Principles and Practices of Plant Genomics

Volume 2: Molecular Breeding

Editors

Chittaranjan Kole

Albert G. Abbott

Department of Genetics and Biochemistry
Clemson University
South Carolina
USA

CRC Press
Taylor & Francis Group
Boca Raton London New York

CRC Press is an imprint of the
Taylor & Francis Group, an **informa** business
A SCIENCE PUBLISHERS BOOK

First published 2008 by Science Publishers

Published 2019 by CRC Press
Taylor & Francis Group
6000 Broken Sound Parkway NW, Suite 300
Boca Raton, FL 33487-2742

ISBN 13: 978-1-57808-537-8 (hbk)

Visit the Taylor & Francis Web site at
http://www.taylorandfrancis.com

and the CRC Press Web site at
http://www.crcpress.com

Library of Congress Cataloging-in-Publication Data

Principles and practices of plant genomics/editors, Chittaranjan Kole, Albert G. Abbott.

p. cm.

Includes bibliographical references and index.

ISBN 978-1-57808-537-8 (hardcover)

1. Plant genomes. 2. Plant genetics. I. Kole, Chittaranjan. II. Abbott, Albert G.

QK981.P75 2008

572.8'62--dc22

2007047428

Dedicated to

Prof. Blannie Bowen

Vice Provost for Academic Affairs
The Pennsylvania State University

The Prologue

Dear Prof. Kole

Having read the preface and contents of your series of volumes on plant genetics, I wish to thank you to have undertaken together with Professor Abbott the heavy work to produce an updated encyclopedia on molecular plant genetics. I am sure that your series on "Principles and Practices of Plant Genomics" will have a tremendous, worldwide impact on plant sciences and their applications.

Our human society has benefited for thousands of years from plants found in their natural environments. Soon people learned to bring about improvements by breeding, as well as by choosing most appropriate soils to grow the plants serving for their nutrition, as well as for medical and ornamental uses.

A true paradigmatic change has started a few decades ago with the progress of scientific investigations at the molecular level. Novel opportunities have become available by a combination of genetic engineering and classical breeding to obtain even more appropriate plant variants for the service to the human population. With novel research strategies and with steadily improving knowledge on gene structure and functions, we can expect that already in the near future many cultivated plants will benefit themselves of an improved health, be more resistant to pests and, what will be of great importance, will provide improved nutritional values to our daily diets. Agriculture will enter a new, more sustainable era.

Your publication series can greatly help the scientists working on this field to reach this goal and to carry out their work, with the required care and responsibility, to the benefit both of the human society and of its environment.

Please take these words as an encouragement to complete and also to steadily update your important project.

With my best regards.

Yours Sincerely,

January 23, 2008

Prof. Werner Arber
Professor Emeritus for Molecular Microbiology
University of Basel
Nobel Laureate in Medicine 1978

Foreword

The elucidation of the double-helix structure of the Deoxyribose Nucleic Acid (DNA) molecule in 1953 by Drs. James Watson, Francis Crick, Maurice Wilkins and Rosalind Franklin marked the beginning of what is now known as *the new genetics*. Research during the last 54 years in the fields of molecular genetics and recombinant-DNA technology has opened up new opportunities in agriculture, medicine, industry, and environment protection.

Availability of DNA based molecular markers since 1980s has facilitated construction of complete genetic maps in several plants of academic and economic importance. Several computer software also have paved the way for handling huge numbers of markers and segregating individuals to frame these maps, also for detection of positions of genes and gene clusters, known as quantitative trait loci, on them.

Molecular marker-based genetic mapping has provided not only comprehensive depictions of the genomes but also facilitated elucidation of phylogenetic relationships and evolutionary pathways, map-based cloning of genes and above all use of molecular markers in various schemes of crop improvement, including germplasm characterization, marker-aided selection and introgression of genes through marker-assisted backcross breeding.

The strategies of construction of BAC and YAC libraries and advent of high-throughput sequencing of DNA have built the platform for physical mapping of chromosomal regions and even whole genomes. The new millennium has already witnessed complete sequencing of the genomes of the model plant *Arabidopsis* and the leading crop plant rice followed by the genomes of a forest tree poplar and a fruit tree peach. Genome initiatives in many more crop plants individually and families comprising important crop plants are indeed progressing fast.

One of the spectacular happening during this period is the merging of information technology and biological science leading to a new discipline called as bioinformatics. This subject is now playing a pivotal role in warehousing biological data on sequences and functions of genes, transcripts, expressed sequence tags and proteins, and their searching to utilize for various purposes. We have now several new branches with the 'omics' sciences such as genomics, transcriptomics, proteomics and metabolomics. However, all of them relates to the genomes and reside under the common roof of genomics.

There is little doubt that the genomics has opened up uncommon opportunities for enhancing the productivity, profitability, sustainability and stability of major cropping systems. It has also created scope for developing crop varieties resistant/tolerant to biotic and abiotic stresses through an appropriate blend of Mendelian and molecular breeding techniques. It has led to the possibility of undertaking anticipatory breeding to meet potential changes in temperature, precipitation and sea level as a result of global warming. There are new opportunities for fostering pre-breeding and farmer-participatory breeding methods in order to continue the merits of genetic efficiency with genetic diversity.

In the coming decades, farm women and men in population-rich but land-hungry countries like India and China will have to produce more food and other agricultural commodities to meet home needs and to take advantage of export opportunities, under conditions of diminishing per capita availability of arable land and irrigation water and expanding abiotic and biotic stresses. The enlargement of the gene pool with which breeders work will be necessary to meet these challenges. Genomics supplemented by transgenics provide breeders with a powerful tool for enlarging the genetic base of crop varieties and to pyramid genes for a wide range of economically important traits.

Genetics can boast of being the subject that progressed with the most rapid speed within the shortest period of time and contributed stupendously in unraveling the nature and function of genes and also towards the genetic improvement of useful microbes, animals and plants in the last century. In fact, the *green revolution* of the sixties was the contribution of classical genetics and conventional breeding. Another revolution in agriculture, medicine and environment is in the wing and we are ready to welcome it as *gene revolution* to be drived by genomics-aided breeding. However, for a meaningful and sustainable application of plant genomics, importance of *orphan* crops which have immense potential for improving food and nutrition security should be taken into account.

The book series "Principles and Practices of Plant Genomics" is thus a timely contribution. It provides an authoritative account of recent progress in plant genomics including structural and functional genomics and their use in molecular- and genomics-assisted breeding. It also provides a road map for genomic research in the present century. I express our gratitude to Profs. Chittaranjan Kole and Albert Abbott for this labor of love in spreading knowledge among both students and scientists on the opportunities as well as challenges opened up by the genomics era.

M S Swaminathan
Chairman
M.S. Swaminathan Research Foundation

Preface to the Series

It is an unequivocal fact that genomics has emerged as the leading discipline in plant sciences. The enormous global efforts in the last two and a half decades have led to a very rapid development of this subject. Currently genetic linkage mapping employing molecular markers has become routine for development of chromosome maps in genetically refractory species, mapping of economic genes and gene clusters, studies on evolution and phylogenetic relationships, map-based cloning of useful genes, and marker-aided breeding. Physical mapping of plant genomes using artificial chromosomes and integration of physical and genetic maps have provided more power of resolution to depict the structural basis of genes and also to elucidate genome colinearity and evolution facilitating exploration of homologous desired genes employing comparative genomics. Whole-genome sequencing of the model plant *Arabidopsis* followed by rice and poplar has already demonstrated the power of structural genomics in elucidating focused understanding of genome organization using sequence data that represents the ultimate level of genetic information. The success stories of genome initiatives in microbes, animals, and plants have inspired 'initiatives' at national and international levels to sequence genomes of an array of other plant genomes. Information on genetic and physical maps, colinearity and whole-genome sequencing serve as the starting point for assignment of biological meaning to putative genes with no known phenotype. For this precise purpose, functional genomics emerged to address the function of genes discovered through sequencing efforts. Employment of concepts and strategies including expressed sequence tags, reverse genetics and transcriptional profiling have facilitated identification and discovery of genes and their expressions in specific tissues and stages.

Plant genomics has already found its deserved place in various courses under agriculture, medicine, and environment sciences. Application of molecular markers for several purposes in crop improvement has fostered the development of another discipline, molecular breeding. Recent developments in functional genomics, transgenics and bioinformatics have vastly enriched the resources of today's plant breeding. Information on a large number of genes handled simultaneously by genomics and thereby genetic integration of diverse processes, tissues, and organisms will highly benefit future plant improvement endeavors. Therefore, future molecular breeding will encompass marker-assisted breeding, transgenic breeding, and genomics-aided breeding. Plant genomics and molecular breeding are going to play the pivotal role in the fields of agriculture, medicine, environment, and ecology in this and hopefully in the coming centuries.

However, through academic and research related interactions with fellow research workers on plant genomics, we recognized the need of a handbook for ready reference. Textbooks explaining the fundamentals and applications of plant genomics are of significant value to students, teachers and scientists of the public and private sectors. Plant genomics is basically the frontier branch of genetics but is complemented strongly by biochemistry, microbiology, computational biology, and bioinformatics; with supplementation from various other disciplines including crop production, statistics, physiology, pathology, entomology, horticulture, just to name a few. Most interestingly in today's world, people from all these fields learn, teach and/or practice plant genomics. Thus, it is essential that academics in these broad disciplines have available the lucid deliberations of the basic concepts and strategies of plant genomics to enhance their own studies, training, and expertise. This was the main driving force for taking up this endeavor to popularize the science and art of plant genomics.

The subject of plant genomics has grown really rapidly within a short period of two and a half decades but with distinct phases. Since the mid-eighties, basic structural genomics emerged comprising construction of genetic linkage maps using molecular markers and mapping of genes and quantitative trait loci. The following phase included application of molecular markers for various purposes of crop improvement, popularly known as molecular breeding. And the third phase involved physical genome mapping with artificial chromosomes culminating in advanced structural genomics including whole-genome mapping, and functional genomics. We have tried to follow almost the same trend while organizing the chapters on different topics under the three volumes of this textbook series.

In the first volume we briefly introduce the historical background and overview on genome mapping in the first chapter. Subsequently, we present 10 chapters deliberating on different types of molecular markers, their detection, relative merits, shortcomings and applications; types of mapping populations, methods of their generation, applications; basic concepts and schematic depiction of construction of genetic linkage maps; concepts and strategies of mapping genes controlling qualitative and quantitative traits on framework genetic linkage maps; rationale, methodologies and implications of comparative mapping; principles, strategies, and outcome of map-based cloning; overviews on the recent advances on plant genomics and genome initiatives; and finally computer strategies and software employed in plant molecular mapping and breeding.

In the second volume, we emphasize areas of application of molecular mapping. We present a general deliberation on the fundamentals of molecular breeding and applications of molecular markers for germplasm characterization; concepts and application of molecular mapping and breeding for yield, quality, and their related component traits; biotic and abiotic stresses; and physiological traits of economic importance. Some plant groups, for example fruit crops, forest trees, fodder, forage, and turf grasses, and polyploid crop plants exhibit unique breeding problems and necessarily require special breeding strategies. We have dedicated independent chapters for them. We include a chapter on transgenic breeding considering its growing popularity and contribution in crop improvement. We also include a chapter on intellectual property rights (IPR) considering the legal implications in using, developing and commercializing plant varieties and genotypes.

In the third volume on advanced genomics, we present an overview on the advances of plant genomics made in the last century; deliberations on the genomics resources; concepts, tools, strategies, and achievements of comparative, evolutionary, and functional genomics and whole-genome sequencing. We also present critical reviews on the already completed genome initiatives and glimpses on the currently progressing genome initiatives. We also have a deliberation on application of the genomics information in genetic improvement of crop plants. We include a chapter on facilitation of the progress of plant genomics researches considering the requirements of generous funding, state-of-the-art facilities and well-trained human resources. Finally, we present depiction of a map of the road for plant genomics to be traveled by us and scientists of our next generations in the twenty-first century.

Articles and reviews on all these concepts, strategies, tools, and achievements in genome analysis and improvement abound but are scattered mainly on the pages of periodicals. Most of them deal with a focus on a particular crop plant or a group of related crop plants. Recently, a seven-volume series on 'Genome Mapping and Molecular Breeding in Plants' was published by Springer that encompasses 82 chapters on more than 100 plant species of economic importance.

Interested readers may enrich their knowledge from this and other resources as well. However, in these three books, we have paid particular attention to the basic principles, strategies, and requirements and applications of various facets of plant genomics useful for students to develop a comprehensive understanding of the subjects and to assist academicians and researchers in disseminating and using the knowledge.

We express our deep gratitude to Prof. Werner Arber, Nobel Laureate in Medicine 1978 for kindly contributing the Prologue for this series and his generous comments about this endeavor. We are grateful to Prof. M.S. Swaminathan for kindly scripting the foreword for this treatise. Views of these science missionaries and social visionaries will be immensely useful for the present and future travelers in the world of plant genomics and will remain as words of inspiration for the present and future generations of plant scientists.

We must mention here that working with the authors of the chapters of these three volumes has been highly enriching, enlightening, and entertaining. We wish to express our thanks and gratitude to these devoted and reputed scientists for their excellent chapters and above all, their affection and cooperation during the pleasant and painful periods of gestation and delivery of these three book volumes. They appreciated the importance of these textbooks and donated their time to contribute these clear and concisely written chapters.

The authors of some chapters of our series have included a few previously published illustrations in original or modified versions. We believe they have obtained permission for such reproduction from the competent authors or publishers of the articles containing these illustrations. However, as editors we believe that it is our moral duty to keep in record our acknowledgement for all these authors, publishers, and sources for the helps that facilitated the improvement of the contents and formats of the concerned deliberations.

We look forward to constructive criticisms and suggestions from all corners for future improvement of the contents and format of these three book volumes.

We must express our thanks and gratitude to the publishers for their constant co-operation extended to us and the authors. We remember that on many occasions we failed to stick to the dead lines of submission of manuscripts, proof corrections and other required inputs due mainly to the conflicts of schedules and other unforeseen assignments and commitments. However, the concerned staff did bear with us with laudable patience.

Chittaranjan Kole
Albert G. Abbott

Preface to the Volume

In plants, the major application of molecular markers and genome maps based on them is in molecular breeding for crop improvement. The concept of 'markers' was conceived as early as in 1923 with the observation of Karl Sax on the association of seed color with seed weight in common bean; and throughout the nineteenth century plant breeders used several simple-inherited morphological or visible traits for indirect selection for improvement in the target traits. Examples that have been used as markers in many plant breeding programs for monitoring agroeconomic traits include traits such as, presence of awn (for insect resistance), shoot pubescence (disease resistance), leaf-rolling (drought tolerance), pigmentation (lycopene content), and others. Additionally, many biochemical compounds such as abscisic acid (drought tolerance), glycine-betane (salt tolerance), phytoalexins (disease resistance), and others have also been used as suggestive indicators for selection of plants with higher tolerance or resistance to abiotic and biotic stresses. However, insufficient coverage of the genomes by visible markers and problems such as long genetic distance from the target loci, linkage drag, dominant nature of inheritance, dependence on age, stage of growth, and above all on environmental factors for expression have limited their use in breeding. The advent of molecular markers in early 1980s has provided solutions to these constraints with several significant advantages. These include the ability to detect theoretically vast numbers of markers to cover the entire genomes in mature plants, seeds or seedlings; the absence of intra-allelic (for most of the markers) and inter-allelic (epistasis) interactions; phenotypic neutrality; and the potential for mitigating linkage drag through high-resolution mapping. A variety of such molecular markers have been developed since 1980s which could be used for specific populations and purposes of genetic augmentation of crop plants.

It is perhaps a paradox that most of the agriculturally desirable traits such as yield, quality, and stress resistance or tolerance are quantitative in nature and are controlled by polygenes. These genes exhibit too small individual effects to be detected by any classical genetic analysis and are highly influenced by environmental fluctuations. In addition, they manifest all sorts of epistatic interaction that confound the ability to assign the precise genetic contribution and localization of these genes employing the traditional genetic and breeding tools including conventional biometrical models. The use of molecular markers has paved the way for development of framework genetic linkage maps with sufficient marker density for detection of the putative chromosomal locations of the polygenic clusters controlling the quantitative traits, designated as quantitative trait loci (QTL) and also for monitoring the QTLs during their introgression into recipient genetic backgrounds. This is perhaps a revolutionary achievement of molecular genetics with potential significant impact on plant genetic improvement. The molecular markers follow Mendelian segregation and the scope of selection for a polygenic trait by using a flanking Mendelian trait is owed again to Karl Sax and thus one could argue that QTL mapping and monitoring are direct descendents of his initial marker-assisted breeding efforts.

There is no denying the fact that scientific plant breeding in the last century has contributed enormously to providing food, feed, clothing, shelter, medicine, and other essentials to the fiercely increasing world population. Innumerable crop varieties were developed using introduction, selection, hybridization, and innovative breeding tools such as heterosis, polyploidy and mutation breeding. However, we are reaching a plateau in productivity in all crops and a dramatic increase of the yield ceiling and value addition are imperative to cope with the population growth particularly in many developing countries. Molecular breeding by employment of markers as a supplement to conventional breeding, development of transgenics, molecular pharming and genomics-assisted breeding are the potential means to respond to the call of the time. The 'green revolution' of the twentieth century must be followed by 'gene revolution' in the twenty-first century to meet the fiercingly increasing demands for higher quantity of quality products from agriculture and forestry.

Molecular breeding has, therefore, emerged as an independent discipline and generated interests in academies and industries, and among policy makers at public and private levels. There is a significant growth in information on plant molecular breeding for individual crop plants or plant groups. However, they are mostly dispersed over the pages of journals, review papers, and reports. A textbook elucidating the concepts and illustrating the strategies and achievements of molecular breeding is indeed timely and would cater to the needs of students, scientists, industries, policy makers, and the media fostering a growth of this subject that could benefit science and mankind. This was the pivotal driving force behind planning this volume.

We are presenting this book volume with thirteen chapters. The first chapter provides an overview on the fundamentals of molecular breeding which has addressed the what, why, and how related to this new discipline.

Establishment, evaluation, and utilization of a germplasm had always been the starting platform for crop improvement as knowledge on availability of desirable genes in the primary, secondary, and tertiary gene pools is a priority for designing breeding schemes for genetic improvement in any plant with regard to the targeted trait(s). These also suggest a requirement for incorporation of or creating new alleles through genetic engineering or mutagenesis. Employment of molecular markers has reinforced the conventional parameters of assessment of available germplasm. Markers have been employed for evaluation of distinctness, uniformity, and stability (DUS) of cultivars which are highly important parameters for breeding as well as for addressing issues of intellectual property rights (IPR). For these reasons, we have dedicated a chapter on application of molecular markers for germplasm conservation.

The major breeding objectives in crop improvement have always been yield, quality, and their component traits. Resistance/tolerance to biotic and abiotic stresses has always attracted attention of plant breeders to mitigate the losses in yield and deterioration of quality of agricultural produce in the field or at post-harvest stages including storage and transportation. Many physiological characters have significant impacts on growth and development of crop plants. Mapping genes or gene clusters controlling these traits and their manipulation require specific strategies. Therefore, we have devoted five chapters on these five groups of traits of high agricultural and industrial importance.

It is unequivocal a fact that the fruit and forest tree species have some unique limitations so far as their genetic studies and breeding are concerned. Particularly, their long juvenile period and inherent heterozygosity invoke special strategies for development of genetic maps, mapping genes and gene clusters and marker-assisted monitoring of target traits with the linked markers. We have included two chapters on these plant groups in this volume. Similarly, the polyploid crop plants have unique meoitic features and invoke for special approach of scoring marker genotypes for linkage mapping and molecular breeding. Hence, we have devoted one chapter on polyploid crop plants to elucidate the unique strategies employed for their mapping and breeding. Recently, the plants used for fodder, forage, and turf grasses are gaining importance in molecular mapping and breeding and thus deserve a chapter for highlighting the extent of work done so far on these crop plants.

Traditional and marker-assisted breeding in crop plants depend mainly on the genes available in the primary gene pool and in certain cases genes available in related species or genera that are cross-compatible with the candidate crop plant. Incorporation of potential alien genes from distant plant taxa and even from bacteria, fungi, and viruses requires the potential employment of transgenic technology. Transgenic crop plants have been developed in an array of species for different purposes. A few examples are resistance to biotic stresses, tolerance to abiotic stresses, reduction of juvenile period, amelioration of quality by adding/increasing desirable chemical components or reducing/eliminating hazardous components, inducing earliness or delay in ripening, and others. Genetic engineering and marker-assisted breeding have complimented each other for the last one decade to reinforce crop improvement strategies. We have therefore included an independent chapter on transgenic breeding.

Intellectual property rights (IPR) have a major role to play in modern plant breeding encompassing the generation of new concepts, strategies, software, genetic stocks, and also new genes with novel products. This is even more relevant for transgenic breeding. A chapter on IPR has been included to provide a thorough understanding on the intricacies and implications of IPR in genome mapping and molecular breeding.

Thirty-nine leading scientists from seven countries have contributed the thirteen chapters of this volume. Their own experience, expertise, and philosophy have made their deliberations elegant in contents, lucid in approach and thought provoking to inspire young students and fellow scientists. We are grateful to all these authors for such contributions and feel happy to get this opportunity to present them to our readers.

The rapid progress in genome science has generated a lot of novel ideas and materials and these have contributed enormously to the genetic improvements in a large number of highly diverse crop plants. This volume will surely miss many such ideas, strategies, success stories, particularly for many traits and for many plants. However, the limitation of the size of this volume precluded inclusion of all these excellent works, and for this we extend our due apology.

We also take responsibility for all other unseen errors and look forward to suggestions for their rectification and also for further improvement of 'quantitative and qualitative traits' of this volume!

Chittaranjan Kole
Albert G. Abbott

Contents

List of Contributors

Albert G. Abbott

Department of Genetics and Biochemistry, 116 Jordan Hall, Clemson University, Clemson, SC 29634, USA; Phone: +1-864-656-3060; Fax: +1-864-656-6879; e-mail: aalbert@clemson.edu

A. Mohan Rao

Department of Genetics and Plant Breeding, University of Agricultural Sciences, Bangalore 560065, Karnataka, India; Phone: +91-80-23624967; Fax: +91-80-23631209; e-mail: amrao8@rediffmail.com

Bharat Singh

Agricultural Research Station, Fort Valley State University, 1005 Stae University Drive, Fort Valley, GA 310, USA; Phone: +1-478-825-6829; Fax: +1-478-825-6376; e-mail: bharats@fvsu.edu

Chittaranjan Kole

Department of Genetics and Biochemistry, 111 Jordan Hall, Clemson University, Clemson, SC 29634, USA; Phone: +1-864-656-3060; Fax: +1-864-656-6879; e-mail: ckole@clemson.edu

Christine A. Hackett

Biomathematics and Statistics Scotland, SCRI, Invergowrie, Dundee DD2 5DA, UK; Phone: ++44-1382-562965; e-mail: Christine@bioss.ac.uk

Christopher J. Lambrides

School of Land, Crop and Food Sciences, The University of Queensland, Brisbane 4072, Australia; Phone: +61-7-3365 1103; Fax: +61-73365 1177; e-mail: chris.lambrides@uq.edu.au

Dan Milbourne

Teagasc, Crops Research Centre, Oak Park, Carlow, Ireland; Phone: ++353-59-9170291; e-mail: Dan.Milbourne@teagasc.ie

David A. Lightfoot

Department of Plant, Soil and General Agriculture, Mailcode 4415, Southern Illinois University at Carbondale, Carbondale IL 62901-4415, USA; Phone: +1-618-453-1797; Fax: +1-618-453-7457; e-mail: ga4082@siu.edu

Eline van Zijll de Jong

Department of Primary Industries, Biosciences Reserach Divison, Victorian AgriBiosciences Centre, La Trobe Research and Development Park, Bundoora, Victoria 3083, Australia; Phone: +61-3-9479-1378; Fax: +61-3-9479-3618; e-mail: eline.vanzijlldejong@dpi.vic.gov.au

Gail Timmerman-Vaughan

New Zealand Institute for Crop and Food Ltd., PO Box 4704, Christchurch, New Zealand; Phone: +64-3-3256400 x 3439; Fax: +64-3-3252074; e-mail: timmermang@crop.cri.nz

Germán C. Spangenberg

Department of Primary Industries, Biosciences Reserach Divison, Victorian AgriBiosciences Centre, La Trobe Research and Development Park, Bundoora, Victoria 3083, Australia; Phone: +61-3-9479-3851; Fax: +61-3-9479-3618; e-mail: german.spangenberg@dpi.vic.gov.au

Gregory J. Rebetzke

CSIRO Plant Industry, ACT, Australia; Phone: +61-2-6246 5153; Fax: +61-2-6246 5399; e-mail: greg.rebetzke@csiro.au

H.C. Lohithaswa

AICRPO on Forage Crops, University of Agricultural Sciences, Bangalore, ZARS, V.C. Farm, Mandya 571405, Karnataka, India; Phone: +91-8232-219671; Fax: +91-8232-277392; e-mail: iohithhc@rediff mail.com

Hari P. Singh

Center for Biotechnology, College of Agriculture, Fort Valley State University, 1005 State University Drive, Fort Valley, GA 31030, USA; Phone: +1-478-825-6887; Fax: +1-478-827-3062; e-mail: singhh@fvsu.edu

Hunter K.C. Laidlaw

School of Land, Crop and Food Sciences, The University of Queensland, Brisbane 4072, Australia; Phone: +61-7-3365 2141; Fax: +61-7-3365 1177; e-mail: laidlaw@uqconnect.net.au

Ian D. Godwin

School of Land, Crop and Food Sciences, The University of Queensland, Brisbane 4072, Australia; Phone: +61-7-3365 2141; Fax: +61-7-3365 1177; e-mail: i.godwin@uq.edu.au

J.B. Power

Plant Sciences Division, School of Biosciences, University of Nottingham, Sutton Bonington Campus, Loughborough, Leicestershire, LE12 5RD, UK;

Phone: +44-(0)115-9513205; Fax: +44-(0)115-9516334; e-mail: brian.power@nottingham.ac.uk

Jaya R. Soneji

University of Florida, IFAS, Citrus Research and Education Center, 700 Experiment Station Road, Lake Alfred, FL 33850, USA; Phone: +1-863-956-1151; Fax: + 1-863-956-4631; e-mail: jrs@crec.ifas.ifas.ufl.edu

Joanne A. Labate

USDA-ARS, Plant Genetic Resources Unit, Geneva, NY 14456, USA; Phone: +1-315-787-2438; Fax: +1-315-787-2339; e-mail: Joanne.Labate@ars.usda.gov

John E. Bradshaw

Genetics Programme, Scottish Crop Research Institute, Invergowrie, Dundee DD2 5DA, UK; Phone: +44-1382-568524; e-mail: John.Bradshaw@scri.ac.uk

John W. Forster

Department of Primary Industries, Biosciences Reserach Divison, Victorian AgriBiosciences Centre, La Trobe University Research and Development Park, Bundoora, Victoria 3083, Australia; Phone: +61-3-9479-5645, Fax: +61-3-9479-3618; e-mail: john.forster@dpi.vic.gov.au

Julie Graham

Scottish Crop Research Institute, Invergowrie, Dundee, DD2 5DA, UK; Phone: +1382-562731; Fax: +1382-562426; e-mail: jgraha@cri.sari.ac.uk

K.C. Lowe

School of Biology, University of Nottingham, University Park, Notingham, Nottinghamshire, NG7 2RD, UK; Phone: +44-(0)115-9513311; Fax: +44-(0)115-9513251; e-mail: kclowe@nottingham.ac.uk

K. Chalapathy Reddy

Mission Biofuels India Pvt. Ltd., 608, Powai Plaza, Hiranandani Business Park, Powai, Mumbai 400076, India; Phone: +91-9323623328; Fax: +91-22-25706215; e-mail: dr.chalapathy@missionbiofuels.com

Kevin F. Smith

Department of Primary Industries, Biosciences Research Divison Hamilton Centre, Hamilton, Victoria 330, Australia; Phone: +61-3-557-30983; Fax: +61-3-557-11523; e-mail: kevin.f.smith@dpi.vic.gov.au

M.J. Iqbal

Institute for Advanced Learning and Research, Institute for Sustainable and Renewable Resources, 150 Slayton Avenue, Danville, VA 24540, USA; Phone: (434) 766-6712; Fax: (434) 791-3279; e-mail: Muhammad.Iqbal@iair.org

Madhugiri Nageswara Rao

University of Florida, IFAS, Citrus Research and Education Center, 700 Experiment Station Road, Lake Alfred, FL 33850, USA; Phone: +1-863-956-1151; Fax: +1-863-956-4631; e-mail: mnrao@crec.ifas.ufl.edu

Marco Maccaferri

Department of Agroenvironmental Sciences and Technology, University of Bologna, Viale Fanin, 44, 40127- Bologna, Italy; Phone: +39-51-2096646; Fax: +39-51-2096241; e-mail: marco.maccaferri@unibo.it

Maria Corinna Sanguineti

Department of Agroenvironmental Sciences and Technology, University of Bologna, Viale Fanin, 44, 40127-Bologna, Italy; Phone: +39-51-2096646; Fax: +39-51-2096241; e-mail: maria.sanguineti@unibo.it

Mark P. Dobrowolski

Department of Primary Industries, Biosciences Research Divison Hamilton Centre, Hamilton, Victoria 330, Australia; Phone: +61-3-557-30918; Fax: +61-3-557-11523; e-mail: mark.dobrowolski@dpi.vic.gov.au

Michael Blakeney

Queen Mary Intellectual Property Research Institute, University of London, 67-69 Lincoln's Inn Fields, London, WC2A 3JB, UK; Phone: +44-115-95163057; Fax: +44-20-7882-6044; e-mail: m.blakeney@qmul.ac.uk

Michael Davey

Plant Sciences Division, School of Biosciences, University of Nottingham, Sutton Bonington Campus, Loughborough, Leicestershire, LE12 5RD, UK; Phone: +44-(0)115-95163057; Fax: +44-(0)115-9516334; e-mail: mike.davey@ nottingham.ac.uk

Milind B. Ratnaparkhe

Division of Plant Sciences, University of Missouri-Columbia, 210 Waters Hall, MO 65211, USA; Phone: +1-573-884-0451; Fax: +1-573-884-1467; e-mail: ratnaparkhem@ missouri.edu

Noel O.I. Cogan

Department of Primary Industries, Biosciences Reserach Division, Victorian AgriBiosciences Centre, La Trobe University Research and Development Park, Bundoora, Victoria 3083, Australia; Phone: +61-3-9479-3249; Fax: 61-3-949-3618; e-mail: noel.cogan@dpi.vic.gov.au

N.V. Hemanth Kumar

Center for Viticulture and Small Fruit Research Station, Division of Plant Biotechnology, Florida Agricultural and Mechanical University, 6505 Mahan Drive, Tallahasscc, FL 32317, USA; Phone: +1-850-412-5191; Fax:+ 1-850-561-2617; e-mail: hemanth_agrihort24@yahoo.co.in

P. Anthony

Plant Sciences Division, School of Biosciences, University of Notttingham, Sutton Bonington Campus, Loughborough, Leicestershire, LE12 5RD, UK; Phone: +44-(0)115-9516364; Fax: +44-(0)115-9516334; e-mail: paul.anthony@ nottingham.ac.uk

Preston R. Aldrich

Department of Biological Sciences, Benedictine University, Birck Hall 341, 5700 College Road, Lisle, IL 60532-0900, USA; Phone: +1-630-829-6519; Fax: +1-630-829-6547; e-mail: paldrich@ben.edu

Roberto Tuberosa

Department of Agroenvironmental Sciences and Technology, University of Bologna, Viale Fanin, 44, 40127-Bologna, Italy; Phone: +39-51-2096646; Fax: +39-51-2096241; e-mail: roberto.tuberosa@unibo.it

R.S. Kulkarni

Department of Genetics and Plant Breeding, University of Agricutural Sciences, Bangalore 560006, Karnataka, India; Phone: +91-80-23352270; Fax: +91-80-23330277; e-mail: kulkarnis@yahoo.com

Wayne Powell

Scottish Crop Research Institute, Invergowrie, Dundee, DD2 5DA, UK Phone: +1382-562731; Fax: +1382-562426; e-mail: wpowel@scri.sari.ac.uk

List of Abbreviations

aadA	Aldehyde-alcohol dehydrogenase (gene)
AB	Advanced backcross
ABA	Abscisic acid
AB-QTL	Advanced backcross QTL
Ac/Ds	Activator/Dissociator (transposable elements)
AFLP	Amplified fragment length polymorphism
ALTM1A	Activated malate transporter
AMOVA	Analysis of molecular variance
AMP	Association mapping panel
ANOVA	Analysis of variance
AP-PCR	Arbitrarily primed PCR
ASI	Anthesis silking interval
ATP	Adenosine tri-phosphate
BAC	Bacterial artificial chromosome
BARC	Beltsville Agricultural Research Center
BDL	Backcross-derived line
BES	BAC-end sequences
BSA	Bulked segregant analysis
Bt	*Bacillus thuriengiensis*
CAD	Cinnamyl alcohol dehydrogenase
CAld5H	5-Hydroxyconiferaldehyde
CaMV	Cauliflower mosaic virus
CAPS	Cleaved amplified polymorphic sequence
CBD	Convention on Biological Diversity
CBF	C-repeat binding factor
Cd	Cadmium
cDNA	Complementary-DNA
CDPK	Calcium dependant protein kinase

CGIAR	Consulting Group for International Agricultural Research
CGN	Center for Genetic Resources, The Netherlands
CIAT	International Center for Tropical Agriculture
CIMMYT	International Maize and Wheat Improvement Center
CIP	International Potato Center
CIV	Chain quadrivalent
COR	Cold-regulated
Cre	Cyclic recombinase
CRT	C-repeat
Cvi	Cape Verdi Islands
DAHP	3-Deoxyheptonate aldolase synthase
DArT	Diversity array technology
DDC	Duplex-duplex coupling
DDM	Duplex-duplex mixed
DDR	Duplex-duplex repulsion
DDRT	Differential display reverse transcription
DH	Doubled haploid
DLE	Delayed light emission
DOE	Department of Energy
DRE	Dehydration-responsive element
DREB	Dehydration-responsive element binding protein
DS	Double simplex
DSR	Duplex-simplex repulsion
dsRNA	Double-stranded RNA
EC	Electrical conductivity
EMS	Ethylmethane sulphonate
EPO	European Patent Office
EPSP	5-Enolpyruvylshikimate-3-phosphate
Epsps	5-Enolpyruvylshikimate-3-phosphate synthase (gene)
eQTL	Expression QTL
ER	*Erecta* (gene)
EST	Expressed sequence tag
ESTP	Expressed sequence tag polymorphism
FAO	Food and Agriculture Organization (United Nations)
FISH	Fluorescent *in situ* hybridization
FLP	Flippase recombinase
FPC	Fingerprint contigs
Fr	Frost tolerance loci
FRT	Flippase recombinase target
G2	Gap 2 phase of the cell cycle

GBSS	Granule-bound starch synthase
GE	Genetic engineering
GEI	Genotype x environment interaction
GFP	Green fluorescent protein
gfp	green fluorescent protein (gene)
GO	Gene ontology
gox	Glyphosate oxidoreductase (gene)
GPC	Grain protein content
GUS	β-Glucuronidase
gus	β-Glucuronidase (gene)
H	Heritability
Ha	Barley hardiness (locus)
H_e	Expected heterozygosity
HI	Harvest index
HICF	High information content fingerprint
HIF	Heterogeneous inbred family
HIV	Human immunodeficiency virus
H_o	Observed heterozygosity
Hpt	Hygromycin phosphotransferase (gene)
HTIRC	Hardwood Tree Improvement and Regeneration Center
ICARDA	International Center for Agriculture in the Dry Areas
ICRISAT	International Crops Research Institute for the Semi-Arid Tropics
ID	Industrial designs
IGER	Institute of Grassland and Environmental Research
IITA	International Institute for Tropical Agriculture
IL	Introgression library
IL	Introgression line
ILGI	International *Lolium* Genome Initiative
ILRI	International Livestock Research Institute
INRA	Institut National de la Recherche Agronomique
IPCC	Inter-government Panel on Climate Change
IPGC	International *Populus* Genome Consortium
IPGRI	International Plant Genetic Resources Institute
IPK	Institut für Pflanzengenetik und Kulturpflanzenforschung
IPR	Intellectual Property Rights
Ipt	Isopentenyl-transferase (gene)
IRRI	International Rice Research Institute
ISSR	Inter-simple sequence repeat
ITIFT	International Transcriptome Initiative for Forage and Turf
L-ABA	Leaf abscisic acid

LD	Linkage disequilibrium
LEA	Late embryogenesis abundant
Ler	Landsberg *erecta*
LG	Linkage group
LIS	Legume Information System
LOD	Logarithm of odds
Lox	Locus of X
luc, lux	Luciferase (gene)
MABC	Marker-assisted backcross
MAS	Marker-assisted selection
MAYG	Mapping as you go
MDS	Multidimensional scaling
MICF	Moderate information content fingerprint
MIM	Multiple interval mapping
MLER	Maximum leaf elongation rate
MR	Magnetic resonance
MRW	Mean ring width
MSE	Managed stress environments
MTA	Material Transfer Agreement
MTP	Minimum tiling path
MUG	4-Methylumbelliferyl-β-D-glucuronide
MYA	Million years ago
NARCH	National Agricultural Research Centre for Hokkaido Region
N_e	Effective population size
NGB	Nordic Gene Bank
NIAS	National Institute of Agrobiological Sciences
NIL	Near-isogenic line
NIRS	Near infra-red reflectance spectroscopy
NMR	Nuclear magnetic resonance
NPGS	National Plant Germplasm System
npt	Neomycin phosphotransferase (gene)
NRC	National Research Council
OA	Osmotic adjustment
ORF	Open reading frame
PAGE	Polyacrylamide gel electrophoresis
PAL	Phenylalanine ammonia-lyase
Pat	Phosphinothricin acetyltransferase (gene)
pAux	Auxillary plasmid
PBR	Plant Breeder's Rights
PCA	Principal component analysis

PCoA	Principal coordinate analysis
PCR	Polymerase chain reaction
PEG	Polyethylene glycol
PHS	Pre-harvest sprouting
PI	Plant introduction
PIC	Polymorphic information content
Pmi	Phosphomannose isomerase (gene)
PNWTIRC	Pacific Northwest Tree Improvement Research Cooperative
PQL	Protein quantitative loci
pSAT	Satellite plasmid
QEI	QTL x environnent interaction
QPM	Quality protein maize
QTL	Quantitative trait loci
QTL × E	QTL × environnent interaction
RAPD	Random(ly) amplified polymorphic DNA
rDNA	Ribosomal-DNA
RDRS	Rice diversity research set
RFLP	Restriction fragment length polymorphism
Rfs	Resistance (gene) against *Fusarium virguliforme*
Rhg	Resistance (gene) against *Heterodera glycines*
RI	Recombinant inbred
Ri	Root-inducing
RIL	Recombinant inbred line
RIV	Ring quadrivalent
RNAi	RNA interference
rol	root loci (gene)
Rps	Resistance (gene) against *Phytophthora sojae* or *Pseudomonas syringae*
Rubisco	Ribulose-bisphosphate carboxylase/oxygenase
S	Synthesis phase of the cell cycle
SAAT	Sonication-assisted *Agrobacterium* transformation
SAMS	S-Adenosylmethionine synthase
SAR	Sodium absorption ratio
SCAR	Sequence characterized amplified region
SCN	Soybean cyst nematode
SD	Short-day
SDS	Soybean sudden death syndrome
SI	Self-incompatibility
SMTA	Standard Materials Transfer Agreement
SNP	Single nucleotide polymorphism

SOS	Salt overly sensitive pathway
SoyGD	Soybean Genome Database
SRAP	Sequence-related amplified polymorphism
SSC	Soluble-solids content
SSC	Simplex-simplex coupling
SSD	Single seed descent
SSR	Simple sequence repeat
SSR	Simplex-simplex coupling
STS	Sequence tagged sites
TAS	Telomeric repeat-associated sequence
T-DNA	Transferred-DNA
TE	Transpiration efficiency
Ti	Tumor-inducing
TILLING	Targeted induced localized lesion in the genome
TmCBF	*Triticum monococcum* C-repeat binding factor
TRIPs	Agreement on Trade-Related Aspects of Intellectual Property Rights 1994
trn	Transportin-like
uidA	β-Glucuronidase (gene)
UHD	Ultra-high density
UNEP	United Nations Environment Program
UPGMA	Unweighted paired group method using arithmetic averages
UPOV	International Convention for the Protection of Plant Varieties
USDA	United States Department of Agriculture
USDA-FS	United States Department of Agriculture – Forest Service
VIGS	Virus-induced gene silencing
vir	Virulence (gene)
VIR	Vavilov Institute of Plant Industry
VPD	Vapor pressure deficit
Vrn	Vernalization loci
WARDA	West Africa Rice Development Association
WIPO	World Intellectual Property Organization
WMO	World Meteorological Organization
WTO	World Trade Organization
X-Gluc	5-Bromo-4-chloro-3-indolyl-β-D-glucuronide
YPDES	Yamanashi Prefectural Dairy Experiment Station

1 Fundamentals of Molecular Breeding

Roberto Tuberosa*, Marco Maccaferri and Maria Corinna Sanguineti

Department of Agroenvironmental Sciences and Technology, Viale Fanin 44, 40127 Bologna, Italy

*Corresponding author: roberto.tuberosa@unibo.it

1 INTRODUCTION

The major challenges faced by plant breeders during the 21st century will be to increase crop production while enhancing the sustainability of agricultural practices and preserving the available biodiversity (Borlaug and Dowswell 2005; Khush 2005; Swaminathan 2005). This challenge is even more daunting in view of the increasing unpredictability of the seasonal trends in temperature and rainfall as a result of global warming. Clearly, a better knowledge of the genetic factors governing harvestable yield and its season-to-season variability will help breeders to more effectively devise strategies for improving yield potential and optimizing yield under a broad range of environmental conditions. Grain yield and the majority of the agronomic traits that breeders contend with are quantitatively inherited. The prevailing assumption in the field of quantitative genetics has been that continuous variation in trait performance is caused by the simultaneous segregation and action of multiple genes with a rather similar effect on the phenotype, together with a major influence of the environment (Falconer 1981). These genes, also referred to as polygenes, are now identified as quantitative trait loci (QTLs; Geldermann 1975). The dissection of quantitative traits improved greatly following the introduction of the molecular platforms that enable us to identify the QTLs governing the relevant genetic variation in cultivated and wild germplasm (Tanksley 1993; Beavis 1994; Ribaut and Hoisington 1998; Koebner and Summers 2003; Varshney et al. 2005; Tuberosa and Salvi 2006). After almost two decades of experiments, it is now clear that although most QTL effects are indeed of small magnitude as originally predicted, a limited number of QTLs have a rather large effect and, as such, can be isogenized and, in some cases, positionally cloned. The list of cloned QTLs has expanded rapidly in the past few years (Salvi and Tuberosa 2005; Price 2006; see Chapter 8 in Volume 1). Therefore, the combination of genomics and genetic engineering offers an array of innovative approaches and unprecedented opportunities for more targeted approaches and for crossing reproductive barriers among species, thus vastly enhancing the repertoire of genes made available to breeders.

In this context, the main objectives of this chapter are to provide basic information on a number of topics important for an effective implementation and exploitation of molecular breeding. In view of the importance of quantitative traits in breeding activities, particular emphasis has been devoted to QTL mapping and the implementation of marker-assisted selection (MAS). Along with this information, some basic details have been provided concerning the methodologies, genetic resources and technological platforms utilized for dissecting the genetic basis of agronomic traits through molecular approaches.

2 THE QTL APPROACH

The basic principles underlying the analysis of quantitative traits were originally provided by Fisher (1918) and were further refined throughout the last century (Falconer 1981; Hallauer and Miranda Fo 1988; Lynch and Walsh 1998; Doerge 2002). The traditional methods (e.g., diallel analysis, generation means analysis, factorial design, etc.) that have been used to investigate the genetic control of quantitative traits provide no information on (i) the chromosomal regions governing naturally-occurring variability of quantitative traits, and (ii) the genetic causes (linkage or pleiotropy) of their association with yield. Conversely, molecular marker technologies allow us to identify at different levels of genetic resolution the relevant loci where the presence of functionally polymorphic alleles influences the variability of quantitatively inherited traits (Stuber et al. 1987, 1999; Tanksley 1993; Lee 1995; Asins 2002; Morgante and Salamini 2003). Importantly, the dissection of quantitative traits into single factors allows breeders to implement MAS on the basis of the polymorphic molecular markers flanking the relevant QTL regions.

2.1 Mapping QTLs

Several reviews provide details and insights on the statistical procedures used to identify QTLs (Tanksley 1993; Beavis 1994, 1998; Churchill and Doerge 1994; Jansen and Stam 1994; Korol et al. 1995, 1998; Stuber et al. 1999; Doerge 2002; Hackett 2002; Broman et al. 2003). The chapters of Doveri et al., Hittalmani et al., and Zeng et al. (Chapters 2, 3 and 6; Volume 1, respectively) provide a detailed description of the materials, approaches and methodologies required for QTL mapping. Here, the basic principles and concepts have been summarized.

QTL detection relies on the occurrence of linkage disequilibrium (LD) between the target sequence and the closely linked markers. The dissection of the phenotypic variation of a quantitative trait into its discrete genetic components is possible through the production of a segregating population, commonly including ca. 150-200 genotypes (e.g., F_2 plants, F_3 families, backcross families, recombinant inbred lines (RILS), doubled haploids (DH), etc.) derived from a cross between two inbred lines differing for the trait(s) of interest (Lee 1995; Quarrie 1996; Prioul et al. 1997). A QTL study includes an accurate phenotypic evaluation and molecular profiling of a segregating population, coupled with a statistical analysis to test the level of significance of the differences between the phenotypic values of the alternative marker alleles, averaged across all genotypes in the population.

The molecular profiling to identify polymorphic loci is carried out using different categories of molecular markers, more commonly RFLPs (restriction fragment length polymorphisms), AFLPs (amplified fragment length polymorphisms) and SSRs (simple sequence repeats), with each category having its own advantages and shortcomings. More recently, a high-throughput platform (DArT: diversity array technology) based on microarray technology has been devised to deliver low-cost marker data at anonymous loci (Wenzl et al. 2004). High-throughput platforms have also been assembled to handle several hundreds of SNPs (single nucletide polymorphisms) simultaneously (Rostoks et al. 2005). When the whole-genome sequence is available, SNPs and indels at each open reading frame (ORF) can be represented on a microarray platform constructed with short oligonucleotides (Mockler and Ecker 2005; Werner et al. 2005). Once all the molecular and phenotypic data are available, a statistical test is applied to verify whether the means of the trait value of the genotypes carrying different alleles at a particular marker are significantly different. A test statistic larger than a threshold value rejects the null hypothesis (i.e. the mean is independent of the genotype at a specific marker locus) and implies a significant association between the investigated marker and a nearby QTL.

The early QTL studies were carried out based on the analysis of the means at single markers using simple test statistics, such as linear regression, *t*-test and analysis of variance (ANOVA). As a genome-wide QTL analysis typically involves a large number of markers, the probability of detecting one or more false positives at the whole-genome level quickly increases unless the threshold of significance is adequately readjusted according to the number of tested markers. Typically, a threshold level of $P_{0.05}$ entails a false-positive discovery rate (i.e. declaring the presence of a QTL when actually there is no QTL) of 5%. Consequently, a mapping experiment based on 100 markers covering the entire genome and tested at $P_{0.05}$ will identify, on average, five markers putatively associated with QTLs even when no real QTL segregates in the population. In order to avoid this problem, the significance threshold is corrected accordingly through a multiple test adjustment (e.g., Bonferroni, Scheffe or Tukey) that will adjust the *P* level according to the number of independent statistical tests that are performed. Additional, and more critical, shortcomings of this single-marker approach are that no information is provided on the most likely position of the QTL and its effects on the phenotype. Due to these major limitations, single-marker analysis is no longer in use for genome-wide QTL searches, which are instead carried out with methods that are based on the estimated order of markers linearly assembled on a genetic map. Arranging the markers into linkage groups, or chromosomes, allows us to produce the genetic framework for positioning QTLs relative to intervals between adjacent pairs of markers. Compared to single-marker analysis, interval mapping provides a much more accurate estimate of the position and genetic effects of each QTL, hence improving the possibility of correctly identifying the most relevant QTLs. With a genetic map available, the QTL is mapped at an interval of defined genetic distance. The estimated genetic distance between markers is a function of the average number of observed recombination events between them. The unit used for expressing the genetic distance between markers on a chromosome is the centiMorgan (cM), which defines the interval along which one recombination event is expected to occur per 100 gametes, per generation. As a density of one marker per ca. 10-15 cM is usually sufficient to detect the presence of QTLs having a major effect on the phenotype, the number of markers required to adequately sample the targeted species varies from as little as 40-50 as in the case of *Arabidopsis*, to well over 300 for large-genomes such as those of oat and wheat.

In interval mapping, each interval between adjacent markers is systematically searched in constant increments (e.g., 2 cM), and statistical methods are applied to test for the likelihood of the presence of a QTL (Lander and Botstein 1989). The results of the likelihood tests carried out at each increment across the ordered markers are expressed as LOD (logarithm of the odds ratio) scores, computed as the $\log_{10}$ of the ratio between the chance of a real QTL being present given the effect measured at that position, divided by the chance of having a similar effect when no QTL is present. The graphical output allows us to compute an empirical confidence interval (usually computed as LOD – 1) around the QTL peak, thus delineating the interval of the most likely QTL position moving away from the peak, under the assumption that the peak is caused by a single QTL and not two closely linked QTLs. In order to avoid declaring false-positive QTLs (i.e. declaring the presence of a QTL when the QTL is actually absent), a reasonably high threshold value for the LOD score should be set (usually > 2.5). Iterative software based upon resampling procedures provides a more accurate estimate of threshold values according to the size of the mapping population and the number of markers (Doerge 2002).

As interval mapping relies on a single-QTL model, LOD peaks can incorrectly identify multiple or 'ghost' QTLs, thus failing to (i) properly account for interactions between multiple QTLs and (ii) unequivocally ascertain which of the many LOD peaks indicate a single QTL. Epistatic interactions occur whenever the combined phenotypic effect of two or more loci exceeds the sum of effects at individual loci. Epistasis can greatly influence the outcome of interval mapping. This problem can be partially overcome with the use of composite interval mapping, a statistical procedure that can account for the effects of other QTLs inherited independently from the interval being considered, hence reducing the possibility of detecting ghost QTLs. Compared to single-QTL interval mapping, statistical approaches for locating multiple QTLs are more powerful because they can differentiate between linked and/or interacting QTLs. As an example, multiple-QTL approaches can identify epistatic QTLs which may otherwise go undetected when using single-QTL interval mapping. Nevertheless, both the computational complexity entailed by multiple-QTL mapping procedures and the difficulty in identifying an appropriate statistical model hinder the routine application of a number of dedicated software for handling multidimensional approaches able to account for the effects due to linked and/or epistatic QTLs (Kao et al. 1999; Sen and Churchill 2001; Stich et al. 2007). Given the potential impact of epistasis on the response to selection, quantifying its influence on target traits is an important component for the design and optimization of any MAS strategy (Podlich et al. 2004). It is likely that the incorporation of epistatic interactions into more properly devised statistical models will play a relevant role in explaining complex regulatory networks governing the expression of quantitative traits.

The accuracy in QTL mapping is influenced by both the individual progenies of the mapping population and the saturation of the genetic map (i.e. the average distance between adjacent markers). Formulas have been devised to estimate how many markers are required to achieve a desired accuracy in estimating QTL location, and how many individuals should be phenotyped given a particular significance level, QTL effect and location (Lynch and Walsh 1998). The main parameters used to define the effects of a QTL are *a* and *d*, i.e. the deviation from the parental mean due to additive and dominant effects, respectively. The additive effect (*a*) equals half of the difference between the homozygotes for one parental allele and the homozygotes for the other parental allele. Values of *d* equal to 0 and *a* indicates additivity and complete dominance (positive or negative according to the sign of the effect), respectively, while absolute values of *d* larger than *a* indicate an overdominant effect (positive or negative according to the sign of the effect). In terms of variance, the general equation $V_P = V_A + V_D + V_E$ describes the contributions of the genetic ($V_A + V_D$) and environmental (V_E) variance components to V_P, the phenotypic variation. The formulation of V_A and V_D varies according to the population being tested; as an example, the dominant component will be absent from populations of homozygous inbred lines.

A major shortcoming of any QTL study is the low accuracy in detecting the real number of QTLs affecting the genetic variation of the investigated traits (Charcosset and Gallais 1996), particularly with populations of less than 150-200 progenies, which is the case in the majority of QTL studies reported so far. In a simulation study applied to experimental data, Beavis (1994) showed that with populations of ca. 100-200 progenies, only a modest fraction of QTLs was identified; furthermore, the effect of each single QTL was usually overestimated. Another study showed that with fewer than 500 progenies and independently from marker density, it is very difficult to detect QTLs of small effect (Beavis 1998). These predictions were substantiated in experiments carried out with maize mapping populations large enough (> 400 progenies) to allow for a meaningful subsampling (Openshaw and Frascaroli 1997; Melchinger et al. 1998). The effectiveness of detecting QTLs using the values of single environments and their mean has been evaluated in a number of studies. Jansen et al. (1995) concluded that the chance of detecting a QTL in several environments is small even in the absence of QTL x Environment (QTL × E) interaction. Inconsistency of QTL detection across environments has been reported by a number of authors (Paterson et al. 1991; Mather et al. 1997; Landi et al. 2002). Additionally, most of the current methods of QTL analysis, in terms of experimental design, population dimension and statistical approach are inadequate for an effective detection of epistatic QTL interactions.

2.2 Comparative QTL Mapping

The availability of QTL data for two or more mapping populations of the same species, enables us to compare the position of QTLs by using 'anchor markers' (usually RFLPs and/or SSRs) common to the investigated maps and/or indirectly through a comparison with a reference map of the same species when the number of markers shared by the mapping populations is limited (see also Chapter 7 by Frary et al. in Volume 1). Reference maps are available for a number of crops (e.g., maize: Davis et al. 1999; barley: Wenzl et al. 2006, Stein et al. 2007, Varshney et al. 2007; wheat: Somers et al. 2004, Lazo et al. 2004; potato: van Os et al. 2006; sorghum: Bowers et al. 2003, Fu et al. 2006). A number of these maps have been subdivided into sectors (bins; for an example on maize, see Davis et al. 1999) flanked by reliable RFLP and/or SSR markers, i.e. markers whose map information can be easily transferred from one segregating population to another. An additional advantage of a reference map is that it allows us to compare the map position of mutants with that of QTLs, thus contributing relevant information for the identification of possible candidate genes causally affecting the investigated trait. On this line, Robertson (1985) suggested that a mutant phenotype at a particular locus may simply be caused by an allele with a much more drastic effect in comparison to that of the QTL alleles at the same locus. In maize, Robertson's hypothesis has been validated for a QTL for plant height co-localized with the mutant *dwarf3* (Touzet et al. 1995). More recently, a mutant at a locus (*ERECTA*) influencing drought tolerance in *Arabidopsis* (Masle et al. 2005) has been shown to co-localize with a QTL for the same trait. Clearly, these results indicate that no real boundary exists between Mendelian and quantitative genetics, and that loci will be classified in one or the other category based on the magnitude and heritability of the effect of the parental alleles segregating in the mapping population. Hence, the information provided by mutants is of great value for QTL studies, particularly due to the availability of methods (e.g., EcoTILLING; Comai et al. 2004) that allow us to gauge the effect that allelic variation at candidate loci has on the phenotypic variability of target traits (Buckler and Thornsberry 2002).

Importantly, the ratio between the genetic and physical distance (i.e. cM vs. base pair) can vary greatly (up to 400-fold) according to the frequency of crossing-over events (i.e. recombination rate) along the chromosome (Boyko et al. 2002; see also Chapter 8 in Volume 1). This factor is of paramount importance for the level of genetic resolution achievable at a particular chromosome region, thus influencing the outcome of MAS and the possibility of positionally cloning a QTL.

2.3 Interpretation of Traits' Association through QTL Analysis

In order to more effectively apply an indirect selection for yield (or other agronomic traits) through the improvement of correlated traits, it is important to verify to what extent the correlation is due to linkage—instead of a cause-effect relationship on a truly biochemical or physiological basis (i.e. pleiotropy). This issue is particularly relevant when morpho-physiological features of the plant limit yield potential, as in the case of crops exposed to different abiotic stresses which negatively impact plant growth (e.g., conditions of limited water supply).

The presence of a genetic association between traits can be ascertained through the evaluation of (i) related families divergently selected from a population segregating for the target trait (Bolaños and Edmeades 1993; Landi et al. 2001; Rebetzke et al. 2002) and/or (ii) a number of unrelated genotypes differing in their mean value for the traits of interest (Condon et al. 1993; Tuberosa et al. 1994; Araus et al. 1997; El Hafid et al. 1998; Sanguineti et al. 2006). Each approach has pitfalls and rarely allows us to determine to what extent the association could be caused by pleiotropy and/or linkage. Conversely, QTL analysis enables us to investigate the genetic basis of trait association with greater precision simply by looking for co-location on the genetic map of the corresponding QTLs and comparing their effects on the investigated traits (Tanksley 1993). In particular, linkage can be distinguished from pleiotropy whenever a higher genetic resolution at the target QTL leads to the identification of progenies showing recombination for the associated traits, i.e. the parental alleles of the two linked genes controlling separately the two traits in a small number of recombinant progenies. Although the opposite finding (i.e. no recombinant progeny) cannot disprove tight association, pleiotropy becomes increasingly more plausible if co-segregation of the two traits is maintained when a larger mapping population is evaluated. An additional way to ascertain the cause of trait association at overlapping QTLs is to analyze the sign of the QTL effects for the investigated traits. In fact, linkage becomes a less plausible cause when a similar association between the genetic effects for the associated traits is present at all relevant QTLs. However, because some QTLs might affect phenotypic variability within a mapping population only at a particular growth stage and not throughout the entire life cycle of the plant, the presence of contrasting traits' associations at regions showing co-location of QTLs (e.g., for root size and grain yield) does not necessarily disprove pleiotropy since the sign of the association could vary according to the environmental conditions (e.g., water availability) present when a particular QTL was expressed or when it influenced phenotypic variability (Sanguineti et al. 1999). The final proof for pleiotropy can only be obtained through the cloning and direct manipulation (e.g., genetic engineering and/or mutagenesis) of the sequence underlying the QTL in question. In this context, a combination of mutants and near-isogenic lines (NILs) were deployed in *Arabidopsis* (i) to reveal a positive genetic correlation between flowering time and carbon-isotope discrimination ($\Delta^{13}C$), an indirect measure of water-use efficiency, and (ii) to demonstrate the presence of pleiotropic effects that controlled both traits (McKay et al. 2003). These results suggested that the correlated evolution of $\Delta^{13}C$ and the flowering time is explained in part by the fixation of pleiotropic alleles that alter both $\Delta^{13}C$ and flowering time.

2.4 Congenic Strains for Mapping and Cloning QTLs

An opportunity for investigating the genetic basis of overlapping QTLs for associated traits in greater detail is offered by the analysis of the traits starting from an adequately large (> 500 F_2 plants) mapping population obtained from the cross of congenic strains (e.g., NILs) contrasted for the parental chromosome regions present at the QTL of interest. NILs can be obtained through repeated selfings (at least six) of one or more individuals heterozygous at the QTL region prior to identifying the individuals homozygous for each one of the two parental segments (Tuinstra et al. 1998). As in this particular case, the NILs are derived from genetically heterogeneous families, they are referred to as heterogeneous inbred families (HIF). Alternatively, each parental line of the original mapping population evaluated for discovering the QTL can be used as a recurrent parent in a backcross scheme in which a single genotype heterozygous at the QTL in question is utilized as a donor of the alternative QTL regions; in this case, the congenic lines are identified as backcrossed-derived lines (BDLs). With NILs it is thus possible to 'Mendelize' major QTLs characterized by a sizeable additive effect.

Unlike genome-wide QTL studies where more than 100 families are usually screened, experiments with NILs involve a much smaller number of genotypes (two as a minimum), thus allowing for a much more refined and detailed analysis of the effects of the QTL (Salvi et al. 2002; Giuliani et al. 2005a, b). However, it should be appreciated that the results of these comparative studies based on NILs could to a certain extent be biased by the action of a number of linked genes affecting the investigated traits, more so when the region flanking the QTL extends for several cM. Importantly, the evaluation of crosses between pairs of NILs at different QTLs allows for accurately testing epistasis.

QTL cloning is an essential entry point for a more effective exploitation of sequence variability at selected target loci and to mine the allelic diversity available from germplasm collections (Tanksley and McCouch 1997). Although QTL cloning in plants is still in its infancy (Salvi and Tuberosa 2005; Price 2006), recent advances in high-throughput profiling of the genome and transcriptome, sequencing and reverse genetics tools (e.g., RNAi; Waterhouse and Helliwell 2003) are expected to streamline the procedures required to identify the sequences governing variation in quantitative traits. Until now, the molecular dissection of a QTL has been prevailingly achieved through positional cloning and association mapping. Both approaches exploit linkage disequilibrium to identify the most promising candidate gene(s) and benefit from the map information of candidate genes and mutants in the species under investigation and in closely related ones. Specific efforts toward enriching linkage maps with function-specific genes, sometimes also described as functional maps, have been undertaken and utilized for QTL analysis (Andersen and Lubberstedt 2003; Stein et al. 2007).

3 MOLECULAR-ASSISTED BREEDING TO IMPROVE CROP PERFORMANCE

Improving crops through conventional breeding has for the most part been achieved with little or no knowledge of the genetic basis of the selected traits, particularly yield, its components and the underlying morpho-physiological determinants. The main obstacle to the success of conventional breeding approaches is represented by the low heritability of yield, particularly under marginal conditions (e.g., low supply of nutrients and/or water) and as a consequence of large Genotype × Environment (G × E) interactions (Ceccarelli 1984; Blum 1988). Compared to empirical breeding, the analytical approach tries to identify heritable attributes, whose selection improves field performance. In this case, additional genetic variability can be acquired through adequate crossing schemes between germplasm accessions contributing new alleles for the trait of interest. If sufficient genetic variation is available for the target trait, other factors to be considered for its use as a selection criterion are its heritability and genetic correlation with yield, which should both be high. Actually, this approach is justified only if the product of heritability for the target trait and its genetic correlation with yield is higher than the heritablity of yield. Selection for a trait characterized by high heritability is feasible at early generations, thus reducing the high costs associated with a breeding scheme where selection for yield is delayed until later generations. Additionally, selecting for highly heritable traits in early generations reduces the number of genotypes to be profiled if MAS is applied at later generations.

3.1 Marker-assisted Selection

The ultimate goal of MAS is to increase the cost-effectiveness of the selection gain per unit time. Although the costs entailed by MAS are still quite high, the sizeable reduction in the time required to release an improved cultivar made possible through MAS can justify its application once suitable genes/QTLs are mapped. On a case-by-case basis, the breeder will evaluate the convenience of adopting MAS to improve the efficiency of the selection process. The effectiveness of MAS will depend on the identification of the relevant QTL alleles and their pyramiding in the correct combinations according also to the genetic background. This approach could be regarded as an evolution of the so-called 'ideotype' breeding (Donald 1968), with the main difference being that we now rely on more powerful tools for (i) dissecting the genetic basis of the phenotype at a much higher level of resolution and (ii) piecing back together the best QTL alleles in a sort of molecular jigsaw puzzle. This concept of 'breeding by design' (Peleman and van der Voort 2003) extends the concept of 'graphical genotypes' first introduced by Young and Tanksley (1989) to portray the parental origin and allelic contribution of each genotype on a genome-wide basis. Although a breeding-by-design approach is technically applicable to all major crops, its impact to improve quantitative traits has so far been marginal. This is due to our poor understanding of the genetic basis of this category of traits, their interaction with environmental factors and, most importantly, the difficulty in predicting the phenotypic value of a new genotype tailored through MAS for several loci. By contrast, more encouraging results have been obtained with Mendelian traits (e.g., disease resistance; Young 1999).

The molecular profiles obtained with molecular markers provide the basic information required to identify the so-called 'haplotype' of each individual at a target chromosome region. As an example, Fig. 1 shows the haplotype composition of a collection of durum wheat genotypes that have been profiled with SSRs to investigate their ancestry relationships. The collection includes several old 'founder' genotypes which have played an important role in modern breeding as parental lines of successful cultivars released in the past decades by durum wheat breeders (Maccaferri et al. 2005). The comparative analysis of the haplotype profiles highlights the contribution of a few chromosome regions which have been preferentially retained throughout the breeding activities of the past decades. It is plausible to postulate that these chromosomal regions are likely to harbor genes important for determining yield and its quality in durum wheat.

3.2 Modeling QTL Effects for Response to Environmental Factors

QTL-based modeling holds promise to allow for a more effective design of 'molecular ideotypes' on the basis of QTL effects for parameters of response curves revealed by exposing mapping populations to varying levels of the main environmental factors that limit growth (Hammer et al. 2006). Additionally, crop modeling can provide useful clues to unravel the genetic basis of G × E interactions and traits' plasticity (Chapman et al. 2003; Reymond et al. 2004; Welcker et al. 2007). An accurate estimate of the consistency of QTL effects in a particular genetic background can be obtained through extensive testing of the materials under varying environmental conditions (e.g., level of irrigation, nutrients, temperature, salinity, etc.) An example on how an ecophysiological model and QTL analysis can be integrated to investigate the genetic basis of leaf growth in response to drought has been provided by Reymond et al. (2003), who have identified QTLs for leaf elongation rate in maize as a function of air humidity, soil water status and meristem temperature. The same approach has also been applied by Welcker et al. (2007) to identify QTLs for the response of silk-elongation rate and anthesis-silking interval in maize exposed to water deficit. Therefore, crop modeling based on QTLs for the response to environmental variables might help us to identify the relevant genetic determinants (Yin et al. 2003, 2004; Welcker et al. 2007) and to more appropriately manage G × E interactions. Clearly, QTLs

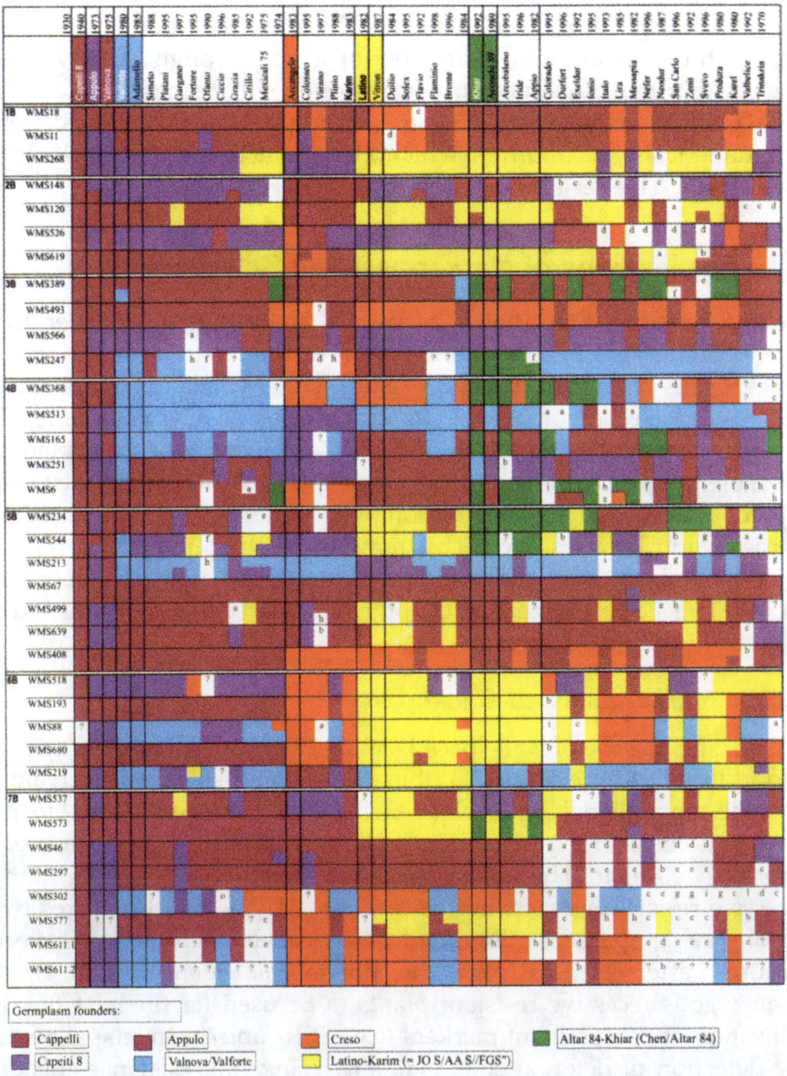

Fig. 1 Graphical genotyping of the B genome across 50 durum wheat cultivars (cvs.) from the Italian germplasm; the molecular profiles are based on 36 SSR loci. Founders (Cappelli, Capeiti 8, Appulo, Valnova, Valforte, Creso, Karim, Latino, Altar 84, and Khiar) included in most of the pedigrees of modern cvs. are represented in chronological order and highlighted by different colors. SSR alleles contributed by each founder are indicated using the corresponding founder-color. Groups of modern cultivars are reported next to the corresponding relevant founder. Cultivars with mixed origin are grouped in the last section.

with a limited QTL × E interaction and with a linear response to a particular environmental factor will provide more predictable opportunities to improve crops' performance through MAS. A major issue is that the crops' performance is often constrained by more than one environmental factor (e.g., drought and heat) occurring simultaneously, a condition which clearly increases the difficulty of predicting QTL effects.

3.3 Applications of Marker-assisted Selection

From a breeding standpoint, MAS can be exploited in order to (i) accelerate the backcross (BC) procedures required to transfer one or more genes/QTLs of interest into an existing cultivar, and (ii) improve the response to selection of target traits within a segregating population. The former is the application that so far has been most widely adopted and has more tangibly impacted breeding programs. MAS has also been used rather extensively to create congenic strains (e.g., NILs, BDLs, introgression libraries, etc.). This notwithstanding, these materials are usually obtained to identify and map genes/QTLs and, as such, usually do not have a direct impact on the outcome of breeding practices and the release of improved cultivars.

3.3.1 Marker-assisted Backcross

On average, the use of well-spaced (ca. 20-25 cM apart) markers can save two to three backcrosses (BCs), thus providing a substantial advantage over the conventional approach (Visscher et al. 1996; Servin and Hospital 2002) and considerably reducing the time required for the release of an improved cultivar. The advantage is greater for the incorporation of recessive resistance genes. In this case, phenotypic selection takes twice as long as compared to dominant genes, since a selfing generation is required after each BC for the phenotypic identification of the homozygous recessive resistant plants to be used for the next BC. The utilization of co-dominant markers (e.g., SSRs and/or indels) allows for the detection of heterozygous plants carrying the resistance-encoding allele, thereby saving one growing season for each BC cycle. The same holds true for dominant markers generating an additional fragment linked to the resistance-encoding allele (Ordon et al. 1999). During the past decade, marker-assisted BC has been routinely used by seed companies to introgress genes/QTLs from wild, sexually compatible relatives of crops and also to introgress transgenes into a broad range of elite materials, as in the case of the *Bt* gene (Ragot et al. 1995). At each generation, individuals heterozygous at the region flanking the target locus and suitable for an additional backcross are identified based exclusively on the results of molecular profiling. This provides the additional advantages of (i) avoiding the vagaries in phenotyping due to abiotic (e.g., rainfall) and/or biotic (e.g., presence of the virulent pathogen/pest) factors when the targeted locus controls a trait whose expression is influenced by such factors (e.g., anthesis-silking interval in maize in response to water deprivation), (ii) reducing the number of individuals to be screened in each selection cycle, and (iii) identifying individuals with a shorter segment introgressed from the donor parental line. An additional benefit is when the phenotyping of the trait under transfer is expensive and/or cumbersome like in the case of genes affecting tolerance to diseases/pests that require artificial inoculation in order to correctly identify those plants carrying the tolerant alleles (e.g., resistance to nematodes; Young 1996; Guo et al. 2005). For traits whose expression is influenced by an environmental variable (e.g., cold, heat, drought, flooding, etc.) accurate phenotyping is only possible under the right environmental conditions. Other cases where marker-assisted BC provides a distinct temporal advantage as compared to conventional procedures is when the phenotypic evaluation of the target trait is destructive or because the trait is expressed after flowering.

3.3.1.1 Recovery of the Recipient Genome

During backcrossing, different rates of recovery of the recipient genome are expected at the target region and the non-target chromosomes. As the percentage of the donor genome at non-target regions is reduced by half at each BC, at least six or seven BCs are required for a satisfactory recovery (ca. 99%) of the recipient genome. However, the number of BCs is frequently higher due to residual linkage drag around the target locus and it is not uncommon that up to nine or ten BCs are implemented before the release of an improved cultivar. Clearly, the long time required to complete the BC procedures reduces the probability of success of the new cultivar.

Selection at the target region. The formulas presented by Hospital (2003) can be used to compute the target control rate (TCR) for concordance between the allelic state at the target locus and the flanking markers. Table 1 reports the TCR value as related to the distance between the target locus and one or two flanking markers. The probability that an individual with the desired genotype at the linked marker(s) has lost the desired genotype at the target is (100 - TCR).

Table 1 Target control ratio (TCR) and minimal population size (N_{min}) when the target is a known gene (modified from Hospital 2003). Values in brackets refer to the utilization of two markers flanking the target gene.

Target marker distance (cM)	*TCR (%)*					
	BC_1	BC_2	BC_3	BC_5	BC_{10}	N_{min}
1	99.0 (100)	98.0 (100)	97.1 (100)	95.1 (100)	90.5 (99.9)	7 (7)
3	97.1 (99.9)	94.3 (99.8)	91.5 (99.7)	86.3 (99.6)	74.4 (99.1)	7 (8)
5	95.2 (99.8)	90.7 (99.5)	86.4 (99.3)	78.4 (98.8)	61.4 (97.5)	7 (8)
10	90.9 (99.0)	82.7 (98.0)	75.2 (97.1)	62.2 (95.2)	38.7 (90.6)	7 (9)
20	83.5 (96.3)	69.7 (92.6)	58.3 (89.2)	40.6 (82.6)	16.5 (68.2)	7 (11)
30	77.4 (92.2)	60.0 (85.0)	46.4 (78.3)	27.9 (66.5)	7.8 (44.3)	7 (12)

The values reported in Table 1 clearly indicate that the level of control made possible with only one marker is insufficient to keep the risk of losing the target allele below 5% throughout BC_5. Conversely, the level of control possible with two flanking markers is considerably higher even when the markers are not as tightly linked to the target locus. If the BC procedure targets a QTL instead of a Mendelian locus, the uncertainty about the exact position of the QTL introduces further complexity. Hospital and Charcosset (1997) have calculated the optimal number and positions of markers to control the backcrossing procedure for a QTL. Table 2 reports the TCR value when targeting a QTL.

Table 2 Target control ratio (TCR) and minimal population size when the target is a QTL (modified from Hospital 2003).

Confidence interval (cM)	*Number of markers*	*Optimal marker positions (cM from marker to expected QTL position)*				*TCR (%)*		
						BC_1	BC_3	N_{min}
10	1	0.0				98.5	95.6	7
	2	-3.6	+3.6			99.9	99.6	8
	3	-4.7	0.0	+4.7		100	99.9	8
	4	-5.4	-1.4	+1.4	+5.4	100	99.9	8
20	1	0.0				97.0	91.5	7
	2	-6.2	+6.2			99.6	98.7	8
	3	-8.5	0.0	+8.5		99.8	99.5	9
	4	-9.9	-2.6	+2.6	+9.9	99.9	99.8	9
40	1	0.0				94.4	84.5	7
	2	-10.4	+10.4			98.7	96.4	9
	3	-14.9	0.0	+14.9		99.5	98.5	10
	4	-17.7	-4.8	+4.8	+17.7	99.7	99.2	11

The values reported in Table 2 indicate that three markers optimally placed are adequate to provide a TCR > 99% even with a rather large confidence interval (40 cM). Additionally, Hospital and Charcosset (1997) showed that when markers are not optimally spaced and the precision on the location of the QTL is low, at least three markers per QTL should be used.

As the quantity of donor genes on the carrier chromosomes decreases much more slowly in comparison to the non-carrier chromosomes, after six BCs the majority of heterozygous loci with undesirable donor alleles are on the carrier chromosome, with the vast majority included in the intact fragment flanking the target locus. On an average, it is expected that up to 40 and 23 cM of the donor genome are retained after BC_6 and BC_{10}, respectively (Naveira and Barbadilla 1992). More importantly, the variance around these expected values can be high as shown by the data reported by Young and Tanksley (1989) following the introgression in tomato of a resistant allele at the *Tm-2* locus. The length of the fragment retained in BC_{11} and BC_{21} reached 50 and 36 cM, respectively. Based on theoretical computations, Hospital (2001) concluded that linkage drag can be reduced effectively only by using markers very close to the target locus, which entails the screening of large populations in order to produce the desired double-recombinant genotype, usually obtained by crossing the single recombinants on each side of the target locus (Young and Tanksley 1989).

Selection at non-target region and optimal population sizes. At the chromosomes not targeted by the BC procedure, it is expected that after n BCs, the probability that any locus remains heterozygous between the donor and the recipient is $(0.5)^n$, which means that each BC halves the residual level of heterozygosity. Based on this, six BCs ensure a level of similarity with the recurrent parent above 99%. Results in different species have shown that there may be a significant deviation from the 75% genomic portion of the recurrent parent expected in the BC_1 generation (Frisch et al. 1999a, b; Servin and Hospital 2002; Uptmoor et al. 2006), thus demonstrating the usefulness of genotype-based selection. The optimal population sizes over successive backcrosses can be defined by means of a computer program designed by Hospital and Decoux (2002) freely available at *http://moulon.inra.fr/~fred/programs*. Using this program, Hospital (2003) has shown that in order to optimize the number of molecular assays, it is preferable to genotype more individuals in advanced BC generations than in early BC generation, a conclusion also reached by Ribaut et al. (2002b). Applying this concept, the number of individuals required to obtain with a global probability of 99% a double recombinant for flanking markers at 2 cM from the target locus drops from ca. 24,000 in BC_1 to ca. 790 in BC_2 and ca. 380 in BC_3. Further details on the optimization of the BC strategy are reported in Hospital (2003).

3.3.1.2 Pyramiding Multiple Loci

With only a few exceptions (e.g., *mlo* resistance in barley) monogenic (Mendelian) resistance based on a single major gene (Leach et al. 2001) is non-durable due to the high mutation rate in plant pathogens (Takken and Joosten 2000), which can lead to the selection of new virulent strains able to overcome the physiological barrier of an individual resistance gene. Consequently, the durability of disease resistance can be increased by screening for new sources of resistance followed by tagging of the relevant genes and their incorporation in adapted cultivars (Huang et al. 1997; Singh et al. 2001). On the other hand, strategies have been developed for extending the useful life-time of single resistance genes in combination with creating cultivars with broader resistance spectra. The procedure for the accumulation of host resistance genes in a single line or cultivar is called pyramiding (Nelson 1978). More broadly, pyramiding can also be implemented for combining favorable alleles at Mendelian loci and QTLs controlling traits other than disease resistance (Langridge 2005). The advantage of pyramiding multiple alleles for resistance are particularly evident with diseases that require repeated inoculation and when phenotypic selection alone does not allow us to reliably combine and detect multiple resistance genes in one genotype, due to epistasis and/or reciprocal masking of genes effects irrespectively of epistatic interactions.

When more loci are targeted in a BC program, the minimum population size increases considerably and rapidly becomes a limiting factor. Hospital and Charcosset (1997) computed the size of the population required when several QTLs are to be backcrossed and concluded that it is futile to plan the simultaneous manipulation of more than three or four QTLs. This number can be increased to five or six if the targets are known loci rather than QTLs selected indirectly through linked markers. When the favorable alleles are contributed by different lines, the easiest strategy is to cross them to produce recombinant progenies and select the desired individuals. Multiple crosses might be required to pyramid all the desired alleles in one single genotype. This strategy can be deployed only with a rather limited number of target loci (up to six-eight; Charmet et al. 1999). Pyramiding alleles at a large number of loci (> 10) is best achieved with a recurrent selection strategy as illustrated by Hospital et al. (2000).

3.3.1.3 Marker-assisted Selection in a Segregating Population

Although MAS has been extensively used for the manipulation of single genes conferring tolerance to diseases/pests (Young 1996, 1999; see also Chapter 5, Volume 2 by Lightfoot and Iqbal) much more limited success has been reported for quantitatively inherited traits (see also Chapter 6, Volume 2 by Lambrides et al. and Chapter 7, Volume 2 by Graham et al.), mainly due to the inconsistency of the QTL effects after their introgression in different genetic backgrounds and the difficulty in identifying major QTLs with a sufficiently large and stable effect for justifying the costs required by MAS (Atlin and Lafitte 2002). While true QTL × E interaction due to variable expression of a trait may cause lack of consistency in QTL detection particularly with traits characterized by low to moderate heritability, the interaction between mapping population of small size (which entails limited power in QTL detection) with variable environments is probably an equally important factor causing inconsistency in QTL detection. This notwithstanding, partial success has been reported in a number of cases (Han et al. 1997; Huang et al. 1997; Lawson et al. 1997; Ribaut et al. 1997, 2002a; Ribaut and Ragot 2007; Toojinda et al. 1998; Hittalmani et al. 2000; Shen et al. 2001; Steele et al. 2006).

The outcome of MAS within a segregating population can be influenced by the genetic make-up of the targeted genetic background in terms of alleles present at other loci which may interact epistatically with the target locus, an aspect which becomes particularly relevant for quantitative traits due to the high number of genes that are usually involved in their control. In this respect, Podlich et al. (2004) have underlined that most evaluations of QTL mapping and MAS strategies have assumed that QTLs act independently, an assumption that has led Bernardo (2001) to suggest that MAS has little, if any, power over traditional phenotypic selection. With maize as a model species, computer simulation showed that gene information is most useful in selection when few loci (e.g., ten) control the trait. With many loci (> 50), the least squares estimates of gene effects become imprecise. Overall, these results led Bernardo (2001) to suggest that the typical reductionist approach pursued through QTL discovery strongly limits the outcome of MAS carried out for traits controlled by many QTLs.

Commonly, the approach to MAS relies on the assumption that the effects of the desirable QTL alleles will not vary after a number of selection cycles. It has been shown that the value of QTL alleles can differ depending on the genetic structure of the current set of germplasm in the breeding program (Wade 2002). To partially overcome this problem, Podlich et al. (2004) have proposed the 'Mapping As You Go' (MAYG) approach, namely a mapping-MAS strategy which partially accounts for the presence of epistasis by implementing MAS such that the estimated values of QTL alleles can evolve as the current germplasm evolves over cycles of selection. The MAYG method operates by cyclically re-estimating the value of QTL alleles each time a new set of germplasm is created during the breeding process. The effectiveness of the MAYG method over conventional MAS for a range of genetic models has been estimated through simulation (Podlich et al. 2004).

In the future, a more routine application of MAS will largely depend on the cost of the molecular analyses. High-throughput genotyping based on the scoring of markers that do not require the use of gels (Salvi et al. 2001) coupled with quick DNA extraction protocols are needed to streamline MAS and lower its cost. In maize, Morris et al. (2003) compared the cost-effectiveness of MAS and conventional breeding for the introgression of an elite allele at a single dominant gene into an elite line. In this particular case, neither method showed clear superiority in terms of both cost and speed: conventional breeding schemes are less expensive, but MAS-based breeding schemes are faster. These findings may question to some extent the validity of integrating MAS with conventional breeding programs. Nonetheless, as agronomically superior QTL alleles characterized by a limited G × E are discovered, MAS will offer increasing opportunities to accelerate the response to selection of quantitative traits.

3.3.1.4 Combining Marker-assisted Selection and Phenotypic Selection

For complex traits, it is likely that the loci targeted by MAS only account for a limited portion of the variability for the selected trait. In this case, it is preferable to control both the variability accounted for by the target QTLs and the background variability unaccounted for by markers, which requires combining MAS with phenotypic selection, hence allowing breeders to exploit also allelic variation at minor QTLs. The early work of Lande and Thompson (1990) produced a selection index combining molecular and phenotypic information weighed by coefficients whose value is chosen so the genetic gain is maximized. Considering molecular score M and phenotype P as the two correlated traits, Lande and Thompson (1990) applied classic selection index theory to compute the coefficients b_M and b_P of the index $I = b_M + b_P P$. For multiple markers, genotype effects can be summed over markers into a single score. This study showed that the method was most efficient with low heritable traits. Later works of Moreau et al. (1998) reached different conclusions:

due to a reduction of the power of detection of the effects associated with markers in a given population, the greatest advantage is achieved with traits with moderate heritability. An aspect that has been debated is to what extent the threshold of significance could be relaxed in order to increase the power to also identify, hence consider for selection, QTLs with small effects (Moreau et al. 1998); however, lowering the significance threshold will inevitably increase the probability of Type I error (i.e. false positive QTLs). Along this line, methods have been proposed that include all marker effects in the index, irrespective of their level of significance, and provide for a greater response to selection (Lange and Whittaker 2001; Meuwissen and Goddard 2001; Meuwissen et al. 2001). Other critical factors to consider when combining molecular and phenotypic selection are the cost of MAS (Moreau et al. 2000) and the overestimation of the effect of the QTLs (Beavis 1994; Melchinger et al. 1998), which may lead to overly emphasizing the MAS component of the selection scheme. Therefore, the application of MAS requires an accurate assessment of the cost-benefit entailed by its deployment on a case-by-case basis.

3.3 Mining Beneficial Alleles in Wild Relatives of Crops

Domestication has drastically reduced the level of genetic variability present in crops compared to their wild counterparts, an aspect probably even more relevant for traits playing a substantial role in survival under natural conditions. To overcome this limitation, Tanksley and Nelson (1996) have implemented advanced backcross QTL (AB-QTL) analysis, an approach that offers the opportunity to quickly discover and exploit beneficial QTL alleles present in wild germplasm. (Grandillo et al. 2007) The AB-QTL approach relies on the evaluation of BC families between an elite variety used as a recurrent parent and a donor accession, usually a wild species sexually compatible with the crop. Preferably, QTL analysis is delayed until the BC_2 generation and after selecting in BC_1 against characteristics with a negative effect on yield (e.g., ear shattering in small-grain cereals). The validity of AB-QTL was first proven in tomato (Tanksley et al. 1996) and rice (Xiao et al. 1998; Moncada et al. 2001) and more recently in barley (Talamè et al. 2004; Yun et al. 2006) and wheat (Huang et al. 2004; Narasimhamoorthy et al. 2006). These results are encouraging as to the possibility of using AB-QTL as a germplasm enhancement strategy for identifying wild progenitor alleles capable of improving the yield of the related crop, particularly under low-input agriculture and marginal environments. Ideally, introgression of such beneficial alleles should bear no negative consequences under more favorable and high-yielding conditions.

In addition to AB-QTL analysis, wild relatives can contribute to the identification of novel alleles for agronomically relevant traits by focusing on loci targeted by selection during both domestication and modern breeding (Peng et al. 2003; Yamasaki et al. 2005, Feuillet et al. 2008). To this end, the comparative analysis of the allelic diversity present in elite accessions, landraces and the undomesticated wild relative of a particular crop allows for the identification of loci devoid of genetic variation within the elite germplasm as a result of domestication and subsequent man-made selection. In this case, it is assumed that the loss of genetic diversity observed from the wild parent to the cultivated crop underscores the strong man-made selection at loci controlling traits of agronomic importance, including those relevant for adaptation to abiotic stress. Therefore, both this 'diversity screen' approach and the AB-QTL approach provide the distinct advantage of identifying agronomically valuable loci which would otherwise go undetected owing to a lack of allelic diversity in the genetic pool presently cultivated. Additionally, the diversity screen approach allows for the identification of candidate genes of potential agronomic importance even without prior knowledge of gene function and the phenotype of interest. However, Yamasaki et al. (2005) have recognized that the applicability of the diversity screen approach is limited by a number of factors, most notably that some of the identified genes may only be hitchhiking with selected genes present in their vicinity. The validity of diversity screening is being tested in maize (Yamasaki et al. 2005), a species particularly suited for this approach due to the extensive allelic richness of teosinte, the wild progenitor of maize, and the landraces presently cultivated.

3.5 *Arabidopsis* as a Model

Due to the relative ease in carrying out extensive phenotyping in *Arabidopsis*, an area worthy of future exploration relates to the identification of genes/QTLs for complex traits which play an important role in the adaptive response to biotic and abiotic stresses (Flavell 2005; Maggio et al. 2006; Zeng et al. 2007). As an example, we have summarized the results of a number of studies that have targeted root architecture and water-use efficiency, two complex traits with an important role in determining plant performance, particularly under water-limited conditions.

The power of applying QTL analysis for complex traits in *Arabidopsis* was recently shown by the identification of a naturally occurring allele at a new type of transcription factor regulating root development (Mouchel et al. 2004, 2006) and by the association of the activity of a sucrose-splitting enzyme and a QTL for root elongation (Sergeeva et al. 2006). The screening of nine *Arabidopsis* accessions grown under rigorously-controlled conditions revealed that one accession was unaffected by water deficit in terms of root growth (Granier et al. 2006). A mapping population including this accession as one of the parents will facilitate the identification of QTLs modulating the response of roots to a decrease in soil moisture. Fitz Gerald et al. (2006) examined the root systems of the closely related *Arabidopsis* ecotypes Landsberg *erecta* (*Ler*) and Columbia (Col) grown under mild osmotic stress conditions and found that *Ler* initiates more lateral root primordia, has an overall larger root system and shows a decreased sensitivity to osmolytes than Col. To understand the genetic basis for these differences, QTLs for root architecture and size under mild osmotic stress were mapped in a *Ler* x Col recombinant inbred population. Two major QTLs were identified and confirmed in NILs. The NILs also allowed for the identification of distinct physiological roles for the gene(s) at each locus (Fitz Gerald et al. 2006). This study provides useful insights to dissect the molecular basis for naturally occurring differences in developmental plasticity of the root system in *Arabidopsis*.

The importance of root architecture in the adaptive response to drought was further investigated by focusing on the inhibition of lateral root development as an adaptive response to drought. Xiong et al. (2006) showed that this response is partly mediated by the sensitivity to abscisic acid (ABA) and devised genetic screens to isolate *drought inhibition of lateral root growth* (*dig*) mutants with altered responses to drought or ABA in lateral root development. Characterization of these *dig* mutants revealed an altered drought stress tolerance, indicating that the response of lateral roots to drought stress is linked to drought adaptation. In *Arabidopsis*, ABA has also been shown to play a central role in mediating the regulatory effects of nitrate on root branching (Signora et al. 2001), thus adding evidence to its role in regulating plasticity of lateral root growth.

Positional cloning has pinpointed the role of the *ERECTA* gene in *Arabidopsis* in the regulation of plant transpiration efficiency, i.e. the ratio between photosynthesis and transpiration rates (Masle et al. 2005). The evaluation of a population of RILs derived from the cross of two ecotypes differing for $\Delta^{13}C$, an indirect measure of water-use efficiency, revealed a major QTL which, depending on growth conditions, accounted for up to 64% of the total phenotypic variation in $\Delta^{13}C$. This QTL (referred to as *transpiration efficiency 1*: *TE1*) peaked on the *ERECTA* gene, a putative leucine-rich-repeat receptor-like kinase known to affect inflorescence development. Subsequent experiments demonstrated a role for the *ERECTA* gene in controlling leaf photosynthetic capacity and balancing biochemical and stomatal limitations on photosynthesis. Additionally, it was shown that the function of *ERECTA* varies according to the genetic background and is modulated by other polymorphisms between the parental ecotypes. The main mechanisms by which *ERECTA* controls $\Delta^{13}C$ include effects on stomatal density, epidermal cell expansion, mesophyll cell proliferation and cell-to-cell contact. *ERECTA* is the first gene to be shown to act on the coordination between transpiration and photosynthesis, and, as such, to be identified as a transpiration efficiency gene, as opposed to simply a gene regulating stomatal density or photosynthesis. *ERECTA* homologues have been identified in several species; in view of this, the *ERECTA* locus is an interesting target for association studies in crops. Phylogenetic analysis has pinpointed that *ERECTA* has evolved during or before early Angiosperms evolution, hence underlining its likely role on plant fitness under the selective pressure of water-limited conditions. From an applicative standpoint, it is worth noting that genetic engineering of *ERECTA* improved transpiration efficiency without detectable penalty in growth, leading Masle et al. (2005) to suggest the potential value of manipulating *ERECTA* as a path for improving crop performance under dry conditions through a decreased stomatal conductance on one hand, and removal of stomatal limitations to improve yield potential under well-watered conditions on the other.

The growing interest in QTL mapping and cloning in *Arabidopsis* and other related model species will provide additional insight into the genetic basis of quantitative traits. To what extent this knowledge will have an impact on the release of better performing crops will depend on our capacity to identify crops' orthologs to *Arabidopsis* at the target QTLs and to single out the most agronomically valuable alleles at these loci (Flavell 2005).

4 TOWARDS HOLISTOMICS?

During the past decade, substantial progress has been achieved in devising platforms able to produce a large amount of data on the dynamics of the transcriptome, proteome and metabolome (see also Chapter 9, Volume 1 by Edwards and Batley). In turn, these 'omics' profiling data facilitate the identification of candidate genes and provide us with a more holistic picture of the molecular events characterizing cellular functions (Tuberosa et al. 2002).

Unlike from the classical QTL positional cloning approach in which we basically 'interrogate' an adequately large mapping population in order to identify the genetic determinants of QTLs, the candidate-gene approach capitalizes on gathering experimental evidence to support and validate the role of an ORF (e.g., glutamine synthetase gene) in governing variation for the putative target trait (e.g., nitrogen-use efficiency). More recently, the candidate-gene approach has also been deployed with no prior knowledge about QTLs for a specific target trait (for an example, see Yamasaki et al. 2005), The role of a candidate gene can be validated through forward- (e.g., genetic engineering, association mapping, etc.) or reverse-genetics (e.g., screening of knockout mutants, RNA interference, TILLING, etc.) approaches.

The major advantage of the candidate-gene approach is that it bypasses the tedious and expensive procedures required by positional cloning. Identifying suitable candidate genes and elucidating their function can be expedited by combining different approaches and high-throughput platforms applied to target crops and/or to model species (Markandeya et al. 2005; Bohnert et al. 2006). From a technical standpoint, combining laser-capture microdissection with the 'omics' platforms offers an unprecedented level of functional resolution at the tissue level, down to a single-cell layer (Nakazono et al. 2003). In maize, a combination of laser-capture microdissection and subsequent microarray analyses applied to the root pericycle of wild type and *rum1* mutant allowed Woll et al. (2005) to identify 19 genes involved in signal transduction, transcription and the cell cycle that are active before lateral root initiation. Among the different platforms available for the mass-scale profiling of the transcriptome, microarrays have been more frequently utilized to gauge the changes in gene expression, particularly in plants exposed to adverse conditions (Ozturk et al. 2002; Zinselmeier et al. 2002; Hazen et al. 2003; Seki et al. 2003; Yu and Setter 2003; Kawaguchi et al. 2004; Schnable et al. 2004; Giuliani et al. 2005a; Rensink 2005).

As the functional basis of a number of plant QTLs has been related to differences in the level of expression (Salvi and Tuberosa 2005; Salvi et al. 2007), QTL cloning could be facilitated through a direct transcriptome profiling approach applied to suitable genetic materials (Wayne and McIntyre 2002; Hazen et al. 2003). In this respect, an interesting application of transcriptome analysis is the identification of the so-called eQTLs, i.e. QTLs able to influence the level of expression (hence the 'e') of a particular gene. Applying QTL analysis to the level of gene expression of each progeny of a mapping population will identify the chromosome regions (eQTLs) that influence the observed variability in mRNA level of the profiled genes (Potokina et al. 2008).

Circumstantial evidence regarding the importance of each ORF in governing variability for the target trait can be obtained by comparing the map position of QTLs for the target trait with the map position of the ORFs and the corresponding eQTLs. In plants, eQTLs were first identified in maize (Schadt et al. 2003). The eQTLs were found to map both at the investigated gene loci for which expression was analyzed and at different chromosome locations, indicating allelic differences for *cis*- and *trans*-acting regulatory factors, respectively. Due to the high cost for profiling RNA samples of an entire mapping population, transcriptome profiling based on microarrays is better suited for studies involving a limited number of samples extracted from congenic strains differing at key genomic regions (e.g., NILs; Giuliani et al. 2005a) and/or bulked RNA samples obtained from the opposite tails (i.e. individuals with an extreme phenotype at target traits of mapping populations. More recently, cDNA-AFLPs have been used as an alternative to microarrays to identify eQTLs in *Arabidopsis* (Vuylsteke et al. 2006). However, distinct drawbacks of using a cDNA-AFLP platform are the limited coverage of the transcriptome and the difficulty to unequivocally identify the genes differentially expressed, a procedure which requires purification and sequencing of individual AFLP fragments. Although the feasibility of the eQTL approach has been demonstrated in the studies conducted so far in animal species and humans (Deutsch et al. 2005; Drake et al. 2006), the small population size of the experimental population usually considered, exacerbates the problem of detection of false-positive eQTLs (Kendziorski and Wang 2006). More specifically, since the same sample/tissue is used for differentiating phenotype and genotype across tens of thousands of ORFs, a number of underlying technical and statistical issues may increase the occurrence of false-positive eQTLs (West et al. 2006). For a more robust implementation of eQTL analysis, Haley and de Koning (2006) have proposed to combine expression studies with the fine-mapping of functional trait loci. Additionally, selective transcriptional profiling based on available information on individual quantitative traits and marker data has been advocated as a means to select a subset of individuals for optimizing the effectiveness and accuracy of RNA profiling (Nettleton and Wang 2006).

The interpretation of the results obtained with profiling experiments of samples collected from plants grown under controlled conditions (e.g., greenhouse and/or growth chambers) and their extrapolation to field-grown plants should take into due consideration the experimental conditions of the applied treatments. As an example, many of the studies aimed to identify genes differentially expressed under drought have been conducted using severe stress-intensity treatments applied in a very short time, commonly only a few hours (Ozturk et al. 2002; reviewed in Hazen et al. 2003). At the cellular level, such experimental conditions will be considerably more damaging compared to similar levels of water deficit that plant tissues experience in the field, where dehydration unfolds over a prolonged period of time (commonly days or weeks), thus allowing the plant to more effectively activate the molecular mechanisms leading to beneficial adaptive responses (e.g., osmotic adjustment, early flowering, thickening of leaf cuticles, etc.) that partially counteract the negative effects of drought. In barley, the changes in gene expression observed in leaves subjected to a rapidly induced dehydration (six hours) have shown a low correlation (r from 0.19 to 0.41) with the changes attained under a slower dehydration regime (ca. one week) which more closely mimicked field conditions (Talamè et al. 2007). Therefore, molecular results obtained under artificially induced experimental conditions should be dealt with cautiously and duly validated with evidence gathered from field studies prior to their utilization in a breeding context. Other factors that limit the effectiveness of profiling experiments to capture the key events triggering important adaptive responses to a stressing agent are the timing of the sampling in relation to the dynamics of the stressing event(s) and the occurrence of post-transcriptional and post-translational modifications.

Additional leads to the changes in cellular metabolism are provided by the profiling of the proteome (Hochholdinger et al. 2005, 2006; Matthiesen and Mutenda 2006; Sauer et al. 2006) and metabolome (Fiehn 2002; Steuer et al. 2003; Morgenthal et al. 2006) which, as compared to the transcriptome, are functionally 'closer' to the phenotype, thus accounting for post-transcriptional and post-translational regulation. Nonetheless, it should be appreciated that proteomics and metabolomics as currently performed report changes for a rather limited portion (ca. 5%) of the ORFs; additionally, proteomics are often unable to detect the changes in gene products (e.g., transcription factors) that despite their low level are more likely to play an important role in pivotal functions (e.g., signal transduction in response to biotic and abiotic stress) and consequently, to underline QTLs. Profiling the proteome of a mapping population offers the opportunity to identify QTLs influencing protein quantity (PQLs, protein quantity loci; de Vienne et al. 1999; Consoli et al. 2002). Co-localization of a PQL with its protein-coding locus would indicate that allelic differences at that locus influence the expression of the protein, whereas co-localization between a PQL and a QTL for a different trait would suggest an association between the candidate protein and trait variation (de Vienne et al. 1999; Pelleschi et al. 1999).

Metabolome profiling applied to a mapping population can also be used to identify QTLs regulating the level of a particular metabolite and verify its coincidence with QTLs for yield and/or genes involved in metabolic pathways. With the present technology, up to ca. 2,000 different metabolites can be profiled in a single sample (Fiehn 2002; Morgenthal et al. 2006; Sakurai and Shibata 2006). The identification and quantification of metabolites in a biological sample can provide important clues for a more comprehensive view of the functional characteristics under investigation. In maize, QTLs for invertase activity have been identified in a population subjected to drought stress (Pelleschi et al. 1999). The number of QTLs for invertase activity detected under drought (nine in total) was more than twice the number detected under well-watered conditions (four in total), an indirect indication of the important role of this enzyme under drought conditions. One QTL common to both treatments was located near *Ivr2*, an invertase-encoding gene on chromosome bin 5.03. More recently, Pelleschi et al. (2006) reported co-location between the activities of three enzymes (invertase, sucrose-P synthase and ADP-glucose pyrophosphorylase) involved in sucrose and starch metabolism and a corresponding structural gene, which can thus be considered as a candidate gene for explaining part of the variability in enzyme activity. These studies imply invertase activity as an important limiting factor for grain yield in maize exposed to drought during the reproductive phase (Boyer and Westgate 2004; McLaughlin and Boyer 2004; Boyer and McLaughlin 2007).

Finally, an area still largely unexplored relates to the so-called 'epigenomics', i.e. the heritable changes in genome functions that occur in the absence of changes to the DNA sequence itself that may provide new insights in our understanding of the variation in gene expression (Martienssen et al. 2005; Richards 2006). The concept of epigenomics postulates that heritable changes that do not involve DNA sequence level but instead modification of the non-DNA portion of the chromatin (e.g., DNA methylation, histone modifications, etc.) regulates growth and its plastic reaction to the environment. Epigenetics may in part account for the lack of reproducibility of QTL effects.

5 CONCLUDING REMARKS

At the dawn of the 21st century, a wealth of genomics and post-genomics platforms promise to expand our understanding of the genetic basis of crops' performance and to improve the efficiency of selection procedures for the release of improved cultivars (Varshney and Tuberosa 2007; Xu and Crouch 2008). While it is easy to predict success in meeting the former objective, reaching the latter appears a much more difficult and daunting undertaking due to the elusive nature of the QTLs that determine yield and its stability. In this respect, QTL cloning will provide valuable clues and will become a more routine practice in the next decade. This, in turn, will facilitate the identification of the best performing QTL alleles and their pyramiding in the most appropriate combinations through MAS. The real challenge will be identifying such combinations. It is under this QTL cloning framework that the molecular basis of quantitative traits will be dissected and better understood, providing important leads for more precisely tailoring the morphology of crops, their metabolism, physiology and, ultimately, productivity.

In terms of experimental materials utilized for QTL studies, growing attention will be devoted to the exploitation of progenies derived from multiparental crosses and adequately large mini-core collections of germplasm accessions with varying levels of linkage disequilibrium (Buckler and Thornsberry 2002; Maccafferri et al. 2006). In the mapping populations so far utilized for QTL discovery, most QTLs go undetected owing to the small size of the population and the presence of functionally monomorphic alleles. The application of linkage mapping to multiparental crosses (Yu et al. 2008) and association mapping to mini-core collections of adequate size (Ersoz et al. 2007) will improve the chances of identifying functional variability at QTLs and selecting the most beneficial alleles in terms of agronomic performance. On a finer scale, high-throughput proteome and metabolome profiling will accelerate the discovery of the causative mechanisms contributing to adaptive responses to adverse environmental conditions (e.g., drought, flooding, heat, etc.) whose frequency and intensity are expected to increase in the near future due to global warming. Nonetheless, the deluge of information originated through the molecular approaches and the 'omics' platforms will not automatically translate into knowledge on how to improve crops' performance. This translation will be facilitated by bridging more systematically the different platforms applied for QTL dissection. A 'systems biology'-like approach (Stephanopoulos et al. 2004) will be instrumental for optimizing the accurate integration and exploitation in breeding terms of all available information. On the molecular side, large EST collections or databases and unigene sets derived from cDNA libraries of different tissues and organs subjected to specific treatments are valuable sources of markers to construct functional maps that could more directly link candidate genes to QTLs (Cushman and Bohnert 2000; Stein et al. 2007). High-throughput DNA profiling based on SNP detection (Rostoks et al. 2005) and other advanced enabling platforms (McCallum et al. 2000; Uetz 2002;

Hardenbol et al. 2003; Wenzl et al. 2004; Mockler and Ecker 2005) will vastly improve our capacity for allele mining and haplotype profiling of the genome, thus streamlining QTL discovery and cloning procedures.

From an applicative standpoint, it is important to emphasize that any molecular approach aiming to discover genes/QTLs and test their effects should be carried out in an experimental context whose results are relevant and applicable to the conditions prevailing in farmers' fields (Hemamalini et al. 2000; Venuprasad et al. 2002; Steele et al. 2006; Reynolds and Tuberosa 2008). The successful exploitation of molecular approaches to enhance crops' performance will also depend on their successful integration with conventional breeding methodologies. Although it is not possible to predict to what extent molecular breeding will be adopted in the next decades, we are confident that the release of improved cultivars will be accelerated and made more efficient through a more systematic and targeted manipulation of the genes/QTLs governing the naturally occurring variation relevant for yield and its stability and sustainability.

References

Andersen JR, Lubberstedt T (2003) Functional markers in plants. Trends Plant Sci 8: 554-560

Araus JL, Amaro T, Zuhair Y, Nachit MM (1997) Effect of leaf structure and water status on carbon isotope discrimination in field-grown durum wheat. Plant Cell Environ 20: 1484-1494

Asins MJ (2002) Present and future of quantitative trait locus analysis in plant breeding. Plant Breed 121: 281-291

Atlin GN, Lafitte HR (2002) Marker-assisted breeding versus direct selection for drought tolerance in rice. Field screening for drought tolerance in crop plants with emphasis on rice. In: Saxena NP, O'Toole JC (eds) Proc Intl Workshop on Field Screening for Drought Tolerance in Rice, Dec 11-14, 2000, ICRISAT, Patancheru, India, pp 71-81

Beavis WD (1994) The power and deceit of QTL experiments: lessons from comparative QTL studies. In: Proc 49th Annu Corn and Sorghum Res Conf, Am Seed Trait Assoc, Washington, DC, USA, pp 250-266

Beavis WD (1998) QTL analysis: power, precision, and accuracy. In: Paterson AH (ed) Molecular Dissection of Complex Traits. CRC Press, Boca Raton, Florida, USA, pp 145-162

Bernardo R (2001) What if we knew all the genes for a quantitative trait in hybrid crops? Crop Sci 41: 1-4

Blum A (1988) Breeding for Stress Environments. CRC Press, Boca Raton, Florida, USA

Bohnert HJ, Gong Q, Li P, Ma S (2006) Unraveling abiotic stress tolerance mechanisms - Getting genomics going. Curr Opin Plant Biol 9: 180-188

Bolaños J, Edmeades GO (1993) Eight cycles of selection for drought tolerance in lowland tropical maize. I. Responses in grain yield, biomass, and radiation utilization. Field Crops Res 31: 233-252

Borlaug NE, Dowswell CR (2005) Feeding a world of ten billion people: a 21st century challenge. In: Tuberosa R, Phillips RL, Gale M (eds) Proc Intl Congr "In the Wake of the Double Helix: From the Green Revolution to the Gene Revolution", May 27-31, 2003, Bologna, Italy, Avenue Media, Bologna, Italy, pp 3-23

Bowers JE, Abbey C, Anderson S, Chang C, Draye X, Hoppe AH, Jessup R, Lemke C, Lennington J, Li Z, Lin YR, Liu SC, Luo L, Marler BS, Ming R, Mitchell SE, Qiang D, Reischmann K, Schulze SR, Skinner DN, Wang YW, Kresovich S, Schertz KF, Paterson AH (2003) A high-density genetic recombination map of sequence-tagged sites for sorghum, as a framework for comparative structural and evolutionary genomics of tropical grains and grasses. Genetics 165: 367-386

Boyer JS, McLaughlin JE (2007) Functional reversion to identify controlling genes in multigenic responses: analysis of floral abortion. J Exp Bot 58: 267-277

Boyer JS, Westgate ME (2004) Grain yields with limited water. J Exp Bot 55: 2385-2394

Boyko E, Kalendar R, Korzun V, Fellers J, Korol A, Schulman AH, Gill BS (2002) A high-density cytogenetic map of the *Aegilops tauschii* genome incorporating retrotransposons and defense-related genes: insights into cereal chromosome structure and function. Plant Mol Biol 48: 767-790

Broman KW, Wu H, Sen S, Churchill GA (2003) QTL mapping in experimental crosses. Bioinformatics 19: 889-890

Buckler ESI, Thornsberry JM (2002) Plant molecular diversity and applications to genomics. Curr Opin Plant Biol 5: 107-111

Ceccarelli S (1984) Plant responses to water stress: a review. Gen Agr 38: 43-73

Chapman S, Cooper M, Podlich D, Hammer G (2003) Evaluating plant breeding strategies by simulating gene action and dryland environment effects, Agron J 95: 99-113

Charcosset A, Gallais A (1996) Estimation of the contribution of quantitative trait loci (QTL) to the variance of a quantitative trait by means of genetic markers. Theor Appl Genet 93: 1193-1201

Charmet G, Robert N, Perretant M, Gay G, Sourdille P, Groos C, Bernard S, Bernard M (1999) Marker-assisted recurrent selection for cumulating additive and interactive QTLs in recombinant inbred lines. Theor Appl Genet 99: 1143-1148

Churchill GA, Doerge RW (1994) Empirical threshold values for quantitative trait mapping. Genetics 138: 963-971

Comai L, Young K, Till BJ, Reynolds SH, Greene EA, Codomo CA, Enns LC, Johnson JE, Burtner C, Odden AR, Henikoff S (2004) Efficient discovery of DNA polymorphisms in natural populations by Ecotilling. Plant J 37: 778-786

Condon AG, Richards RA, Farquhar GD (1993) Relationships between carbon isotope discrimination, water use efficiency and transpiration efficiency for dryland wheat. Aust J Agri Res 44: 1693-1711

Consoli L, Lefévre CL, Zivy M, de Vienne D, Damerval C (2002) QTL analysis of proteome and transcriptome variations for dissecting the genetic architecture of complex traits in maize. Plant Mol Biol 48: 575-581

Cushman JC, Bohnert HJ (2000) Genomic approaches to plant stress tolerance. Curr Opin Plant Biol 3: 117-124

Davis GL, McMullen MD, Baysdorfer C, Musket T, Grant D, Staebell M, Xu G, Polacco M, Koster L, Melia-Hancock S, Houchins K, Chao S, Coe EHJ (1999) A maize map standard with sequenced core markers, grass genome reference points and 932 expressed sequence tagged sites (ESTs) in a 1736 locus map. Genetics 152: 1137-1172

de Vienne D, Leonardi A, Damerval C, Zivy M (1999) Genetics of proteome variation for QTL characterization: application to drought-stress responses in maize. J Exp Bot 50: 303-309

Deutsch S, Lyle R, Dermitzakis ET, Attar H, Subrahmanyan L, Gehrig C, Parand L, Gagnebin M, Rougemont J, Jongeneel CV, Antonarakis SE (2005) Gene expression variation and expression quantitative trait mapping of human chromosome 21 genes. Hum Mol Genet 14: 3741-3749

Doerge RW (2002) Mapping and analysis of quantitative trait loci in experimental populations. Nat Rev Genet 3: 43-52

Donald CM (1968) The breeding of crop ideotypes. Euphytica 17: 385-403

Drake TA, Schadt EE, Lusis AJ (2006) Integrating genetic and gene expression data: application to cardiovascular and metabolic traits in mice. Mamm Genom 17: 466-479

El Hafid R, Smith DH, Karrou M, Samir K (1998) Physiological attributes associated with early-season drought resistance in spring durum wheat cultivars. Can J Plant Sci 78: 227-237

Ersoz ES, Yu J, Buckler ES (2007) Applications of linkage disequilibrium and association mapping. In: Varshney RK, Tuberosa R (eds) Genomics-Assisted Crop Improvement Vol 1: Genomics Approaches and Platform. Springer, Dordrecht, pp 97-120

Falconer DS (1981) Introduction to Quantitative Genetics, 2nd edn. Longman, London, UK

Feuillet C, Langridge P, Waugh R (2008) Cereal breeding takes a walk on the wild side. Trends Genet 24: 24-32

Fiehn O (2002) Metabolomics - the link between genotypes and phenotypes. Plant Mol Biol 48: 155-171

Fisher (1918) The correlation between relatives on the supposition of Mendelian inheritance. Trans Roy Soc Edinb 52: 399-433

Fitz Gerald JN, Lehti-Shiu MD, Ingram PA, Deak KI, Biesiada T, Malamy JE (2006) Identification of quantitative trait loci that regulate Arabidopsis root system size and plasticity. Genetics 172: 485-498

Flavell R (2005) Model plants, with special emphasis on Arabidopsis thaliana, and crop plant improvement. In: Tuberosa R, Phillips RL, Gale M (eds) Proc Intl Congr "In the Wake of the Double Helix: From the Green Revolution to the Gene Revolution", May 27-31, 2003, Bologna, Italy, Avenue media, Bologna, Italy, pp 365-378

Frisch M, Bohn M, Melchinger AE (1999a) Minimum sample size and optimal positioning of flanking markers in marker-assisted backcrossing for transfer of a target gene. Crop Sci 39: 967-975

Frisch M, Bohn M, Melchinger AE (1999b) Comparison of selection strategies for marker-assisted backcrossing of a gene. Crop Sci 39: 1295-1301

Fu Y, Wen TJ, Ronin YI, Chen HD, Guo L, Mester DI, Yang Y, Lee M, Korol AB, Ashlock DA, Schnable PS (2006) Genetic dissection of intermated recombinant inbred lines using a new genetic map of maize. Genetics 174: 1671-1683

Geldermann H (1975) Investigation on inheritançe of quantitative characters in animals by gene markers. I. Methods. Theor Appl Genet 46: 319-330

Giuliani S, Clarke J, Kreps JA, Sanguineti MC, Salvi S, Landi P, Zhu T, Tuberosa R (2005a) Microarray analysis of backcrossed-derived lines differing for *root-ABA1*, a major QTL controlling root characteristics and ABA concentration in maize. In: Tuberosa R, Phillips RL, Gale M (eds) Proc Intl Congr "In the Wake of the Double Helix: From the Green Revolution to the Gene Revolution", 27-31 May 2003, Bologna, Italy, Avenue media, Bologna, Italy, pp 463-490

Giuliani S, Sanguineti MC, Tuberosa R, Bellotti M, Salvi S, Landi P (2005b) *Root-ABA1*, a major constitutive QTL, affects maize root architecture and leaf ABA concentration at different water regimes. J Exp Bot 56: 3061-3070

Grandillo S, Tanksley SD, Zamir D (2007) Exploitation of natural biodiversity through genomics. In: Varshney RK, Tuberosa R (eds) Genomics-Assisted Crop Improvement, Vol 1: Genomics Approaches and Platforms, Springer, Dordrecht, pp 121-150

Granier C, Aguirrezabal L, Chenu K, Cookson SJ, Dauzat M, Hamard P, Thioux JJ, Rolland G, Bouchier-Combaud S, Lebaudy A, Muller B, Simonneau T, Tardieu F (2006) PHENOPSIS, an automated platform for reproducible phenotyping of plant responses to soil water deficit in *Arabidopsis thaliana* permitted the identification of an accession with low sensitivity to soil water deficit. New Phytol 169: 623-635

Guo B, Sleper DA, Arelli PR, Shannon JG, Nguyen HT (2005) Identification of QTLs associated with resistance to soybean cyst nematode races 2, 3 and 5 in soybean PI 90763. Theor Appl Genet 111: 965-971

Hackett C (2002) Statistical methods for QTL mapping in cereals. Plant Mol Biol 48: 585-599

Haley C, de Koning DJ (2006) Genetical genomics in livestock: potentials and pitfalls. Anim Genet 37: 10-12

Hallauer AR, Miranda Fo JB (1988) Quantitative Genetics in Maize Breeding, 2nd edn. Iowa State University Press, Ames, Iowa, USA

Hammer G, Cooper M, Tardieu F, Welch S, Walsh B, van Eeuwijk F, Chapman S, Podlich D (2006) Models for navigating biological complexity in breeding improved crop plants. Trends Plant Sci 11:587-593

Han F, Romagosa I, Ullrich SE, Jones BL, Hayes PM, Wesenberg D (1997) Molecular marker-assisted selection for malting quality traits in barley. Mol Breed 3: 427-437

Hardenbol P, Baner J, Jain M, Nilsson M, Namsaraev EA, Karlin-Neumann GA, Fakhrai-Rad H, Ronaghi M, Willis TD, Landegren U, Davis RW (2003) Multiplexed genotyping with sequence-tagged molecular inversion probes. Nat Biotechnol 21: 673-678

Hazen SP, Wu YJ, Kreps JA (2003) Gene expression profiling of plant responses to abiotic stress. Funct Integr Genom 3: 105-111

Hemamalini GS, Shashidhar HE, Hittalmani S (2000) Molecular marker assisted tagging of morphological and physiological traits under two contrasting moisture regimes at peak vegetative stage in rice (*Oryza sativa* L.). Euphytica 112: 69-78

Hittalmani S, Parco A, Mew TV, Zeigler RS, Huang N (2000) Fine mapping and DNA marker-assisted pyramiding of the three major genes for blast resistance in rice. Theor Appl Genet 100: 1121-1128

Hochholdinger F, Sauer M, Dembinsky D, Hoecker N, Muthreich N, Saleem M, Yan L (2006) Proteomic dissection of plant development. Proteomics 6: 4076-4083

Hochholdinger F, Woll K, Guo L, Schnable PS (2005) The accumulation of abundant soluble proteins changes early in the development of the primary roots of maize (*Zea mays* L.). Proteomics 5: 4885-4893

Hospital F (2001) Size of donor chromosome segments around introgressed loci and reduction of linkage drag in marker-assisted backcross programs. Genetics 158: 1363-1379

Hospital F (2003) Marker-assisted breeding. In: Newbury HJ (ed) Plant Molecular Breeding. Blackwell Scientific, London, UK, pp 30-56

Hospital F, Charcosset A (1997) Marker-assisted introgression of quantitative trait loci. Genetics 147: 1469-1485

Hospital F, Decoux G (2002) Popmin: a program for the numerical optimization of population sizes in marker-assisted backcross programs. J Hered 93: 383-384

Hospital F, Goldringer I, Openshaw S (2000) Efficient marker-based recurrent selection for multiple quantitative trait loci. Genet Res 75: 357-368

Huang B, Angeles ER, Domingo J, Magpantay G, Singh S, Zhang G, Kumaravadivel N, Benett J, Khush GS (1997) Pyramiding of bacterial blight resistance genes in rice: marker-assisted selection using RFLP and PCR. Theor Appl Genet 95: 313-320

Huang XQ, Kempf H, Ganal MW, Roder MS (2004) Advanced backcross QTL analysis in progenies derived from a cross between a German elite winter wheat variety and a synthetic wheat (*Triticum aestivum* L.). Theor Appl Genet 109: 933-943

Jansen RC, Stam P (1994) High resolution of quantitative traits into multiple loci via interval mapping. Genetics 136: 1447-1455

Jansen RC, Van Ooijen JW, Stam P, Lister C, Dean C (1995) Genotype by environment interaction in genetic mapping of multiple quantitative trait loci. Theor Appl Genet 91: 33-37

Kao CH, Zeng ZB, Teasdale RD (1999) Multiple interval mapping for quantitative trait loci. Genetics 159: 1203-1216

Kawaguchi R, Girke T, Bray EA, Bailey-Serres J (2004) Differential mRNA translation contributes to gene regulation under non-stress and dehydration stress conditions in *Arabidopsis thaliana*. Plant J 38: 823-839

Kendziorski C, Wang P (2006) A review of statistical methods for expression quantitative trait loci mapping. Mamm Genom 17: 509-517

Khush GS (2005) Green Revolution: challenges ahead. In: Tuberosa R, Phillips RL, Gale M (eds) Proc Intl Congr "In the Wake of the Double Helix: From the Green Revolution to the Gene Revolution", May 27-31, 2003, Bologna, Italy. Avenue media, Bologna, Italy, pp 37-51

Koebner RMD, Summers RW (2003) 21st century wheat breeding: plot selection or plate detection? Trends Biotechnol 21: 59-63

Korol AB, Ronin YI, Kirzhner VM (1995) Interval mapping of quantitative trait loci employing correlated trait complexes. Genetics 140: 1137-1147

Korol AB, Ronin YI, Nevo E (1998) Approximate analysis of QTL-environment interaction with no limits on the number of environments. Genetics 148: 2015-2028

Lande R, Thompson R (1990) Efficiency of marker-assisted selection in the improvement of quantitative traits. Genetics 124: 743-756

Lander ES, Botstein D (1989) Mapping Mendelian factors underlying quantitative traits using RFLP linkage maps. Genetics 121: 185-199

Landi P, Sanguineti MC, Conti S, Tuberosa R (2001) Direct and correlated responses to divergent selection for leaf abscisic acid concentration in two maize populations. Crop Sci 41: 335-344

Landi P, Sanguineti MC, Darrah LL, Giuliani MM, Salvi S, Tuberosa R (2002) Detection of QTLs for vertical root pulling resistance in maize and overlaps with QTLs for root traits in hydroponics and for grain yield at different water regimes. Maydica 47: 233-243

Lange C, Whittaker J (2001) On prediction of genetic values in marker-assisted selection. Genetics 159: 1375-1381

Langridge P (2005) Molecular breeding of wheat and barley. In: Tuberosa R, Phillips RL, Gale M (eds) Proc Intl Congr "In the Wake of the Double Helix: From the Green Revolution to the Gene Revolution", May 27-31, 2003, Bologna, Italy, Avenue media, Bologna, Italy, pp 279-286

Lawson DM, Lunde CF, Mutschler MA (1997) Marker-assisted transfer of acylsugar-mediated pest resistance from the wild tomato, *Lycopersicon pennellii*, to the cultivated tomato, *Lycopersicon esculentum*. Mol Breed 3: 307-317

Lazo GR, Chao S, Hummel DD, Edwards H, Crossman CC, Lui N, Matthews DE, Carollo VL, Hane DL, You FM, Butler GE, Miller RE, Close TJ, Peng JH, Lapitan NLV, Gustafson JP, Qi LL, Echalier B, Gill BS, Dilbirligi M, Randhawa HS, Gill KS, Greene RA, Sorrells ME, Akhunov ED, Dvorák J, Linkiewicz AM, Dubcovsky J, Hossain KG, Kalavacharla V, Kianian SF, Mahmoud AA, Miftahudin, Ma X-F, Conley EJ, Anderson JA, Pathan MS, Nguyen HT, McGuire PE, Qualset C. O, Anderson OD (2004) Development of an Expressed Sequence Tag (EST) resource for wheat (*Triticum aestivum* L.): EST generation, unigene analysis, probe selection and bioinformatics for a 16,000-locus bin-delineated map. Genetics 168: 585-593

Leach JE, Leung H, Wang GL (2001) Dissection of defence response pathways in rice. In: Rice Biotechnology: Improving Yield, Stress Tolerance and Grain Quality, No 236. Novartis Foundation, Basel, Switzerland, pp 190-204

Lee M (1995) DNA markers and plant breeding programs. Adv Agron 55: 265-344

Lynch M, Walsh B (1998) Genetics and Analysis of Quantitative Traits. Sinuaer Assoc, Sunderland, MA, USA

Maccaferri M, Sanguineti MC, Corneti S, Araus-Ortega JL, Ben Salem M, Bort J, DeAmbrogio E, Garcia del Moral LF, Demontis Montis A, El-Ahmed A, Maalouf F, Machlab H, Martos V, Moragues M, Motawaj J, Nachit M, Nserallah N, Ouabbou H, Royo C, Slama A, Tuberosa R (2008) Quantitative trait loci for grain yield and adaptation of durum wheat (*Triticum durum* Desf.) across a wide range of water availability. Genetics 178: 489-511

Maccaferri M, Sanguineti MC, Natoli V, Araus-Ortega JL, Ben Salem M, Bort J, Chenenaoui S, DeAmbrogio E, Garcia Del Moral L, De Montis A, El-Ahmed A, Maalouf F, Machlab H, Moragues M, Motawaj J, Nachit M, Nserallh N, Ouabbou H, Royo C, Tuberosa R (2006) A panel of elite accessions of durum wheat (*Triticum durum* Desf.) suitable for association mapping studies. Plant Genet Res 4: 79-85

Maccaferri M, Sanguineti MC, Noli E, Tuberosa R (2005) Population structure and long-range linkage disequilibrium in a durum wheat elite collection. Mol Breed 15: 271-290

McLaughlin JE, Boyer JS (2004) Sugar-responsive gene expression, invertase activity, and senescence aborting maize ovaries at low water potentials. Ann Bot 94: 675-689

Maggio A, Zhu JK, Hasegawa PM, Bressan RA (2006) Osmogenetics: Aristotle to Arabidopsis. Plant Cell 18: 1542-1557

Markandeya G, Babu PR, Lachagari VBR, Feltus FA, Paterson AH, Reddy AR (2005) Functional genomics of drought stress response in rice: transcript mapping of annotated unigenes of an indica rice (*Oryza sativa* L. cv. Nagina 22). Curr Sci 89: 496-514

Martienssen RA, Doerge RW, Colot V (2005) Epigenomic mapping in Arabidopsis using tiling microarrays. Chrom Res 13: 299-308

Masle J, Gilmore SR, Farquhar GD (2005) The ERECTA gene regulates plant transpiration efficiency in *Arabidopsis*. Nature 436: 866-870

Mather DE, Tinker NA, LaBerge DE, Edney M, Jones BL, Rossnagel BG, Legge WG, Briggs KG, Irvine RB, Falk DE, Kasha KJ (1997) Regions of the genome that affect grain and malt quality in a North American two-row barley cross. Crop Sci 37: 544-554

Matthiesen R, Mutenda KE (2006) Introduction to proteomics. Meth Mol Biol 367: 1-36

McCallum CM, Comai L, Greene EA, Henikoff S (2000) Targeted screening for induced mutations. Nat Biotechnol 18: 455-457

McKay JK, Richards JH, Mitchell-Olds T (2003) Genetics of drought adaptation in *Arabidopsis thaliana*: I. Pleiotropy contributes to genetic correlations among ecological traits. Mol Ecol 12: 1137-1151

Melchinger AE, Utz HF, Schon CC (1998) Quantitative trait locus (QTL) mapping using different testers and independent population samples in maize reveals low power of QTL detection and large bias in estimates of QTL effects. Genetics 149: 383-403

Meuwissen TH, Goddard ME (2001) Prediction of identity by descent probabilities from marker-haplotypes. Genet Sel Evol 33: 605-34

Meuwissen T, Hayes B, Goddard M (2001) Prediction of total genetic value using genome-wide dense marker maps. Genetics 157: 1819-1829

Mockler TC, Ecker JR (2005) Applications of DNA tiling arrays for whole-genome analysis. Genomics 85: 1-15

Moncada P, Martinez C, Borrero JMC, Gauch HJ, Guimaraes E, Tohme J, McCouch SR (2001) Quantitative trait loci for yield and yield components in an *Oryza sativa* × *Oryza rufipogon* BC_2F_2 population evaluated in an upland environment. Theor Appl Genet 102: 41-52

Moreau L, Charcosset A, Hospital F, Gallais A (1998) Marker-assisted selection efficiency in populations of finite size. Genetics 148: 1353-1365

Moreau L, Lemarié S, Charcosset A, Gallais A (2000) Economic efficiency of one cycle of marker-assisted selection. Crop Sci 40: 329-337

Morgante M, Salamini F (2003) From plant genomics to breeding practice. Curr Opin Biotechnol 14: 214-219

Morgenthal K, Weckwerth W, Steuer R (2006) Metabolomic networks in plants: Transitions from pattern recognition to biological interpretation. Biosystems 83: 108-117

Morris M, Dreher K, Ribaut J, Khairallah M (2003) Money matters (II): costs of maize inbred line conversion schemes at CIMMYT using conventional and marker-assisted selection. Mol Breed 11: 235-247

Mouchel CF, Briggs GC, Hardtke CS (2004) Natural genetic variation in *Arabidopsis* identifies *BREVIS RADIX*, a novel regulator of cell proliferation and elongation in the root. Gen Dev 18: 700-714

Mouchel CF, Osmont KS, Hardtke CS (2006) BRX mediates feedback between brassinosteroid levels and auxin signalling in root growth. Nature 443: 458-461

Nakazono M, Qiu F, Borsuk LA, Schnable PS (2003) Laser-capture microdissection, a tool for the global analysis of gene expression in specific plant cell types: identification of genes expressed differentially in epidermal cells or vascular tissues of maize. Plant Cell 15: 583-596

Narasimhamoorthy B, Gill BS, Fritz AK, Nelson JC, Brown-Guedira GL (2006) Advanced backcross QTL analysis of a hard winter wheat × synthetic wheat population. Theor Appl Genet 112: 787-796

Naveira H, Barbadilla A (1992) The theoretical distribution of lengths of intact chromosome segments around a locus held heterozygous with backcrossing in a diploid species. Genetics 130: 205-209

Nelson RR (1978) Genetics of horizontal resistance to plant diseases. Annu Rev Phytopathol 16: 359-378

Nettleton D, Wang D (2006) Selective transcriptional profiling for trait-based eQTL mapping. Anim Genet 37: 13-17

Openshaw S, Frascaroli E (1997) QTL detection and marker-assisted selection for complex traits in maize. In: Proc 52nd Annu Corn and Sorghum Res Conf, Am Seed Trade Assoc, Washington, DC, USA, pp 44-53

Ordon F, Schiemann A, Pellio B, Dauck V, Bauer E, Streng S, Friedt W, Graner A (1999) Application of molecular markers in breeding for resistance to the barley yellow mosaic virus complex. J Plant Dis Protect 106: 256-264

Ozturk ZN, Talamè V, Deyholos M, Michalowski CB, Galbraith DW, Gozukirmizi N, Tuberosa R, Bohnert HJ (2002) Monitoring large-scale changes in transcript abundance in drought- and salt-stressed barley. Plant Mol Biol 48: 551-573

Paterson AH, Damon S, Hewitt JK, Zamir D, Rabinowitch HD, Lincoln SE, Lander ES, Tanksley SD (1991) Mendelian factors underlying quantitative traits in tomato: comparison across species, generations, and environments. Genetics 127: 181-197

Peleman JD, van der Voort, JR (2003) Breeding by design. Trends Plant Sci 8: 330-334

Pelleschi S, Guy S, Kim JY, Pointe C, Mahe A, Barthes L, Leonardi A, Prioul JL (1999) *Ivr2*, a candidate gene for a QTL of vacuolar invertase activity in maize leaves. Gene-specific expression under water stress. Plant Mol Biol 39: 373-380

Pelleschi S, Leonardi A, Rocher JP, Cornic G, de Vienne D, Thévenot C, Prioul JL (2006) Analysis of the relationships between growth, photosynthesis and carbohydrate metabolism using quantitative trait loci (QTLs) in young maize plants subjected to water deprivation. Mol Breed 17: 21-39

Peng J, Ronin Y, Fahima T, Roder MS, Li Y, Nevo E, Korol A (2003) Domestication quantitative trait loci in *Triticum dicoccoides*, the progenitor of wheat. Proc Natl Acad Sci USA 100: 2489-2494

Podlich DW, Winkler CR, Cooper M (2004) Mapping as you go: an effective approach for marker-assisted selection of complex traits. Crop Sci 44: 1560-1571

Potokina E, Druka A, Luo ZW, Wise R, Waugh R, Kearsey M (2008) Gene expression quantitative trait locus analysis of 16,000 barley genes reveals a complex pattern of genome-wide transcriptional regulation. Plant J 53: 90-101

Price AH (2006) Believe it or not, QTLs are accurate! Trends Plant Sci 11: 213-216

Prioul JL, Quarrie S, Causse M, de Vienne D (1997) Dissecting complex physiological functions through the use of molecular quantitative genetics. J Exp Bot 48: 1151-1163

Quarrie SA (1996) New molecular tools to improve the efficiency of breeding for increased drought resistance. Plant Growth Reg 20: 167-178

Ragot M, Biasiolli M, Delbut MF, Dell'Orco A, Margarini L, Thevenin P, Vernoy J, Vivant J, Zimmermann R, Gay G (1995) Marker-assisted backcrossing: a practical example. In: Techniques et Utilisations des Marqueurs Moléculaires, Les Colloques, No 72. INRA, Paris, France, pp 45-46

Rebetzke GJ, Condon AG, Richards RA, Farquhar GD (2002) Selection for reduced carbon isotope discrimination increases aerial biomass and grain yield of rainfed bread wheat. Crop Sci 42: 739-745

Rensink WA (2005) Microarray expression profiling resources for plant genomics. Trends Plant Sci 10: 603-609

Reymond M, Muller B, Leonardi A, Charcosset A, Tardieu F (2003) Combining quantitative trait loci analysis and an ecophysiological model to analyze the genetic variability of the responses of maize leaf growth to temperature and water deficit. Plant Physiol 131: 664-675

Reymond M, Muller B, Tardieu F (2004) Dealing with the genotype × environment interaction via a modelling approach: a comparison of QTLs of maize leaf length or width with QTLs of model parameters. J Exp Bot 55: 2461-2472

Reynolds M, Tuberosa R (2008) Translational research impacting on crop productivity in drought-prone environments. Curr Opin Plant Biol 11: 171-179

Ribaut JM, Hoisington D (1998) Marker-assisted selection: new tools and strategies. Trends Plant Sci 3: 236-239

Ribaut JM, Ragot M (2007) Marker-assisted selection to improve drought adaptation in maize: the backcross approach, perspectives, limitations, and alternatives. J Exp Bot 58: 351-360

Ribaut JM, Jiang C, Gonzalez-de-Leon D, Edmeades G, Hoisington DA (1997) Identification of quantitative trait loci under drought conditions in tropical maize. 2. Yield components and marker-assisted selection strategies. Theor Appl Genet 94: 887-896

Ribaut JM, Banziger M, Betran J, Jiang C, Edmeades GO, Dreher K, Hoisington D (2002a) Use of molecular markers in plant breeding: drought tolerance improvement in tropical maize In: Kang MS (ed) Quantitative Genetics, Genomics, and Plant Breeding. CABI Publishing, Wallingford, UK, pp 85-99

Ribaut JM, Jiang C, Hoisington D (2002b) Efficiency of a gene introgression experiment by backcrossing. Crop Sci 42: 557-565

Richards EJ (2006) Inherited epigenetic variation - revisiting soft inheritance. Nat Rev Genet 7: 395-401

Robertson DS (1985) A possible technique for isolating genic DNA for quantitative traits in plant. J Theor Biol 117: 1-10

Rostoks N, Borevitz JO, Hedley PE, Russell J, Mudie S, Morris J, Cardle L, Marshall DF, Waugh R (2005) Single-feature polymorphism discovery in the barley transcriptome. Genom Biol 6(6): R54. Epub 2005 May 11

Rostoks N, Mudie S, Cardle L, Russell J, Ramsay L, Booth A, Svensson JT, Wanamaker SI, Walia H, Rodriguez EM, Hedley PE, Liu H, Morris J, Close TJ, Marshall DF, Waugh R (2005) Genome-wide SNP discovery and linkage analysis in barley based on genes responsive to abiotic stress. Mol Genet Genom 274: 515-527

Sakurai N, Shibata D (2006) KaPPA-View for integrating quantitative transcriptomic and metabolomic data on plant metabolic pathway maps. J Pesti Sci 31: 293-295

Salvi S, Tuberosa R (2005) To clone or not to clone plant QTLs: present and future challenges. Trends Plant Sci 10: 297-304

Salvi S, Tuberosa R, Phillips RL (2001) Development of PCR-based assays for allelic discrimination in maize by using the 5-nuclease procedure. Mol Breed 8: 169-176

Salvi S, Tuberosa R, Chiapparino E, Maccarerri M, Veillet S, Van Beuningen L, Isaac P, Edwards K, Phillips RL (2002) Toward positional cloning of *Vgt1*, a QTL controlling the transition from the vegetative to the reproductive phase in maize. Plant Mol Biol 48: 601-613

Salvi S, Sponza G, Morgante M, Tomes D, Xiaomu Nui X, Fengler KA, Meeley R, Ananiev EV, Svitashev S, Bruggemann E, Li B, Haney CF, Radovic S, Zaina G, Rafalski JA, Tingey SV, Miao GH, Phillips RL, Tuberosa R (2007) Conserved non coding genomic sequences controlling flowering time differences in maize. Proc Natl Acad Sci USA 104: 11376-11381

Sanguineti MC, Tuberosa R, Landi P, Salvi S, Maccaferri M, Casarini E, Conti S (1999) QTL analysis of drought-related traits and grain yield in relation to genetic variation for leaf abscisic acid concentration in field-grown maize. J Exp Bot 50: 1289-1297

Sanguineti MC, Duvick DN, Smith S, Landi P, Tuberosa R (2006) Effects of long term selection on seedling traits and ABA accumulation in commercial maize hybrids. Maydica 51: 329-338

Sauer M, Jakob A, Nordheim A, Hochholdinger F (2006) Proteomic analysis of shoot-borne root initiation in maize (*Zea mays* L.). Proteomics 6: 2530-2541

Schadt EE, Monks SA, Drake TA, Lusis AJ, Che N, Colinayo V, Ruff TG, Milligan SB, Lamb JR, Cavet G, Linsley PS, Mao M, Stoughton RB, Friend SH (2003) Genetics of gene expression surveyed in maize, mouse and man. Nature 422: 297-302

Schnable PS, Hochholdinger F, Nakazono M (2004) Global expression profiling applied to plant development. Curr Opin Plant Biol 7: 50-56

Seki M, Kamei A, Yamaguchi-Shinozaki K, Shinozaki K (2003) Molecular responses to drought, salinity and frost: common and different paths for plant protection. Curr Opin Biotechnol 14: 194-199

Sen S, Churchill GA (2001) A statistical framework for quantitative trait mapping. Genetics 159: 371-387

Sergeeva LI, Keurentjes JJ, Bentsink L, Vonk J, van der Plas LH, Koornneef M, Vreugdenhil D (2006) Vacuolar invertase regulates elongation of *Arabidopsis thaliana* roots as revealed by QTL and mutant analysis. Proc Natl Acad Sci USA 103: 2994-2999

Servin B, Hospital F (2002) Optimal positioning of markers to control genetic background in marker-assisted backcrossing. J Hered 93: 214-217

Shen L, Courtois B, McNally KL, Robin S, Li Z (2001) Evaluation of near-isogenic lines of rice introgressed with QTLs for root depth through marker-aided selection. Theor Appl Genet 103: 75-83

Signora L, De Smet I, Foyer CH, Zhang H (2001) ABA plays a central role in mediating the regulatory effects of nitrate on root branching in Arabidopsis. Plant J 28: 655-662

Singh S, Sidhu JS, Huang N, Vikal Y, Li Z, Brar DS, Dhaliwal HS, Khush GS (2001) Pyramiding three bacterial blight resistance genes (*xa5, xa13* and *Xa21*) using marker-assisted selection into indica rice cultivar PR106. Theor Appl Genet 102: 1011-1015

Somers DJ, Isaac P, Edwards K (2004) A high-density microsatellite consensus map for bread wheat (*Triticum aestivum* L.). Theor Appl Genet 109: 1105-1114

Steele KA, Price AH, Shashidhar HE, Witcombe JR (2006) Marker-assisted selection to introgress rice QTLs controlling root traits into an Indian upland rice variety. Theor Appl Genet 112: 208-221

Stein N, Prasad M, Scholz U, Thiel T, Zhang H, Wolf M, Kota R, Varshney RK, Perovic D, Grosse I, Graner A (2007) A 1,000-loci transcript map of the barley genome: new anchoring points for integrative grass genomics. Theor Appl Genet 114: 823-839

Stephanopoulos G, Alper H, Moxley J (2004) Exploiting biological complexity for strain improvement through system biology. Nat Biotechnol 22: 1261-1267

Steuer R, Kurths J, Fiehn O, Weckwerth W (2003) Observing and interpreting correlations in metabolomic networks. Bioinformatics 19: 1019-1026

Stich B, Yu J, Melchinger AE, Piepho HP, Utz P, Maurer HP, Buckler (2007) Power to detect higher-order epistatic interactions in a metabolic pathway using a new mapping strategy. Genetics 176: 563-570

Stuber CW, Edwards MD, Wendel JF (1987) Molecular marker facilitated investigations of quantitative trait loci in maize. II. Factors influencing yield and its component traits. Crop Sci 27: 639-648

Stuber CW, Polacco M, Senior ML (1999) Synergy of empirical breeding, marker-assisted selection, and genomics to increase crop yield potential. Crop Sci 39: 1571-1583

Swaminathan MS (2005) Towards an Ever-Green Revolution. In: Tuberosa R, Phillips RL, Gale M (eds) Proc Intl Congr "In the Wake of the Double Helix:

From the Green Revolution to the Gene Revolution", May 27-31, 2003, Bologna, Italy. Avenue media, Bologna, Italy, pp 25-36

Takken FLW, Joosten MHAJ (2000) Plant resistance genes: their structure, function and evolution. Eur J Plant Pathol 106: 699-713

Talame V, Sanguineti MC, Chiapparino E, Bahri H, Ben Salem M, Forster BP, Ellis RP, Rhouma S, Zoumarou W, Waugh R, Tuberosa R (2004) Identification of *Hordeum spontaneum* QTL alleles improving field performance of barley grown under rainfed conditions. Ann Appl Biol 144: 309-319

Talamè V, Ozturk NZ, Bohnert HJ, Tuberosa R (2007) Barley transcript profiles under dehydration shock and drought stress treatments: a comparative analysis. J Exp Bot 58: 229-240

Tanksley SD (1993) Mapping polygenes. Annu Rev Genet 27: 205-233

Tanksley S, Nelson J (1996) Advanced backcross QTL analysis: a method for the simultaneous discovery and transfer of valuable QTLS from unadapted germplasm into elite breeding lines.Theor Appl Genet 92: 191-203

Tanksley SD, McCouch SR (1997) Seed banks and molecular maps: unlocking genetic potential from the wild. Science 277: 1063-1066

Tanksley SD, Grandillo S, Fulton T, Zamir D, Eshed T, Petiard V, Lopez J, Beck BT (1996) Advanced backcross QTL analysis in a cross between an elite processing line of tomato and its wild relative *L. pimpinellifolium*. Theor Appl Genet 92: 213-224

Toojinda T, Baird E, Booth A, Broers L, Hayes P, Powell W, Thomas W, Vivar H, Young G (1998) Introgression of quantitative trait loci (QTLs) determining stripe rust resistance in barley: an example of marker assisted line development. Theor Appl Genet 96: 123-131

Touzet P, Winkler RG, Helentjaris T (1995) Combined genetic and physiological analysis of a locus contributing to quantitative variation. Theor Appl Genet 91: 200-205

Tuberosa R, Sanguineti MC, Landi P (1994) Abscisic acid concentration in the leaf and xylem sap, leaf water potential, and stomatal conductance in drought-stressed maize. Crop Sci 34: 1557-1563

Tuberosa R, Gill B, Quarrie S (2002) Cereal genomics: ushering in a brave new world. Plant Mol Biol 48: 445-449

Tuberosa R, Salvi S (2006) Genomics approaches to improve drought tolerance in crops.Trends Plant Sci 11: 415-412

Tuinstra MR, Ejeta G, Goldbrough P (1998) Evaluation of near-isogenic sorghum lines contrasting for QTL markers associated with drought tolerance. Crop Sci 38: 835-842

Uetz P (2002) Two-hybrid arrays. Curr Opin Chem Biol 6: 57-62

Uptmoor R, Wenzel W, Ayisi K. Donaldson G, Gehringer A, Friedt W, Ordon F (2006) Variation of the genomic proportion of the recurrent parent in BC1 and its relation to yield performance in sorghum (*Sorghum bicolor*) breeding for low-input conditions. Plant Breed 125: 532-534

van Os H, Andrzejewski S, Bakker E, Barrena I, Bryan GJ, Caromel B, Ghareeb B, Isidore E, de Jong W, van Koert P, Lefebvre V, Milbourne D, Ritter E, van der Voort JN, Rousselle-Bourgeois F, van Vliet J, Waugh R, Visser RG, Bakker J, van Eck HJ (2006) Construction of a 10,000-marker ultra-dense genetic recombination map of potato: providing a framework for accelerated gene isolation and a genomewide physical map. Genetics 173: 1075-1087

Varshney RK, Tuberosa R (2007) Genomics-assisted Crop Improvement: an overview. In Varshney RK, Tuberosa R (eds) Genomics Approaches and Platforms. Springer, Dordrecht, pp 1-12

Varshney RK, Graner A, Sorrells ME (2005) Genomics-assisted breeding for crop improvement. Trends Plant Sci 10: 621-630

Varshney RK, Marcel TC, Ramsay L, Russell J, Roder MS,Stein N, Waugh R, Langridge P, Niks RE, Graner A (2007) A high density barley microsatellite consensus map with 775 SSR loci. Theor Appl Genet 114: 1091-1103

Venuprasad R, Shashidhar HE, Hittalmani S, Hemamalini GS (2002) Tagging quantitative trait loci associated with grain yield and root morphological traits in rice (*Oryza sativa* L.) under contrasting moisture regimes. Euphytica 128: 293-300

Visscher P, Haley C, Thompson R (1996) Marker-assisted introgression in backcross breeding programs. Genetics 144: 1923-1932

Vuylsteke M, van den Daele H, Vercauteren A, Zabeau M, Kuiper M (2006) Genetic dissection of transcriptional regulation by cDNA-AFLP. Plant J 45: 439-446

Wade MJ (2002) A gene's eye view of epistasis, selection, and speciation. J Evol Biol 15: 337-346

Waterhouse PM, Helliwell CA (2003) Exploring plant genomes by RNA-induced gene silencing. Nat Rev Genet 4: 29-38

Wayne ML, McIntyre L (2002) Combining mapping and arraying: an approach to candidate gene identification. Proc Natl Acad Sci USA 99: 14903-14906

Welcker C, Boussuge B, Bencivenni C, Ribaut JM, Tardieu F (2007) Are source and sink strengths genetically linked in maize plants subjected to water deficit? A QTL study of the responses of leaf growth and of Anthesis-Silking Interval to water deficit. J Exp Bot 58: 339-349

Wenzl P, Carling J, Kudrna D, Jaccoud D, Huttner E, Kleinhofs A, Kilian A (2004) Diversity Arrays Technology (DArT) for whole-genome profiling of barley. Proc Natl Acad Sci USA 101: 9915-9920

Wenzl P, Li H, Carling J, Zhou M, Raman H, Paul E, Hearnden P, Maier C, Xia L, Caig V, Ovesna J, Cakir M, Poulsen D, Wang J, Raman R, Smith KP, Muehlbauer GJ, Chalmers KJ, Kleinhofs A, Huttner E, Kilian A (2006) A high-density consensus map of barley linking DArT markers to SSR, RFLP and STS loci and agricultural traits. BMC Genomics 7: 206. doi:10.1186/1471-2164-7-206

Werner JD, Borevitz JO, Warthmann N, Trainer GT, Ecker JR, Chory J, Weigel D (2005) Quantitative trait locus mapping and DNA array hybridization identify an FLM deletion as a cause for natural flowering-time variation. Proc Natl Acad Sci USA 102: 2460-2465

West MA, Kim K, Kliebenstein DJ, van Leeuwen H, Michelmore RW, Doerge RW, St Clair DA (2006) Global eQTL mapping reveals the complex genetic architecture of transcript level variation in Arabidopsis. Genetics 175: 1441-1450

Woll K, Borsuk LA, Stransky H, Nettleton D, Schnable PS, Hochholdinger F (2005) Isolation, characterization, and pericycle-specific transcriptome analyses of the novel maize lateral and seminal root initiation mutant *rum1*. Plant Physiol 139: 1255-1267

Xiao JH, Li J, Grandillo S, Ahn SN, Yuan L, Tanksley SD, McCouch SR (1998) Identification of trait-improving quantitative trait loci alleles from a wild rice relative, *Oryza rufipogon*. Genetics 150: 899-909

Xiong L, Wang R-G, Mao G, Koczan JM (2006) Identification of drought tolerance determinants by genetic analysis of root response to drought stress and abscisic acid. Plant Physiol 142: 1065-1074

Xu Y, Crouch JH (2008) Marker-assisted selection in plant breeding: From publications to practice. Crop Sci 48: 391-407

Yamasaki M, Tenaillon MI, Bi IV, Schroeder SG, Sanchez-Villeda H, Doebley JF, Gaut BS, McMullen MD (2005) A large-scale screen for artificial selection in maize identifies candidate agronomic loci for domestication and crop improvement. Plant Cell 17: 2859-2872

Yin XY, Stam P, Kropff MJ, Schapendonk AHCM (2003) Crop modeling, QTL mapping, and their complementary role in plant breeding. Agron J 95: 90-98

Yin X, Struik PC, Kropff MJ (2004) Role of crop physiology in predicting gene-to-phenotype relationships. Trends Plant Sci 9: 426-432

Young ND (1996) QTL mapping and quantitative disease resistance in plants. Annu Rev Phytopathol 34: 479-501

Young ND (1999) A cautiously optimistic vision for marker-assisted breeding. Mol Breed 5: 505-510

Young ND, Tanksley SD (1989) Restriction fragment length polymorphism maps and the concept of graphical genotypes. Theor Appl Genet 77: 95-101

Yu LX, Setter TL (2003) Comparative transcriptional profiling of placenta and endosperm in developing maize kernels in response to water deficit. Plant Physiol 131: 568-582

Yu JM, Holland JB, McMullen MD, Buckler ES (2008) Genetic design and statistical power of nested association mapping in maize. Genetics 178: 539-551.

Yun SJ, Gyenis L, Bossolini E, Hayes PM, Matus I, Smith KP, Steffenson BJ, Tuberosa R, Muehlbauer GJ (2006) Validation of quantitative trait loci for multiple disease resistance in barley using advanced backcross lines developed with a wild barley. Crop Sci 46: 1179-1186

Zeng H, Luo L, Zhang W, Zhou J, Li Z, Liu H, Zhu T, Feng X, Zhong Y (2007) PlantQTL-GE: a database system for identifying candidate genes in rice and *Arabidopsis* by gene expression and QTL information. Nucl Acids Res 35 (Database issue): D879-D882

Zinselmeier C, Sun Y, Helentjaris T, Beatty M, Yang S, Smith H, Habben J (2002) The use of gene expression profiling to dissect the stress sensitivity of reproductive development in maize. Field Crops Res 75: 111-121

2 Molecular Markers in Germplasm Conservation

Joanne A. Labate
USDA-ARS, Plant Genetic Resources Unit, Geneva, NY 14456, USA
E-mail: Joanne.Labate@ars.usda.gov

1 INTRODUCTION

Plant breeders require large and diverse pools of genetic variation for crop improvement, particularly for novel traits (McCouch 2004). Major challenges which breeders face include global warming, increased environmental contamination, disease and pest epidemics, depleted soils, and fresh water shortages. Consumers in nations with bountiful supplies of food, such as the USA, are increasingly interested in foods perceived to be healthy (Weimer 1999). This trend has led to the production of leaner meats and, concurrently, fruits and vegetables with enhanced levels of phytonutrients.

Although most new mutations are harmful (Crow 1992), a large fraction of the DNA variation observable within a species, so-called 'standing genetic variation', does not effect an individual's fitness (survival and reproduction) (Kimura 1983). These neutral variations include synonymous changes, changes in noncoding regions such as introns, and amino acid substitutions that are functionally equivalent within a given environment. Neutral polymorphisms can provide a reservoir of raw material for adaptive change under new selection pressure (Gardner and Latta 2006).

A large supply of plant genetic variation is held in the form of more than six million accessions stored *ex situ* in global collections (Scarascia-Mugnozza and Perrino 2002). According to the Food and Agriculture

Organization of the United Nations (FAO), 15 national collections hold about one third of the world's *ex situ* plant genetic resources. An additional 12% are held by the genebanks of the Consultative Group on International Agricultural Research (CGIAR) Centers (Table 1) (FAO 1998). As of May 2008, over one-million total accessions were held by CGIAR (*http://www.fao.org/*) and United States Department of Agriculture (USDA) National Plant Germplasm System (NPGS) (*http://www.ars-grin.gov/npgs/*) genebanks alone. This included 219 families, 2,022 genera and 12,495 species in NPGS.

Wild relatives of crops, landraces, cultivars, and advanced breeding lines represent a continuum of germplasm types available to end users of germplasm collections. In general the total amount of genetic variation progressively decreases in this series as the potential for agronomic performance increases (Fig. 1). By crossing highly selected material back to wild progenitors or landraces, novel favorable alleles can be reincorporated into the cultivated gene pool with great success (McCouch 2004 cites examples).

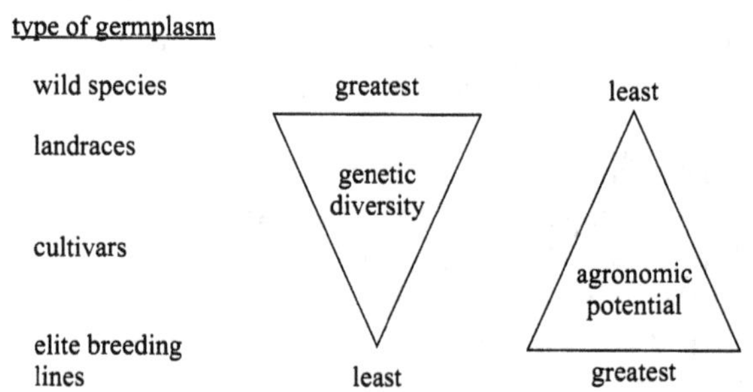

Fig. 1 Relative amount of genetic diversity decreases while agronomic potential increases as the continuum of germplasm advances from wild crop relatives through elite breeding lines.

Improved phenotypes originate from new combinations of already available alleles rather than from spontaneous mutations. For efficient breeding through artificial selection, an allele cannot be so rare (e.g., a single copy) in a population so as to completely escape selection. This has raised questions about what should be conserved in genebanks, allelic richness (identities of alleles) or allelic evenness (moderate frequencies of alleles, i.e. heterozygosity). A recognition of the value of genetic resources as collections of alleles per se follows from the analytical result that heterozygosity is proportional to the amount of genetic variance and

Table 1 Holdings at Consultative Group on International Agricultural Research (CGIAR) Centers.

Name	*Location*	*Principal Crop(s)*[1]	*Number of Accessions*[2]
International Center for Tropical Agriculture (CIAT)	Cali, Colombia	Bean (*Phaseolus* spp.)	31,718
		Cassava (*Manihot esculentum* Crantz)	5,728
		Forages (*Aeschynomene* spp., *Desmodium* spp., *Stylosanthes* spp.)	18,138
International Maize and Wheat Improvement Center (CIMMYT)	El Batan, Mexico	Maize (*Zea mays* L.)	20,411
		Wheat (*Triticum aestivum* L.)	95,113
International Potato Center (CIP)	Lima, Peru	Andean roots and tubers (*Oxalis tuberosa* M., *Ullucus tuberosus* C., *Tropaeolum tuberosum* R. & P., *Arracacia xanthorrhiza* B, *Smallanthus sonchifolius* (P & E) H. Rob., *Canna indica* L., *Pachyrhizus ahipa* (Wedd.) Parodi, *P. tuberosus* (Lam.) Spreng., *Lepidium meyenii* W., *Mirabilis expansa* R. & P.)	1,112
		Potato (*Solanum tuberosum* L.)	5,057
		Sweet potato (*Ipomoea batatas* L.)	6,413
International Center for Agriculture in the Dry Areas (ICARDA)	Aleppo, Syria	Barley (*Hordeum vulgare* L.)	24,218
		Chickpea (*Cicer arietinum* L.)	9,116
		Faba bean (*Vicia faba* L.)	9,074
		Forage legumes (*Lathyrus* spp., *Vicia* spp.)	24,581
		Lentil (*Lens culinaris* Medik.)	7,827
		Wheat (*Triticum aestivum* L., *T. durum* Desf.)	30,270
International Crops Research Institute for	Patancheru, India	Chickpea (*Cicer arietinum* L.)	16,961
		Groundnut (*Arachis hypogaea* L.)	14,357

Contd.

Contd.

the Semi-Arid Tropics (ICRISAT)		Pearl millet (*Pennisetum glaucum* L.) Pigeonpea (*Cajanus cajan* L.) Sorghum (*Sorghum bicolor* L.)	21,250 12,698 35,780
International Institute for Tropical Agriculture (IITA)	Ibadan, Nigeria	Bambara groundnut (*Vigna subterranea* L.) Cassava (*Manihot esculentum* Crantz) Cowpea (*Vigna unguiculata* L.) Soybean (*Glycine max* L.) Wild *Vigna* Yam (*Dioscorea* spp.)	2,029 2,158 15,001 1,909 1,634 2,878
International Livestock Research Institute (ILRI)	Nairobi, Kenya	Forages (*Crotalaria kirkii* Bak., *Pseudarthria hookeri* Wight & Arn., *Stylosanthes fruticosa* (Retz.) Alston, *Lotononis angolensis* Welw. ex Baker *Aeschynomene americana* L., *Zornia glabra* Desv., *Cajanus cajan* (L.) Millsp., *Stylosanthes scabra* Vog.)	11,537
International Plant Genetic Resources Institute (IPGRI)	Maccarese, Italy	Banana (*Musa* spp.)	931
International Rice Research Institute (IRRI)	Los Banos, Philippines	Rice (*Oryza* spp.)	80,617
West Africa Rice Development Association (WARDA)	Bouaké, Cote d'Ivoire	Rice (*Oryza* spp.)	14,917
World Agroforesty Centre	Nairobi, Kenya	*Sesbania* spp.	25
TOTAL			523,458

[1]Numerous additional species including other crops and wild crop relatives are not listed here.
[2]Data from *http://www.cgiar.org* as of May 2008.

therefore related to the immediate response to selection (natural or artificial), whereas the limit of response to selection over generations is determined by allelic composition regardless of allelic frequencies (James 1971; reviewed by Petit et al. 1998).

Many user-friendly software programs are available for estimating genetic diversity parameters from molecular marker data (Labate 2000). Table 2 summarizes various population measures that can be estimated from markers. Descriptive statistics include mean number of alleles per locus, effective number of alleles, percent polymorphic loci, observed heterozygosity, and expected heterozygosity. The partitioning of diversity within and among various hierarchical levels such as individuals, accessions, and groups of accessions is also of great interest when utilizing resources. This population structuring can be estimated using *F*-, *G*-, or *Rho*-statistics. Additional methods for describing the structure of diversity include analysis of molecular variance (AMOVA), tests of homogeneity, measures of migration and isolation-by-distance.

Matrices of genetic distances or co-ancestry between pairs of populations provide input data for clustering methods such as neighbor-joining and an unweighted paired group method using arithmetic averages (UPGMA). Bootstrapped dendrograms produced by clustering are an efficient way to visualize relationships among taxa. A multivariate method such as principal component analysis (PCA), principal coordinate analysis (PCoA), and multidimensional scaling (MDS) is sometimes preferable over bifurcating dendrograms for visualizing clusters.

Hardy-Weinberg, two-locus, and multilocus equilibrium reflect population-level processes such as random mating and recombination (among others) over one or more generations. A comparison of the relative utility and merits of various statistical methods is beyond the scope of this chapter, an extensive review was published by Mohammadi and Prasanna (2003).

Molecular markers are attractive tools for genetically characterizing germplasm because, unlike morphological traits, they are not influenced by the environment. The extensive sampling needed to effectively study large germplasm collections (multiple accessions, multiple plants per accession, and multiple loci), often times combined with a lack of laboratory infrastructure have previously made molecular studies by genebanks somewhat cost-prohibitive. This is rapidly changing, there is now a multitude of published studies using molecular markers as tools to help solve problems in genebank management. Markers are not seen as a panacea, and are usually most effective when interpreted in

Table 2 Various population measures that can be estimated using molecular genetic markers.

Measure	*Reference*
Diversity	
Mean number of alleles per locus	
Effective number of alleles[1]	Petit et al. 1998
Percent polymorphic loci	
Observed heterozygosity	
Expected heterozygosity[2]	Nei 1973
Partitioning of diversity	
F-statistics	Weir 1990
G-statistics	Weir 1990
Rho-statistics[3]	Rousset 1996
Analysis of Molecular Variance (AMOVA)	Excoffier et al. 1992; Michalakis and Excoffier 1996
Homogeneity test	Goudet et al. 1996; Raymond and Rousset 1995
Migration	Barton and Slatkin 1986; Slatkin 1985
Isolation-by-distance	Rousset 1997
Genetic distance	
Nei's	Nei 1972, 1978
Rogers'	Rogers 1972
Co-ancestry	Reynolds et al. 1983
Clustering	
Neighbor-joining	Saitou and Nei 1987
Unweighted Paired Group Method using Arithmetic Averages (UPGMA)	Sokal and Michner 1958
Principal Component Analysis (PCA)	Manly 2004
Principal Coordinate Analysis (PCoA)	Manly 2004
Multidimensional Scaling (MDS)	Manly 2004
Equilibrium	
Hardy-Weinberg	Weir 1990
Two-locus	Weir 1990
Multilocus	Weir 1990

[1]Number of alleles corrected for sample size.

[2]Probability that two randomly sampled alleles are different.

[3]Applicable to microsatellite allele length.

combination with phenotypic and passport data. This chapter reviews the diverse and creative applications of markers to help solve problems in conservation and utilization of genetic resources, with emphasis on *ex situ* collections. It is organized according to the primary activities associated with germplasm conservation: acquisition, preservation, evaluation, documentation, and utilization.

2 ACQUISITION

Genetic diversity within and among populations is influenced by the mating system, population size, mutation, migration, demography, and selection (Brown 1978). Given the extensive range in these parameters among plant species, simplifying assumptions are needed to implement a general strategy for maximizing genetic variation collected for *ex situ* conservation. The simplistic strategy is to collect 50 to 100 random individuals per site from as many sites as possible, and to sample sites from as broad a range of environments as possible (Brown 1978). This method assumes that all populations contain similar levels of genetic variation, populations are differentiated from each other with respect to the identity of alleles they contain, and that allelic differentiation can potentially translate into desirable differences in phenotypes.

Strategies for sampling representative extant genetic diversity for *ex situ* conservation have been refined with the aid of molecular markers. If populations lack genetic differentiation, collection and subsequent maintenance can be made more efficient by sampling fewer populations and more individuals per population. This is illustrated by a study of the endangered aquatic grass Texas wild rice (*Zizania texana* Hitchc.) which grows in small stands in the upper four km of the San Marcos River in south central Texas. This geographically restricted distribution permitted extensive sampling and genotyping over the entire species' range for multiple years (Richards et al. 2007). Six simple sequence repeat (SSR) markers were genotyped for 471 plants from 23 stands. The authors showed that sampling of large stands efficiently captured the genetic diversity within the species, while stands of areas less than two m^2 contained no unique alleles. It was concluded that large stands provided stable reservoirs of genetic diversity for this species. Guidelines for sampling strategies for *ex situ* conservation and recommendations for *in situ* conservation were proposed based on the results.

In a primarily self-fertilizing species, sampling large numbers of plants with few seeds per plant can be highly efficient in capturing allelic diversity. This was concluded from a demographic study of five wild

Lima bean (*Phaseolus lunatus* L.) populations in the Central Valley of Costa Rica using six enzyme loci (Bi et al. 1998). Wild plant populations in this region are at risk of extinction due to increased urbanization and agriculture. The authors estimated that sampling one to four seeds from 200 to 300 plants per population would capture eight to ten alleles across loci, including rare (frequency less than 0.01) alleles. For targeting alleles at frequencies of a least 0.05, sampling one to two seeds from 10 to 80 plants would yield four to eight alleles across loci.

For vegetatively propagated crops, molecular markers are sometimes used to help verify trueness-to-type during acquisition. Genebank acceptance of clonal bud wood carries the risk of introducing graft-transmissible pathogens; this has been circumvented by using asexual (nucellar) seed and seedlings in citrus (*Citrus* spp.) (Krueger and Roose 2003). In this study, sexual (zygotic) seedlings needed to be distinguished from nucellar because many citrus varieties produce both. To identify nucellar seedlings of polyembryonic types, 1,340 seedlings from 88 seed sources were fingerprinted using inter-simple sequence repeat (ISSR) markers. Success rate was high; seedlings from 72 sources were categorized as nucellar and were chosen for potential addition to the collection. The investigators planned to use morphological data to ensure that the selected nucellar individuals were identical to the named type. This technique improved the efficiency of acquiring new citrus germplasm by seed.

Germplasm requiring *in vitro* culture is relatively costly to conserve. For example, cultivated potato (*Solanum tuberosum* L.) accessions require high maintenance as slow growing cultures that must be periodically transferred and kept virus-free. It is particularly desirable to avoid acquiring duplicate genotypes given these requirements. An example where this situation was dealt with at the Nordic Gene Bank (NGB) was reported by Vetelainen et al. (2005). NGB recognized the need to acquire additional Nordic landraces of potato in its repository based on survey data. Four accessions donated by a private individual as potentially representing the landrace 'Tärendö' were morphologically indistinguishable from NGB accession 'Mandel' as well as from each other. To examine this problem within the context of diversity within the larger collection, 32 Nordic landraces were scored for 57 morphological traits and genotyped using amplified fragment length polymorphisms (AFLPs). All but the above-mentioned five accessions were distinguishable based on morphology. The five morphologically indistinguishable accessions were found to comprise two distinct AFLP genotypes, namely 'Tärendö' and 'Mandel'. Among the three identical samples of the landrace 'Tärendö', the one with the most reliable and longest known history according to information provided by the donor was chosen for long-term conservation.

Ex situ collections can be compared to their *in situ* counterparts to help develop guidelines for population sampling during germplasm acquisition. For example, the maize (*Zea mays* L. subsp. *mays*) race Mexican Jala was sampled from eight populations in farmers' fields and compared to samples from eight accessions held by CIMMYT using 22 SSR loci (Rice et al. 2006). Total genetic diversity in the *ex situ* and *in situ* populations was similar (expected heterozygosity, H_e = 0.62 and 0.61, respectively), and the two groups were not genetically distinct from each other (F_{ST} = 0.01). For both *in situ* and *ex situ* material, populations collected in the same year were genetically similar to each other, implying panmixia within the Jala valley. The authors recommended collecting many individuals per field, rather than collecting a few individuals from many farmers' fields, for *ex situ* conservation of this landrace.

Home gardens potentially represent substantial reservoirs of *in situ* genetic diversity in the tropics, where families traditionally cultivate and maintain small populations of plants for culinary and medicinal purposes. To investigate diversity in home gardens, 34 accessions of chili pepper (*Capsicum* spp.) were collected from home gardens throughout the Department of Alta Verapaz, Guatemala (Guzman et al. 2005). These were compared in their AFLP profiles to 40 accessions conserved in the *ex situ* collection of the Universidad de San Carlos. The *ex situ* germplasm was selected to be representative of the national diversity found in Guatemala. Remarkably, the amount of variation found in the home gardens approximated the total genetic diversity of *Capsicum* in Guatemala. Only a small fraction of the total surveyed genetic variation was partitioned between the two groups (G_{ST} = 0.04). In addition, the *in situ* group contained more unique AFLP bands (three) than the *ex situ* group (one), possible evidence for the maintenance of rare alleles in home gardens. The authors pointed out that pepper consumption in Alta Verapaz is the highest among all departments in Guatemala owing to its importance in local indigenous culture. Molecular markers strongly reinforced Alta Verapaz as a region for potential discovery of additional pepper types for *ex situ* conservation, as well as being a highly suitable region for *in situ* conservation efforts.

3 PRESERVATION

To preserve diversity in *ex situ* populations of species conserved as seed, regeneration and maintenance protocols must minimize genetic changes within accessions. In clonally propagated species gross mutations (bud sports) or somaclonal mutation during *in vitro* culturing can alter trueness-to-type. In the case of accession mishandling (and subsequent misidentification) either in the field, during *in vitro* culturing, or during cryo-preservation, markers can often provide firm evidence of identity.

Seed stocks become depleted through distribution, and embryos eventually die during cold storage. Periodic regenerations of accessions are necessary, often under field conditions. Regeneration protocols for cross-pollinated species can require isolation of accessions from each other in large mesh cages in the field. Pollinators, such as honey bees (*Apis mellifera* L.), are introduced into the cages during flowering. Loss of genetic identity can occur through contamination by foreign pollen, seed adulteration during harvesting, threshing, and packaging, and through mutation (Steiner et al. 1997). Genetic diversity of an accession can be altered or lost by genetic drift (random changes in allele frequencies caused by random sampling of gametes during sexual reproduction) and unintentional selection. Extensive measures are taken to minimize loss of integrity of genebank accessions. Many studies have examined allele frequency changes in accessions during preservation, storage, and regeneration so that protocols for *ex situ* conservation can be optimized.

Most new mutations are rare and random. The large sample sizes needed to screen for these make them difficult to study, although there have been some attempts (e.g., Chwedorzewska et al. 2002a). Some studies indicated that major chromosomal aberrations occurred during seed storage and senescence, especially under desiccating conditions, but these gross mutations did not persist in regenerated populations (Wu et al. 1998).

Inadvertent selection for more adapted genotypes or genotypes with relatively more viable seed can occur during regeneration (Wu et al. 1998). Selection acts on specific loci. Without knowing which loci these might be or even in the case of identified candidate loci for certain traits, selection as a causal agent in allele frequency change is difficult to verify.

For practical purposes evidence of selection can be studied in conjunction with evidence for drift. A good example comes from a study of Texas wild rice (Richards et al. 2004). Seeds of some species cannot survive the desiccation used in standard methods of preservation so are cryo-preserved. For Texas wild rice embryo mortality associated with cryo-preservation was quite high (25 to 45%) (Walters et al. 2002). To study the impact of cryo-preservation-associated mortality on genetic diversity, allele frequencies at five SSR loci were compared for cryo-preserved versus untreated (control) samples of embryos post-germination (Richards et al. 2004). Allele frequency changes at each locus among treatments were not significantly different from zero; initial allelic richness and gene diversity remained intact. The authors concluded that no detectable drift or selection had been imposed on the five loci, and that cryogenic protocols were adequate for *ex situ* conservation of this genetically diverse species.

During the reproductive cycle of a plant, the seedling stage seems particularly vulnerable to natural or artificial selection. Seed is typically sown in a greenhouse and seedlings are later transplanted to the field. Workers tend to select vigorous appearing seedlings for transplant, as they are more likely to survive. Random environmental influences can lead to non-uniformity in vigor of seedlings. However, if certain seedlings exhibit poor vigor due to genetic differences, it may be worthwhile to conserve them in the interest of preserving intrapopulation diversity. This reasoning was stated as the impetus for a study of potential loss of diversity in wild potato (*Solanum* spp.) accessions due to variation in seedling vigor (Bamberg and del Rio 2006). Seed lots of 11 inbreeding wild potato species were studied using random amplified polymorphic DNA (RAPD) markers to assess the potential of losing diversity at the seedling transplant stage. Vigorous versus non-vigorous seedlings within lots were genotyped in bulk samples. Genetic similarity between paired vigorous and non-vigorous bulked siblings ranged from 98.4% to 100%, and in no case was statistically significantly different. The authors concluded that the extra effort needed to conserve low-vigor seedlings of inbreeding wild potato (given their associated lower survival rate, slower growth, and reduced fertility and seed set) was not warranted.

A study of rye (*Secale cereale* L.) seedling populations after long-term seed storage and regeneration showed significant alterations in AFLP frequencies (Chwedorzewska et al. 2002b). In this study, all four seedling populations traced back to a single parent population of 'Dańkowskie Złote'. The greatest genetic differentiation between populations was found between seed stored for four years with high (91%) viability versus seed stored for 14 years with low (5%) viability. After three regenerations, accessions showed significant differences in 6 to 8% of the 231 polymorphic markers. Observed genetic alterations were consistent with previous studies on seed aging and regeneration in rye and other crops where significant changes in morphology and protein markers were observed (Chwedorzewska et al. 2002b and references therein).

Genetic drift, contamination, and misidentification are easier to detect than mutation and selection; many studies have found evidence for one or more of these (del Rio et al. 1997; Wu et al. 1998; Börner et al.

2000; Parzies et al. 2000). Genetic alterations of accessions are sometimes extreme. For seven oat (*Avena sativa* L.) lines maintained for 124 years, electrophoresis of storage proteins showed results ranging from no contamination of a line to complete replacement of a line by a foreign phenotype (Steiner et al. 1997). Both mishandling of seed and pollen contamination during maintenance were implicated.

The rate of drift depends on effective population size (N_e), which is defined as the number of individuals in a theoretically ideal population having the same magnitude of drift as the actual population (Hartl and Clark 1989). N_e is usually smaller than census size and will be substantially so if an accession undergoes a genetic bottleneck.

Molecular markers showed evidence of genetic drift during regeneration in one of six wild potato (*Solanum jamesii*) (del Rio et al. 1997) and one of eight wheat (*Triticum aestivum* L.) (Börner et al. 2000) accessions. In two barley (*Hordeum vulgare* L.) landraces that had been maintained for over 70 years, N_e was estimated to be 4.7 using morphological and isozyme markers, even though census sizes in regeneration plots were routinely 600 plants (Parzies et al. 2000). The authors stated that either intense directional selection or a single bottleneck event could explain the extreme loss of genetic diversity. Four of six accessions of rye (*Secale cereale* L.) regenerated up to 14 times over 45 years showed significant changes in SSR allele frequencies (Chebotar et al. 2003). Field notes showed that strong winter damage had reduced plant numbers from several hundred to a few dozen in regeneration populations. High frequencies of new alleles were also detected, probably from cross-pollination.

To prevent drift during regeneration, theoretical studies have shown that carefully sampling equal numbers of seeds from as many seed parents as possible is highly effective (Crossa and Vencovsky 1997; Vencovsky and Crossa 1999). However, this method is inefficient and labor intensive. An alternative strategy is to sample a single inflorescence per plant rather than the whole plant when bulking seed of outcrossing plants (Johnson 1998). This will improve N_e providing the relative variation in seeds per spike is less than that of seeds per plant. In annual ryegrass (*Lolium multiflorum* L.) balanced (equal numbers of seeds per plant combined), spike (one inflorescence per plant combined), and bulk (seeds combined proportionally according to seeds per plant) regeneration samples were compared (Johnson 1998). The first method was clearly superior at maintaining genetic integrity of accessions. The latter two methods maintained diversity of eight isozymes as estimated by heterozygosity and mean numbers of alleles per locus, but allele frequencies shifted using those two methods.

Misclassification of accessions is often confirmed using molecular markers. Sixteen of 124 wheat (*Triticum* spp.) accessions were suspected to be misclassified based on morphology (Cao et al. 1999). RAPD markers were used to correctly identify the accessions; RAPD results were further supported by cytogenetic analyses. For 52 accessions of Spanish common wheat (*Triticum aestivum* L. ssp. *vulgare*) 25% of samples were misidentified when assessed by comparing gliadin allele profiles between original spike collections and genebank samples (Ruiz et al. 2002). In some instances a probable cause, such as erroneous labeling, was inferred. Misidentification of accessions has also been reported in bamboo (Barkley et al. 2005) and species of *Crotolaria* (Wang et al. 2006).

Molecular markers have also been used to explore regeneration and germplasm management strategies. Two examples (van Treuren and van Hintum 2001; Bamberg and del Rio 2004) involving the management of intrapopulation heterogeneity (a mixture of homozygous genotypes within accessions) will be described to illustrate this.

The mating system of primarily self-pollinating crops results in substantial homozygosity within lines. However, there may be intrapopulation heterogeneity of selfing species that is desirable to conserve. Conversion of intra-accession heterogeneity into inter-accession variation, to avoid loss of diversity during regeneration, was explored for six accessions of selfing barley using AFLPs (van Treuren and van Hintum 2001). Genetic distances between genotypes within accessions were found to be small with the exception of one landrace that contained 12 genotypes. AMOVA showed that this accession could be converted into eight distinct lines with 88.5% of the total variation distributed among lines. Additional advantages of transforming intraaccession to interaccession diversity for a selfing species include requiring smaller populations for regeneration, easier characterization, and easier identification of contamination of an accession. The authors concluded that these factors must be weighed against maintaining an increased number of accessions in a collection.

Using RAPDs, mean intrapopulation heterogeneity was estimated to be 6.6% for selfing (*Solanum fendleri* and *S. verrucosum*) compared to 36.8% in outcrossing (*S. jamesii* and *S. sucrense*) species of wild potato (Bamberg and del Rio 2004). To avoid sampling error caused by heterogeneity, sample sizes of 25 to 30 plants were recommended from populations of both selfing and outcrossing wild potato species during collection, regeneration, and evaluation.

4 EVALUATION

Molecular markers reflect pedigrees, which may not be significantly correlated with gross morphology. In the Brazilian cassava (*Manihot esculentum* Crantz) collection, cultivars with similar agronomic traits were very heterogeneous at the molecular level (Carvalho and Schaal 2001). For four major Sorghum races, grouping of accessions based on agronomic descriptors did not correlate with groupings produced from RAPD markers (Dahlberg et al. 2002). The authors pointed out that such correlations should not be expected because in most cases genes underlying an agronomic trait represent a very small fraction of the genome. Molecular markers have been shown to be highly valuable when interpreted alongside morphological data during evaluation of collections. Reasons for evaluating collections include reducing redundancies, identifying gaps, and assembling core subsets. Examples of these will be described in turn.

4.1 Reduce Redundancy

Many empirical studies have addressed the question of unintentional duplication of conserved germplasm by examining subsets of collections. Various types of duplication have been defined - identical, common, partial, compound, and parental (van Hintum and Visser 1995) (Table 3). Although precise language such as this can lead to refined studies of redundancy the use of terms such as these has not become widespread.

Table 3 Types of duplication of germplasm accessions.

Type	*Definition*
Identical	Genetically identical
Common	Derived from the same ancestral population and sharing all alleles
Partial	Derived from the same ancestral population and sharing a proportion of alleles
Compound	All alleles of an accession are a subset of alleles in another accession
Parental	Parents and offspring are included in the collection, all alleles in offspring are found in one or more parents

A frequently applied approach to analyzing redundancy is to compare identically or similarly named accessions within a species using molecular marker, passport, and phenotypic data. An estimated 800,000 accessions of wheat (*Triticum* spp.) are held globally, 28,000 of which are maintained in the Institut für Pflanzengenetik und Kulturpflanzenforschung (IPK). To develop a strategy for reducing such extensive collections, 70 potential duplicate accessions comprising 12 groups were chosen for SSR genotyping and analysis (Dobrovolskaya et al. 2005). Results showed that cultivar name or accession number were clearly insufficient for identifying duplicates. The authors proposed a comprehensive strategy for addressing this problem - i) group accessions by name and/or accession number; ii) SSR genotype and sequentially bootstrap the data to detect both statistically homogeneous groups and distinct accessions; iii) compare passport data for accessions of statistically homogeneous groups for further separation; iv) phenotypically characterize accessions from homogeneous groups; iv) identify duplicates. Based on historical information, the most original accession can be maintained and the others eliminated. Alternatively, the authors recommended pooling accessions of homogeneous groups to reduce holdings.

Estimates of redundancy from molecular data should be interpreted within the context of *in situ* distribution of variation, sampling during collection, regeneration protocols, sample sizes for genotyping, and the resolution of the marker system, as was done in successive studies of the 314 accessions of wild potato (series *Acaulia)* collection at the Center for Genetic Resources, The Netherlands (CGN). AFLP genotyping of at least two plants per accession with a minimum of two primer combinations identified 36 accessions as potentially redundant (McGregor et al. 2002). Additional sampling and genotyping of the 36 accessions plus one additional accession revealed 15 accessions (5% of total holdings) to be redundant (van Treuren et al. 2004). AMOVA showed 91% of the total variation to be partitioned among the 37 accessions. This variance component did not change when the 15 redundant accessions were excluded from analysis.

Duplication may refer to accessions held by different repositories that originated from distributed progeny of a single original germplasm collection. These samples are assumed to be highly genetically similar. This assumption was tested by comparing 22 pairs of reputed duplicate accessions of 18 different potato species held by International Potato Center (CIP) and NPGS (del Rio et al. 2006). Mean genetic similarity of reputed duplicates was quite high as revealed by RAPDs, although some pairs with significant genetic differences were found. This study also reported evidence for inter-genebank genetic drift based on SSR data.

Duplicates may be unintentionally conserved when they originate from different sources under different names. SSRs were used in European hazelnut (*Corylus avellana* L.) to uncover instances where morphologically identical cultivars represented such cases (Gokirmak et al. 2005). Clonally propagated garlic (*Allium sativum* L.) is highly phenotypically plastic; because of its morphological diversity it is not unusual for varieties to be renamed by collectors and growers. To identify potential clones with unique cultivar names, 211 *Allium sativum* and *Allium longicuspis* accessions from NPGS and commercial sources were genotyped using AFLPs (Volk et al. 2004). Extensive duplications, 64% and 41% of the surveyed accessions from NPGS and commercial sources, respectively, were identified.

In a study of date palm (*Phoenix dactylifera* L.) varieties collected in Egypt, AFLP profiling showed that similarity coefficients for pairs of same-named varieties ranged from 46% to 100% (El-Assar et al. 2005). The authors interpreted this finding in relation to commercial propagation and production in this crop, where sexual seedlings rather than clonal propagules are frequently used. Unintentional outcrossing could not be distinguished from mislabeling for cases in which varieties had the same name but were highly genetically dissimilar. Somaclonal mutation during *in vitro* culture was suspected for two accessions with 75% and 77% similarity to other individuals in their respective groups.

Studies of duplication often address pooling of duplicate material and estimate how much genetic variation will be lost by doing so. One strategy for pooling is to maximize the ratio of the similarity within groups to the dissimilarity between groups (van Hintum et al. 1996) and many studies have applied this model. Examples include all accessions of sorghum [*Sorghum bicolor* (L.) Moench] named 'Orange' in NPGS (Dean et al. 1999), Dutch landraces of *B. oleracea* accessions in CGN (van Hintum et al. 1996), flax (*Linum usitatissimum*) accessions designated as 'breeder's lines' at CGN (van Hintum and van Treuren 2001), all Peruvian sweet potato (*Ipomoea batatas* L.) accessions at CIP (Huamán et al. 1999), and all accessions of cabbage named Golden Acre (*B. oleracea* var. *capitata* L.) in NPGS (Phippen et al. 1997). In general these studies recommended pooling duplicate accessions based on AMOVA in a way that would retain greater than 90% of the among-group component of molecular genetic variation.

4.2 Identify Gaps

Molecular markers can be used to assess what diversity is being conserved and to fill in gaps if necessary. To ensure a wide representation of conserved diversity the geographical distribution of genetic variation within species must be understood (Ferguson et al. 1998).

The vegetatively propagated crop jackfruit (*Artocarpus heterophyllus* Lam., Moraceae) originated in India and is widely produced in Southeast Asia for consumption by humans and livestock. In response to an increasing demand for improved germplasm in the USA and tropical America, a collection of 26 accessions held by the Fairchild Tropical Botanic Garden in Florida was studied (Schnell et al. 2001). AFLP data were consistent with phenotypic evaluations; both revealed limited diversity in the collection. Acquisition of additional accessions from the center of diversity in India was recommended.

In a similar study, the 450-accession common vetch (*Vicia sativa* L.) collection of IPK was compared to 673 accessions (landraces and local cultivars) from different geographical regions of Russia held by the Vavilov Institute of Plant Industry (VIR) (Potokina et al. 2002). This economically important fodder crop has been cultivated in Russia since ancient times. The 43 accessions of the IPK collection originating from the area formerly known as the USSR were found to be a narrow, nonrandom sample of available diversity when compared to the VIR collection. This study provided a framework for the acquisition of additional common vetch germplasm from VIR by IPK.

4.3 Assemble Core Collections

Relatively large numbers of accessions within collections provide increased opportunities for mining genetic diversity, but collections can become so large as to be intractable with respect to characterization and utilization (Nienhuis 2001). Core subsets are tools for users to efficiently work with a large portion of the total genetic diversity from a collection. Typically the number of accessions in a core collection approximates 10% of the entire collection and retains more than 70% of alleles (Brown 1989). Random sampling methods are applied to develop core subsets. In order to broadly capture diversity, sampling may be from subsets of accessions initially grouped according to phenotype or ecogeographic origin (Li et al. 2002). Phenotypic traits are often poor indicators of genetic variation; they are influenced by the environment and/or can arise from independent genetic bases. Molecular markers can help overcome limitations in phenotypic and provenance data when assembling core collections.

Molecular markers have been used i) to compare various techniques used to generate a core (e.g., *Sorghum bicolor*; Grenier et al. 2000), ii) as criteria to decide which accessions to include in a core (e.g., cashew, *Anacardium occidentale* L.; Dhanaraj et al. 2002) Andean potato (*Solanum phureja*; Ghislain et al. 1999), and iii) to validate that a core is representative of a given base collection (e.g., Mexican common bean, *Phaseolus vulgaris* L.; Skroch et al. 1998). Computer simulations, statistical techniques such as PCA, and genetic descriptors such as numbers of alleles, expected heterozygosity, and percentage of rare alleles, are frequently used to establish and validate core subsets.

Storing accessions in the form of seed rather than planted trees is both economical and secure, this provided impetus for developing core sets of seed for wild apple (*Malus sieversi* Ledeb.) accessions (Volk et al. 2005). Core sets representing more than 90% of the genetic diversity from two Kazakhstan sites were assembled based on seven unlinked SSR loci and 19 quantitative traits. The 35 trees within each of the two core sets were later used as parents in crosses to capture the diversity as seed for long-term *ex situ* conservation.

As an alternative to a core the concept of a rice (*Oryza sativa* L.) diversity research set (RDRS) was implemented as a means to provide users with a manageable number (less than 100) of accessions for in-depth agronomical and physiological studies (Kojima et al. 2005). The National Institute of Agrobiological Sciences (NIAS) maintains 32,000 rice accessions; 332 of these were randomly sampled based on geographical representation and genotyped using 179 RFLP markers. An RDRS of 67 accessions retained 91% of the alleles and 80% to 100% of the range of agro-morphological variation found in the original 332 accessions. Genotypic and phenotypic data on this set of germplasm can continue to accumulate and be shared by users.

Another alternative to a core has been to apply markers to identify a manageable number of accessions that maximizes the number of alleles not found in a set of elite germplasm (Charcosset and Moreau 2004). The authors describe a preliminary study in which 18 SSR loci were genotyped in samples from 276 maize populations. The populations contained 28 alleles that were not found in a set of inbred lines already containing 86 alleles at the loci. A set of 10 populations containing 25 out of 28 alleles was identified. By crossing this relatively small set of populations with the inbred lines, new breeding populations with 30% more available genetic variation relative to the current inbred gene pool could be formed.

5 DOCUMENTATION

Molecular genetic markers aid in classification of accessions at inter- and intra-specific levels. Randomly distributed, selectively neutral markers are indicative of historical relationships within and among species;

accessions often cluster according to their geographic origin and subspecific status. Examples include 25 African accessions of Bambara groundnut (*Vigna subterranea* L.) (Amadou et al. 2001) and 113 accessions of the diploid grass *Aegilops tauschii* Coss. originating from the Caucasus and Central Asia (Pestsova et al. 2000).

Studies of the family Solanaceae provide good examples of applying molecular markers to problems in germplasm classification. As a medium-sized family of angiosperms, it is comprised of approximately 90 genera and 3,000 to 4,000 species, nearly half of which are in the large and highly variable genus *Solanum* (Knapp et al. 2004). The family is economically important, including species used for food (e.g., *Solanum lycopersicum* L., tomato; *Solanum tuberosum* L., potato; *Solanum melongena* L., eggplant), drugs (e.g., *Nicotiana* spp., tobacco) and ornamentals (e.g., *Petunia hybrida* hort., petunia). Black nightshades (*Solanum nigrum* L. complex) are valuable for *ex situ* conservation due to their potential use as medicinal, forage, and nutritional plants, as well as sources for disease resistance in potato and eggplant breeding. Black nightshades encompass a complex of highly variable, closely related subtaxa sometimes originating through hybridization and with no discernable interspecific boundaries (Dehmer and Hammer 2004 and references therein).

In a case study of black nightshade taxonomy, 44 of 56 accessions held by IPK were genotyped using AFLPs (Dehmer and Hammer 2004). Classification of 23 accessions was uncertain, and for 24 accessions provenance was not reliably recorded. AFLP data aided in correcting the classification of one accession, and in successfully classifying 13 accessions formerly listed only as *Solanum* sp. Taxonomic identities were further confirmed using morphology. AFLP and morphological data were inconclusive in classifying two accessions. The authors noted that AFLP data grouped accessions of known provenance according to geographic origin. The study illustrated the potential for applying AFLPs to define centers of diversity for black nightshades, and for identifying taxonomic and geographic gaps in the collection.

A second *Solanum* example is illustrated by a taxonomic study of eggplants and their close relatives held by NPGS (Furini and Wunder 2004). Eggplant taxonomy has traditionally been based on morphology, crossability and hybrid fertility. However, this can be problematic because *S. melongena* is highly variable morphologically and can form fertile hybrids with distantly related species (Furini and Wunder 2004 and references therein). Morphological and AFLP data were used to conclusively assign to specific taxa eight out of nine accessions of *Solanum* that had not been previously classified. In addition, 14 accessions were discovered to have been misnamed when they were received. The authors recommended applying molecular markers to characterize and manage germplasm collections from the time of receipt.

Molecular markers can provide evidence of gene flow at the subspecies level within germplasm collections. The value of this to breeding is illustrated in studies of the two major common bean (*P. vulgaris*) gene pools, Andean and Middle American. The Andean gene pool has a narrow genetic base within which it is difficult to make genetic gains. Severe limitations (e.g., F_1 hybrid lethality) of inter-gene pool crosses have been observed during attempts to overcome this (Gepts and Bliss 1985). However, evidence of natural introgression from Middle American into Andean races (presumably originating in the farmers' fields) was observed in about 10% of the Andean accessions from the CIAT common bean core collection (Beebe et al. 2001). In a follow-up study Islam et al. (2004) i) confirmed the level of and ii) quantified the effects of introgression on landrace disease resistance, morphology, and nutritional attributes. Results of RAPD markers and phenotypic data identified 58 of 426 Andean accessions as introgressed; these shared qualitative traits typical of Middle American types. These naturally introgressed accessions are valuable sources of new alleles for the Andean gene pool. The authors noted that they could be particularly beneficial as sources of disease resistance alleles because *P. vulgaris* genotypes are often relatively more resistant to pathogen races that they did not co-evolve with.

6 UTILIZATION

Ultimately, the value of germplasm collections is to be found in the exploitation of new, favorable alleles. Large numbers of accessions, lack of descriptive information, and the ever increasing gap in agronomical performance between germplasm resources and elite material has traditionally made utilization of *ex situ* collections a laborious process for breeders (Charcosset and Moreau 2004). However, there are numerous examples where molecular markers have aided in the process.

For example, cultivars can be compared to breeding germplasm to discover sources of new alleles without placing strong emphasis on a specific trait. This was done in a study of sorghum in Japan, where the crop has been cultivated for 700 to 1,100 years and for which pedigrees of current cultivars are largely unknown (Anas and Yoshida 2004). Cluster analysis based on 9 SSR loci showed 12 accessions of sorghum breeding material from ICRISAT, USA, and Japan to be genetically distinct from 19 sorghum lines cultivated in Japan. The germplasm can be more efficiently utilized through making crosses among rather than within clusters.

Cluster analysis has been similarly used to infer potentially independent genetic bases for a key trait. Drought stress in wheat (*Triticum aestivum* L.) is prevalent and crop improvement under such stress is difficult because of lower heritability of selection (Trethowan et al. 2002). To study this trait 14 wheat cultivars from Iran and 14 developed or obtained by CIMMYT were evaluated for drought tolerance and genotyped using AFLPs (Moghaddam et al. 2005). Clear differences between the Iranian and CIMMYT gene pools were observed and distinct genotypes with favorable agronomic performance under drought stress were identified. The authors proposed to increase efficiency of breeding for drought tolerance in wheat by combining independent sources of favorable alleles for improvement.

Germplasm collections are excellent sources of alleles for improvement of crop disease resistance (e.g., Castano et al. 2005; Iwaro et al. 2006; Serdani et al. 2006). New resistance sources are continually being sought as pathogen races overcome resistance alleles via natural selection. A general strategy for isolating a novel disease resistance gene (Fig. 2) is exemplified in the case of leaf rust resistance in wheat (*Triticum aestivum* L.; Obert et al. 2005). Forty NPGS accessions (landraces) from Afghanistan and Iran previously categorized as leaf rust resistant were evaluated in the field at multiple locations. A gene from accession PI 289284 was found to be highly resistant to the predominant races of leaf rust in the Great Plains region of the USA. Based on genetic map position and virulence assays, the authors concluded that they had identified a new gene rather than a new allele of a known gene. AFLP and DNA sequence analyses led to the development of sequence tagged site (STS) marker linked to the gene at a distance of 1.4 cM. This effort can help advance an economically sound and environmentally responsible method of disease control in this host-pathogen system via multilocus genetic resistance.

Increasingly sophisticated molecular genetics tools are being employed for efficacious utilization of *ex situ* collections. At CIAT, broad-based germplasm resources provide a key component in crop improvement strategies that combine traditional breeding with genomics tools (Ishitani et al. 2004). For example, to study abiotic stress resistance, physiological, biochemical, and phenotypic assays are used to screen a gene pool and measure germplasm response to a particular stress (e.g., drought or aluminum resistance). Microarrays are then used to dissect genetic variation in the gene pool; each array can analyze the expression of tens of thousands of genes simultaneously. Analyses of microarray

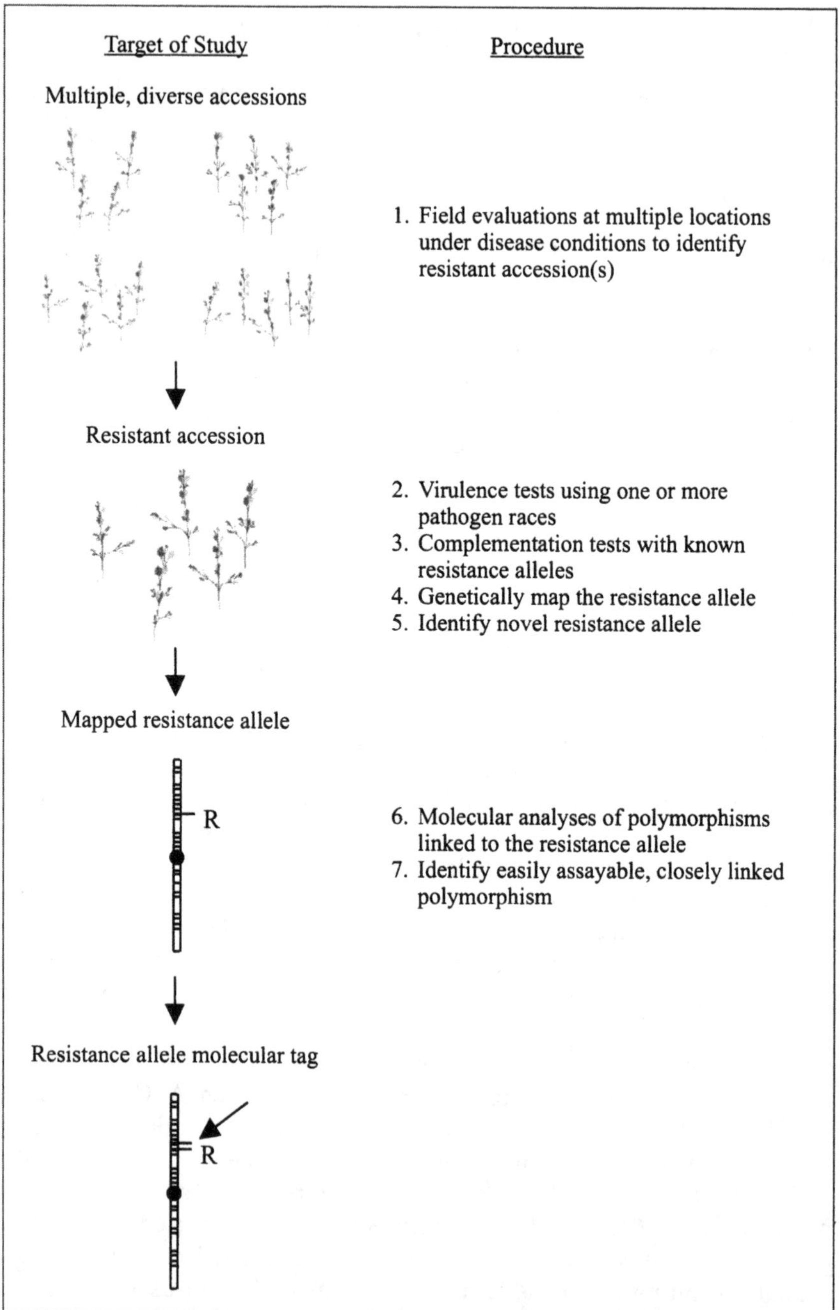

Fig. 2 Strategy for isolating a new disease resistance gene marker.

data combined with bioinformatics can lead to the identification of candidate genes responding to the stress. Finally, alleles at candidate genes can be tested through genetic transformation either in the crop species or in a model system such as *Arabidopsis* for proof of function *in planta*. This strategy is being applied at CIAT to a diverse array of species such as common bean (*P. vulgaris*), cassava (*Manihot esculenta* Crantz), tropical grasses, and upland rice (*O. sativa*). For traits with low heritability and complex genetic bases, these combined methods may prove more effective over traditional breeding alone.

Another tool recently adopted in plant breeding, association mapping, makes use of variable populations with an extensive history of genetic recombination, as is typical of germplasm collections. In association mapping DNA polymorphisms that correlate with phenotypes are identified and interpreted within the context of population structure. As crosses to create mapping populations are not required these methods are much faster than alternative linkage mapping; they also give higher resolution (Buckler et al. 2006).

Linkage disequilibrium (LD) is a critical parameter when designing association-based studies. Germplasm collections are often comprised of gene pools (e.g., groups of accessions) varying in their extent and magnitude of LD due to different evolutionary histories. Association mapping studies can be designed to utilize these gene pools in a complementary fashion, as was suggested in a report on barley (*Hordeum* spp.) LD (Caldwell et al. 2006). In a study of 74 cultivated, 23 landrace, and 34 wild barley accessions, inter- and intra-genic associations were estimated across four loci within 215 kb of the barley hardiness (*Ha*) locus. Significantly different patterns were observed for different barley gene pools, with LD extending to at least 212 kb in elite barley lines and rapidly decaying in ancestral populations. These results demonstrate the opportunity for low-resolution whole-genome scanning in cultivated barley germplasm followed by high-resolution LD mapping in ancestral landrace and wild species populations for discovery of alleles for crop improvement.

As a final note, high-throughput DNA sequencing (Margulies et al. 2005) and genotyping (Gunderson et al. 2005) methods offer the promise of rapidly advancing the progress of mining valuable alleles from germplasm collections. The technique of 'transcriptome shadowing' – the rapid sequencing of large populations of cDNA to a deep level of coverage for a diverse set of germplasm, can be used for discovery of large numbers of single nucleotide polymorphisms (SNPs) in expressed genes (Paterson 2006). This will boost the number of molecular markers available to assay, typically numbering in the hundreds, by up to several orders of magnitude depending on the species. Association mapping studies incorporating many thousands of SNPs, hundreds of genotypes, and multiple environments offer a powerful approach to quantifying function of alleles. Molecular markers will be key not only for exploiting standing genetic variation from germplasm collections, but for cataloging its vast potential.

References

Amadou H, Bebeli P, Kaltsikes P (2001) Genetic diversity in Bambara groundnut (*Vigna subterranea* L.) germplasm revealed by RAPD markers. Genome 44: 995-999

Anas Yoshida T (2004) Genetic diversity among Japanese cultivated sorghum assessed with simple sequence repeats markers. Plant Prod Sci 7: 217-223

Bamberg J, del Rio A (2004) Genetic heterogeneity estimated by RAPD polymorphism of four tuber-bearing potato species differing by breeding system. Am J Potato Res 81: 377-383

Bamberg J, del Rio AH (2006) Seedling transplant selection does not cause genetic shifts in Genebank populations of inbred potato species. Crop Sci 46: 424-427

Barkley NA, Newman ML, Wang ML, Hotchkiss MW, Pederson GA (2005) Assessment of the genetic diversity and phylogenetic relationships of a temperate bamboo collection by using transferred EST-SSR markers. Genome 48: 731-737

Barton NH, Slatkin M (1986) A quasi-equilibrium theory of the distribution of rare alleles in a subdivided population. Heredity, UK 56: 409-415

Beebe S, Rengifo J, Gaitan E, Duque MC, Tohme J (2001) Diversity and origin of Andean landraces of common bean. Crop Sci 41: 854-862

Bi IZ, Maquet A, Degreef J, Wathelet B, Baudoin JP (1998) Sample size for collecting seeds in germplasm conservation: the case of the Lima bean (*Phaseolus lunatus* L.). Theor Appl Genet 97: 187-194

Börner A, Chebotar S, Korzun V (2000) Molecular characterization of the genetic integrity of wheat (*Triticum aestivum* L.) germplasm after long-term maintenance. Theor Appl Genet 100: 494-497

Brown AHD (1978) Isozymes, plant population genetic structure and genetic conservation. Theor Appl Genet 52: 145-157

Brown AHD (1989) Core collections: a practical approach to genetic resources management. Genome 31: 818-824

Buckler ES, Gaut BS, McMullen MD (2006) Molecular and functional diversity of maize. Curr Opin Plant Biol 9: 172-176

Caldwell KS, Russell J, Langridge P, Powell W (2006) Extreme population-dependent linkage disequilibrium detected in an inbreeding plant species, *Hordeum vulgare*. Genetics 172: 557-567

Cao W, Scoles G, Hucl P, Chibbar R (1999) The use of RAPD analysis to classify *Triticum* accessions. Theor Appl Genet 98: 602-607

Carvalho L, Schaal B (2001) Assessing genetic diversity in the cassava (*Manihot esculenta* Crantz) germplasm collection in Brazil using PCR-based markers. Euphytica 120: 133-142

Castano F, Gulya T, Re J, Echeverria M, Rodriguez R (2005) Reaction of some sunflower accessions to *Albugo tragopogonis* and *Sclerotinia sclerotiorum* infections. Helia 28: 25-31

Charcosset A, Moreau L (2004) Use of molecular markers for the development of new cultivars and the evaluation of genetic diversity. Euphytica 137: 81-94

Chebotar S, Roder MS, Korzun V, Saal B, Weber WE, Boerner A (2003) Molecular studies on genetic integrity of open-pollinating species rye (*Secale cereale* L.) after long-term genebank maintenance. Theor Appl Genet 107: 1469-1476

Chwedorzewska KJ, Bednarek PT, Puchalski J (2002a) Studies on changes in specific rye genome regions due to seed aging and regeneration. Cell Mol Biol Lett 7: 569-576

Chwedorzewska KJ, Bednarek PT, Puchalski J, Krajewski P (2002b) AFLP-profiling of long-term stored and regenerated rye genebank samples. Cell Mol Biol Lett 7: 457-463

Crossa J, Vencovsky R (1997) Variance effective population size for two-stage sampling of monoecious species. Crop Sci 37: 14-26

Crow JF (1992) Genetic load. In: Keller EF, Lloyd EA (eds) Keywords in Evolutionary Biology. Harvard Univ Press, Cambridge, Massachusetts, pp 132-136

Dahlberg J, Zhang X, Hart G, Mullet J (2002) Comparative assessment of variation among sorghum germplasm accessions using seed morphology and RAPD measurements. Crop Sci 42: 291-296

Dean R, Dahlberg J, Hopkins M, Mitchell S, Kresovich S (1999) Genetic redundancy and diversity among 'Orange' accessions in the U.S. National *Sorghum* collection as assessed with simple sequence repeat (SSR) markers. Crop Sci 39: 1215-1221

Dehmer KJ, Hammer K (2004) Taxonomic status and geographic provenance of germplasm accessions in the *Solanum nigrum* L. complex: AFLP data. Genet Resour Crop Evol 51: 551-558

del Rio AH, Bamberg JB, Huaman Z (2006) Genetic equivalence of putative duplicate germplasm collections held at CIP and US potato genebanks. Am J Potato Res 83: 279-285

del Rio AH, Bamberg JB, Huaman Z (1997) Assessing changes in genetic diversity of potato gene banks. 1. Effects of seed increase. Theor Appl Genet 95: 191-198

Dhanaraj A, Rao E, Swamy K, Bhat M, Prasad T, Sondur S (2002) Using RAPDs to assess the diversity in Indian cashew (*Anacardium occidentale* L.) germplasm. J Hort Sci Biotechnol 77: 41-47

Dobrovolskaya O, Saleh U, Malysheva-Otto L, Roeder MS, Boerner A (2005) Rationalising germplasm collections: a case study for wheat. Theor Appl Genet 111: 1322-1329

El-Assar AM, Krueger RR, Devanand PS, Chao CCT (2005) Genetic analysis of Egyptian date (*Phoenix dactylifera* L.) accessions using AFLP markers. Genet Resour Crop Evol 52: 601-607

Excoffier L, Smouse PE, Quattro JM (1992) Analysis of molecular variance inferred from metric distances among DNA haplotypes: application to human mitochondrial DNA restriction data. Genetics 131: 479-491

FAO (1998) The state of the world's plant genetic resources for food and agriculture. FAO, Rome, Italy

Ferguson M, Ford-Lloyd B, Robertson L, Maxted N, Newbury H (1998) Mapping the geographical distribution of genetic variation in the genus *Lens* for the enhanced conservation of plant genetic diversity. Mol Ecol 7: 1743-1755

Furini A, Wunder J (2004) Analysis of eggplant (*Solanum melongena*)-related germplasm: Morphological and AFLP data contribute to phylogenetic interpretations and germplasm utilization. Theor Appl Genet 108: 197-208

Gardner KM, Latta RG (2006) Identifying loci under selection across contrasting environments in *Avena barbata* using quantitative trait locus mapping. Mol Ecol 15: 1321-1333

Gepts P, Bliss FA (1985) F1 hybrid weakness in the common bean: differential geographic origin suggests two gene pools in cultivated bean germplasm. J Hered 76: 447-450

Ghislain M, Zhang D, Fajardo D, Huaman Z, Hijmans R (1999) Marker-assisted sampling of the cultivated Andean potato *Solanum phureja* collection using RAPD markers. Genet Resour Crop Evol 46: 547-555

Gökirmak T, Mehlenbacher SA, Bassil NV (2005) Investigation of genetic diversity among European hazelnut (*Corylus avellana*) cultivars using SSR markers. Acta Hort 686: 141-147

Goudet J, Raymond M, deMeeus T, Rousset F (1996) Testing differentiation in diploid populations. Genetics 144: 1933-1940

Grenier C, Deu M, Kresovich S, Bramel-Cox P, Hamon P (2000) Assessment of genetic diversity in three subsets constituted from the ICRISAT sorghum collection using random vs non-random sampling procedures B. Using molecular markers. Theor Appl Genet 101: 197-202

Gunderson KL, Steemers FJ, Lee G, Mendoza LG, Chee MS (2005) A genome-wide scalable SNP genotyping assay using microarray technology. Nat Genet 37: 549-554

Guzman FA, Ayala H, Azurdia C, Duque MC, deVicente MC (2005) AFLP assessment of genetic diversity of *Capsicum* genetic resources in Guatemala: home gardens as an option for conservation. Crop Sci 45: 363-370

Hartl D, Clark A (1989) Principles of Population Genetics. Sinauer Associates, Sunderland, MA, USA

Huamán Z, Aguilar C, Ortiz R (1999) Selecting a Peruvian sweetpotato core collection on the basis of morphological, eco-geographical, and disease and pest reaction data. Theor Appl Genet 98: 840-844

Ishitani M, Idupulapati R, Wenzl P, Beebe S, Tohme J (2004) Integration of genomics approach with traditional breeding towards improving abiotic stress adaptation: drought and aluminum toxicity as case studies. Field Crops Res 90: 35-45

Islam FMA, Beebe S, Munoz M, Tohme J, Redden RJ, Basford KE (2004) Using molecular markers to assess the effect of introgression on quantitative attributes of common bean in the Andean gene pool. Theor Appl Genet 108: 243-252

Iwaro AD, Butler DR, Eskes AB (2006) Sources of resistance to *Phytophthora* pod rot at the International Cocoa Genebank, Trinidad. Genet Resour Crop Evol 53: 99-109

James JW (1971) The founder effect and response to artificial selection. Genet Res 16: 241-250

Johnson R (1998) Genetic structure of regeneration populations of annual ryegrass. Crop Sci 38: 851-857

Kimura M (1983) The Neutral Theory of Molecular Evolution. Cambridge Univ Press, Cambridge, UK

Knapp S, Bohs L, Nee M, Spooner DM (2004) Solanaceae - a model for linking genomics with biodiversity. Comp Funct Genom 5: 285-291

Kojima Y, Ebana K, Fukuoka S, Nagamine T, Kawase M (2005) Development of an RFLP-based rice diversity research set of germplasm. Breed Sci 55: 431-440

Krueger RR, Roose ML (2003) Use of molecular markers in the management of citrus germplasm resources. J Am Soc Hort Sci 128: 827-837

Labate JA (2000) Software for population genetic analyses of molecular marker data. Crop Sci 40: 1521-1528

Li Z, Zhang H, Zeng Y, Yang Z, Shen S, Sun C, Wang X (2002) Studies on sampling schemes for the establishment of core collection of rice landraces in Yunnan, China. Genet Resour Crop Evol 49: 67-74

Manly BFJ (2004) Multivariate Statistical Methods: A Primer. 3rd edn. CRC Press, Boca Raton, Florida

Margulies M et al. (2005) Genome sequencing in microfabricated high-density picolitre reactors. Nature 437: 376-380

McCouch S (2004) Diversifying selection in plant breeding. PLoS Biol 2: e347-e306

McGregor CE, van Treuren R, Hoekstra R, van Hintum TJL (2002) Analysis of the wild potato germplasm of the series *Aculia* with AFLPs: Implications for *ex situ* conservation. Theor Appl Genet 104: 146-156

Michalakis Y, Excoffier L (1996) A generic estimation of population subdivision using distances between alleles with special reference to microsatellite loci. Genetics 142: 1061-1064

Moghaddam ME, Trethowan RM, William HM, Rezai A, Arzani A, Mirlohi AF (2005) Assessment of genetic diversity in bread wheat genotypes for tolerance to drought using AFLPs and agronomic traits. Euphytica 141: 147-156

Mohammadi SA, Prasanna BM (2003) Analysis of genetic diversity in crop plants - salient statistical tools and considerations. Crop Sci 43: 1235-1248

Nei M (1972) Genetic distance between populations. Am Naturalist 106: 283-292

Nei M (1973) Analysis of gene diversity in subdivided populations. Proc Natl Acad Sci USA 70: 3321-3323

Nei M (1978) Estimation of average heterozygosity and genetic distance from a small number of individuals. Genetics 89: 583-590

Nienhuis J (2001) Molecular marker-based characterization of genetic diversity in core collections. HortScience 36: 427

Obert D, Fritz A, Moran J, Sukhwinder S, Rudd J, Menz M (2005) The identification and molecular tagging of a gene from PI 289824 conferring resistance to leaf rust (*Puccinia triticina*) in wheat. Theor Appl Genet 110: 1439-1444

Parzies H, Spoor W, Ennos R (2000) Genetic diversity of barley landrace accessions (*Hordeum vulgare* ssp. *vulgare*) conserved for different lengths of time in *ex situ* gene banks. Heredity 84: 476-486

Paterson AH (2006) Leafing through the genomes of our major crop plants: strategies for capturing unique information. Nat Rev Genet 7: 174-184

Pestsova E, Korzun V, Goncharov N, Hammer K, Ganal M, Röder M (2000) Microsatellite analysis of *Aegilops tauschii* germplasm. Theor Appl Genet 101: 100-106

Petit RJ, El Mousadik A, Pons O (1998) Identifying populations for conservation on the basis of genetic markers. Conserv Biol 12: 844-855

Phippen W, Kresovich S, Candelas F, McFerson J (1997) Molecular characterization can quantify and partition variation among genebank holdings: a case study with phenotypically similar accessions of *Brassica oleracea* var. *capitata* L. (cabbage) 'Golden Acre'. Theor Appl Genet 94: 227-234

Potokina E, Blattner FR, Alexandrova T, Bachmann K (2002) AFLP diversity in the common vetch (*Vicia sativa* L.) on the world scale. Theor Appl Genet 105: 58-67

Raymond M, Rousset F (1995) An exact test for population differentiation. Evolution 49: 1280-1283

Reynolds J, Weir BS, Cockerham CC (1983) Estimation of the coancestry coefficient: basis for a short-term genetic distance. Genetics 105: 767-779

Rice EB, Smith ME, Mitchell SE, Kresovich S (2006) Conservation and change: A comparison of *in situ* and *ex situ* conservation of Jala maize germplasm. Crop Sci 46: 428-436

Richards CM, Reilley A, Touchell D, Antolin MF, Walters C (2004) Microsatellite primers for Texas wild rice *(Zizania texana)* and a preliminary test of the impact of cryogenic storage on allele frequency at these loci. Conserv Genet 5: 853-859

Richards CM, Antolin MF, Reilley A, Poole J, Walters C (2007) Capturing genetic diversity of wild populations for *ex situ* conservation: Texas wild rice (*Zizania texana*) as a model. Genet Resour Crop Evol 54: 837-848

Rogers J (1972) Measures of genetic similarity and genetic distance. Studies in Genetics VII. Univ Texas Publ 7213, Univ Texas, Austin, Texas

Rousset F (1996) Equilibrium values of measures of population subdivision for stepwise mutation processes. Genetics 142: 1357-1362

Rousset F (1997) Genetic differentiation and estimation of gene flow from F-statistics under isolation by distance. Genetics 145: 1219-1228

Ruiz M, Rodriguez-Quijano M, Metakovsky EV, Francisco-Vazquez J, Carrillo JM (2002) Polymorphism, variation and genetic identity of Spanish common wheat germplasm based on gliadin alleles. Field Crops Res 79: 185-196

Saitou N, Nei M (1987) The neighbor-joining method: A new method for reconstructing phylogenetic trees. Mol Biol Evol 4: 406-425

Scarascia-Mugnozza GT, Perrino P (2002) The history of *ex situ* conservation and use of plant genetic resources. In: Engels J, Rao VR, Brown A, Jackson M (eds) Managing Plant Genetic Diversity. CABI Publ, Wallingford, pp 1-22

Schnell RJ, Olano CT, Campbell RJ, Brown JS (2001) AFLP analysis of genetic diversity within a jackfruit germplasm collection. Sci Hort (Amsterdam) 91: 261-274

Serdani M, Spotts RA, Calabro JM, Postman JD, Qu AP (2006) Evaluation of the USDA National Clonal Pyrus Germplasm collection for resistance to *Podosphaera leucotricha*. HortScience 41: 717-720

Skroch P, Nienhuis J, Bebee S, Tohme J, Pedraza F (1998) Comparison of Mexican common bean (*Phaseolus vulgaris* L.) core and reserve germplasm collections. Crop Sci 38: 488-496

Slatkin M (1985) Rare alleles as indicators of gene flow. Evolution 39: 53-65

Sokal R, Michner C (1958) A statistical method for evaluating systematic relationships. Univ Kansas Sci Bull 28: 1409-1438

Steiner A, Ruckenbauer P, Goecke E (1997) Maintenance in genebanks, a case study: contaminations observed in the Nürnberg oats of 1831. Genet Resour Crop Evol 44: 533-538

Trethowan RM, van Ginkel M, Sanjaya R (2002) Progress in breeding wheat for yield and adaptation in global drought affected environments. Crop Sci 42: 1441-1446

van Hintum TJL, Visser DL (1995) Duplication within and between germplasm collections. Genet Resour Crop Evol 42: 135-145

van Hintum TJL, van Treuren R (2001) Improving PGR management with molecular markers. In: Broad Variation and Precise Characterization-Limitation for the Future. Proc XVI EUCARPIA Genet Resour Sect Workshop, Poznan, Poland, pp 194-197 van Hintum TJL, Boukema IW, Visser DL (1996) Reduction of duplication in a *Brassica oleracoa* germplasm collection. Genet Resour Crop Evol 43: 343-349

van Treuren R, van Hintum TJL (2001) Identification of intra-accession genetic diversity in selfing crops using AFLP markers: Implications for collection management. Genet Resour Crop Evol 48: 287-295

van Treuren R, Magda A, Hoekstra R, van Hintum TJL (2004) Genetic and economic aspects of marker-assisted reduction of redundancy from a wild potato germplasm collection. Genet Resour Crop Evol 51: 277-290

Vencovsky R, Crossa J (1999) Variance effective population size under mixed self and random mating with applications to genetic conservation of species. Crop Sci 39: 1282-1294

Vetelainen M, Gammelgard E, Valkonen JPT (2005) Diversity of Nordic landrace potatoes (*Solanum tuberosum* L.) revealed by AFLPs and morphological characters. Genet Resour Crop Evol 52: 999-1010

Volk GM, Henk AD, Richards CM (2004) Genetic diversity among U.S. garlic clones as detected using AFLP methods. J Am Soc Hort Sci 129: 559-569

Volk GM, Richards CM, Reilley AA, Henk AD, Forsline PL, Aldwinckle HS (2005) *Ex situ* conservation of vegetatively propagated species: development of a seed-based core collection for *Malus sieversii*. J Am Soc Hort Sci 130: 203-210

Walters C, Touchell D, Power P, Wesley-Smith J, Antolin M (2002) A cryopreservation method for embryos of the endangered species *Zizania texana*. Cryo Lett 23: 291-298

Wang M, Mosjidis J, Morris J, Dean R, Jenkins T, Pederson G (2006) Genetic diversity of *Crotolaria* germplasm assessed through phylogenetic analysis of EST-SSR markers. Genome 49: 707-715

Weimer J (1999) Chapter 19 Accelerating the trend toward healthy eating: Public and private efforts. In: E Frazao (ed) America's Eating Habits: Changes and Consequences. Agri Info Bull 750, USDA/ERS, pp 385-401

Weir B (1990) Genetic Data Analysis. Sinauer Assoc, Sunderland, MA, USA

Wu XM, Wu NF, Qian XZ, Li RG, Huang FH, Zhu L (1998) Phenotypic and genotypic changes in rapeseed after 18 years of storage and regeneration. Seed Sci Res 8 (Suppl 1): 55-64

3 Molecular Mapping and Breeding for Yield and Yield-related Traits

Gail Timmerman-Vaughan
New Zealand Institute for Crop and Food Research Ltd., PO Box 4704, Christchurch, New Zealand
E-mail: TimmermanG@crop.cri.nz

1 INTRODUCTION

Yield is arguably the most valuable trait for arable, horticultural or fiber crops, but physiologically and genetically it one of the most complex traits. During the twentieth century, incredible increases in grain yield occurred - either expressed as yield per unit area, which doubled in the 40 years from 1940 to 1980 and continues to increase (Richards 2000), or in total food grain production, which went from approximately one billion tons in 1960 to two billion tons in 2000 (Khush 2001). For example, during the post-green revolution years from 1966 to 1999, world rice production increased by 132% and wheat production by 91% (Khush 2001). The increases seen in yield are attributable to genetic increases as a result of plant breeding and to agronomic practices, such as the use of irrigation and inorganic fertilizers. Into the future, world population is expected to reach about nine billion by 2050 (*www.census.gov/ipc/www/worldpop.html*). Therefore, further yield increases will be required, with a greater reliance on genetic rather than agronomic gain as resources such as available land, water for irrigation, energy for production of inorganic N fertilizers, or natural sources of inorganic P become more scarce and expensive.

Depending on the crop and its intended use, the yield trait of direct economic interest may be, for example, biomass, grain or seed yield, fruit yield, or the yield of tubers or rhizomes. While much of the improvement in yield observed in the past 40-plus years has been due to genetic gains in adaptation (e.g., tolerance or resistance to biotic or abiotic stresses, including manipulation of phenology to match climatic conditions), the focus of this chapter is predominantly on the genetics of yield *per se* or attributes that contribute to yield.

2 YIELD: PHYSIOLOGICAL COMPLEXITY

Yield of the harvested product of an economic plant species is a complex trait that is the result of interactions between plant genome and environment and a product of the expression of most processes and pathways in the plant. In a recent review, Richards (2000) attributes historical increases in yield to two components - increased photosynthesis per unit land area, and increased partitioning of biomass into the harvested component - and further credits the major gains in the former to agronomic practice while the latter (biomass partitioning) is mainly attributable to genetic improvement. In wheat, yield improvement has been attributed to increased harvest index (HI = seed dry weight/total plant dry weight) rather than biomass, while the converse is true in maize, where yield improvement has been attributed to increases in biomass/unit land area. QTL mapping of physiological traits is discussed in Chapter 7, Volume 2. Chapter 7 discusses three topics of particular relevance to yield determination - QTL for photosynthesis, nutrient absorption, and flowering time.

Breeders have long sought traits related to biomass partitioning that can be used as indirect selection tools to improve yield, so consequently have dissected yield into its component traits. In cereals, for example, yield per unit area can be described as the product of the number of plants per unit area, the number of spikes per plant, the number of spikelets per spike, the number of grains per spikelet, and the mean grain weight (Slafer 2003). More simply, for any seed or fruit yielding plant, yield per unit area is the mean seed or fruit weight × the total seed or fruit number per unit area. Typically, yield has low heritability because of its sensitivity to the environment, while traits like grain weight have higher heritability and therefore are more attractive as breeding targets, if their relevance to yield as an alternate selection measure can be demonstrated. However, the relationships between yield *per se* and yield component traits are not straightforward, and positive or negative correlations are often observed. For example, an inverse relationship between mean seed weight and seed number is commonly observed (Timmerman-Vaughan et al. 2005). QTL mapping of yield *per se* and of well chosen yield-related traits has the potential to elucidate the genetic basis of these correlations by determining coincidence of QTLs, and the magnitude and direction of QTL effects, for related traits in the same genomic region.

Biomass can also be expressed in terms of its component parts as the product of the solar radiation over the crop period duration (*Q*), the amount of light intercepted by the crop canopy (*I*), and the captured light energy converted to plant dry matter (*E*), expressed as:

$$\text{Biomass} = \sum_{\text{sowing}}^{\text{harvest}} Q \times I \times E$$

As discussed by Richards (2000), genetic approaches have been most successful at manipulating the crop canopy (*I*). For example, yield in maize is affected by leaf number, blade area and leaf angle, as well as by tassel architecture due to the ability of the tassel to shade the plant significantly (reviewed by Mickelson et al. 2002). Crop period duration (*Q*) is also amenable to breeding and was manipulated very successfully to improve yield and yield stability in 'Green Revolution' plant cultivars (reviewed by Khush 2001). In this case, however, crop period duration was shortened and segregants were selected that would result in acceptable yields (usually as a result of early vigor) when combined with shorter plant types that are more suitable to intensive cropping. Finally, as discussed by Richards (2000), substantial genetic differences in photosynthetic rates per unit of leaf area exist in crop species so there is potential to manipulate photosynthetic rates to enhance biomass production. However, pleiotropic effects exist between photosynthesis, leaf nitrogen, specific leaf area (leaf area/leaf weight) and 'sink strength'. Therefore, the effects of selecting for high photosynthetic rate on yield may also have unexpected results.

3 POPULATION DEVELOPMENT

The generation of experimental populations suitable for QTL mapping has been discussed in Chapter 3 of Volume 1. To map QTLs for yield and yield-related traits it is important that population sizes are large enough to allow loci with small effects on phenotypic variation (perhaps even less than 5%) to be detected. Another important reason to use large populations is to detect significant epistatic interactions, which are commonly observed with yield and yield-related traits. Since yield is subject to G × E interactions as well, it is important that QTL detection is conducted in suitably replicated trials in multiple environments.

Furthermore, plant densities in plots used to estimate yield should resemble conventional agronomic practice, and plots should be large enough to minimize edge effects or allow inside rows to be sampled. The implication of this is that relatively large amounts of seed of each segregating progeny line or family are required. In practice, the population sizes used for QTL mapping of yield and yield-related traits are a compromise between the statistical power desired from the experiment and the time and cost involved in population development, linkage map construction and trait evaluation. Typically, yield QTLs are most commonly characterized using around 200 lines, but populations can vary from less than 100 lines (e.g., Nichols et al. 2006) to nearly 1,000 lines (e.g., Schön et al. 2004).

Classical approaches to population development for QTL mapping depend on the breeding system of the plant, and in particular its tolerance for inbreeding. Crosses for QTL mapping population development may be between adapted or elite genotypes. However, wide crosses produced by crossing an adapted with a wild genotype have also been used quite widely. Likewise, most yield QTL mapping experiments have been carried out using balanced populations - populations that have equal representation of the parental alleles, such as F_2-derived families, recombinant inbred (RI) or doubled haploid (DH) lines. An alternate approach that uses unbalanced populations, and is particularly suited for discovering QTL from wild germplasm, is development of advanced backcross (AB) populations (Tanksley and Nelson 1996), discussed below. The AB-QTL mapping approach leads to the development of introgression lines (ILs) containing marker-delimited genomic regions derived from wild germplasm introgressed into an elite genetic background.

The majority of yield QTL mapping research has been conducted in inbreeding crops like cereals (wheat, barley, rice), legumes (soybeans, beans, peas), or crops that tolerate inbreeding (maize, Brassicas), enabling the development of RI or DH populations where adequate seed can be produced for trait evaluation. For outcrossing species that are usually grown from hybrid seed, such as maize or Brassicas, it is common for the trait estimation phase of the experiment to be conducted using testcross seed developed by crossing a tester line with the individual RI or DH lines (e.g., Schön et al. 2004; Quijada et al. 2006).

4 QTL MAPPING OF YIELD USING BALANCED POPULATIONS IN ADAPTED GERMPLASM

QTL mapping has characterized the genetics of yield and yield-related traits in a wide range of economically important plant species, including cereals, legumes, vegetables, oilseeds, forages, and fruits. Most experiments to map QTL for yield have a number of common design elements. These include the development of segregating populations using adapted parental lines (breeding lines or cultivars) that nonetheless show variation in yield for, use of 'balanced' segregating populations (e.g., F_2-derived families, RI lines or DH lines) in which alleles from both parents are present in approximately equal frequencies, construction of molecular linkage maps of moderate saturation, and analysis of yield *per se* and related traits in field trials where progeny families are replicated (at least) and conducted in multiple environments (sites and/or years). The yield-related traits that are assessed are often components of yield, as discussed above, or phenological traits related to yield. In major crops, QTLs have been mapped in a number of pedigrees, offering the opportunity to compare QTL genomic regions and phenotypic effects. Most studies examine a number of traits that show either positive or negative correlation with yield *per se*. Very often, QTLs for such related or correlated traits coincide on the molecular linkage maps being used. This coincidence suggests either that the QTL detects a complex genetic locus containing a cluster of genes that may co-ordinatedly regulate the associated traits or that the gene underlying the QTL acts pleiotropically resulting in variation in a number of traits.

A recently published study in hexaploid wheat demonstrates many aspects of this experimental approach (Quarrie et al. 2005). The population of 95 doubled haploid (DH) lines was developed by crossing Chinese Spring (a wheat variety that is widely used in genetic research) × SQ1 (a breeding line high in drought induced abscisic acid). The linkage map of the 21 wheat chromosomes was built using 567 molecular markers, mostly SSRs. The map had a mean distance between markers of approximately 6 cM, therefore could be considered to be of moderate saturation. Traits (yield *per se*, grains/ear, ears/plant, thousand-grain weight) were measured at four sites and over a number of years (a total of 24 environments) but also included estimation of yield in the presence of a number of abiotic stresses - drought, salt stress, nitrogen and nutrient stress. Based on position on linkage maps, QTLs for yield were associated with one gene for dwarfing (*Rht*-B1), two genes for vernalization (*Vrn*-A1, *Vrn*-B1), as well as with 14 other genomic regions. In most cases, one or more of the yield component QTLs also coincided with a yield QTL. Both parents contributed high value alleles at the detected QTL. The identification of beneficial alleles from germplasm that performs poorly for the trait of interest is another common feature of QTL mapping experiments in general, including those for characterizing the genetics of yield. Many of the QTLs identified in the Quarrie et al. (2005) study were also found in previously published studies (e.g., Börner et al. 2002; Groos et al. 2003). Furthermore, comparison of the wheat and rice linkage maps led Quarrie et al. (2005) to suggest that their wheat yield QTLs coincided with rice yield QTLs in 10 genomic regions. Since an aim - stated or unstated - of many linkage mapping studies is eventually to understand the genes underlying the trait of interest, their regulation and interaction with other genes, the discovery of wheat and rice yield QTLs in syntenic genomic regions is promising.

The question of QTL co-localization was also examined by ter Steege et al. (2005) who examined the basis of early vigor in *Aegilops tauschii,* which is the donor of the D genome of wheat. For this study, an F_4 population of 134 families was developed and many traits were measured. These traits contributed to three components of early vigor - leaf length, relative growth rate and plant mass. To characterize the correlation among the traits contributing to early vigor, path maps were constructed for leaf length, relative growth rate and plant mass, and the correlations (positive and negative) among the traits underlying these components of vigor were displayed. QTLs were found on all the seven D genome homeologous chromosomes. The authors did not find any significant QTL for plant mass and shoot mass - and consequently hypothesized that these traits are governed by a large number of loci of small effect. Comparison of the results of the ter Steege et al. (2005) study with other QTL mapping studies in rice and wheat suggested that wheat chromosomes 4 and 7 (rice chromosomes 3 and 6) are important for determining seed mass, yield, and growth.

5 NEW SOURCES OF YIELD VARIATION FROM WILD GERMPLASM

The domesticated germplasm of crop species is generally less genetically diverse than that of wild accessions, largely as a result of bottlenecks introduced by the selection of desirable traits or plant forms by humans over the past 10,000 years. Therefore, the potential exists to discover favorable alleles in wild accessions and eventually to deploy these for crop improvement. In this way, genetic gains may be obtained that are difficult if not impossible using the narrower genetic base of domesticated germplasm. Particular successes in using wild germplasm have been made in introducing major genes, such as for resistance to biotic stresses or for plant form. For example, Hoisington et al. (1999) summarized a range of pest and disease resistance genes that have been introgressed into wheat from a number of related species. In general, wild species are agronomically less desirable than modern cultivars. Characteristics of wild species include low yield, indeterminacy, pod or head shattering, seed hardness and growth habits that do not suit mechanical harvesting. Therefore, a problem with using wild germplasm to improve quantitatively inherited traits in a crop breeding program is the presence of alleles or epistatic effects from wild germplasm that have a deleterious effect on phenotypes of value for agricultural species (Tanksley and McCouch 1997). This has been a particular problem when the purpose of breeding is to improve quantitatively inherited traits like yield. In spite of this, QTL mapping approaches can be used to discover and characterize favorable QTL alleles present in wild germplasm. In this section, the discovery of favorable QTL alleles for yield and yield-related traits originating in wild germplasm will be discussed. In addition, the advanced backcross QTL (AB-QTL) mapping approach for discovering and deploying favorable QTL alleles from wild germplasm will be described.

In rice, the potential of *Oryza rufipogon* to contribute alleles related to yield to a cultivated *O. sativa* background was explored (Xiao et al. 1998). A BC_2 testcross population for QTL identification was developed by crossing the *O. rufipogon* parent (♂) with the *O. sativa* parent (V20B, ♀) to produce F_1 plants. These plants showed strong vegetative heterosis. BC_1 plants were produced by backcrossing with the *O. sativa* parent (V20A, ♂). After selection of 10 BC_1 plants with a desirable plant type, maturity and fertility, BC_2 plants were produced. The final BC_2 testcross families were produced by crossing 300 BC_2 plants with elite rice line Ce64. Notably, the parents V20A and Ce64 are the parents of a commercial F_1 hybrid. The BC_2 testcross families were genotyped with 122 markers spanning most of the rice genome. Yield *per se* and 11 yield-related traits were measured in only a single environment. The association of the traits with marker genotypes was analyzed by ANOVA. Genotyping of the BC_2 testcross families revealed that the mean *O. rufipogon* DNA content averaged 4.5% (range from 1.5 to 11.5%). Trait analysis using this population showed that grain yield was strongly and positively correlated with plant height, panicles per plant, spikelets per plant, grains per panicle and grains per plant (r = 0.968!), but not with seed weight, heading date or days to maturity.

Of the 68 QTLs detected, the favorable alleles for a remarkable 35 loci (51%) were provided by *O. rufipogon*. This high proportion of discovery of favorable alleles from *O. rufipogon* may be due to the plant type selection that took place at BC_1 during population development. As is commonly observed in QTL mapping studies of yield and yield-related traits, coincident QTLs were observed in association with a number of genomic regions. For these coincident QTLs, it was often the case that a favorable *O. rufipogon* allele for one trait was associated with deleterious alleles for other traits. For example, in two cases where the *O. rufipogon* allele increased yield, there was also an increase in plant height associated with the QTL. In rice, increased plant height is not a favorable trait because of increased susceptibility to lodging, and consequently yield loss. At a QTL degree of resolution, this illustrates the difficulty of using wild germplasm in a conventional breeding program without use of breeding strategies designed to isolate the valuable wild alleles in an elite germplasm background. Still, 19 of the 35 *O. rufipogon*-derived QTLs detected had no negative effects of associated traits so may provide a valuable source of novel genetic diversity for rice improvement. In particular, these authors found promising wild alleles at two loci that resulted in shortened growth duration without decreasing grain yield, at least in this single environment.

6 ADVANCED BACKCROSS (AB) QTL MAPPING

The advanced backcross (AB) method was first suggested by Tanksley and Nelson (1996) as a way to identify beneficial QTL alleles from wild relatives of economically important plant species while developing germplasm that is suitable for inclusion in a cultivar breeding program. The use of wild relatives in plant breeding is accompanied by a range of problems that can be very time consuming to overcome, including low hybridization rates and fertility of progeny, non-Mendelian segregation of loci, suppression of recombination, and linkage drag. The AB-QTL method was first described in tomato, but has been used more widely to identify wild QTL alleles for yield and yield-related traits in, for example, bean (Blair et al. 2006), barley (von Korff et al. 2006), wheat (Narasimhamoorthy et al. 2006), and rice (Yoon et al. 2006).

The steps in AB-QTL mapping are: 1) cross an elite × a wild genotype to develop hybrid plant(s); 2) backcross hybrids to the elite parent to produce BC_1 and BC_2 populations; 3) conduct selection against known deleterious alleles by phenotypic selection or using markers; 4) continue backcrossing to the elite parent; 5) using molecular markers genotype either BC_2 or BC_3 populations; 6) using BC_3 or BC_4 populations, conduct agronomic evaluation of traits of interest; 7) carry out QTL mapping to identify QTL(s) and discover beneficial wild alleles using molecular marker and trait information; 8) develop NILs for selected genomic regions that carry beneficial wild alleles by marker-based selection; and 9) validate agronomic performance of NILs carrying the wild alleles in comparison with elite parental line.

An example of an AB-QTL population development and mapping strategy implemented in tomato based on a *Lycopersicon esculentum* × *L. parviflorum* cross (Fulton et al. 2000) is presented in Fig. 1. Development of the AB-QTL population proceeded using 8 BC_1 lines retained after marker-assisted selection from a total of 50 BC_1 lines produced. In this case, the retained BC_1 lines were selected for homozygous *L. esculentum* alleles for a single trait, determinate growth, which was required for agronomic characters to be assessed. At this stage, it was possible that the BC_1 lines used would only carry a portion of the wild parental

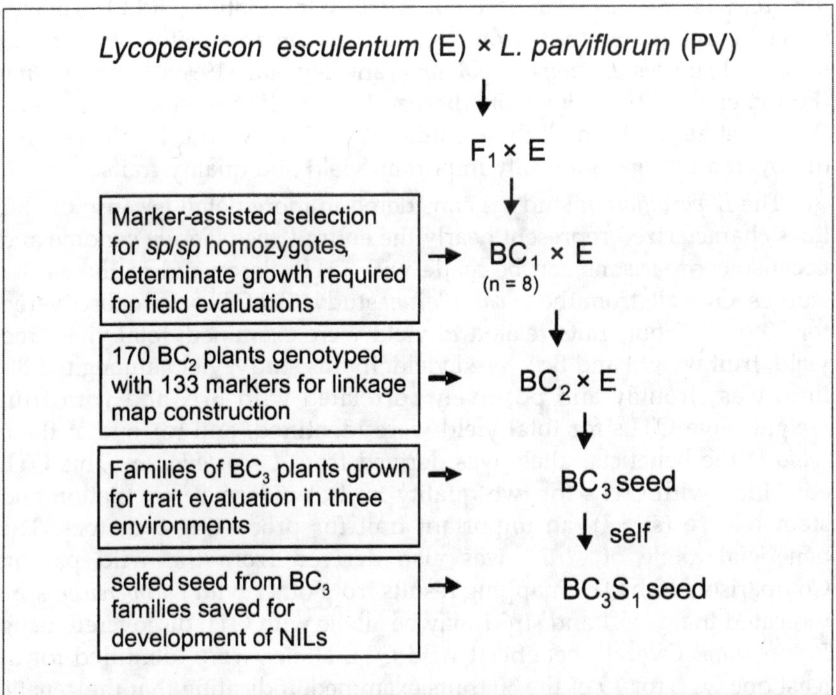

Fig. 1 Schematic diagram of an AB-QTL mapping strategy based on Fulton et al. (2000). Marker genotyping of BC_2 plants showed that *L. parviflorum* alleles were present in BC_2 lines for nearly all genomic regions, and that the average number of E/PV heterozygotes was 19% per locus, close to the expected value for a BC_2 of 25%.

alleles. For example, the AB-QTL population developed by Fulton et al. (1997) by crossing an elite tomato line with *L. peruvianum* contained alleles representing only about 67% of the *L. peruvianum* genome. The original elite × *L. peruvianum* cross experienced poor fertility and as a result only five fertile BC_1 lines were obtained. Therefore, loss of *L. peruvianum* genome segments would be expected due to the very small size of the BC_1 population as well as the fertility problems. Then in the *Lycopersicon esculentum* × *L. parviflorum* cross (Fig. 1), linkage maps were constructed using BC_2 plants. The BC_2 plants were backcrossed to the elite parent to produce BC_3 seeds for agronomic trials. The consequence of planting BC_3 families is that one-half of the individuals in a BC_3 plot will be heterozygous for the wild parental QTL or marker allele while the remainder will be homozygous for the respective *L. esculentum* alleles. This must be taken into account when calculating the estimated QTL allelic effects.

AB-QTL mapping has been applied most extensively to-date in tomato, where it has been used successfully to identify wild QTL alleles for yield and quality traits from crosses between an elite tomato line and four wild species, *L. pimpinellifolium* (Tanksley et al. 1996), *L. peruvianum* (Fulton et al. 1997), *L. hirsutum* (Bernacchi et al. 1998a) and *L. parviflorum* (Fulton et al. 2000). In all these studies, beneficial wild QTL alleles were discovered for agronomically important yield and quality traits.

The *L. parviflorum* study is considered in more detail because the BC lines characterized represent nearly the entire *L. parviflorum* genome and because comparisons can be made with QTL discovered in the earlier studies. Overall, from the *L. parviflorum* study, 199 QTLs were discovered for 30 traits. Four traits related to yield were examined: total yield, red yield, fruit weight and Brix × red yield. In this study, yield among the BC lines was strongly and positively correlated with Brix and with fruit weight. Five QTLs for total yield were identified, and for one of them (*yld6.1*) the beneficial allele was derived from *L. parviflorum*. This QTL coincided with QTLs for two quality traits, epidermal reticulation and stem release (*str6.1*), an important trait for processing tomatoes. The beneficial allele of *str6.1* was also derived from the wild parent. Comparison with QTL mapping results from other wild *Lycopersicon* spp. indicated that *yld6.1* and *str6.1* may be allelic with QTL discovered using *L. hirsutum*. Overall, beneficial wild QTL alleles were identified for at least one QTL for 19 of the 30 traits examined, indicating that the genetic diversity offered by *L. parviflorum* has the potential to make a significant contribution to development of future tomato varieties. Likewise, comparison of the *L. parviflorum* AB-QTL mapping results with the other three wild species indicated that possible orthologous QTLs were discovered for 16 of the 30 traits examined, and for these 16 traits, 39% of the QTLs were potentially orthologous. The identification of this number of potentially common QTLs among four tomato relatives confirms that the QTLs detected are indeed real and are especially not due to Type I error (false declaration of a QTL where one does not exist). By using the AB-QTL strategy, the authors have taken the first step toward developing the QTL NILs that can be intercrossed to pyramid QTL alleles and, with the use of marker-assisted selection, to obtain rapid and substantial progress towards breeding objectives.

7 LINKAGE DISEQUILIBRIUM (LD) MAPPING: BACKGROUND

As discussed above, the conventional method for dissecting the genetics of a complex trait and related traits using QTL mapping begins with the development of an experimental population developed from a cross between two parental lines. Linkage disequilibrium (LD) mapping (also called association mapping) provides an alternative approach that is based on broader genetic variation that may be found in individuals, germplasm or cultivars representative of the species being examined (Gaut and Long 2003). LD mapping has attained considerable popularity in human genetics and has been applied successfully to identify mutations associated with a number of disease-causing genes. Mapping in humans is outside the scope of this chapter, but was reviewed recently by Cardon and Bell (2001).

LD is the non-random association of alleles due to genetic linkage. Consequently, LD mapping is conceptually deceptively simple – the aim is to discover markers that are so tightly linked to the trait that the marker and trait alleles are very strongly associated, even in the presence of segregating variation in a diverse collection of individuals or plant genotypes, including natural populations. LD mapping has a number of advantages over conventional linkage or QTL mapping. LD mapping examines a wider range of genetic variation in a more diverse genetic background. It, therefore, has the potential to detect marker-trait associations that may be missed in experimental populations because of lack of segregation at marker and/or trait alleles. LD mapping is capable of attaining a higher resolution than linkage mapping because LD populations are likely to contain many more informative meioses than conventional mapping populations. Finally, LD mapping can use collections of germplasm or cultivars for which trait data have been gathered over many trials and stored in databases. Trait evaluation is time consuming and expensive, so LD has the potential to take advantage of the investment made in characterization of cultivars or germplasm. For example, genetic loci for yield and yield stability were mapped using LD in spring barley cultivars using trait data gathered from official Danish variety trials over eight years (Kraakman et al. 2004). For other species, data can be found in publicly available databases (cf. www.grin.ars.gov).

LD is most commonly measured using the metrics D, D′, and r^2. A description of the concept of LD can be found in Fig. 2 and its legend. More detailed reviews of the use of LD in plants were published recently (Flint-Garcia et al. 2003; Gaut and Long 2003).

Three common metrics are used to measure LD quantitatively, D, D′, and r^2.

$$D' = P_{A(green)\,B(red)} - (p_{A(green)} \times p_{B(red)})$$

where $P_{A(green)B(red)}$ is the frequency of sequences with the green allele at locus A and the red allele at locus B, $p_{A(green)}$ is the marginal frequency of the green allele at locus A and $p_{B(red)}$ is the marginal frequency of the red allele at locus B. Therefore, for Case 1, complete LD, D = (0.5) - (0.5)(0.5) = 0.25. The value of D is dependent on marginal allele frequencies, the value of D obtained will vary substantially over different SNPs, even if the sites are in complete LD.

The LD metric r^2 is the covariance between alleles at different loci and can also be thought of as the square of the correlation coefficient between the two loci.

$$r^2 = \frac{D_{AB}^2}{p_A p_a p_B p_b}$$

r^2 can only reach a value of 1.0 when the allele frequencies at the two loci being compared are equal.

D′ is a measure of LD that is normalized with respect to allele frequencies. D′ has a value of 1.0 if two alleles are in complete LD (as in Case 1) and also if only three gametic phases are present (as in Case 2).

$$|D'| = \frac{(D_{AB})^2}{\min(p_A p_b, p_a p_B)} \quad \text{for } D_{ab} < 0, \text{ or}$$

$$|D'| = \frac{(D_{AB})^2}{\min(p_A p_B, p_a p_b)} \quad \text{for } D_{ab} > 0$$

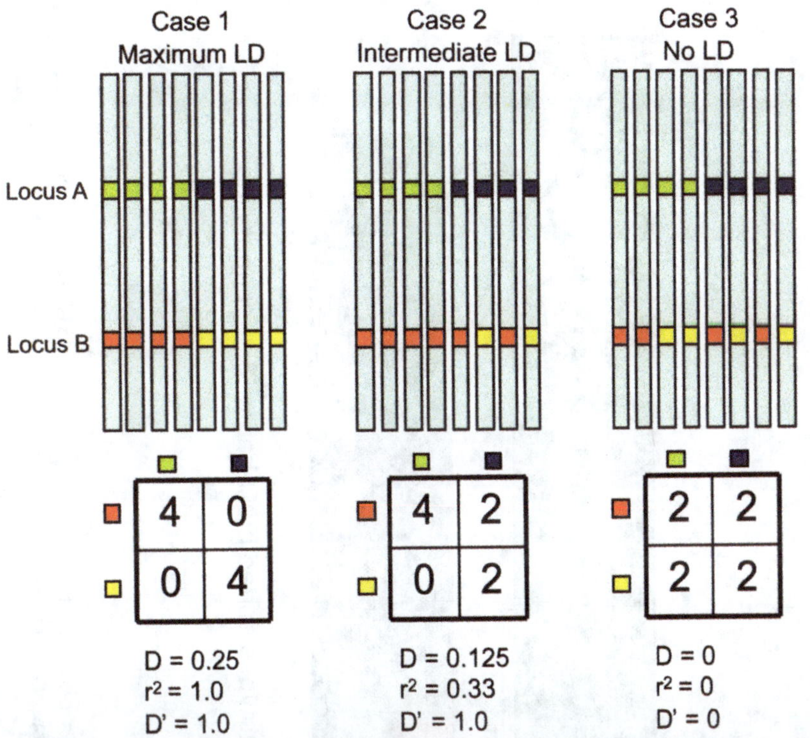

Fig. 2 Estimation of LD. In eight individuals, patterns of allelic variation at two loci (A and B) are shown for three Cases. 2 × 2 contingency tables for allelic segregation are shown below the hypothetical chromosome segregation patterns. Values computed for D, r^2, and D' are shown below the contingency tables. In Case 1, alleles are in complete LD since there is no evidence of historical recombination. In Case 2, alleles are in intermediate LD, and only three gametic phases are present. In Case 3, alleles show no LD.

The statistical significance (*P* value) of LD is calculated either using Fisher's exact test if two alleles are present or multifactorial permutation testing if more than two alleles are present.

The use of LD mapping for QTL discovery and characterization can be confounded because LD can be due to sources other than genetic linkage. In particular, LD can be affected by historical forces such as multilocus selection (especially directional selection involving epistasis), random drift in small populations, population migration or blending (admixture), population bottlenecks, and mutation. These evolutionary forces have the potential to increase LD above the rates expected from

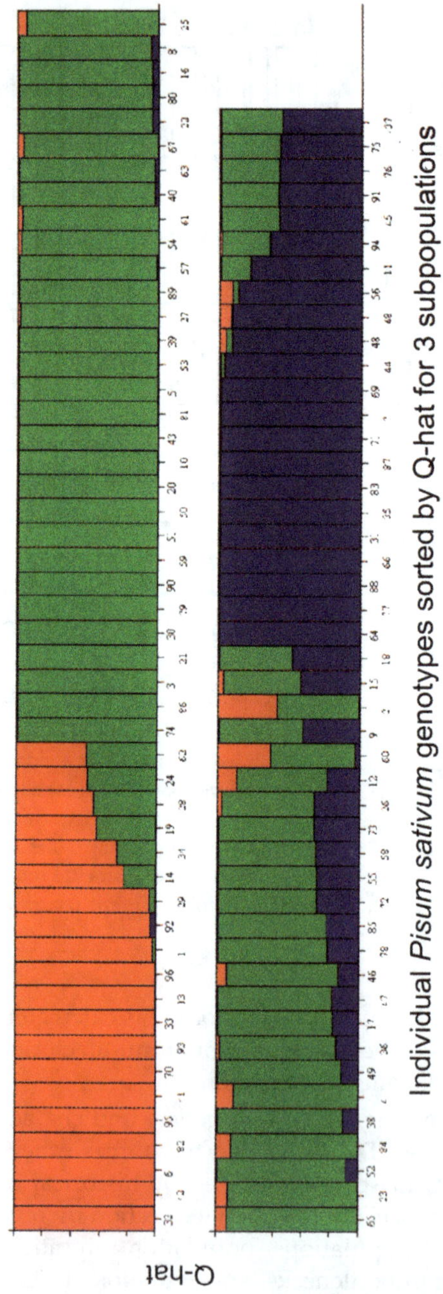

Fig. 3 Population structure for 96 *Pisum sativum* germplasm accessions estimated using the software STRUCTURE (Pritchard 2000), solved for a model of k = 3, 3 subpopulations. Q-hat – the proportion of the individual's genome that is estimated to originate in a single subpopulation – is plotted against the individual lines.

chance. If present, therefore, it will result in an increase in the rate of false-positive associations found. The mating system also has a significant effect on LD. Selfing species seem to have increased recombination rates per meiosis but have high amounts of homozygosity, which reduces the opportunities for alleles to be shuffled due to a lack of double heterozygotes. Outbreeding species have higher amounts of heterozygosity so have a greater opportunity per meiosis to develop informative crossovers than do inbreeding species. The effect of these considerations is that alleles can remain correlated over relatively long distances in selfing species, so LD can be observed over long physical distances. In contrast, LD decays more rapidly in outbreeding species, so LD is maintained over much shorter distances.

In crop plants, the history of domestication (introducing genetic bottlenecks, resulting in effective population sizes that are generally small) and of breeding (resulting in non-random mating, introducing admixture and selection), as well as the prevalence of inbreeding species, has resulted in the potential for LD to be conserved, with an increased risk of obtaining false-positive associations between marker and trait alleles when using LD mapping. A common approach is to correct for population stratification, particularly when characterizing LD in crop species populations prior to carrying out LD mapping, using molecular marker polymorphisms that have been scored on the germplasm being investigated. Recent research characterizing LD in maize (Stich et al. 2006), barley (Caldwell et al. 2006) and potato (Simko et al. 2006) describes the statistical approaches taken for characterizing population stratification. Generally two approaches are taken: statistical methods like cluster analysis and/or principle coordinates analysis, employing measures of genetic diversity based on nucleotide substitution rates, or a Bayesian model-based approach using the software package STRUCTURE (Pritchard et al. 2000; Pritchard and Wen 2004). Using the results of these analyses, individuals or lines can be assigned to subpopulations or further divided into subgroups, and this information can be considered when testing marker-trait associations. Figure 3 shows the results of analysis of population structure in 96 *Pisum sativum* germplasm accessions that have been genotyped using a collection of 36 molecular markers, including SSRs, sequenced tagged sites and RAPDs using the STRUCTURE software. In this case, the model being tested was that three subpopulations existed among the lines (k = 3). This is represented graphically by plotting the statistic Q-hat against the individual lines, as shown in Fig. 3. The Q-hat statistic estimates the fraction of the genome of the individual line that is estimated to have originated in the subpopulation. In the example in Fig. 3, many lines are estimated to reside almost completely within a single subpopulation, which is represented by a bar of just one color. Lines whose genomes are likely to have resulted from admixture between two or even three populations are shown with two or three colors.

Two approaches are taken to LD mapping. The first is LD mapping in a defined genomic region based on prior information gained from linkage mapping studies. This approach is often taken as a means of identifying nucleotide sequence polymorphisms in the candidate genomic region that may be responsible for trait determination. It leads to the identification of the gene or sequence variants underlying the genetic locus. Examples from recent literature include: the use of LD mapping to characterize sequence variation in and around the *Dwarf8* locus of maize and its association with flowering time (Thornsberry et al. 2001); and characterization of haplotype variation in the vicinity of the Cryptochrome2 (*CRY2*) photoperiod receptor in *Arabidopsis thaliana* (Olsen et al. 2004), resulting in identification of an amino acid variant associated with an early flowering phenotype.

In the second approach, the entire genome is 'scanned' and LD mapping is conducted using molecular markers that span the entire genome. Examples include LD mapping of a gene for bolting in sea beet (Hansen et al. 2001) and genome-wide LD mapping of major flowering time (*FRI*) and pathogen resistance (*Rpm1, Rps2,* and *Rps5*) genes in *A. thaliana* (Aranzana et al. 2005).

As described by Aranzana et al. (2005), the success of genome scanning via LD mapping depends on the magnitude of the gene's effect on phenotypes, as well as the gene's allelic frequencies; the size and underlying stratification of the population sample; and the density of the marker map. The extent of LD varies widely in different genomic regions within a species, therefore the marker densities required to provide reasonable power for detecting a significant marker-trait association also vary. Aranzana et al. (2005) estimated that a 20 kbp marker spacing may be appropriate for *A. thaliana.* This would require approximately 6,000 polymorphisms that have been selected to reflect haplotype structure in the local genomic regions. Therefore, development of marker resources for genome scanning via LD mapping with high power will require substantial investment, particularly for 'pre-genome' species that have not yet been sequenced in full.

8 LD MAPPING OF YIELD AND YIELD-RELATED TRAITS

In barley, a genome scanning LD mapping approach has been taken to identify AFLP markers significantly associated with yield, yield stability and adaptability using a collection of 148 spring barley cultivars (Kraakman et al. 2004). Attempts to identify population structure did not succeed using either cluster analysis or model-based Bayesian population structure analysis so the authors concluded there was an absence of population structure. Analysis of LD among the 123 AFLP markers that were able to be placed on an integrated barley genetic map revealed many instances of LD among markers that mapped more than 10 cM distance apart. Only markers with band-present frequencies between 5 and 95% were used, and a total of 236 AFLP markers were examined. Significant marker-trait correlations were observed with AFLP markers - including both mapped and unmapped markers. The authors were careful to control for multiple testing and controlled for a false discovery fate of more than 0.20. The mapped AFLP markers tagged 10 genomic regions on chromosomes 2, 3, 4, 5, 6, and 7 that had previously been shown to contain QTLs for yield in various mapping populations, from seven published studies (summarized in Kraakman et al. 2004). This result confirms the potential value of LD mapping for identifying genetic loci for yield and similarly genetically complex and economically valuable traits. Two markers were associated with yield stability as well as yield, and QTL for yield G × E interactions associated with these genomic regions were previously reported (Voltas et al. 2001; Malosetti et al. 2004).

Days to heading and plant height are two phenological traits with high relevance to production. A number of conventional QTL mapping studies have focussed on these traits, identifying many QTLs. Again, in modern spring barley cultivars, LD mapping was employed to identify markers associated with these traits, as well as morphological and disease resistance characteristics (Kraakman et al. 2006). In this study, the 148 spring barley cultivars that were the subject of the study discussed above were genetically sequenced using a combination of AFLP and SSR markers resulting in 286 polymorphic markers. Of these markers, 128 AFLP and 11 SSR markers were located on a combined barley map. In the case of this study, only two markers were associated with days to heading, much less than expected given the preponderance of QTLs described in the literature (summarized by Kraakman et al. 2006). The authors rationalized this result by proposing that many QTLs for days to heading operate in the cultivars examined, and therefore, it may be difficult to detect LD between one marker and one of the many QTL affecting the trait. The results for plant height were more expected. LD mapping identified 11 markers (three mapped and eight unmapped) associated with plant height. Association of heading date with the three mapped markers had all been reported before. Regarding plant height, the authors reasoned that success with this trait may reflect the contribution of fewer QTLs in cultivated germplasm than for days to heading, therefore marker-trait associations were less obscured by numerous other QTLs.

Kernel weight and dimensions in wheat are other yield-related traits that have been examined by LD mapping (Breseghello and Sorrells 2006). The research focussed on markers mapped to chromosomes 2D and 5A/5B because previous unpublished research indicated that QTL affecting kernel weight occurred on these chromosomes. Therefore, these authors carried out a more focussed LD mapping experiment rather than a genome scan. In this case a population of 149 soft winter wheat cultivars was characterized using 93 SSR loci from 19 wheat chromosomes. Population structure was characterized using the STRUCTURE software (Pritchard et al. 2000), resulting in division of the germplasm into four subpopulations. Of the 93 SSR loci examined, 36 were unlinked, 33 mapped to chromosome 2D, 20 to 5A and 9 to 5B. Significant marker-trait associations were found for kernel length and width on chromosome 2D, and for all traits (kernel weight, length, area, and width) on chromosomes 5A and 5B. Among the chromosome 2D markers associated with kernel width, marker *Xgwm515* has previously been related to kernel weight by QTL mapping (Dholakia et al. 2003). The results from this research, therefore, demonstrate the power of LD mapping to confirm and enhance the results of QTL mapping, with the potential to improve the resolution of QTL mapping results as well as to identify alleles that are potentially useful for selection.

9 CASE STUDY: DISCOVERY AND MANIPULATION OF QTL FOR YIELD AND RELATED TRAITS IN SOYBEAN

Soybean (*Glycine max* L.) is the world's major legume crop and is valued for its protein and oil content. Like many other economically important crops, soybean cultivars have a narrow genetic base. For example, North American soybean cultivars can trace 80% of their gene pool back to 13 ancestral cultivars (Gizlice et al. 1994). In spite of this narrow genetic base, soybean has achieved yield increases averaging 1% per year (Specht et al. 1999), mostly through crosses involving elite genotypes. Understanding the genetic basis of yield via QTL studies will benefit yield improvement in soybean by increasing the understanding of the genetic basis of yield, identifying marker-QTL associations, and identifying high value alleles from within the *G. max* primary gene pool, including the wild species *G. soja*. In this section, we discuss yield QTL discovery in cultivated soybean as well as in wider sources of genetic diversity, including plant introduction (PI) lines as well as *G. soja*. The progress and pitfalls to date in marker-assisted introgression of yield QTL in soybean are discussed in the 'marker-assisted selection' section below.

While a number of published studies have mapped QTLs for agronomic traits in cultivated soybean (summarized in Zhang et al. 2004), relatively fewer of them have mapped QTLs for yield *per se* (for example, Mansur et al. 1996; Orf et al. 1999; Specht et al. 2001; Yuan et al. 2002). In these studies, conventional QTL mapping has been carried out using RIL populations. Genomic regions that contain QTLs determining a number of yield and yield-related traits have been identified through a number of studies, particularly on linkage groups C2 (U9), F (U13), L (U14) and M (U11). These have been summarized graphically in Fig. 4, based on information from the above papers and linkage maps from the Soybean Breeders Toolbox (*http://soybeanbreederstoolbox.org/*).

The observed coincidence of QTL for a number of traits related to yield undoubtedly provides valuable information about the genetics of

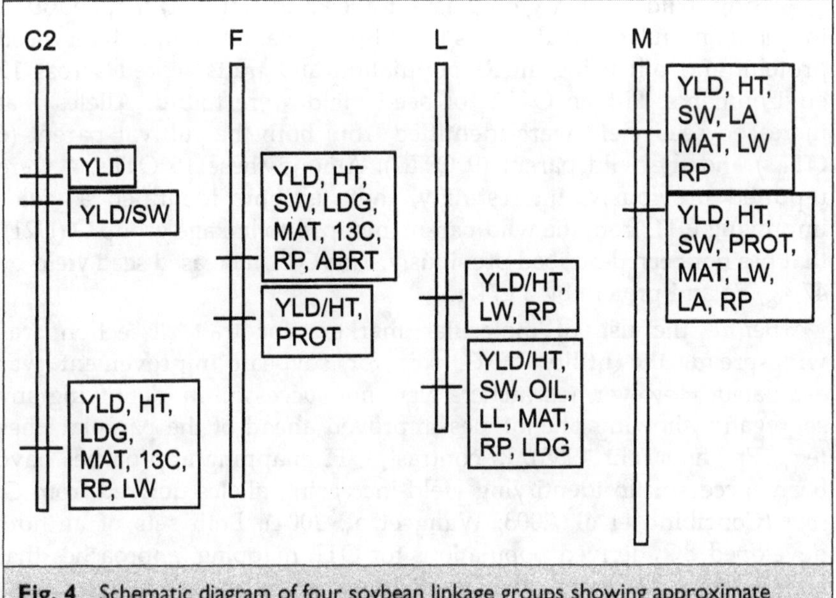

Fig. 4 Schematic diagram of four soybean linkage groups showing approximate locations of QTL for seed yield (YLD) and coincidence with other yield-related traits: seed weight (SW), height (HT), lodging (LDG), maturity (MAT), stable carbon isotope ratio $\delta^{13}C$ (13C), reproductive traits (RP), leaf width (LW), leaf area (LA), leaf length (LL), seed abortion (ABRT), seed protein (PROT), seed oil (OIL).

yield, for example by providing genetic explanations for the correlations that are observed between traits. For example, QTLs for yield and plant height are positively correlated in most studies and also coincide at most of these QTL regions. However, this association may be deleterious in some populations and for the QTL regions on linkage groups C2, F and L because QTLs for lodging are also associated. There are two possible genetic explanations for QTL coincidence - either the pleiotropic actions of a single gene or the actions of a cluster of genes that determine the individual traits. Since the resolution of QTL mapping experiments is generally low, it is usually not possible to distinguish these two hypotheses unless more research is conducted to fine-map the QTL regions.

To access novel sources of genetic variation that may not be available in modern soybean cultivars, QTLs for yield and yield-related traits have been identified using either 'wild' *G. max* plant introduction (PI) lines (Kabelka et al. 2004) or *G. soja* lines (Concibido et al. 2003; Wang et al. 2004) as parents in mapping population development. As will be described below, these studies have successfully identified genetic loci from wild germplasm that may improve yield.

Using 'wild' *G. max* germplasm, Kabelka et al. (2004) mapped QTL for yield and related traits (days to maturity, plant height, lodging, seed protein, and oil) using an RI population and traits scored across 12 environments. Fifteen QTLs for seed yield were found. Alleles that increased seed yield were identified from both the cultivar parent (6 QTLs) and the wild parent (9 QTLs). Among these 15 QTLs, 4 were reported previously. Interestingly, these authors identified a yield-improving QTL from the wild parent, mapped to linkage group O (U21), that has not been described previously. This QTL increased seed yield by 47 kg/ha and protein by 3 g/kg.

Before the use of molecular markers for trait dissection was widespread, the utility of *G. soja* for soybean improvement was evaluated. However, researchers were not successful in identifying any segregants showing phenotypes improved ahead of the parental lines (e.g., Ertl and Fehr 1985). In contrast, QTL mapping approaches have been successful in identifying yield-increasing alleles derived from *G. soja* (Concibido et al. 2003; Wang et al. 2004). Both sets of authors developed BC_2-derived populations for QTL mapping, approaches that are similar in concept to the development of advanced backcross (AB) lines for developing interspecific mapping populations, but that have been tailored to soybean, in which it is difficult to generate the large number of F_1 progeny required for the AB-QTL mapping approach.

Concibido et al. (2003) identified a *G. soja*-derived QTL that improved yield by 9.4%, mapped to linkage group B2 (U26). Lines carrying the *G. soja* allele at the marker associated with this QTL also had increased plant height and more pods on the main stem. This QTL may be orthologous to a QTL discovered in cultivated germplasm, since it occurred in the same genomic region as the linkage group B2 QTL discovered by Orf et al. (1999). Wang et al. (2004) identified 4 QTLs on linkage groups C2, E, K, and M that exceeded the experimental significance threshold. In these cases, however, the yield-improving alleles were all associated with the adapted soybean genotype rather than with *G. soja*. When Wang et al. (2004) reduced the significance threshold for QTL detection to $P < 0.05$ (but thereby increased the likelihood of false QTL declaration), they did succeed in identifying 4 yield QTLs where a positive effect was obtained from the *G. soja* allele. However, all these QTLs were previously described in *G. max*. A significant drawback to the experiment that was carried out by Wang et al. (2004) was that only approximately 22.8% of the *G. soja* genome was segregating in the BC_2-based populations examined. Therefore, only a fraction of the potential QTL for yield improvement could be discovered. Regardless, Wang et al. (2004) concluded that it will be "difficult to unlock positive allele diversity from *G. soja*".

10 TOWARD MARKER-ASSISTED SELECTION FOR YIELD AND RELATED TRAITS

Marker-trait associations discovered by LD mapping may have promise for selecting improved genotypes. Building on the markers associated with yield and yield-related traits discovered by LD mapping, Kraakman et al. (2004) took a multiple regression approach to ask whether markers could predict phenotypic responses. Using stepwise regression, selection of 18-20 markers from the total AFLP markers examined successfully explained significant proportions of the trait variation (for yield, yield stability, and adaptability), ranging from 40 to 58%. For all three traits, the minimum to maximum values predicted from using these subsets of markers vastly exceeded the observed range, suggesting that selection on these markers might produce genotypes that are superior for yield, yield stability, or adaptability. In contrast, stepwise regression using only those markers that were previously identified as being associated with the yield and yield stability explained much less of the variation, ranging from 22 to 34%.

The marker-assisted development of NILs carrying desirable QTL alleles derived from wild germplasm is one outcome of AB-QTL mapping. As a result of the AB-QTL mapping process, the association of these QTL alleles with linked markers is known, therefore valuable resources - both germplasm and marker - are available for marker-assisted breeding. In tomato, Bernacchi et al. (1998b, c) evaluated NILs carrying introgressions from two wild species, *L. hirsutum* and *L. pimpinellifolium*, for agronomic performance to test the applicability of the AB-QTL method to plant improvement. These authors examined 23 BC_3S_1 NILs developed for 15 genomic regions that contained QTLs for seven agronomic traits including yield, red yield and brix × red yield. Overall, 88% of the NILs derived from the *L. hirsutum* or *L. pimpinellifolium* AB-QTL experiments resulted in superior performance in at least one trial environment. Important issues that were examined included the ability of the AB-QTL mapping results to predict NIL performance, the magnitude of the NIL performance in comparison with QTL predictions, comparison of the significance levels for detecting effects in NILs versus the significance obtained from QTL detection, and the stability of the effects across trial environments. In general, there were modest relationships between the values obtained from BC_3 lines during QTL mapping and the NILs. Regarding the magnitude of NIL performance versus the BC_3 lines, generally similar results were obtained but the allelic effects of the NILs were usually slightly less than with the BC_3 lines. The significance of the effects detected in the NILs also generally mirrored the results obtained with the BC_3 lines - meaning that QTL detected at high significance levels in BC_3 lines also usually gave highly significant effects in the NILs. The effect of trial location on QTL effects in the BC_3 lines versus the NILs was quite variable, indicating the likelihood of genotype × environment effects at many of the QTL loci. In spite of these difficulties, the magnitude of the QTL effects can be relatively large, therefore opportunities exist to make many improvements in agronomic traits. For example, the *L. hirsutum* allele at the total yield QTL *yld4.1* was predicted by AB-QTL mapping to improve yield by 14.8% on average - and in the NIL carrying this introgression an average yield gain of 12% was gained. However, the magnitude of the yield gain varied between environments, and in two environments yield was reduced slightly versus two other environments where yield was increased more substantially. Improvements in red yield and brix × red yield were also associated with this NIL, and the same variable response across environments was observed. Historically, the yield gain in processing tomatoes from a conventional breeding has been 1% per annum. AB-QTL takes about four years, therefore the results of this study suggest that a rate of yield improvement of 3% may be obtained. Similar genetic gains may also be obtained using other wild *Lycopersicon* species or different accessions of *L. hirsutum*.

The potential for marker-assisted introgression of yield QTL alleles in soybean cultivar development has also been examined, but mixed results have been obtained from two studies published to-date. Reyna and Sneller (2001) obtained no significant yield improvement (a negative result) when three QTL alleles for improved yield (derived from northern *G. max* cultivar Archer) were introgressed into two different southern soybean cultivars. In contrast, Concibido et al. (2003) introgressed the linkage group B2 *G. soja* QTL discussed above into six soybean cultivars, and consistently obtained an yield advantage (up to 9%) in two of the six genetic backgrounds.

The success of Concibido et al. (2003) in obtaining yield increases from *G. soja* alleles in modern soybean cultivars is encouraging, and the limitations to their results may also explain the negative results obtained by Reyna and Sneller (2001). Clearly the adapted cultivar genetic background affected the expression of the increased yield phenotype, either due to epistatic interactions of the introduced QTL allele with background loci or because the adapted cultivar allele at that QTL position is similar in value to the yield-increasing allele being introduced.

11 SUMMARY

QTL genetic analysis has tremendous potential to contribute to increasing yield into the future through a number of routes. There are many examples of QTLs being identified for phenological traits, biomass partitioning traits, and yield *per se*. Significant opportunities may exist to discover the basis of genetic variation in relatively unexplored physiological traits, particularly related to photosynthesis. QTL analysis of yield and yield-related traits at the very least establishes the minimum number of genetic loci involved in trait determination, and provides estimates of the potential magnitude of effect of the alleles derived from particular germplasm sources – in themselves very valuable information for plant breeders. Placement of QTLs on a linkage map or demonstration of association with particular molecular marker loci have the potential to identify coincident or closely linked QTLs for either related or unrelated traits. Knowledge of the genetic linkage of loci determining plant productivity or produce quality and resistance to biotic or abiotic stresses should be of particular interest to plant breeders, since such linkages may need to be broken during the breeding process to develop high yielding, high quality cultivars containing stress resistance genes from less agronomic genotypes.

There are many examples of the discovery of coincident QTLs for correlated traits from the published literature. Such examples give the breeder clues about the nature of the genetic locus that has been detected and its effect on a range of traits. Coincident QTLs can be due either to pleiotropic interactions or to close linkage of a number of genes. Usually the resolution of QTL mapping experiments is too poor to distinguish these two possibilities and further research is required, usually fine-mapping of the QTL locus.

Of course, a common aim of experiments to map QTLs for yield and yield-related traits is to identify marker-trait associations that can be exploited in a breeding program through marker-assisted selection to produce rapid gains in these genetically complex traits. Through newer approaches such as AB-QTL and LD mapping, wider sources of genetic variation than the elite or domesticated germplasm base are examined to identify QTLs and associated markers. First published reports of marker-assisted selection based on these approaches have produced promising results with impressive gains in agronomic trait values.

References

Aranzana MJ, Zim S, Zhao K, Bakker E, Horton M, Jakob K, Lister C, Molitor J, Shindo C, Tang C, Toomajian C, Traw B, Zheng H, Bergelson J, Dean C, Marjoram P, Nordborg M (2005) Genome-wide association mapping in Arabidopsis identifies previously known flowering time and pathogen resistance genes. PLoS Genet 1: e60

Blair MW, Iriarte G, Beebe S (2006) QTL analysis of yield traits in an advanced backcross population derived from a cultivated Andean × wild common bean (*Phaseolus vulgaris* L.) cross. Theor Appl Genet 112: 1149-1163

Bernacchi D, Beck-Bunn T, Eshed Y, Lopez J, Petiard V, Uhlig J, Zamir D, Tanksley SD (1998a) Advanced backcross QTL analysis of tomato. I. Identification of QTLs for traits of agronomic importance from *Lycopersicon hirsutum*. Theor Appl Genet 97: 381-397

Bernacchi D, Beck-Bunn T, Emmatty D, Eshed J, Inai S, Lopez J, Petiard V, Sayama H, Uhlig J, Zamir D, Tanksley SD (1998b) Advanced backcross QTL analysis of tomato. II. Evaluation of near-isogenic lines carrying single donor introgression desirable wild QTL-alleles derived from *Lycopersicon hirsutum* and *L. pimpinellifolium*. Theor Appl Genet 97: 170-180

Bernacchi D, Beck-Bunn T, Emmatty D, Eshed J, Inai S, Lopez J, Petiard V, Sayama H, Uhlig J, Zamir D, Tanksley SD (1998c) ERRATUM Advanced backcross QTL analysis of tomato. II. Evaluation of near-isogenic lines carrying single donor introgression desirable wild QTL-alleles derived from *Lycopersicon hirsutum* and *L. pimpinellifolium*. Theor Appl Genet 97: 1191-1196

Börner A, Schumann E, Furste A, Coster H, Leithold B, Röder MS, Weber WE (2002) Mapping of quantitative trait loci determining agronomic important characters in hexaploid wheat (*Triticum aestivum* L.). Theor Appl Genet 105: 921-936

Breseghello F, Sorells ME (2006) Association mapping of kernel size and milling quality in wheat (*Triticum aestivum* L.) cultivars. Genetics 172: 1165-1177

Caldwell KS, Russell J, Langridge P, Powell W (2006) Extreme population-dependent linkage disequilibrium detected in an inbreeding plant species, *Hordeum vulgare*. Genetics 172: 557-567

Cardon LR, Bell JI (2001) Association study designs for complex diseases. Nature Rev Genet 2: 91-99

Concibido VC, La Vallee B, Mclaird P, Pineda N, Meyer J, Hummel L, Yang J, Wu K, Delannay X (2003) Introgression of a quantitative trait locus for yield from Glycine soja into commercial soybean cultivars. Theor Appl Genet 106: 575-582

Dholakia BB, Amiraju JSS, Singh H, Lagu MD, Roder MS (2003) Molecular marker analysis of kernel size and shape in bread wheat. Plant Breed 122: 392-395

Ertl DS, Fehr WR (1985) Agronomic performance of soybean genotypes from *Glycine max* × *Glycine soja* crosses. Crop Sci 25: 589-592

Flint-Garcia SA, Thornsberry JM, Buckler ES (2003) Structure of linkage disequilibrium in plants. Annu Rev Plant Biol 54: 357-374

Fulton TM, Beck-Bunn T, Emmatty D, Eshed Y, Lopez J, Petiard V, Uhlig J, Zamir D, Tanksley SD (1997) QTL analysis of an advanced backcross of *Lycopersicon peruvianum* to the cultivated tomato and comparisons with QTLs found in other wild species. Theor Appl Genet 95: 881-894

Fulton TM, Grandillo S, Beck-Bunn T, Fridman E, Frampton A, Lopez J, Petiard V, Uhlig J, Zamir D, Tanksley SD (2000) Advanced backcross QTL analysis of a *Lycopersicon esculentum* × *L. parviflorum* cross. Theor Appl Genet 100: 1025-1042

Gaut BS, Long AD (2003) The lowdown on linkage disequilibrium. Plant Cell 15: 1502-1506

Gizlice Z, Carter TE, Burton JW (1994) Genetic base for North American public soybean cultivars released between 1947 and 1988. Crop Sci 34: 1143-1151

Groos C, Robert N, Bervas E, Charmet G (2003) Genetic analysis of grain protein-content, grain yield and thousand-kernel weight in bread wheat. Theor Appl Genet 106: 1032-1040

Hansen M, Kraft T, Ganestam S, Säll T, Nilsson N-O (2001) Linkage disequilibrium mapping of the bolting gene in sea beet using AFLP markers. Genet Res Camb 77: 61-66

Hoisington D, Khairallah M, Reeves T, Ribaut J-M, Skovmand B, Taba S, Warburton M (1999) Plant genetic resources: what can they contribute toward increased crop productivity? Proc Natl Acad Sci USA 96: 5937-5943

Kabelka EA, Diers BW, Fehr WR, LeRoy AR, Baianu IC, You T, Neece DJ, Nelson RL (2004) Putative alleles for increased yield from soybean plant introductions. Crop Sci 44: 784-791

Khush GS (2001) Green revolution: the way forward. Nat Rev Genet 2: 815-822

von Korff M, Wang H, Léon J, Pillen K (2006) AB-QTL analysis in spring barley: II. Detection of favourable exotic alleles for agronomic traits introgressed from wild barley (*H. vulgare* ssp. *spontaneum*). Theor Appl Genet 112: 1221-1231

Kraakman ATW, Niks RE, van den Berg PMMM, Stam P, van Eeuwijk FEV (2004) Linkage disequilibrium mapping of yield and yield stability in modern spring barley cultivars. Genetics 168: 435-446

Kraakman ATW, Martínez F, Mussiraliev B, van Eeuwijk FA, Niks RE (2006) Linkage disequilibrium mapping of morphological, resistance and other agronomically relevant traits in modern spring barley cultivars. Mol Breed 17: 41-58

Malosetti M, Voltas J, Romagosa I, Ullrich SE, van Eeuwijk FA (2004) Mixed models including environmental variables for studying QTL by environment interaction. Euphytica 137: 139-145

Mansur LM, Orf JH, Chase K, Jarvik T, Cregan PB, Lark KG (1996) Genetic mapping of agronomic traits using recombinant inbred lines of soybean. Crop Sci 36: 1327-1336

Mickelson SM, Stuber CS, Senior L, Kaeppler SM (2002) Quantitative trait loci controlling leaf and tassel traits in a B73 × Mo17 population of maize. Crop Sci 42: 1902-1909

Narasimhamoorthy B, Gill BS, Fritz AK, Nelson JC, Brown-Guedira GL (2006) Advanced backcross QTL analysis of a hard winter wheat × synthetic wheat population. Theor Appl Genet 112: 787-796

Nichols DM, Glover KD, Carlson SR, Specht JE, Diers BW (2006) Fine mapping of a seed protein QTL on soybean linkage group I and it correlated effects on agronomic traits. Crop Sci 46: 834-839

Olsen KM, Halldorsdottir SS, Stinchcombe JR, Weinig C, Schmitt J, Purugganan MD (2004) Linkage disequilibrium mapping of Arabidopsis *CRY2* flowering time alleles. Genetics 167: 1361-1369

Orf JH, Chase K, Jarvik T, Mansur LM, Cregan PB, Adler FR, Lark KG (1999) Genetics of soybean agronomic traits: I. Comparison of three related recombinant inbred populations. Crop Sci 39: 1642-1651

Pritchard JK, Stephens M, Donnelly P (2000) Inference of population structure using multilocus genotype data. Genetics 155: 945-959

Pritchard JK, Wen W (2004) Documentation for Structure Software. Univ Chicago Press, Chicago, Illinois, USA

Quarrie SA, Steed A, Calestani C, Semikhodskii A, Lebreton C, Chinoy C, Steele N, Pljevlakusiæ D, Waterman E, Weyen J, Schondelmaier J, Habash DZ, Farmer P, Saker L, Clarkson DT, Abugalieva A, Yessimbekova M, Turuspekov Y, Abugalieva S, Tuberosa R, Sanguineti M-C, Hollington PA, Aragués R, Royo A, Dodig D (2005) A high density genetic map of hexaploid wheat (*Triticum aestivum* L.) from the cross Chinese Spring × SQ1 and its use to compare QTLs for grain yield across a range of environments. Theor Appl Genet 110: 865-880

Quijada PA, Udall JA, Lambert B, Osborn TC (2006) Quantitative trait analysis of seed yield and other complex traits in hybrid spring rapeseed (*Brassica napus* L.): 1. Identification of genomic regions from winter germplasm. Theor Appl Genet 113: 549-561

Richards RA (2000) Selectable traits to increase crop photosynthesis and yield of grain crops. J Exp Bot 51: 447-458

Reyna N, Sneller CH (2001) Evaluation of marker-assisted introgression of yield QTL alleles into adapted soybean. Crop Sci 41: 1317-1321

Schön CC, Utz HF, Groh S, Truberg B, Openshaw S, Melchinger AE (2004) Quantitative trait locus mapping based on resampling in a vast maize testcross experiment and its relevance to quantitative genetics for complex traits. Genetics 167: 485-498

Simko I, Haynes KG, Jones RW (2006) Assessment of linkage disequilibrium in potato genome with single nucleotide polymorphism markers. Genetics 173: 2237-2245

Slafer GA (2003) Genetic basis of yield as viewed from a crop physiologist's perspective. Ann Appl Biol 142: 117-128

Specht JE, Hume DJ, Kumudini SV (1999) Soybean yield potential – a genetic and physiological perspective. Crop Sci 39: 1560-1570

Specht JE, Chase K, Macrander M, Graef GL, Chung J, Markwell JP, Germann M, Orf JH, Lark KG (2001) Soybean response to water: a QTL analysis of drought tolerance. Crop Sci 41: 493-509

Stich B, Maurer HP, Melchinger AE, Frisch M, Heckenberger M, van der Voort JR, Peleman J, Sorensen AP, Reif JC (2006) Comparison of linkage disequilibrium in elite European maize inbred lines using AFLP and SSR markers. Mol Breed 17: 217-226

Tanksley SD, Grandillo S, Fulton TM, Zamir D, Eshed Y, Petiard V, Lopez J, Beck-Bunn T (1996) Advanced backcross QTL analysis in a cross between an elite processing line of tomato and its wild relative *L. pimpinellifolium*. Theor Appl Genet 92: 213-224

Tanksley SD, Nelson JC (1996) Advanced backcross QTL analysis: a method for the simultaneous discovery and transfer of valuable QTLs from unadapted germplasm into elite breeding lines. Theor Appl Genet 92: 191-203

Tanksley SD, McCouch SR (1997) Seed banks and molecular maps: Unlocking genetic potential from the wild. Science 277: 1063-1066

ter Steege MW, den Ouden FM, Lambers H, Stam P, Peeters AJM (2005) Genetic and physiological architecture of early vigor in *Aegilops tauschii*, the D-genome donor of hexaploid wheat. A quantitative trait loci analysis. Plant Physiol 139: 1078-1094

Thornsberry JM, Goodman MM, Doebley J, Kresovich S, Nielsen D (2001) *Dwarf8* polymorphisms associate with variation in flowering time. Nat Genet 28: 286-289

Timmerman-Vaughan GM, Mills A, Whitfield C, Frew T, Butler R, Murray S, Lakeman M, McCallum J, Russell A, Wilson D (2005) Linkage mapping of

QTL for seed yield, yield components and developmental traits in pea. Crop Sci 45: 1336-1344

Voltas J, Romagosa I, Ullrich SE, van Eeuwijk FA (2001) Identification of adaptive patterns in the 'Steptoe × Morex' barley mapping population integrating genetic, phenotypic and environmental information. Presented at the 7th Quantitative Trait Locus Mapping and Marker-assisted Selection Workshop, Valencia, Spain

Wang D, Graef GL, Procopiuk AM, Diers BW (2004) Identification of putative QTL that underlie yield in interspecific soybean backcross populations. Theor Appl Genet 108: 458-467

Xiao J, Li J, Grandillo S, Ahn SA, Yuan L, Tanksley SD, McCouch SR (1998) Identification of trait-improving quantitative trait loci alleles from a wild rice relative, *Oryza rufipogon*. Genetics 150: 899-909

Yoon D-B, Kang K-H, Kim H-J, Ju H-G, Kwon S-J, Suh J-P, Jeong O-Y, Ahn S-N (2006) Mapping quantitative trait loci for yield components and morphological traits in an advanced backcross population between *Oryza grandiglumis* and the *O. sativa japonica* cultivar Hwaseongbyeo. Theor Appl Genet 112: 1052-1062

Yuan J, Njiti VN, Meksem K, Iqbal JJ, Triwitayakorn K, Kassem MA, Davis GT, Schmidt ME, Lightfoot DA (2002) Quantitative trait loci in two soybean recombinant inbred line populations segregating for yield and disease resistance. Crop Sci 42: 271-277

Zhang W-K, Wang Y-J, Luo G-Z, Zhang J-S, He C-Y, Wu X-L, Gai J-Y, Chen S-Y (2004) QTL mapping of ten agronomic traits on the soybean (*Glycine max* L. Merr.) genetic map and their association with EST markers. Theor Appl Genet 108: 1131-1139

4 Molecular Mapping and Breeding for Quality

Hari P. Singh[1], N.V. Hemanth Kumar[2], Madhugiri Nageswara Rao[3], Jaya R. Soneji[3], Bharat Singh[4], H.C. Lohithaswa[5] and Chittaranjan Kole[6*]

[1]Center for Biotechnology, College of Agriculture, Fort Valley State University, 1005 State University Drive, Fort Valley, Georgia 31030, USA

[2]Center for Viticulture and Small Fruit Research, Florida A and M University, 6505 Mahan Drive, Tallahassee, Florida 32317, USA

[3]University of Florida, IFAS, Citrus Research and Education Center, 700 Experiment Station Road, Lake Alfred, Florida 33850, USA

[4]Agricultural Research Station, Fort Valley State University, 1005 State University Drive, Fort Valley, Georgia 31030, USA

[5]AICRIPO on Forage Crops, University of Agricultural Sciences, Bangalore, ZARS, V.C. Farm, Mandya 571405, Karnataka, India

[6]Clemson University, Department of Genetics and Biochemistry, III Jordan Hall, Clemson, SC 29634, USA

*Corresponding author: ckole@clemson.edu

1 INTRODUCTION

Quality is one of the principal targets in plant breeding which will always remain a major challenge to the plant breeders. Quality, the degree of excellence or superiority, is a product trait that is related to human appreciation. It is a combination of attributes, properties or characteristics that add value to each crop in terms of its intended use. Crop quality attributes, most often than not, are the resultant of actions and interactions of the genotype of a crop, the environment in which it is grown and the interaction between genotype and environment. Crop quality may be classified by attributes like nutritional, anti-nutritional, nutraceutical, consumer preference, pre- and post-harvest characteristics, mechanized harvesting related, transportation and storage related, and industrial characteristics. The importance given to a specific quality attribute may vary with the commodity concerned and with the individual (producer or consumer) or market concerned with the quality assessment.

Nutritional value is one of the most important traits for crop quality improvement. It deals with the enhancement of the protein content, micronutrients, vitamins, minerals, dietary fiber, composition of amino acids, sugars, carbohydrates, starch, lipids, oils, etc., which are essential for a healthy diet. A recent World Food Program (WFP) report states that 842 million people worldwide suffer from chronic malnutrition (WFP 2006). The increase in nutritional value of food crops can help in improving human health. Reduction or elimination of anti-nutritional chemical compounds such as glucosides, neurotoxins, tannins, lectins, alkaloids, etc., will also help in the quality improvement of certain crops. Nutraceuticals (natural bioactive compounds such as flavonoids, carotenoids, isoflavones, antioxidants, phytosterols, etc.) are health promoting, disease preventing or medicinal properties. They can be enhanced to improve the quality of the crop which in turn will lead to the improvement of human health. The integration of nutraceutical crops into agriculture and manufacturing will impact the economic development and can replace or complement traditional crops.

Quality can also be enhanced for consumer preference by manipulating appearance, texture, and flavor. Appearance quality factors may include size, shape, color, gloss, freedom from decays, and defects. Firmness, juiciness, crispiness, toughness, and mealiness form the textural quality factors and are important in horticultural crops for their cooking and eating quality. Flavor quality involves perception of the tastes (such as sweetness, sourness, acidity, astringency, and bitterness) and aromas. Pre- and post-harvest characteristics involve attaining and maintaining good flavor, nutritional quality, resistance to physiological disorders, prevention of defects and decays, and improved storage or shelf-life. Maturity at harvest is the most important determinant of storage-life and final quality especially in fruits. Prevention of mechanical injuries during harvesting can increase the quality. Several traits are desirable for mechanized agriculture (e.g., jointless fruits in tomato, uniform ripening in pineapple, indehiscence in cereals, pulses and oilseeds, etc.), industrial and processing qualities (e.g., bread-making and malting quality in cereals, firmness and texture in fruits, etc.), and transportation and storage (e.g., fiber traits, sucrose content, storage quality, etc.). Besides, fiber quality in crops including cotton, jute, sunhemp, flax; wood quality of forest trees; and polysaccharide contents and composition of biofuel crops deserve attention of plant breeders. Table 1 lists some quality-related traits in a few representative crops.

Table 1 Important quality traits targeted during crop improvement.

S.No.	Crop	Targeted quality traits
Cereals		
1	Rice	Proteins, carbohydrates, vitamins
2	Wheat	Grain color, grain hardiness, flour proteins, starch, flour pigments, dough strength, dough extensibility
3	Maize	Protein concentration, amino acid, kernel size/weight/uniformity, kernel hardiness (density), kernel breakage susceptibility (brittleness), stress cracking, extractable starch, oil concentration, essential nutrient level, toxins
4	Oat	Grading, test weight, groat percentage (proteins, lipids, β-glucan, starch, antioxidants), milling yield
5	Barely	Malting quality
Pulses		
6	Legumes	Seed size, color, shape and seed coat integrity, starch, protein (methionine), vitamins, minerals Antinutritional factors: lectins, tannins, trypsin, alkaloids, RFO (raffinose oligosaccharides), phytic acids, vicine, corvicine, DOPA (dihydroxyphenylalanine), antigenecity/allergenic factors
7	Soybean	Edible oil, protein, isoflavones
8	Mungbean	Seed size, shape, testa color, luster, weathering resistance
9	Common bean	Protein, minerals, vitamin content
Oilseed crops		
10	Turnip	Oil content, quality
11	Linseed	Fat and fatty acid composition
12	Oilseed rape	Alkenyl and indolyl glucosinolates content, fiber, fatty acid content
13	Castor	Hydroxy fatty acid content
14	Sunflower	High oil percentage, protein, and husk content
15	Indian mustard	Erucic acid content
16	Sesame	Oil, oleic and linoleic acid content
17	Peanut	Oil, sugar and starch characters, oleic and linoleic acid content

Contd.

Contd.

Fiber crops		
18	Cotton	Fiber strength, fiber length, fiber fineness
19	Hemp	Fiber quality (bast, woody core), cannabinoid content
20	Jute	Fiber strength, firmness, lignin content
Sugar crops		
21	Sugarbeet	Sucrose
22	Sugarcane	Fructose, sucrose, starch, and cellulose content
Tuber crops		
23	Potato	Biological trait (proteins, carbohydrates and minerals); sensorial traits (flavor, texture), industrial traits (tuber shape, cold sweetening, starch quality)
24	Yam	Tuber quality, starch content
25	Sweet potato	Tuber quality, beta carotene content
26	Cassava	Carbohydrate and starch content, high dry matter content, easy peeling
Narcotic Crops		
27	Opium Poppy	Opiumless, alkaloid free, latex content, morphine, and thebaine content
28	Tobacco	High quality cured leaves (tar potential, leaf body, grain color, and chemical balance), and nicotine content
Fruit Crops		
29	Mango	Fruit color, flavor, size, fruit weight,
30	Citrus	Fruit size, color, flavor, acidity, seedlessness, easy peeling
31	Apple	Fruit size, color, appearance, storage characteristics, flavor
32	Apricot	Fruit size, sugar content, sweet kernel
33	Avocado	Fruit weight, ripening time, seed size
34	Peach	Fruit firmness, texture, color, season of maturity

With the rise in demand for crops having better nutritional and industrial traits, quality improvement is gaining unprecedented importance in both developed and developing countries. It is difficult to improve the quality through conventional breeding, as the traits are usually complex and are controlled by the actions and interactions of several genes which are docile to environmental fluctuations. Products with improved quality will affect various agricultural product-based sectors, improve human health, increase farm income, and also provide good quality raw materials for industries.

An important development during the last decade in genetics was the introduction of DNA-based molecular markers which provided researchers with the ability to identify putative genomic (chromosomal) regions called quantitative trait loci (QTL) that are responsible for variation in a trait (Paterson et al. 1988). QTLs represent the polygenic clusters underlying the quantitative traits and have been identified for various quality related traits. Marker-assisted selection (MAS) makes it possible to 'tag' alleles conferring desirable quality traits (Mohan et al. 1997). Development of markers tightly linked to quality traits or resistance to quality-affecting stresses should enable breeders to select on a genotypic basis. Despite their great potential, DNA-based markers have not yet been widely implemented in breeding for quality, particularly in developing countries, due to economic constraints and cost effectiveness. Thus, the improvement of quality characters in crop plants has great potential to alleviate problems caused by poverty and malnutrition through both direct (food quality and quantity) and indirect effects (income stability, etc). The improvement of the nutritional quality of crops is a problem-orientated challenge which will be solved by exploiting the most recent developments in genomics and biotechnology. This chapter deals with the various aspects of quality improvement through molecular mapping and breeding.

2 BREEDING FOR QUALITY IMPROVEMENT

2.1 Traditional Breeding

2.1.1 Selection

Breeding has been practiced since the early human civilization and selection was the first method of breeding, adding the criteria of suitability for man's use (e.g., larger seed, better taste, easier harvestability, etc.) to those of natural adaptation, fitness, and offspring. It can be defined as systems for selection of superior genotypes from genetically variable populations derived from sexual recombination. The system is powerful because it is evolutionary; progress can be cumulative, with improved individuals continually serving as parents for subsequent cycles of breeding. For quality improvement, this may be accomplished by selecting naturally occurring variants or by selecting superior parents to create improved variants exhibiting better quality traits as compared to the parents. In many cases, the selection process has been unconscious as was seen in wild wheat where the grains scattered by disarticulation (separation of the seed from the seed head) (Poehlman

1986). When these grains were harvested by cutting the heads with a sickle, an unconscious selection occurred for 'non-shattering' types that were then continually replanted. For some crops maintained by vegetative propagation, a clear conscious selection must have occurred, as in the case of seedless banana, by planting offshoots of the plant (Janick and Moore 1996). Selection from natural variation has also been carried out in tomato for carotenoid content (Liu et al. 2003), in maize for protein content (Moose et al. 2004) and in potato and rice for starch content (Fernie and Willmitzer 2004).

2.1.2 Hybridization

Interspecific hybridization has been attempted to produce yellow seeded mutants having thinner seed coats, higher oil contents, higher protein and lower fiber contents in the meal than black seeds with the same genetic background in *Brassica napus* germplasm (Liu et al. 2005). Intervarietal hybridization has been carried out between soft and hard wheat as a strategy for improvement of quality traits (Campbell et al. 2001). Selection is either carried out by bulking desirable types (mass selection) or by recurrent selection (cyclical selection). In maize breeding programs, recurrent selection was used to assess changes in the genetic structure of grain yield, grain moisture and broken stalks (Doerksen et al. 2003). Interspecific hybridization of *B. juncea* with *B. carinata* has been carried out to develop *B. carinata* with low 2-propenyl glucosinolate (Getinet et al. 1997). Recurrent selection was also evaluated for four seed quality traits, viz., protein content, dye binding capacity, tryptophan content, and specific weight using four selection cycles in a modified opaque-2 population of maize. The material from each cycle and from test crosses of each cycle with a homozygous opaque-2 tester were grown for two consecutive years and showed a nearly linear rate of progress without altering any other agronomic traits (Pani et al. 1985). In durum wheat, selection is aimed at improving major quality characters such as kernel size, kernel vitreousness, protein content, gluten strength, and pigment concentration (Abdalla et al. 1992). Considerable advances have been achieved in selection for quality-related traits such as texture, color, shape and storage-ability of edible parts of Chinese cabbage and quality improvement of hot pepper (Yoon et al. 1999). The eating and cooking quality of 'Zhenshan 97' was improved by introgressing the *Waxy* gene region from Minghui 63 (wx-MH), a restorer line that has medium amylase content, soft gel consistency and high gelatinization temperature. The wx-MH fragment was transferred to 'Zhenshan 97B' by three backcrosses and one selfing, then from 'Zhenshan 97B' to 'Zhenshan 97A' by a cross and a backcross (Zhou et al. 2003). A population of 300 $F_{3:4}$ families were generated from the cross between two maize inbred lines *F2* and *F252*. Each $F_{3:4}$ family was crossed with an inbred line tester, *MBS847* and the testcross progenies were selected on an index combining grain yield and grain moisture at harvest (Moreau et al. 2004). A population consisting of 150 F_4-derived single-seed descent (SSD) lines from a cross between the Australian wheat cultivars Schomburgk and Yarralinka showed segregation for protein, milling yield and flour color (Parker et al. 1999).

2.1.3 Mutation Breeding

Another technique available to breeders in their quest to develop new cultivars for the improvement of crop quality is mutation breeding (Muggleston 1995). Mutations can result in valuable new traits. Spontaneous mutations occur in nature at a relatively frequent rate. A large number of spontaneous mutations have been recorded in fruit trees and have been used either directly as new cultivars or in their breeding programs. In apples, increases in the amount of anthocyanin in the outer cell layers of the fruit skin are quite common (Janick et al. 1996). The apples 'Royal Gala' and 'Imperial Gala' are natural mutations of 'Gala'. Spontaneous, nearly seedless mutants ('Washington' navel orange and 'Marsh' grapefruit) have occurred in citrus (Soost and Roose 1996). Spontaneous mutations in pears (Bell et al. 1996) have resulted in completely red fruits (cv. Starkrimson from cv. Clapp's Favorite) and completely russeted fruits (cv. Russet Barlett from cv. Barlett).

Recent decades have witnessed intensive work on the induction of mutations by using physical and chemical mutagens. The frequency of induced mutations almost doubles those occurring naturally and they have been looked on as a powerful tool for the development of new cultivars. Chemical mutagens (e.g., ethylmethane sulfonate (EMS), sodium azide, etc.) have been used to generate non-lethal mutants expressing decreased levels of phytic content in seeds of maize (Raboy et al. 2000; Pilu et al. 2003) and barley (Larson et al. 1998). Several high-oleic mutants with oleic acid (C18:1) contents of 77 to 80% in the seed oil were produced by EMS treatment of a population of oilseed rape (*B. napus* cv. Wotan; Rucker and Robbelen 1995). Eight high-oleic mutant lines from these were selected to study the inheritance of high oleic acid mutations (Schierholt et al. 2001). EMS was applied to seeds of the Ethiopian mustard (*B. carinata* A. Braun) line C-101 and bulk samples of M_3 seeds were evaluated for the fatty acid composition in their oil. A putative mutant was obtained which showed very low oleic acid content while in M_5 generation, a high erucic acid line was obtained (Velasco et al. 1998). Doubled haploid (DH) lines of *B. carinata* with modified erucic acid content have been obtained by EMS treatment of microspores (Barro et al. 2001). An opiumless and alkaloid-free non-narcotic var. Sujata of opium poppy has been developed by induced mutagenesis in an otherwise narcotic (with opium and opium alkaloids both) strain Mass-2B of opium poppy (Sharma et al. 1999). The var. Sujata is also attributed with high yield of seeds which are rich in both protein and oil content.

In apricots, an early-ripening cultivar ('Early Blenheim') was selected following thermal neutron irradiation (Layne et al. 1996). When gamma irradiation was used in rice, two mutants, one lethal and one non-lethal, were identified for decreased levels of phytic content (Larson et al. 2000). Doubled haploid lines of *B. carinata* with modified glucosinolate and erucic acid content have been obtained by UV treatment of microspores (Barro et al. 2002). Gamma irradiation of dormant budwood has been used in New Zealand in an attempt to induce useful mutations in 'Hayward' kiwifruit. A line lacking all the seed lipoxygenase isozymes in soybean was obtained by γ-ray irradiation (Hajika et al. 1991). In bambara groundnut, the M_3 generation, obtained after gamma irradiation, offered the best opportunity for selection for the number of pods per plant and seed size (Adu-Dapaah and Sangwan 2004).

2.1.4 Polyploidy Breeding

Many crop plants are polyploids and some genomes have been duplicated more recently than others (Udall and Wendel 2006). Polyploid plants may offer considerable potential for crop improvement through the exploitation of their horticulturally important and useful characteristics. Polyploidy provides genome buffering, increased allelic diversity and heterozygosity, and permits novel phenotypic variation to be generated. Genome buffering for the *waxy* locus that encodes granule bound starch synthase I (GBSS) was first characterized in a population of potato obtained by crossing two genotypes that were duplex for the null *GBSS* allele (Flipse et al. 1996). The amylose content in tubers was not linearly correlated with dosage suggesting that there was 'buffering' against the mutant alleles. In cotton, tetraploid *Gossypium barbadense* and *G. hirsutum* (AD genome species) have been domesticated for their epidermal seed trichomes (cotton fiber) (Applequist et al. 2001) as they produce longer, finer and stronger fibers as compared to their diploid relatives. The Hardness (*Ha*) locus controls grain hardiness in hexaploid wheat (*Triticum aestivum*) and its relatives (*Triticum* and *Aegilops* species) and is a classical example of a trait whose variation arose from gene loss after polyploidization (Chantret et al. 2005).

Tetraploidy has been induced with colchicines in the 'Fuerte' and 'Mexicola' cultivars of avocado (Bergh and Lahav 1996). In pineapple, tetraploidy increases the fruitlet width (Leal and d'Eeckenbrugge 1996). Spontaneous tetraploids occur in pears and can be recognized by their larger than normal fruits (Bell et al. 1996). Polyploid plants can also be used in breeding programs aimed at producing lines for seedlessness. A tetraploid plant can be bred with a diploid to obtain a triploid progeny. These triploids are sterile as they cannot form gametes with a balanced assortment of chromosomes. This has been used effectively in citrus breeding programs. 'Shasta Gold®'or 'TDE2' (late maturing), 'Tahoe Gold®'or 'TDE3' (mid-season maturing) and 'Yosemite Gold®'or 'TDE4' (mid-late season maturing) are triploid mandarin hybrids of tetraploid female parent ('Temple' tangor x 4*n* 'Dancy' mandarin) x diploid male parent 'Encore' mandarin that combines large fruit size, attractive deep orange rind color, rich fruit flavor, and the virtual absence of seeds (Roose and Williams 2006). Seedless watermelons and bananas are other examples of triploidy.

Polyploid breeding has also been used to produce a new species by crossing a cabbage with a radish. Most of the hybrids produced by this cross were mostly infertile. Only a few fertile hybrids were produced, probably due to spontaneous doubling of the chromosome number in the somatic cells that went on to form gametes. Fruit size and flower bud setting of cranberry (Zeldin and McCown 2003), black chokeberry and blueberried honeysuckle (Katsuji et al. 2003) was improved through polyploidy. Polyploid plants not only have larger cells but the plants themselves are often larger. This has led to the deliberate creation of polyploid varieties of plants such as marigolds and snapdragons.

2.2 Limitations of Traditional Breeding

The quality traits are usually controlled by the action of several genes, and are also influenced by the environment. The complex genetics of quality traits makes them difficult to improve through traditional plant breeding tools. These limitations are mainly due to the sexual system. It is usually not possible to incorporate genes from non-related species or to incorporate small changes without disturbing the particular combination of genes that make a particular type unique. With traditional breeding, it is also not possible to improve a unique genotype by adding a single character. Mutations may be induced, but they are often deleterious or connected with undesirable effects. Limitations of traditional breeding are particularly apparent when a needed character (such as quality improvement) is unavailable in populations that can be incorporated by sexual crosses. Often, when the trait is available it is highly polygenic and has a low heritability making progress extremely slow. Biological limitations associated with degrees of sexual sterility and/or incompatibility prevents the utilization of certain breeding parents and exploitation of specific parental combinations. Traditional breeding also has technical or economic limitations to detect infrequent or rare recombinants, the lack of sufficient time to generate cycles of recombination, space to grow necessary populations to recover superior recombinants, or resources to be able to select, identify, evaluate, and fix desired recombinants. Polyploids have high heterozygosity, severe inbreeding depression and require exceptionally large populations of seedlings to be screened in order to recover individuals for evaluation as potential clones making initial selection for desirable quality characters inefficient and time consuming. Retaining the essential characteristics and achieving significant improvements in quality traits demand the application of genetic improvement techniques other than traditional breeding through sexual hybridization and selection.

The last decade has seen a lot of important developments in the field of quantitative genetics. This has facilitated identification of the genome regions responsible for variation of a trait using molecular markers (Paterson et al. 1988) and tag alleles conferring desirable quality traits using QTL and MAS. For many crop plants large numbers of molecular markers of different types (RAPD, RFLP, AFLP, SSR, SNP, etc.) are now available (see Chapter 2, Volume 1). These vary widely in their cost of development and deployment. The choice of the markers to be used depends largely on the breeder's objectives and the facilities available. Markers based on the use of PCR will predominantly be used as they have many advantages over other methods (amount of DNA required, speed, cost, ease of use, etc.) and co-dominant markers, such as SSRs, offer considerable advantages in determining allelic composition in heterozygous material. Development of markers tightly linked to quality traits or resistance to stresses that affect the quality should enable the breeders to select on a genotypic basis.

3 EVALUATION OF CROP QUALITY PARAMETERS

Methods used for evaluating quality of food crops generally make a measurement of a chemical or physical parameter of the food crop associated or correlated with quality and does not measure the quality directly. In the case where the method requires destruction and/or sacrifice of the sample, a random sample is used for predicting quality of the surrounding population. The nondestructive method makes an analysis of the food crop without altering its physical objective or chemical condition due to which that food crop can be stored or marketed even after the analysis (Wafada 1992). If the quality has to be monitored over a period of time then nondestructive methods are preferred. The type of method used depends on the food crop. Some of the methods of evaluating quality attributes are based on human sensory evaluation. The quality of a number of agricultural products such as fruits and vegetables, cereals, legumes, and their processed products is judged by the colors they display or fail to display as well as by their flavor and texture. Sensory evaluation of breadfruit includes rating cooked fruit for palatability, maturity, flavor, sweetness, texture, and discoloration at six stages of development to ascertain the preferred stage of maturity for consumption (Worrell et al. 1998) as well as to assess it for crispness, flavor, color, and overall preference (Omobjuwajo 2003). Such an evaluation, however, is not adequate to be a general standard on a large scale, and for a wide range of operations. Factors such as eye fatigue, lack of color memory, variations in color discrimination between individuals, personal bias, and variations in lighting conditions greatly influence the decision of an individual in determining the quality of any given product. This has led to considerable research in developing instrumentation sensitive to a broad range of electromagnetic spectrum and establishing quality standards for several agricultural products (Deshpande et al. 1984).

Reflectance, the light reflected from below the skin of an illuminated food crop or few millimeters below the surface of an illuminated ground sample, is used to grade tomatoes, oranges, lemons, and apples in the packing house. Reflectance is also used to detect blight, soft rot and scab in potato tubers (Porteous et al. 1981). It is used in grains and oil seeds to predict starch, protein, oil, and moisture contents (Norris 1983). Spectrophotometer is used to detect internal defects like hollow heart in potato tubers, core breakdown in pears (Wang and Worthington 1979), water core in apples, freeze damage in citrus and smut content in wheat (Wafada 1992). Maturity of fruit crops such as apples, peaches, blueberries, etc. is also detected by spectrophotometer. Surface defects like scab on dried prunes and potatoes can also be detected by spectrophotometer (Gunasekaran et al. 1985). Spectroscopy is used for measuring the firmness, skin and flesh color, and dry matter content of pickling cucumbers (Kavdir et al. 2007). Spectroscopy has been used to determine the total carotenoids and sugars in different carrot varieties (Schulz et al. 1998).

Maturity of papaya is predicted by determining the chlorophyll, carotenoid and soluble solids content using interactance spectroscopy

(Birth et al. 1984). Interactance spectroscopy has also been used to determine the flesh color in clingstone peaches (Slaughter and Crisosto 2006). Fluorescence is used to measure chilling injury in tomatoes (Kamps et al. 1987). In pears, apples, watermelons, and tomatoes, acoustic vibration is used to measure the firmness of the fruits (Shmulevich et al. 1996). Fruit ripening in pear is also measured using a laser doppler vibrometer (Murayama et al. 2006). Delayed light emission (DLE) is used to sort tomatoes and persimmons according to their color or maturity (Chuma et al. 1982a, b). DLE is also used for early detection of chilling stress in cucumbers (Abbott and Massie 1985). Imaging technology is used in separating peaches based on their color development (Miller and Delwiche 1988) and good dried prunes from defective ones (Delwiche et al. 1988). It is also used to detect defects in apples caused by scab, hail damage, bird pecks and insect stings (Rehkugler and Throop 1987). Nuclear magnetic resonance (NMR) has been used to detect the development of water core in apples (Wang et al. 1988) and core breakdown in pears (Wang and Wang 1989). Magnetic resonance (MR) has been used to study the moisture and oil content of agricultural products (Choi et al. 2005). X-rays have been used to detect hollow hearts in potatoes (Finney and Norris 1978). Resonance and vibration transmissibility have been used to predict anthocyanin content in raspberries (Wafada 1992). Hand densimeter has been used to evaluate the firmness of olives (Garcia and Yousfi 2005). High-performance liquid chromatography (HPLC) has been used to evaluate the carotenoids present in durum wheat grain (Hentschel et al. 2002).

4 APPLICATIONS OF MOLECULAR BREEDING TOOLS FOR QUALITY IMPROVEMENT

Molecular breeding utilizes the molecular markers associated with specific genes and has many applications in the improvement of crop quality. When tightly linked to genes of interest, the molecular markers can be used to select indirectly for the desirable quality trait. This represents the simplest form of MAS. MAS can be used either to accelerate backcrossing of a trait or in pyramiding several desirable traits. Markers can also be used for scrutinizing polygenic traits into their Mendelian components or QTL. QTLs increase the understanding of the inheritance and gene action for the specific trait and allow the use of MAS as a complement to conventional breeding and selection procedures.

4.1 Quantitative Trait Loci (QTL)

Qulaity related QTLs have been detected in a number of crops (Table 2; Fig. 1). Their inheritance pattern has also been detected and they have been manipulated by MAS for quality improvement purposes.

Table 2 QTLs detected for quality traits in different crops.

Target Trait	*Mapping Population*	*Marker Loci #*	*Reference*
	Wheat		
Bread-making quality: Zeleny sedimentation values, protein, kernel hardness, and 1000-kernel weight	226 RIL*	187	Zanetti et al. 2001
Bread making quality-hardness, protein, dough strength	187DH*	380	Perretant et al. 2000
Pre-harvest sprouting	204 RIL	183	Zanetti et al. 2000
Grain weight	100 RIL	BSA*	Varshney et al. 2000
Grain protein	100 RIL 97 RIL	BSA 35	Prasad et. al. 1999 Snape et al. 2005
Milling yield	150 RIL	178	Parker et al. 1999
Kernel morphology and texture: kernel length, width, area, perimeter, test weight, 1000-kernel weight	78 RIL		Campbell et al. 1999
Flour color	150 RIL	147	Parker et al. 1998
Water absorption, flour color, dough strength, dough extensibility, dough development time and sponge and dough baking quality	180 DH	140	Schmidt et al. 2004
	Soybean		
Seed quality: protein content, oil content, and seed weight	82 F_2	120	Csanadi et al. 2001
Seed quality: protein, oil, and seed size	131 RIL	100	Hyten et al. 2004
	Barley		
Malting quality: malt extract percentage, α-amylase activity, diastatic power, and malt β-glucan content	92 DH	-	Han et al. 1997

Contd.

Contd.

Grain α-amylase content	120 DH	355	Li et al. 2003
	Rice		
Grain Quality: amylose content, alkali spreading score, gel consistency, percentage of grain with a white core, and the square of the white core	132 DH	243	He et al. 1999
Moisture content, filled rough/total rough rice ratio, grain density, de-husked grain, green grain, damaged-yellow rice, red grain, milling ratio, head rice grain, broken rice grain, crushed rice, chalkiness, amylose content, and gel consistency	285 BC_2F_2	131	Septiningsih et al. 2003
Milling quality	165 RIL	289	Dong et al. 2004
	Tomato		
Fruit quality: fruit color and fruit firmness	23NIL	BSA	Bernacchi et al. 1998
Fruit quality: lycopene content	123 BC_1S_1	151	Chen et al. 1999
	Pea		
Seed protein concentration	88 RIL	207	Taran et al. 2004
	Sunflower		
Oil quality: stearic acid	-	99	Perez-Vich et al. 2004
	Cotton		
Fiber quality	F_2	566	Lin et al. 2005

*RIL= Recombinant inbred line; DH = Doubled haploid; BSA = Bulked Segregant Analysis

A lot of work has been done to identify QTLs associated with fiber quality traits in cotton. Lin et al. (2005) identified 13 QTLs associated with fiber traits of which two QTLs were for fiber strength, four for fiber length and seven for micronaire value using SRAP, SSR and RAPD markers. RFLPs have been used to detect QTLs in nine linkage groups for fiber properties with three QTLs for fiber strength, two for fiber length, four for micronaire and four for fiber perimeter (Ulloa and Meredith 2000). Kohel et al. (2001) used RAPD and RFLP markers to develop a framework map and found four QTLs to influence bundle fiber strength, three to influence fiber length, and six to influence fiber fineness.

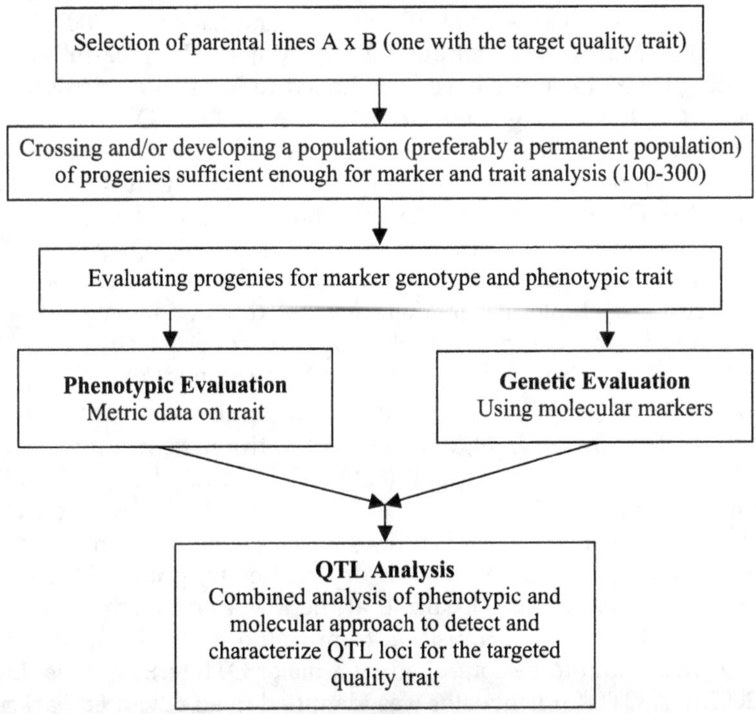

Fig. 1 A generalized flow chart depicting steps in QTL mapping for quality improvement.

Paterson et al. (2003) mapped a total of 76 QTLs controlling six fiber quality traits under well-watered vs. water-limited conditions. They reported that fiber length, length uniformity, elongation, strength, fineness, and color (yellowness) were influenced by 6, 7, 9, 21, 25, and 11 QTLs, respectively, and these QTLs could be detected in one or more treatments. Seven QTLs for six fiber-related traits were detected using AFLPs, SSRs and RFLPs, of which five were distributed on the chromosomes of the A-subgenome (Mei et al. 2003). Most of the QTLs influencing fiber quality were located on the D subgenome which was derived from an ancestor that does not produce spinnable fiber (Jiang et al. 1998). SSRs and RAPDs have been used to identify two QTLs for fiber strength (Zhang et al. 2003). Interval mapping of a map based on RFLP, SSR, and AFLP markers led to the detection of 15, 12, 21, and 16 QTLs, respectively controlling fiber length, strength, fineness, and color (Lacape et al. 2005).

In wheat, QTLs governing flour yellow pigment have been identified on chromosome 7 (Parker et al. 1998; Parker and Langridge 2000; Elouafi et al. 2001; Mares and Campbell 2001). Major QTLs for pre-harvest sprouting (PHS) tolerance have been linked to RFLP and microsatellites markers (Anderson et al. 1993; Roy et al. 1999). Three QTLs, located on chromosome arms 2AL, 2DL, and 4AL, have been found to be associated with the differences in germination index in wheat grains (Mares and Mrva 2001). QTLs associated with grain polyphenol oxidase (PPO) have been located on 2A and 2D (Mares and Campbell 2001). QTLs controlling late maturity α-amylase (LMA) in wheat have also been identified (Mrva and Mares 2001). Molecular markers associated with milling yield, water absorption, flour color, and dough properties for Australian hexaploid bread wheat have been identified using QTLs (Schmidt et al. 2004).

A total of 130 QTLs associated with a number of quality traits in tomato such as fruit weight, diameter, color, firmness, meltiness, aroma volatiles, etc. have been mapped (Saliba-Colombani et al. 1998; Saliba-Colombani et al. 2000; Causse et al. 2002). QTLs with strong effect for fruit weight, fruit elasticity, color, sourness, aroma intensity, candy aroma, mealiness, and contents of dry matter weight, soluble solids, sugars, and eugenol were localized at the lower end of chromosome 2 (Causse et al. 2002). The transition from round to pear-shaped fruit in tomato was found to be controlled by a major QTL termed *ovate* (Liu et al. 2002a). A QTL for fruit color was identified in an advanced backcross QTL study in *Lycopersicon hirsutum* (Bernacchi et al. 1998). QTLs involved in the evolution and domestication of tomato from small berries to large fruit have been mapped (Frary et al. 2000). Paterson et al. (1991) identified seven QTLs influencing soluble solids concentration, nine QTLs influencing fruit pH, 11 QTLs influencing mass per fruit and four QTLs influencing fruit diameter and pH, in a cross between a cultivated tomato (*L. esculentum* Mill.) variety and a closely-related wild species, *L. cheesmanii*.

A number of QTLs have been identified in rice. Three QTLs have been found to be associated with grain length elongation in cooked rice (chromosomes 2, 6, and 11), six with width expansion (chromosomes 1, 2, 3, 6, 9, and 11) and two with water absorption (chromosomes 2 and 6). A single QTL located near the waxy (*wx*) gene on chromosome 6 seemed to influence all the traits tested for the cooked rice quality (Ge et al. 2005). Four QTLs for grain length, five QTLs for grain width and three QTLs for the ratio of grain length to width have been identified and found to be closely linked on chromosomes 1, 2, 3, and 10 (Huang et al. 1997). QTLs have been detected for milling traits in rice (Fig. 2). Two QTLs for brown rice percentage have been identified and mapped to chromosomes 9 and

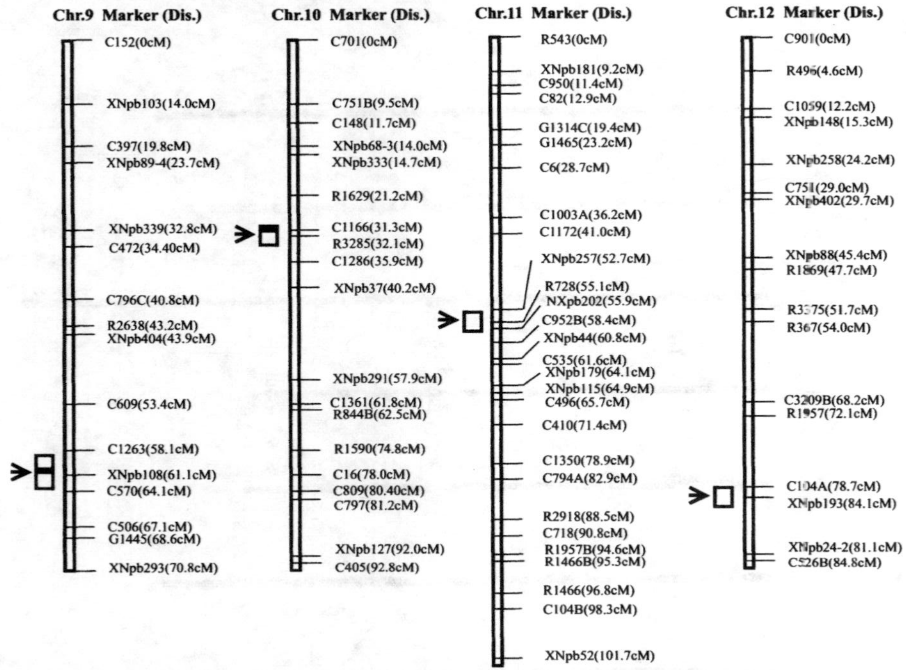

Fig. 2 Chromosomal locations of QTLs for milling quality in an RI population of rice from the cross between Asominori and IR24. Shaded (Chr.9 and 10), white (Chr.11 and 12) and black (Chr.1, 3, and 5) bars genomic regions of putative QTLs for brown rice percentage, milled rice percentage, and milled head rice percentage, respectively, based on composite interval mapping method. Black arrowheads indicate the location of peak LOD for putative QTLs detected (Dong et al. 2004).

10, two QTLs governing milled rice percentage have been detected on chromosomes 11 and 12, while three QTLs controlling milled head rice percentage have been mapped to chromosomes 1, 3, and 5 (Dong et al. 2004).

Molecular markers have been used to detect QTLs contributing to the phenotypic expression of grain and malt quality in barley (Igartua et al. 2000; Fig. 3). Four QTLs linked to the grain and malting quality have

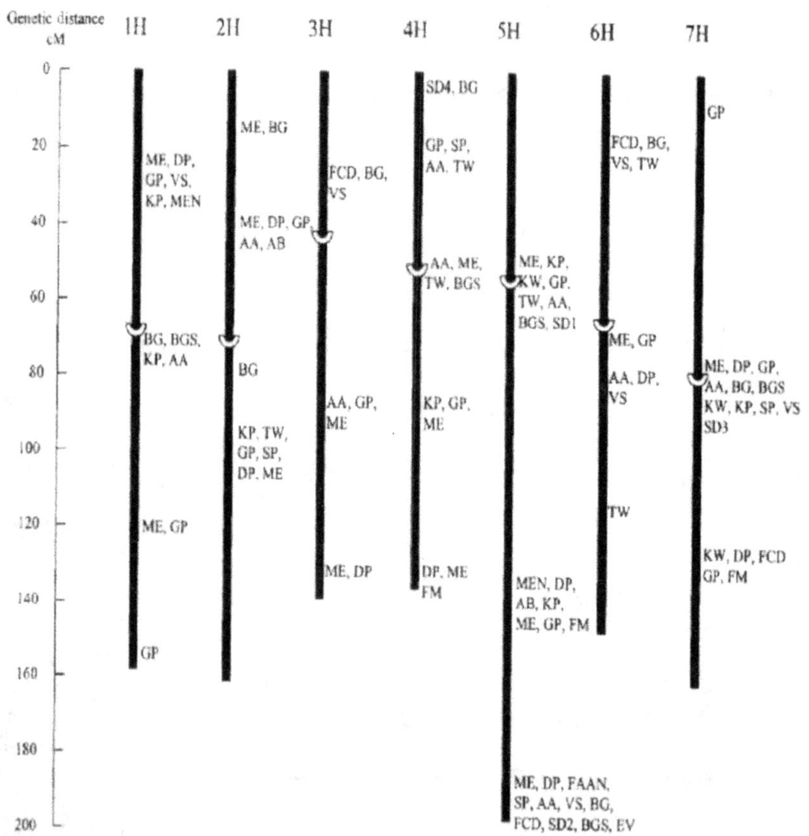

Fig. 3 Chromosomal locations of mapped QTLs controlling malting quality related to the barley consensus linkage map (Qi et al. 1996). AA, α-amylase; AB, starch granule A/B; BG, β-glucan; BGS, β-glucanase; DP, diastatic power; EV, early vigor; FAAN, free amino acid nitrogen; FCD, fine-coarse difference; FM, fermentability; GP, grain protein; KP, kernel plumpness; ME, malt extract; MEN, milling energy; SD, seed dormancy; SP, soluble protein; TW, test weight; VS, wort viscosity.

been detected (Qi et al. 1996). These QTLs have effects on extract β-glucan, fine-coarse difference, kernel weight, kernel plumpness, grain protein, soluble protein, diastatic power, α-amylase activity, and fine-grind extract in the Harrington/TR306 population (Igartua et al. 2000). Five QTLs related to seed dormancy have been identified in barley (Vanhala and Stam 2006). Two major QTLs have also been identified for malt extract percentage, α-amylase activity, diastatic power, and malt β-glucan content on chromosomes 1 and 4 (Han et al. 1997).

In *B. juncea*, Cheung et al. (1998) identified QTLs for various seed quality traits such as oil content and also concentrations of erucic acid (Fig. 4), linolenic acid. total alkenyl glucosinolates, and individual classes of alkenyl glucosinolates. Two QTLs underlying the variation of seed erucic acid content were assigned to two linkage groups of a *B. juncea* map using a doubled haploid (DH) mapping population derived from a high×low erucic acid F_1 hybrid (Gupta et al. 2004). QTL controlling the tocopherol content in winter oilseed rape (*B. napus*) has also been mapped (Marwede et al. 2005).

Three QTLs controlling fatty acid composition have been identified in sunflower. These would be useful in improving the stearic acid content in the seeds of sunflower (Perez-Vich et al. 2004). QTLs for seed size, color, storage proteins, and pod color have been identified in bean and cowpea (Kelly et al. 2003). Several components of grain quality, such as visual quality, nutritional quality including digestibility and availability of nutrients, anti-nutritional factors, milling characteristics, cooking quality, consumer acceptability, and storage stability have been addressed using molecular breeding in sorghum *(Sorghum bicolor* L. *Moench),* pearl millet *(Pennisetum americanum* L.*),* pigeon-pea *(Cajanus cajan* L. *Millsp.),* and chick-pea *(Cicer arietinum* L.*)* (Jambunathan et al. 1984). QTL analysis for the creation of poppy cultivars with high alkaloid content as well as free of alkaloids is in progress (Nemeth and Petheo 2003). Research is in progress to identify QTLs associated with latexless/partial latex and to improve the quality of poppy seeds that contain high proportion of (edible) vegetable oil which is rich in unsaturated fatty acids, particularly oleic acid and linoleic acid (Anon 1966). Ming et al. (2001) analyzed genetic control of sugar content in sugarcane using QTL mapping. QTLs have been identified in oat for groat protein and oil content (Zhu et al. 2004). Csanadi et al. (2001) identified four markers for protein content, three for oil content and eight for seed weight associated with QTLs influencing seed quality characters in a cross of two early maturing soybean (*Glycine max* L. Merr.) cultivars (Ma.Belle and Proto) adapted to the short growing seasons of Central Europe. Twenty-one QTLs have been mapped for

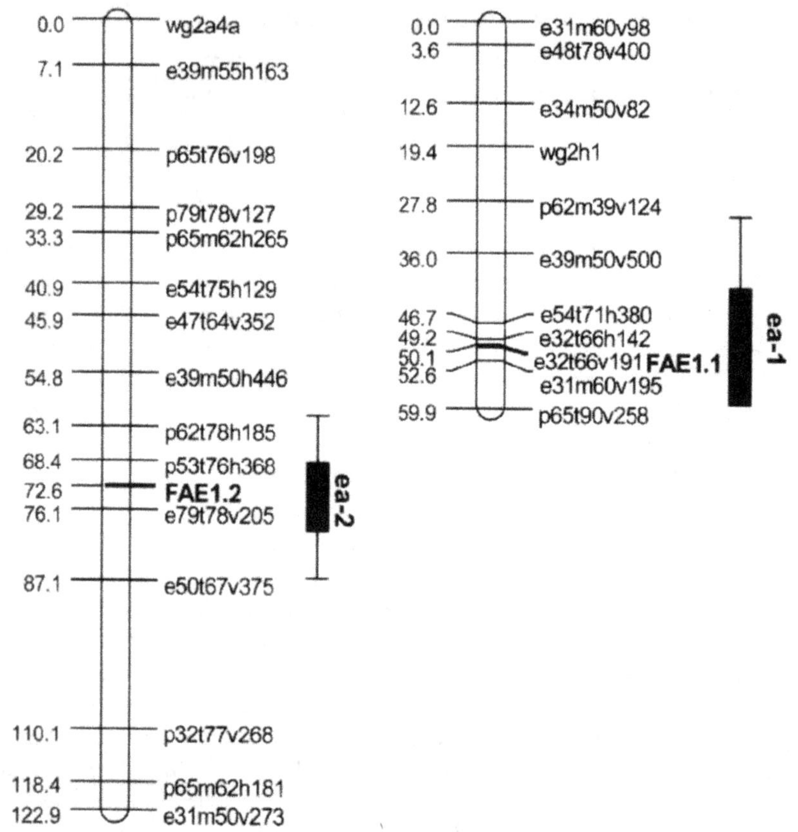

Fig. 4 Molecular mapping and tagging of erucic acid trait in *Brassica juncea* by QTL mapping and single nucleotide polymorphisms in *FAE1* gene. (Two QTLs underlying the variation of seed erucic acid content were assigned to two linkage groups of a *B. juncea* map using a doubled haploid (DH) mapping population derived from high × low erucic acid F_1 hybrid. Single Nucleotide Polymorphism (SNP) marker system was utilized for genotyping. The position of two QTLs, *ea-1* and *ea-2* (black rectangular boxes), controlling the seed erucic acid content on LG 17 and LG 3, respectively, of *B. juncea* map are shown. The *FAE1.1* and *FAE1.2* also mapped to LG17 and LG 3, coinciding with *ea-1* and *ea-2* QTLs, respectively (Gupta et al. 2004).

sucrose content and other quality parameters such as amino nitrogen, potassium and sodium content in a sugar beet population (Schneider et al. 2002; Fig. 5).

QTLs have been identified for fresh weight, color, pH, titratable acidity, soluble-solids content (SSC), acid (malic, citric, and quinic acids),

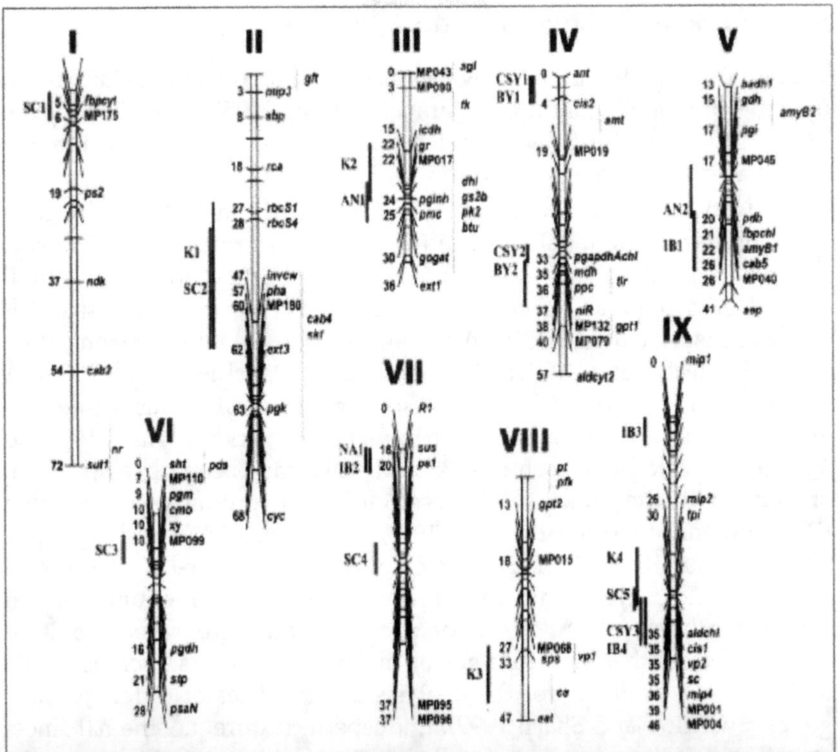

Fig. 5 Genetic map of sugar beet with chromosomal positions of functional genes and QTLs for sugar yield and quality traits. The genetic map comprises 75 genes for which symbols and names are reported on *http://www.mpiz-koeln.mpg.de/sugarbeet.* AFLP markers are indicated by dashes. QTL regions for the traits CSY, corrected sugar yield; BY, beet yield; SC, sugar content; AN, content of amino nitrogen; K, potassium; NA, sodium and IB, ion balance are indicated by bars in the figure (Schneider et al. 2002).

and sugar (sucrose, glucose, fructose, sorbitol) contents in peach (Dirlewanger et al. 1999). QTL analysis for fruit color in peach has also been carried out (Quilot et al. 2005). Nine QTLs for earliness, eight for fruit shape, six for fruit weight, and five for sugar content have been identified for melon fruit quality improvement (Monforte et al. 2004). Molecular markers have been developed linked to QTLs controlling fruit acidity in citrus (Fang et al. 1997). QTL analysis has been successfully used to tag fruit quality traits in mango (Vasanthaiah et al. 2007) where fruit color and yield were the main aspects.

4.2 Marker-assisted Selection (MAS)

Though most quality traits show continuous variation and are influenced by environmental conditions (Francia et al. 2005), they have been monitored using MAS in several major crops such as wheat, tomato, barley, rice, soybean, maize, cotton, etc. (Table 3).

In wheat, MAS has been used for a number of quality traits. The high grain protein content (GPC) gene (*Gpc-B1*) from *Triticum turgidum* ssp. *dicoccoides* has been introduced into durum wheat (Joppa and Cantrell 1990; Deckard et al. 1996). RFLP, SSR, and sequence-specific PCR markers associated with GPC have been developed and are being used in MAS (Mesfin et al. 1999; Khan et al. 2000; Distelfield et al. 2006). A molecular marker loci (*Xwmc41*) associated with GPC was located on 2DL (Varshney et al. 2001). Two molecular markers (*Xwmc104* and *Xmst101*), located one each on 6BS and 7DL, have been developed that are linked to the major QTLs associated with pre-harvest sprouting (PHS) tolerance (Anderson et al. 1993; Roy et al. 1999; Varshney et al. 2001) for use in MAS. Three genes, viz., *Wx-A1*, *Wx-D1* and *Wx-B1* located on 7AS, 7DS and 4AL, respectively code for granule-bound starch synthase (GBSS) or waxy protein which are responsible for amylase biosynthesis. A number of molecular markers such as CAPS and SSRs are available for the analysis of the alleles of different waxy genes (Shariflou and Sharp 1999). Endosperm texture, i.e. the hardiness or softness of the grain, is controlled by the *Hardiness* (*Ha*) locus on the short arm of chromosome 5D. The *Pina-D1* and *Pinb-D1* genes code for lipid binding proteins puroindolines a (PINA) and b (PINB). These two genes act together in the determination of grain texture and are tightly linked to the *Ha* locus on chromosome 5DS (Morris 2002; Turnbull and Rahman 2002). PCR-specific primers and CAPs markers have been developed for use in MAS (Gautier et al. 1994; Tranquilli et al. 1999). High Molecular Weight-Gluteinin Subunit (HMW-GS) *Glu-A1*, *Glu-B1* and *Glu-D1* are coded by three different loci located on the long arms of group 1 chromosomes (Payne 1987), while the gene families located on the short arms of the same chromosomes code for Low Molecular Weight-Gluteinin Subunit (LMW-GS). It is a classic example of MAS where a non-DNA based assay is used. The electrophoretic profiles are obtained from crude seed protein extracted by Sodium Dodecyl Sulfate-Polyacrylamide Gel Electrophoresis (SDS-PAGE) to exploit the correlation between bread-making quality and allelic status at the *Glu-1* loci. PCR- based markers have also been developed for MAS of HMW and LMW glutenin alleles (D'Ovidio et al. 1992; D'Ovidio and Anderson 1994). Molecular markers such as SSR, RFLP, and STS have been

Table 3 Quality traits for which MAS has been proved to be reliable approach for selection.

Trait	*Marker type*	*Reference*
	Wheat	
Milling yield	RFLP	Parker et al. 1999
Grain protein content	SSR	Prasad et al. 1999
Waxy	SSR	Shariflou and Sharp 1999
High grain protein	RFLP	Mesfin et al. 1999
HMW glutenins	STS	Ahmad 2000
Dough strength	SCAR	Radovanovic and Cloutier 2003
Flour color	STS	Parker and Langridge 2000
Kernel size and shape	-	Dholakia et al. 2003
	Rice	
Eating and cooking quality	AFLP and SSR	Zhou et al. 2003
Fragrance of cooked rice	RFLP, SSR, and SNP	Ahn et al. 1992; Cordeiro et al. 2002
Grain length, width, and ratio of length to width	RFLP and RAPD	Huang et al. 1997
	Barley	
Malting quality	RFLP	Igartua et al. 2000
Malting quality	RFLP	Han et al. 1997
	Cotton	
Fiber strength	SSR and RAPD	Zhang et al. 2003
	Tomato	
Fruit size	RFLP and CAPS	Frary et al. 2000
Total soluble solids (sugar and acids)	RFLP, SCAR, CAPS	Fridman et al. 2000
Total soluble solids (sugar and acids)	RFLP and SCAR	Frary et al. 2003
Fruit elongation and neck constriction	RFLP, SCAR, CAPS	Liu et al. 2002a
	Brassica	
2-propenyl glucosinolate content	SCAR	Ripley and Roslinsky 2005
Oil content	RFLP	Cheung et al. 1998
Linoleic acid desaturation	RAPD	Somers et al. 1998
	Soybean	
Seed composition	RFLP	Qiu et al. 1999

identified associated with flour yellow pigment on chromosome 7 for use in MAS (Parker et al. 1998; Parker and Langridge 2000; Elouafi et al. 2001; Mares and Campbell 2001). RFLP markers were used to analyze grain quality traits in a population of 78 recombinant inbred lines (RILs) derived from the cross of the soft white wheat NY6432-18 and the hard white wheat 'Clark's Cream'. Multiple regression models constructed from significant RFLP markers explained the following percentage of the phenotypic variance: PHS, 35%; flour protein concentration, 40%; alkaline water retention capacity, 28%; adjusted flour yield, 14%; and kernel hardness, 9% (Anderson et al. 1994).

Molecular markers such as RFLP, AFLP and RAPD (Saliba-Colombani et al. 1998, 2000) have been identified, for use in MAS, associated with tomato quality traits (physical, chemical, and sensory) using a range of 144 recombinant inbred (RI) populations derived from a cross between a cherry tomato chosen for its good flavor and a line with bigger fruits but poor taste (Causse et al. 2001, 2002). Marker-assisted backcross scheme was applied for the introgression of five QTLs controlling fruit quality traits derived from *L. esculentum* var. *cerasiforme* into three different genetic backgrounds of cultivated tomato (Lecomte et al. 2004).

An RFLP marker, RG28, on chromosome 8 was identified as a candidate to produce a PCR-based marker capable of discriminating between fragrant and non-fragrant rice cultivars (Ahn et al. 1992; Garland et al. 2000). SSR and SNP markers have also been identified linked to the fragrance genes in rice (Corderio et al. 2002; Christopher et al. 2004). A single QTL linked to RFLP and microsatellite markers has been detected for grain extension on cooking (Ahn et al. 1993; Christopher et al. 2004). Three traits, viz., amylose content (AC), a hard gel consistency (GC) and low gelatinization temperature (GT), for cooking and eating quality are controlled by the single *waxy* locus and/or the tightly linked genomic region (Tan et al. 1999). SSR (Bligh et al. 1995) and RFLP markers linked to the *waxy* locus have been used for MAS while AFLP markers were used to recover the genetic background (Zhou et al. 2003). RFLP mapping of isozymes and RAPDs have been used to identify marker loci associated with a grain shape and will be of great use in MAS (Huang et al. 1997).

Three SSR and six RAPD markers were identified linked to two loci for fiber strength in cotton and MAS revealed that these DNA markers could be used in increasing the fiber strength of commercial cotton cultivars (Zhang et al. 2003). The efficiency of MAS for increasing fiber strength was enhanced when SSR markers were used as they could identify the homozygous genotype. RAPDs and RFLPs have also been used for identification of loci associated with bundle fiber strength, fiber length, and fiber fineness and can be used in MAS for cotton fiber quality improvement (Kohel et al. 2001). Marker-assisted backcross introgression scheme was undertaken for analysis of cotton fiber qualities using a combined RFLP-SSR-AFLP genetic map of a *G. hirsutum/G. barbadense* (Lacape et al. 2005). SCAR marker linked to a major QTL for high fiber strength has been developed and has been used for MAS in upland cotton (Guo et al. 2003).

RAPD, RFLP, AFLP, ISSR, STS, SCAR, SSR and SNP marker analyses have been used in MAS in various *Brassica* crops (Cardoza and Stewart 2004; Snowdon and Freidt 2004). RFLP markers have been used to localize genes for seed quality traits in *B. juncea* and used for MAS (Cheung et al. 1998). In *B. napus*, RAPD markers associated with linoleic acid desaturation have been identified and found that the gene *FAD3* localized near one of the QTLs identified for this trait (Somers et al. 1998). RAPD marker associated with linolenic acid content has been converted to SCAR marker for use in MAS in *B. napus* (Hu et al. 1999). Tanhuanpaa et al. (1998) identified an association between the *FAD2* gene, which codes 18:1 desaturase, and a QTL for oleic acid in *B. rapa*. An ISSR marker has been identified linked to the genes controlling contents of 2-propenyl glucosinolates and has been converted to SCAR marker. This marker is of great use in MAS in canola *B. juncea* to select lines having low concentration of glucosinolate per gram of seed and in mustard *B. juncea* to select lines having high concentration of glucosinolate per gram of seed (Ripley and Roslinsky 2005). SNP markers have been identified linked to the low and high erucic acid types in two different types of segregating populations and representative samples of *B. juncea* germplasm. The accurate allele-specific tagging of the erucic acid trait with these SNP markers could be effectively used for marker-assisted transfer of the low erucic acid trait in *B. juncea* (Gupta et al. 2004).

SSR-based MAS has been used to transfer the quality protein maize (QPM) phenotype, controlled in part by a mutant allele of a gene called *opaque2* which doubles the lysine content of the endosperm, from one elite maize inbred line to another elite inbred line (Dreher et al. 2000) and thereby improving the protein quality of the grain. Two major loci for malt extract percentage, α-amylase activity, diastatic power, and malt β-glucan content have been identified and effectively used in MAS for malting quality traits in barley (Han et al. 1997). Two RFLP markers (B072 and B148) have been found to be associated with loci controlling protein concentration and the marker B072 was also found to be associated with loci controlling seed oil concentration in soybeans. These markers will be of great use in MAS for improving the quality of soybeans (Qiu et al. 1999). RFLP markers have also been used to identify five other independent markers associated with soybean seed protein and four markers associated with seed oil content (Lee et al. 1996). Markers associated with seed weight QTLs associated with soybean seed quality showed effects large enough to be useful in a marker-assisted breeding program (Csanadi et al. 2001).

Linkage map of sunflower has been developed using 80 RFLP and 19 SSR markers for improving its oil quality by increasing the stearic acid content in the seeds. Loci controlling fatty acid composition have been identified and have high potential for use in MAS to improve the seed quality of sunflower (Perez-Vich et al. 2004). In bean and cowpea, genetic linkage maps have been developed associated with seed quality traits such as size, color and storage proteins as well as for pod color and are in use for MAS experiments (Kelly et al. 2003). MAS is also being used to improve the feed quality of the crop residues in pearl millet and sorghum (Hash et al. 2003). Some of the important quality traits like seed characteristics, and oil and protein quality parameters have been improved through molecular mapping in sunflower (Tavoljanskiy et al. 2004), castor (Bhardwaj et al. 1996), *Lesquerella fendleri* L. (Roetheli et al. 1991) and peanut (Isleib et al. 2004).

4.3 RNA-Interference Technology

Recent attempts in using RNA-interference (RNAi) to further improve conventionally developed QPM can be used to illustrate the conventional and molecular breeding interface. *Opaque-2*, the gene behind the QPM is recessive in nature and therefore hard to monitor during breeding and after release. RNAi, a mechanism where the presence of certain fragments of double-stranded RNA (dsRNA) interferes with the expression of a particular gene which shares a homologous sequence with dsRNA (Guru 2000), shows promise in suppressing the production of seed proteins of low nutritional value in maize (Segal et al. 2003). The use of antisense gene to create complementary and reverse orientation mRNA that form unstable RNA-RNA complex more difficult to transport out of the nucleus has been harnessed to slow down ripening of tomatoes after harvest (Gray et al. 1992). 'Flavr Savr' was the first whole food commercially sold in commerce using antisense RNA to regulate the expression of enzyme polygalacturonase thought to be responsible for softening in the ripe tomato (Kramer and Redenbaugh 1994).

RNAi technology has also been used in several other plants to improve their nutritional quality. In coffee plants, the caffeine content was reduced by RNAi-mediated suppression of the caffeine synthase gene (Ogita et al. 2003). In another study, RNAi was successfully used to generate a dominant high-lysine maize variant by knocking out the expression of the 22-kD maize zein storage protein, a protein that is poor in lysine content (Segal et al. 2003). RNAi technology has also been successful in genetic modification of the fatty acid composition of cotton seed oil. RNAi mediated by a hairpin RNA was used to knocked-own two key fatty acid desaturase genes leading to an increase in nutritionally improved high-oleic and high-stearic cottonseed oils (Liu et al. 2002b).

5 CONCLUSIONS

Traditional breeding has been used in improving the quality of various crop plants. However, it is a time-consuming process and the limited genetic resources of most crops have left little room for continued improvement by these means (Davies 2003). Recent advances in genomics and bioinformatics offer great potential to revolutionize the field of breeding and genetics through targeted manipulations of quality traits by greatly enhancing our understanding of structural and functional aspects of plant genomes leading to the integration of basic knowledge in ways that can enhance our ability to improve crop plants (Varshney et al. 2005).

Over the past few decades, the possibilities for improvement have been broadened by the development of various molecular marker technologies for gene mapping and identification. This has led to the investigations in gene expression studies, helped in the construction and integration of fine genetic maps for various quality traits and whole-genome sequencing of model plants and crops. These genetic maps will be useful for the crop breeders as they will allow the initial early selection of the desired quality traits in a young progeny population, thereby improving the efficiency of traditional breeding. These maps will also be useful in MAS breeding programs for desirable quality trait improvement. MAS will enhance the efficiency of the breeding process by using DNA markers along with traditional hybridization and selection processes. RNAi triggered by dsRNA has great advantages over antisense and co-suppression approaches, owing to its higher gene silencing efficiency and the shorter time needed to screen the targeted plants (Tang and Galili 2004).

At present, genomes of rice (both *Oryza sativa* var. *indica* and *O. sativa* var. *japonica*) and *Arabidopsis thaliana* have been sequenced. The genetic maps of several other crop species are at different levels of completion. As a result, knowledge regarding location of genomic regions controlling quality traits in various crops will be available for exploitation. Technologies such as microarrays enable scientists to uncover or generate information on thousands of genes and expressed sequence tags (ESTs) in their search for various quality traits. These ESTs can be efficiently used in the molecular breeding approaches for exploiting the quality traits related gene(s). However, the usefulness of new genetic combinations or genotypes must be proved by its performance in field trials or target environment(s) to verify the function and productivity of the quality traits as well as acceptance by farmers, growers, quality managers, and consumers.

References

Abbott A, Massie DR (1985) Delayed light emission for the early detection of chilling stress in cucumber and bell pepper. J Am Soc Hort Sci 110: 42-47

Abdalla O, Dieseth JA, Singh RP (1992) Breeding durum wheat at CIMMYT. In: Rajaram S, Saari EE, Hettel GP (eds) Durum Wheats: Challenges and opportunities. Wheat Special Report No 9, CIMMYT, Mexico DF, pp 1-13

Adu-Dapaah HK, Sangwan RS (2004) Improving bambara groundnut productivity using gamma irradiation and *in vitro* techniques. Afr J Biotech 3(5): 260-265

Ahmad M (2000) Molecular marker-assisted selection of HMW glutenin alleles related to wheat bread quality by PCR-generated DNA markers. Theor Appl Genet 101: 892-896

Ahn SN, Bollich CN, Tanksley SD (1992) RFLP tagging of a gene for aroma in rice. Theor Appl Genet 84: 825-828

Ahn SN, Bollich CN, McClung AM, Tanksley SD (1993) RFLP analysis of genomic regions associated with cooked-kernel elongation in rice. Theor Appl Genet 87: 27-32

Anderson JA, Sorrells ME, Tanksley SD (1993) RFLP analysis of genomic regions associated with resistance to preharvest sprouting in wheat. Crop Sci 33(3): 453-459

Anderson JA, Sorrells ME, Finney PL, Tanksley SD (1994) Molecular mapping of grain quality traits in wheat. In: Plant and Animal Genome II Conf, San Diego, CA, USA

Anonymous (1966) Wealth of India: Raw Materials 7: 234- 246

Applequist WL, Cronn R, Wendel JF (2001) Comparative development of fiber in wild and cultivated cotton. Evol Dev 3: 3-17

Barro F, Fernández-Escobar J, De La Vega M, Martín A (2001) Doubled haploid lines of *Brassica carinata* with modified erucic acid content through mutagenesis by EMS treatment of isolated microspores. Plant Breed 120(3): 262-264

Barro F, Fernández-Escobar J, De La Vega M, Martín A (2002) Modification of glucosinolate and erucic acid contents in double haploid lines of *Brassica carinata* by UV treatment of isolated microspores. *Euphytica* 129: 1-6

Bell RL, Quamme HA, Layne REC, Skirvin R (1996) Pears. In: Janik JJ, Moore JN (eds) Fruit Breeding. John Wiley, New York, USA, pp 441-514

Bergh BO, Lahav E (1996) Avocados. In: Janik JJ, Moore JN (eds) Fruit Breeding. John Wiley, New York, USA, pp 79-111-166

Bernacchi D, Beck-Bunn T, Emmatty D, Eshed Y, Inai S, Lopez J, Petiard V, Sayama H, Uhlig J, Zamir D, Tanksley S (1998) Advanced backcross QTL analysis of tomato. II. Evaluation of near-isogenic lines carrying single-donor introgressions for desirable wild QTL-alleles derived from *Lycopersicon hirsutum* and *L. pimpinellifolium*. Theor Appl Genet 97: 1191-1196

Bhardwaj HL, Mohamed AI, Webber CL, Lovell GR (1996) Evaluation of castor germplasm for agronomic and oil characteristics. In: Janick JJ (ed) Progress in New Crops. ASHS Press, Alexandria, VA, USA, pp 342-346

Birth GS, Dull GG, Magee JB, Chan HT, Cavaletto CG (1984) An optical method for estimating papaya maturity. J Am Soc Hort Sci 109: 62-66

Bligh HFJ, Till RI, Jones CA (1995) A microsatellite sequence closely linked to the *waxy* gene of *Oryza sativa*. Euphytica 86: 83-85

Campbell KG, Bergman CJ, Gualberto DG, Anderson JA, Giroux MJ, Hareland G, Fulcher RG, Sorrells ME, Finney PL (1999) Quantitative trait loci associated with kernel traits in a soft × hard wheat cross. Crop Sci 39: 1184-1195

Campbell KG, Finney PL, Bergman CJ, Gualberto DG, Anderson JA, Giroux MJ, Siritunga D, Zhu J, Gendre F, Roue C, Verel A, Sorrells ME (2001) Quantitative trait loci associated with milling and baking quality in a soft hard wheat cross. Crop Sci 41: 1275-1285

Cardoza V, Stewart CN Jr. (2004) Brassica biotechnology: Progress in cellular and molecular biology. In Vitro Cell Dev Biol-Plant 40: 542-551

Chantret N, Salse J, Sabot F, Rahman S, Bellec A, et al. (2005) Molecular basis of evolutionary events that shaped the hardness occurs in diploid and polyploid Wheat Species (*Triticum* and *Aegilops*). Plant Cell 17: 1033-1045

Chen FQ, Foolad MR, Hyman J, St Clair DA, Beelaman RB (1999) Mapping of QTLs for lycopene and other fruit traits in a *Lycopersicon esculentum* and *L. pimpinellifolium* cross and comparison of QTLs across tomato species. Mol Breed 5: 283-299

Cheung WY, Landry BS, Raney P, Rakow GFW (1998) Molecular mapping of seed quality traits in *Brassica juncea* L. Czern. and Coss. Acta Hort 459: 139-148

Choi KH, Lee KJ, Kim G (2005) Nondestructive quality evaluation technology for fruits and vegetables using near-infrared spectroscopy. In: 7th Intl Fruit/ Vegetable/Food Process Technol Exposition and Symp (IFVE), 21-26 Sept 2005, Yantai, Shandong Province, China

Causse M, Lecomte L, Baffert N, Duffé P, Hospital F (2001) Marker-assisted selection for the transfer of QTLs controlling fruit quality traits into tomato elite lines. Acta Hort 546: 557-564

Causse M, Saliba-Colombani V, Lecomte L, Duffe P, Rousselle P, Buret M (2002) QTL analysis of fruit quality in fresh market tomato: a few chromosome regions control the variation of sensory and instrumental traits. J Exp Bot 53(377): 2089-2098

Cordeiro GM, J Christopher M, Robert HJ, Russell RF (2002) Identification of microsatellite markers for fragrance in rice by analysis of the rice genome sequence. Mol Breed 9: 245-250

Christopher M, Cordeiro G, Waters D, Henry R (2004) Marker assisted selection in rice improvement. Rural Indust Res Devt Corp, No 04/011: 1-15

Chuma Y, Nakaji K, McClure WF (1982) Delayed light emission as a means of automatic color sorting of persimmon fruit, DLE fundamental characteristics of persimmon fruits. J Fac Agri 27(1-2): 13-20

Chuma Y, Nakaji K, Takagawa A (1982) Delayed light emission as a means of automatic color sorting of tomatoes. J Fac Agri 26(4): 221-234

Csanadi G, Vollmann J, Stift G, Lelley T (2001) Seed quality QTLs identified in a molecular map of early maturing soybean. Theor Appl Genet 103: 912-919

Davies WP (2003) An historical perspective from the green revolution to the gene revolution. Nutr Rev 61: S124-S134

Deckard EL, Joppa LR, Hammond JJ, Hareland GA (1996) Grain protein determinants of the Langdon durum-dicoccoides chromosome substitution lines. Crop Sci 36(6): 1513-1516

Delwiche MJ, Tang S, Thompson JF (1988) Prune defect detection by line-scan imaging. ASAE 87-3024

Deshpande SS, Gunasekaran S, Paulsen MR, Cheryan M, Salunkhe DK (1984) Nondestructive optical methods of food quality evaluation. CRC Crit Rev Food Sci Nutr 21(4): 323-379

Dholakia BB, Ammiraju JSS, Singh H, Lagu MD, Roder MS, Rao VS, Dhaliwal HS, Ranjekar PK, GuptaVS (2003) Molecular marker analysis of kernel size and shape in bread wheat. Plant Breed 122: 392-395

Dirlewanger E, Moing A, Rothan C, Svanella L, Pronier V, Guye A, Plomion C, Monet R (1999) Mapping QTLs controlling fruit quality in peach (*Prunus persica* (L.) Batsch). Theor Appl Genet 98: 18-31

Distelfeld A, Uauy C, Fahima T, Dubcovsky J (2006) Physical map of the wheat high-grain protein content gene *Gpc-B1* and development of a high-throughput molecular marker. New Phytol 169: 753-763

Doerksen TK, Kannenberg LW, Lee EA (2003) Effect of recurrent selection on combining ability in maize breeding populations. Crop Sci 43: 1652-1658

Dong Y, Tsuzuki E, Lin D, Kamiunten H, Terao H, Matsuo M, Cheng S (2004) Molecular genetic mapping of quantitative trait loci for milling quality in rice (*Oryza sativa* L.). J Cereal Sci 40: 109-114

D'Ovidio R, Anderson OD (1994) PCR analysis to distinguish between alleles of a member of a multigene family correlated with wheat bread-making quality. Theor Appl Genet 88: 759-763

D'Ovidio R, Tanzarella OT, Masci S, Lafiandra D, Porceddu E (1992) RFLP and PCR analyses at *Gli-1, Gli-2, Glu-1* and *Glu-3* loci in cultivated and wild wheats. Hereditas 116: 79-85

Dreher K, Morris M, Khairallah M, Ribaut JM, Pandey S, Srinivasan G (2000) Is marker-assisted selection cost-effective compared to conventional plant breeding methods? The case of quality protein maize. In: 4th Annu Conf Intl Consort on Agri Biotechnol Res (ICABR), Econ Agri Biotechnol, Aug 24-28, Ravello, Italy

Elouafi I, Nachit MM, Martin LM, (2001) Identification of a microsatellite on chromosome 7B showing a strong linkage with yellow pigment in durum wheat (*Triticum turgidum* L. var. *durum*). Hereditas 135(2-3): 255-261

Fang DQ, Federici CT, Roose ML (1997) Development of molecular markers linked to gene controlling fruit acidity in citrus. Genome 40: 841-849

Fernie AR, Willmitzer L (2004) Carbohydrate metabolism. In: Christou P, Klee H (eds) Handbook of Plant Biotechnology. John Wiley, Chichester, UK, pp 1230-1262

Finney EE, Norris KH (1978) X-ray scans for detecting hallow heart in potatoes. Am Potato J 55: 95-105

Flipse E, Keetels CJAM, Jacobsen E, Visser RGF (1996) The dosage effect of the wildtype *GBSS* allele is linear for *GBSS* activity but not for amylose content: Absence of amylose has a distinct influence on the physico-chemical properties of starch. Theor Appl Genet 92: 121-127

Francia E, Tacconi G, Crosatti C, Barabaschi D, Bulgarelli D, Dall'Aglio E, Vale G (2005) Marker assisted selection in crop plants. Plant Cell Tiss Org Cult 82: 317-342

Frary A, Nesbitt TC, Frary A, Grandillo S, van der Knaap E, Cong B, Liu JP, Meller J, Elber R, Alpert KB, Tanksley SD (2000) A quantitative trait locus key to the evolution of tomato fruit size. Science 289: 85-88

Frary A, Doganlar S, Frampton A, Fulton T, Uhlig J, Yates H, Tanksley S (2003) Fine mapping of quantitative trait loci for improved fruit characteristics from *Lycopersicon chmielewskii* chromosome 1. Genome 46: 235-243

Fridman E, Pleban T, Zamir D (2000) A recombination hotspot delimits a wild-species quantitative trait locus for tomato sugar content to 484 bp within an invertase gene. Proc Natl Acad Sci USA 97: 4718-4723

Garcia JM, Yousfi K (2005) Non-destructive and objective methods for the evaluation of the maturation level of olive fruit. Eur Food Res Technol 221(3-4): 538-541

Garland S, Lewin L, Blakeney A, Reinke R (2000) PCR-based molecular markers for the fragrance gene in rice (*Oryza sativa* L.). Theor Appl Genet 101: 364-371

Gautier MF, Aleman ME, Guirao A, Marion D, Joudrier P (1994) *Triticum aestivum* puroindolines, two cystine-rich seed proteins: cDNA sequence analysis and developmental gene expression. Plant Mol Biol 25: 43-57

Ge XJ, Xing YZ, Xu CG, He YQ (2005) QTL analysis of cooked rice grain elongation, volume expansion, and water absorption using a recombinant inbred population. Plant Breed 124: 121-126

Getinet A, Rakow G, Raney JP, Downey RK (1997) Glucosinolate content in interspecific crosses of *Brassica carinata* and *B. juncea* and *B. napus*. Plant Breed 116: 39-46

Gray J, Picton S, Shabbeer J, Schuch W, Grierson D (1992) Molecular biology of fruit ripening and its manipulation with antisense genes. Plant Mol Biol 19: 69-87

GuoW, Zhang T, Shen X, Yu JZ, Kohel RJ (2003) Development of SCAR marker linked to a major QTL for high fiber strength and its usage in molecular-marker assisted selection in upland cotton. Crop Sci 43: 2252-2256

Gunasekaran S, Paulsen MR, Shove GC (1985) Optical methods for nondestructive quality evaluation of agricultural and biological materials. J Agri Engg Res 32(3): 209-241

Gupta V, Mukhopadhyay A, Arumugam N, Sodhi YS, Pental D, Pradhan AK (2004) Molecular tagging of erucic acid trait in oilseed mustard (*Brassica juncea*) by QTL mapping and single nucleotide polymorphisms in *FAE1* gene. Theor Appl Genet 108: 743-749

Guru T (2000) A silence that speaks volumes. Nature 404: 804-808

Hajika M, Igita K, Kitamura K (1991) A line lacking all the seed lipoxygenase isozymes in soybean (*Glycine max* (L.) Merrill) induced by gamma-ray irradiation. Jap J Breed 41: 507-509

Han F, Romagosa I, Ullrich SE, Jones BL, Hayes PM, Wesenberg DM (1997) Molecular marker-assisted selection for malting quality traits in barley. Mol Breed 3: 427-437

Hash CT, Bhasker Raj AG, Lindup S, Sharma A, Beniwal CR, Folkertsma RT, Mahalakshmi V, Zerbini E, Blummel M (2003) Opportunities for marker-assisted selection (MAS) to improve the feed quality of crop residues in pearl millet and sorghum. Field Crops Res 84: 79-88

He GY, Rooke L, Steele S, Békés F, Gras P, Tatham AS, Fido R, Barcelo P, Shewry PR, Lazzeri PA (1999) Transformation of pasta wheat (*Triticum turgidum* L. var. *durum*) with high-molecular-weight glutenin subunit genes and modification of dough functionality. Mol Breed 5: 377-386

Hentschel V, Kranl K, Hollmann J, Lindhauer MG, Bohm V, Bitsch R (2002) Spectrophotometric determination of yellow pigment content and evaluation of carotenoids by high-performance liquid chromatography in durum wheat grain. J Agri Food Chem 50(23): 6663-6668

Hu J, Li G, Struss D, Quiros CF (1999) SCAR and RAPD markers associated with 18-carbon fatty acids in rapeseed, *Brassica napus*. Plant Breed 118: 145-150

Huang N, Parco A, Mew T, Magpantay G, McCouch S, Guiderdoni E, Xu J, Subudhi P, Angeles ER, Khush GS (1997) RFLP mapping of isozymes, RAPD and QTLs for grain shape, brown planthopper resistance in a doubled haploid rice population. Mol Breed 3: 105-113

Hyten DL, Pantalone VR, Sams CE, Saxton AM, Landau-Ellis D, Stefaniak TR, Schmidt ME (2004) Seed quality QTL in a prominent soybean population. Theor Appl Genet 109: 552-561

Igartua E, Edney M, Rossnagel BG, Spaner D, Legge WG, Scoles GJ, Eckstein PE, Penner GA, Tinker NA, Briggs KG, Falk DE, Mather DE (2000) Marker-based selection of QTL affecting grain and malt quality in two-row barley. Crop Sci 40: 1426-1433

Isleib TG, Pattee HE, Giesbrecht FG (2004) Oil, sugar, and starch characteristics in peanut breeding lines selected for low and high oil content and their combining ability. Agri Food Chem 52 (10): 3165-3168

Jambunathan R, Singh U, Subramanian V (1984) Grain quality of sorghum, pearl millet, pigeonpea and chickpea. In: Achaya KT (ed) Interfaces Between Agriculture, Nutrition and Food Science. Proc Workshop Biotechnol Trop Crop Improv, Patancheru, India, pp 47-60

Janick JJ, Cummins JN, Brown SK, Hemmat M (1996) Apples. In: Janik JJ, Moore JN (eds) Fruit Breeding. John Wiley, New York, USA, pp 1-77

Janick JJ, Moore JN (eds) (1996) Fruit Breeding. 3 vols. John Wiley, New York, USA

Jiang CX, Wright RJ, El-Zik K, Paterson AH (1998) Polyploid formation created unique avenues for response to selection in *Gossypium* (cotton). Proc Natl Acad Sci USA 95: 1419-1424

Joppa LR, Cantrell RG (1990) Chromosomal location of genes for grain protein content of wild tetraploid wheat. Crop Sci 30(5): 1059-1064

Kamps TL, Isleib TG, Herner RC, Sink KS (1987) Evaluation of techniques to measure chilling injury in tomato. HortScience 22(6): 1309-1312

Katsuji O, Takashi S, Tetsuo M (2003) Polyploidy breeding of functional northern small fruit trees by utilizing *in vitro* cultures treated with colchicines. Urakami Found Mem 11: 24-31

Kavdir I, Lu R, Ariana D, Ngouajio M (2007) Visible and near-infrared spectroscopy for nondestructive quality assessment of pickling cucumbers. Postharvest Biotechnol 44(2): 165-174

Kelly JD, Gepts P, Miklas PN, Coyne DP (2003) Tagging and mapping of genes and QTL and molecular marker-assisted selection for traits of economic importance in bean and cowpea. Field Crops Res 82: 135-154

Khan IA, Procunier JD, Humphreys DG, Tranquilli G, Schlatter AR, Marcucci-Poltri S, Frohberg R, Dubcovsky J (2000) Development of PCR-based markers for a high grain protein content gene from *Triticum turgidum* ssp. *dicoccoides* transferred to bread wheat. Crop Sci 40(2): 518-524

Kohel RJ, Yu J, Park YH, Lazo GR (2001) Molecular mapping and characterization of traits controlling fiber quality in cotton. Euphytica 121(2): 163-172

Kramer MG, Redenbaugh K (1994) Commercialization of a tomato with antisense polygalacturonase gene: The FLVR SAVR^{TM} tomato story. Euphytica 79: 293-297

Lacape JM, Nguyen TB, Courtois B, Belot JL, Giband M, Gourlot JP, Gawryziak G, Roques S, Hau B (2005) QTL analysis of cotton fiber quality using multiple *G. hirsutum* x *G. barbadense* backcross generations. Crop Sci 45: 123-140

Larson SR, Young KE, Cook A, Blake TK, Raboy V (1998) Linkage mapping two mutations that reduce phytic acid content of barley grain. Theor Appl Genet 97: 141-146

Larson SR, Rutger JN, Young KA, Raboy V (2000) Isolation and genetic mapping of a non-lethal rice (*Oryza sativa* L.) *low phytic acid 1* mutation. Crop Sci 40: 1397-1405

Layne REC, Bailey CH, Hough LF (1996) Apricots. In: Janik JJ, Moore JN (eds) Fruit Breeding. John Wiley, New York, USA, pp 79-111

Leal F, d'Eeckenbrugge GC (1996) Pineapple. In: Janik JJ, Moore JN (eds) Fruit Breeding. John Wiley, New York, USA, pp 515-557

Lecomte L, Duffe P, Buret M, Servin B, Hospital F, Causse M (2004) Marker-assisted introgression of five QTLs controlling fruit quality traits into three tomato lines revealed interactions between QTLs and genetic backgrounds. Theor Appl Genet 109: 658-668

Lee SH, Bailey MA, Mian MAR, Carter TE Jr, Shipe ER, Ashley DA, Parrott WA, Hussey RS, Boerma HR (1996) RFLP loci associated with soybean seed protein and oil content across populations and locations. Theor Appl Genet 93: 649-657

Li CD, Tarr A, Lance RCM, Harasymow S, Uhlmann J, Westcor S, Young KJ, Grime CR, Cakir M, Broughton S, Apples R (2003) A major QTL controlling seed dormancy and pre-harvest sprouting/grain α-amylase in two-rowed barley (*Hordeum vulgare* L.). Aust J Agri Res 54: 1303-1313

Lin Z, He D, Zhang X, Nie Y, Guo X, Feng C, Stewart J McD (2005) Linkage map construction and mapping QTL for cotton fibre quality using SRAP, SSR and RAPD. Plant Breed 124: 180-187

Liu JP, Van Eck J, Cong B, Tanksley SD (2002a) A new class of regulatory genes underlying the cause of pear-shaped tomato fruit. Proc Natl Acad Sci USA 99: 13302-13306

Liu Q, Singh S, Green A (2002b) High-oleic and high-stearic cottonseed oils: nutritionally improved cooking oils developed using gene silencing. J Am Coll Nutr 21(3): S205-S211

Liu YS, Gur A, Ronen G, Causse M, Damidaux R, Buret M, Hirschberg J, Zamir D (2003) There is more to tomato fruit color than candidate genes. Plant Biotechnol J 1: 195-207

Liu ZW, Fu TD, Tu JX, Chen BY (2005) Inheritance of seed color and identification of RAPD and AFLP markers linked to the seed color gene in rapeseed (*Brassica napus* L.). Theor Appl Genet 110: 303-310

Mares DJ, Campbell AW (2001) Mapping components of flour and noodle color in Australian wheat. Aust J Agri Res 52: 1297-1309

Mares DJ, Mrva K (2001) Mapping quantitative trait loci associated with variation in grain dormancy in Australian wheat. Aust J Agri Res 52: 1257-1265

Marwede V, Gul MK, Becker HC, Eckle W (2005) Mapping of QTL controlling tocopherol content in winter oilseed rape. Plant Breed 124(1): 20-26

Mei M, Syed NH, Gao W, Thaxton PM, Smith CW, Stelly DM, Chen ZJ (2003) Genetic mapping and QTL analysis of fibre-related traits in cotton (*Gossypium*). Theor Appl Genet 108: 280-291

Mesfin A, Frohberg RC, Anderson JA (1999) RFLP markers associated with high grain protein from *Triticum turgidum* L. var. dicoccoides introgressed into hard red spring wheat. Crop Sci 39: 508-513

Miller BK, Delwiche MJ (1988) A color vision system for peach grading. ASAE 88-6025

Ming R, Liu SC, Moore PH, Irvine JE, Paterson AH (2001) QTL analysis in a complex autopolypoid: Genetic control of sugar content in sugarcane. Genome Res 11: 2075-2084

Mohan M, Nair S, Bhagwat A, Krishna TG, Yano M, Bhatia CR, Sasaki T (1997) Genome mapping, molecular markers and marker assisted selection in crop plants. Mol Breed 3: 87-103

Monforte A J, Oliver M, Gonzalo M J, Alvarez JM, Dolcet-Sanjuan R, Arffls P (2004) Identification of quantitative trait loci involved in fruit quality traits in melon (*Cucumis melo* L.) Theor Appl Genet 108: 750-758

Moose SP, Dudley JW, Rocheford TR (2004) Maize selection passes the century mark: a unique resource for 21st century genomics. Trends Plant Sci 9: 358-364

Moreau L, Charcosset A, Gallais A (2004) Experimental evaluation of several cycles of marker-assisted selection in maize. Euphytica 137: 111-118

Morris CF (2002) Puroindolines: the molecular basis of wheat grain hardness. Plant Mol Biol 48: 633-647

Mrva K, Mares DJ (2001b) Quantitative trait locus analysis of late maturity α-amylase in wheat using the doubled haploid population Cranbrook x Halberd. Aust J Agri Res 52: 1267-1273

Muggleston S (1995) What is involved in plant breeding (Part I). Orchardist NZ 68(9): 49

Murayama H, Konno I, Terasaki S, Yamamoto R, Sakurai N (2006) Nondestructive method for measuring fruit ripening of 'La France' pears using a laser doppler vibrometer. J Jap Soc Hort Sci 75(1): 79-84

Nemeth BÉ, Petheõ F (2003) Alkaloid accumulation in capsules of the selfed and cross-pollinated poppy. Plant Breed 122(3): 263-267

Norris KH (1983) Instrumental techniques for measuring quality of agricultural crops. In: Lieberman M (ed) Post-harvest Physiology and Crop Preservation. Plenum press, New York, USA, p 471

Ogita S, Uefuji H, Yamaguchi Y, Koizumi N, Sano H (2003) Producing decaffeinated coffee plants. Nature 423: 823

Omobjuwajo TO (2003) Compositional characteristics and sensory quality of biscuits, prawn crackers and fried chips produced from breadfruit. Innov Food Sci Emerg Technol 4: 219-225

Pani F, Gentinetta E, Verderio A, Lorenzoni C, Motto M (1985) Recurrent selection for seed quality traits in an opaque-2 maize population. Euphytica 34(3): 857-863

Parker GD, Chalmers KJ, Rathjen AJ, Langridge P (1998) Mapping loci associated with flour colour in wheat (*Triticum aestivum* L.) Theor Appl Genet 97: 238-245

Parker GD, Chalmers KJ, Rathjen AJ, Langridge P (1999) Mapping loci associated with milling yield in wheat (*Triticum aestivum* L.). Mol Breed 5: 561-568

Parker GD, Langridge P (2000) Development of a STS marker linked to a major locus controlling flour colour in wheat (*Triticum aestivum* L.). Mol Breed 6: 169-174

Paterson AH, Lander ES, Hewitt JD, Peterson S, Lincoln SE, Tanksley SD (1988) Resolution of quantitative factors by using a complete linkage map of restriction fragment length polymorphisms. Nature 335: 721-726

Paterson AH, Damon S, Hewitt JD, Zamir D, Rabinowitch HD, Lincoln SE, Lander ES, Tanksley SD (1991) Mendelian factors underlying quantitative traits in tomato: comparison across species, generations, and environments. Genetics 127: 181-197

Paterson AH, Saranga Y, Menz M, Jiang CX, Wright RJ (2003) QTL analysis of genotype x environment interactions affecting cotton fiber quality. Theor Appl Genet 106: 384-396

Payne PI (1987) Genetics of wheat storage proteins and the effect of allelic variation on bread-making quality. Annu Rev Plant Physiol 38: 141-153

Pérez-Vich B, Knapp SJ, Leon AJ, Fernández-Martínez JM, Berry ST (2004) Mapping minor QTL for increased stearic acid content in sunflower seed oil. Mol Breed 13: 313-322

Perretant MR, Cadalen T, Charmet G, Sourdille P, Nicolas P, Boeuf C, Tixier MH, Branlard G, Bernard S, Bernard M (2000) QTL analysis of bread-making quality in wheat using a doubled haploid population. Theor Appl Genet 100: 1167-1175

Pilu R, Panzeri D, Gavazzi G, Rasmussen S, Consonni G, Nielsen E (2003) Phenotypic, genetic and molecular characterization of a maize low phytic acid mutant (*lpa241*). Theor Appl Genet 107: 980-987

Poehlman JM (1986) Breeding Field Crops. 3rd edn. AVI Publ, Westport, Connecticut, USA

Porteous RL, Muir AV, Wastie RL (1981) The identification of diseases and defects in potato tubers from measurements of optical spectral reflectance. J Agri Engg Res 26: 151

Prasad M, Varshney RK, Kumar A, Balyan HS, Sharma PC, Edwards KJ, Singh H, Dhaliwal HS, Roy JK, Gupta PK (1999) A microsatellite marker associated with a QTL for grain protein content on chromosome arm 2DL of bread wheat. Theor Appl Genet 99: 341-345

Qi X, Stam P, Lindhout P (1996) Comparisons and integration of four barley genetic maps. Genome 39: 379-394

Qiu BX, Arelli PR, Sleper DA (1999) RFLP markers associated with soybean cyst nematode resistance and seed composition in a 'Peking' × 'Essex' population. Theor Appl Genet 98: 356-364

Quilot B, Kervella J, Genard M, Lescourret F (2005) Analyzing the genetic control of peach fruit quality through an ecophysiological model combined with a QTL approach. J Exp Bot 56(422): 3083-3092

Raboy V, Gerbasi P, Young KA, Stoneberg S, Pickett SG, Bauman AT, Murthy PPN, Sheridan WF, Ertl DS (2000) Origin and seed phenotype of maize *low phytic acid 1-1* and *low phytic acid 2-1*. Plant Physiol 124: 355-368

Radovanovic N, Cloutier S (2003) Gene-assisted selection for high molecular weight glutenin subunits in wheat doubled haploid breeding programs. Mol Breed 12: 51-59

Rehkugler G, Throop JA (1987) Image processing algorithm for apple defect detection. ASAE 87-3041

Ripley VL, Roslinsky V (2005) Identification of an ISSR marker for 2-propenyl glucosinolate content in *Brassica juncea* L. and conversion to a SCAR marker. Mol Breed 16: 57-66

Roetheli JC, Carlson KD, Kleiman R, Thompson AE, Dierig DA, Glaser LK, Blasé MG, Goodell J (1991) Lesquerella as a Source of Hydroxy Fatty Acids for Industrial Products. An Assessment. USDA/CSRS Off Agri Mater, Washington, DC, USA

Roose ML, Williams TE (2006): *http://www.citrusvariety.ucr.edu/citrus/mandarins.html#Shasta*

Roy JK, Prasad M, Varshney RK, Balyan HS, Blake TK, Dhaliwal HS, Singh H, Edwards KJ, Gupta PK (1999) Identification of a microsatellite on chromosomes 6B and a STS on 7D of bread wheat showing an association with preharvest sprouting tolerance. Theor Appl Genet 99(1-2): 336-340

Rucker B, Robbelen G (1995) Development of high oleic acid rapeseed. In: 9th Proc Intl Rapeseed Congr, July 4-7, Cambridge, UK, pp 389-391

Saliba-Colombani V, Causse M, Philouze J, Buret M, Issanchou S, Lesschaeve I (1998) QTLS for organoleptic quality in fresh market tomato. In: Proc XV EUCARPIA Congr on Genet Breed for Crop Quality and Resistance, Sept 20-25, Viterbo, pp 291-299

Saliba-Colombani V, Causse M, Gervais L, Philouze J (2000) Efficiency of AFLP, RAPD and RFLP markers for the construction of an intraspecific map of the tomato genome. Genome 43: 29-40

Schierholt A, Rucker B, Becker HC (2001) Inheritance of high oleic acid mutations in winter oilseed rape (*Brassica napus* L.). Crop Sci 41: 1444-1449

Schmidt AL, Liu C, Martin D, Kelly A, McIntyre L (2004) Molecular markers for selected quality traits in Australian hexaploid bread wheat. In: 4th Intl Crop Sci Congr, Brisbane, Australia (*http://www.cropscience.org.au/icsc2004/*)

Schneider K, Schäfer-Pregl R, Borchardt DC, Salamini F (2002) Mapping QTLs for sucrose content, yield and quality in a sugar beet population fingerprinted by EST-related markers. Theor Appl Genet 104: 1107-1113

Schulz H, Drews HH, Quilitzsch R, Kruger H. (1998) Application of near infrared spectroscopy for the quantification of quality parameters in selected vegetables and essential oil plants. J Near Infrared Spectroscopy 6: A125-A130

Segal G, Song R, Messing J (2003) A new opaque variant of maize by a single dominant RNA-interference-inducing transgenic. Genetics 165: 387-397

Septiningsih EM, Trijatmiko KR, Moeljopawiro S, McCouch SR (2003) Identification of quantitative trait loci for grain quality in an advanced backcross population derived from the *Oryza sativa* variety IR64 and the wild relative *O. rufipogon*. Theor Appl Genet 107: 1433-1441

Shariflou MR, Sharp PJ (1999) A polymorphic microsatellite in the 32 end of 'waxy' genes of wheat, *Triticum aestivum*. Plant Breed 118: 275-277

Sharma JR, Lal RK, Gupta AP, Misra HO, Pant V, Singh NK, Pandey V (1999) Development of non-narcotic (opiumless and alkaloid-free) opium poppy, *Papaver somniferum*. Plant Breed 118 (5): 449-452

Shmulevich I, Galili N, Rosebfeld D (1996) Detection of fruit firmness by frequency analysis. Transaction ASAE 39: 1047-1055

Slaughter DC, Crisosto CH (2006) Nondestructive determination of internal quality in clingstone peaches. In: ASAE Annu Meet, Pap no 066180

Snape J, Fish L, Leader D, Bradburne R, Turner A (2005) The impact of genomics and genetics on wheat quality improvement. Turk J Agri For 29: 97-103

Snowdon RJ, Friedt W (2004) Molecular markers in *Brassica* oilseed breeding: current status and future possibilities. Plant Breed 123: 1-8

Somers DJ, Friesen KRD, Rakow G (1998) Identification of molecular markers associated with linoleic acid desaturation in *Brassica napus*. Theor Appl Genet 96: 897-903

Soost RK, Roose ML (1996) Citrus. In: Janik JJ, Moore JN (eds) Fruit Breeding. John Wiley, New York, USA, pp 257-323

Tan YF, Li JX, Yu SB, Xing YZ, Xu CG, Zhang Q (1999) The three important traits for cooking and eating quality of rice grains are controlled by a single locus in an elite rice hybrid, Shanyou 63. Theor Appl Genet 99: 642-648

Tang G, Galili G (2004) Using RNAi to improve plant nutritional value: from mechanism to application. Trends Biotechnol 22(9): 463-469

Tanhuanpaa P, Vilkki J, Vihinen M (1998) Mapping and cloning of *FAD2* gene to develop allele-specific PCR for oleic acid in spring turnip rape (*Brassica rapa* ssp. *oleifera*). Mol Breed 4: 543-550

Taran B, Warkentin T, Somers DJ, Miranda D, Vandenberg A, Blade S, Bing D (2004) Identification of quantitative trait loci for grain yield, seed protein concentration and maturity in field pea (*Pisum sativum* L.). Euphytica 136: 297-306

Tavoljanskiy NP, Chiryaev PV, Scherstyuk SV, Altinnikova VI, Tikhomirov VT (2004) Development of original material for sunflower breeding for seed characteristics, oil and protein quality in the conditions of Central-Chernozem region. Helia 27(40): 117-122

Tranquilli G, Lijavetzky D, Muzzi G, Dubcovsky J (1999) Genetic and physical characterization of grain texture-related loci in diploid wheat. Mol Gen Genet 262: 846-850

Turnbull KM, Rahman S (2002) Endosperm texture in wheat. J Cereal Sci 36: 327-337

Udall JA, Wendel JF (2006) Polyploidy and crop improvement. Crop Sci 46: 3-14

Ulloa M, Meredith WR (2000) Genetic linkage map and QTL analysis of agronomic and fibre quality traits in an intraspecific population. J Cotton Sci 4: 161-170

Vanhala TK, Stam P (2006) Quantitative trait loci for seed dormancy in wild barley (*Hordeum spontaneum* C. Koch). Genet Resour Crop Evol 53: 1013-1019

Varshney RK, Prasad M, Roy JK, Harjit-Singh NK, Dhaliwal HS, Balyan HS, Gupta PK (2000) Identification of eight chromosomes and a microsatellite marker on 1AS associated with QTL for grain weight in bread wheat. Theor Appl Genet 100: 1290-1294

Varshney RK, Prasad M, Roy JK, Roeder MS, Balyan HS, Gupta PK (2001) Integrated physical maps of 2DL, 6BS and 7DL carrying loci for grain protein content and pre-harvest sprouting tolerance in bread wheat. Cereal Res Commun 29(1-2): 33-40

Varshney RK, Graner A, Sorrells ME (2005) Genomics-assisted breeding for crop improvement. Trends Plant Sci 10(12): 621-630

Vasanthaiah HKN, Ravishankar KV, Mukanda GK (2007) Mango. In: Kole C (ed) Genome Mapping and Molecular Breeding in Plants. Vol 4: Fruits and Nuts. Springer, Berlin, Heidelberg, New York, pp 303-323

Velasco L, Fernandez-Martinez JM, deHaro A (1998) Increasing erucic acid content in Ethiopian mustard through mutation breeding. Plant Breed 117(1): 85-87

Wafada (1992) Food crops: nondestructive quality evaluation. In: Hui YH (ed) Encyclopedia of Food Science and Technology. John Wiley New York, USA, pp 1016-1019

Wang CY, Worthington JT (1979) A non-destructive method for measuring ripeness and detecting core break-down in 'Bartlett' pears. J Am Soc Hort Sci 104: 629-931

Wang SY, Wang PC (1989) Non-destructive detection of core break-down in 'Bartlett' pears with nuclear magnetic resonance imaging. HortScience 24: 106-109

Wang SY, Wang PC, Faust M (1988) Non-destructive detection of watercore in apple with Nuclear Magnetic Resonance imaging. Sci Hort 35: 227-234

World Food Program (2006) Facts and figures (*http://www.wfp.org/food_aid/doc/39149_Nutrition.pdf*)

Worrell DB, Carrington CM, Huber DJ (1998) Growth, maturation and ripening of breadfruit, *Artocarpus altilis* (Park.). Fosb Sci Hort 76: 17-28

Yoon JY, Lee WM, Woo JG (1999) Quality improvement of major kimchi vegetables through plant breeding and biotechnology. Acta Hort 483: 49-55

Zanetti S, Winzeler M, Keller M, Keller B, Messmer M (2000) Genetic analysis of pre-harvest sprouting resistance in a wheat × spelt cross. Crop Sci 40: 1406-1417

Zanetti S, Winzeler M, Feuillet C, Keller B, Messmer M (2001) Genetic analysis of bread making quality in wheat and spelt. Plant Breed 120: 13-19

Zeldin EL, McCown BH (2003) Application of polyploidy to cranberry breeding and biotechnology. Acta Hort 626: 133-139

Zhang TZ, Yuan YL, Yu J, Guo WZ, Kohel RJ (2003) Molecular tagging of a major QTL for fiber strength in upland cotton and its marker-assisted selection. Theor Appl Genet 106: 262-268

Zhou PH, Tan YF, He YQ, Xu CG, Zhang Q (2003) Simultaneous improvement for four quality traits of Zhenshan 97, an elite parent of hybrid rice, by molecular marker-assisted selection. Theor Appl Genet 106: 326-331

Zhu S, Rossnagel BG, Kaeppler HF (2004) Genetic analysis of quantitative trait loci for groat protein and oil content in oat. Crop Sci 44: 254-260

5 Molecular Mapping and Breeding for Biotic Stress Resistance

D.A. Lightfoot* and M.J. Iqbal**
Illinois Soybean Center of Excellence, Department of Plant, Soil and Agricultural Systems, 176 Agriculture Building, MC 4415, Southern Illinois University at Carbondale, Carbondale, Illinois 62901-4415, USA
*Corresponding author: ga4082@siu.edu
**Present Address: Institute for Advanced Learning and Research, Institute for Sustainable and Renewable Resources, 150 Slayton Ave, Danville, VA 24540, USA

1 INTRODUCTION

Crop geneticists and breeders aim to improve harvestable yield, reduce crop losses, and reduce grower inputs by re-assortment among favorable haplotypes (Iqbal and Lightfoot 2004; Stefaniak et al. 2005; Karangula et al. 2007) and the incorporation of new genes for new traits (Anand 1992; Arelli 1994). Rapid genetic gains can be made by incorporating genes for resistance to biotic stress. The positive effect of biotic stress resistance on seed yield is so reliable that journals rarely accept yield data (Hnetkovsky et al. 1996). However, each disease has an economic threshold, a point at which significant yield loss occurs (Wrather et al. 2003). For each disease and pest, it is important to determine the threshold and use resistance genes only when the threshold is exceeded for two reasons. First, many biotic stress resistance genes reduce yield potential in the absence of disease (Yuan et al. 2002). Second, the use of resistance genes selects for new more virulent mutants or populations of pathogens.

Molecular mapping and marker-assisted selection (MAS) are cost and labor intensive endeavors. Therefore, biotic stress resistance traits with clear easily scored phenotypes are often selected without markers. Traits with a bi-geneic or oligo-geneic inheritance pattern are most efficiently selected with DNA markers (Meksem et al. 2001a, b). Traits with a low to moderate heritability due to significant interactions with the environment are most efficiently selected with DNA markers (Kassem et al. 2006).

1.1 Intellectual Property and Due Diligence

A major use of molecular markers is to define regions encompassing resistance genes that can be used to claim and protect intellectual property (Webb et al. 1995, 1996). Early patents in this area were quite broad, claiming large tracts of genome because they contained a single resistance gene. More recent actions by the patent office try to restrict the claims to single markers for single genes making this type of patent more difficult and expensive to obtain. Some companies holding broad early patents have sought to broaden the claims still further to encompass selection for ANY locus in the region where a biotic stress gene is found (Webb et al. 2000, 2003). Here they seek to limit "unintentional" selection of linked traits. Therefore, each new project should start with due diligence, an examination of intellectual property related to the trait and regions to be selected.

1.2 Marker Development

For most crops there is a continued distributed community effort to add new markers to the existing linkage maps. In soybean this has expanded to include satellites and single nucleotide polymorphisms (SNPs; Figs. 1-3; Song et al. 2004; Shultz et al. 2006) and has been focused on regions that do not have many markers currently. The current work relies heavily on the physical map (Wu et al. 2004; Shultz et al. 2006). In future, the development of new markers will rely on the genome sequence and the haplotype map that will be developed from it (Zhu et al. 2003).

For map development renewable genetic resources are necessary. In soybean there are many renewable collections of recombinant inbred lines (RILs). However, most lie in private hands and are not widely available. Recently released was 100 lines derived from the cross of Essex and Forrest (Lightfoot et al. 2005) that are available from SIUC on request. In addition there is an immortal collection of 100 near-isogenic line (NIL) pairs, each pair capturing the heterogeneous regions derived from a single RIL (Njiti et al. 1998, 2003), that are available.

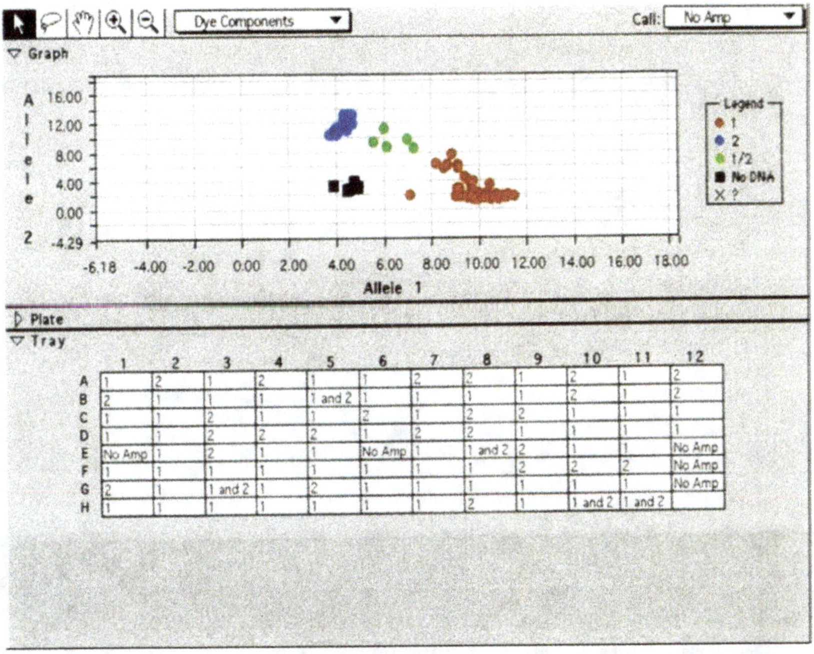

	1	2	3	4	5	6	7	8	9	10	11	12
A	1	2	1	2	1	1	2	2	1	2	1	2
B	2	1	1	1	1 and 2	1	1	1	1	2	1	2
C	1	1	2	1	1	2	1	2	2	1	1	1
D	1	1	2	2	2	1	2	2	1	1	1	1
E	No Amp	1	2	1	1	No Amp	1	1 and 2	2	1	1	No Amp
F	1	1	1	1	1	1	1	1	2	2	2	No Amp
G	2	1	1 and 2	1	2	1	1	1	1	1	1	No Amp
H	1	1	1	1	1	1	1	2	1	1 and 2	1 and 2	

Fig. 1 Detection of the rhg1 marker polymorphism by the TaqMan method allows allelic discrimination of soybean genotypes with manual selection of genotypes (from Meksem et al. 2001b with permission). Data for 90 individuals from an F_5-derived population of RILs from the cross Essex × Forrest that segregate for resistance to SCN are shown. **Panel 1:** The fluorescent signals were viewed under the "Dye Component" field of the sequence detection software and the genotypes were manually selected based on the ratio of FAM to TET signals. Allele 1, homozygous Forrest type: TET/FAM greater than 2. Allele 2, homozygous Essex type: FAM/TET greater than 2. If the TET level was less than twofold greater or lesser than the FAM level, the individual was scored as heterozygous Essex/Forrest (1/2). **Panel 2:** The Excel spreadsheet shows the scores for the samples as they were arranged in the 96-well plate. There was no DNA in wells H11 and H12. Essex DNA was placed in well A1, Forrest DNA in well B1, The RIL DNA was arrayed in the rest of the 96-well plate (C1-G12). *1* Resistant allele, *2* susceptible allele, *1* and 2 heterozygous lines.

1.3 Scoring of Phenotypes

In order to avoid a 'garbage in garbage out' (GIGO) map, biotic stress resistance traits must be measured under controlled and/or highly replicated assays with clear easily scored phenotypes. Traits with a bigeneic or oligo-geneic inheritance pattern are most efficiently scored on a

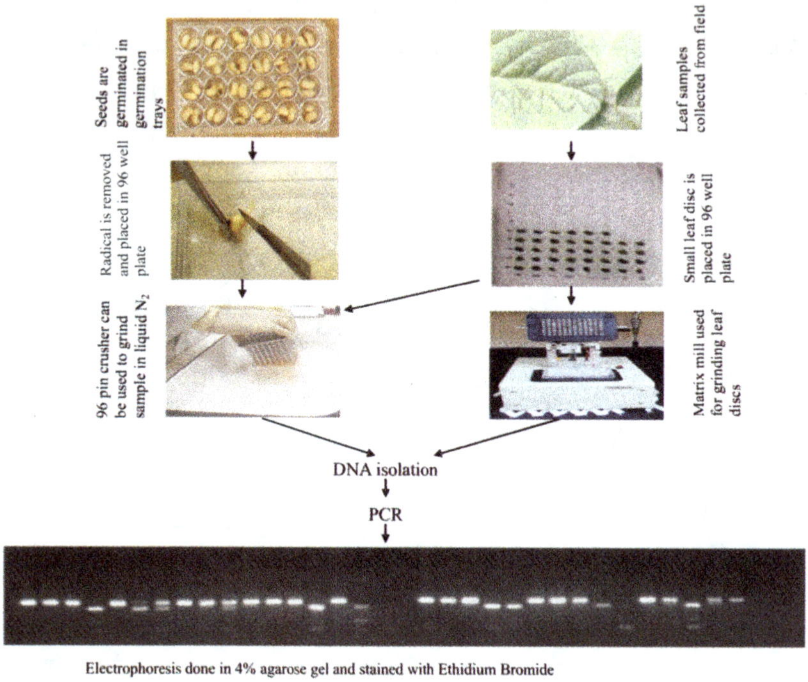

Fig. 2 Steps involved in MAS. Gel electrophoresis can be replaced by other capillary-based electrophoresis techniques, a fluorescent-based assay like TaqMan or invader or MALDI-TOF can be used for scoring polymorphism at the target allele.

percentage scale (Njiti et al. 2001; Kassem et al. 2006). Traits with a low to moderate heritability due to significant interactions with the environment are most efficiently selected in greenhouse assays, IF the assay can be shown to correspond to the field results. The use of low doses of standardized inocula was particularly important to many assays (Arelli 1994; Njiti et al. 2001). Trait distributions should be analyzed before mapping to determine skewness and kurtosis, indicators of major gene effects. Major morphological effects like maturity dates and determinacy MUST be measured and correlated with the disease resistance traits to avoid selection of loci underlying such confounding traits (Prabhu et al. 1999; Njiti and Lightfoot 2006).

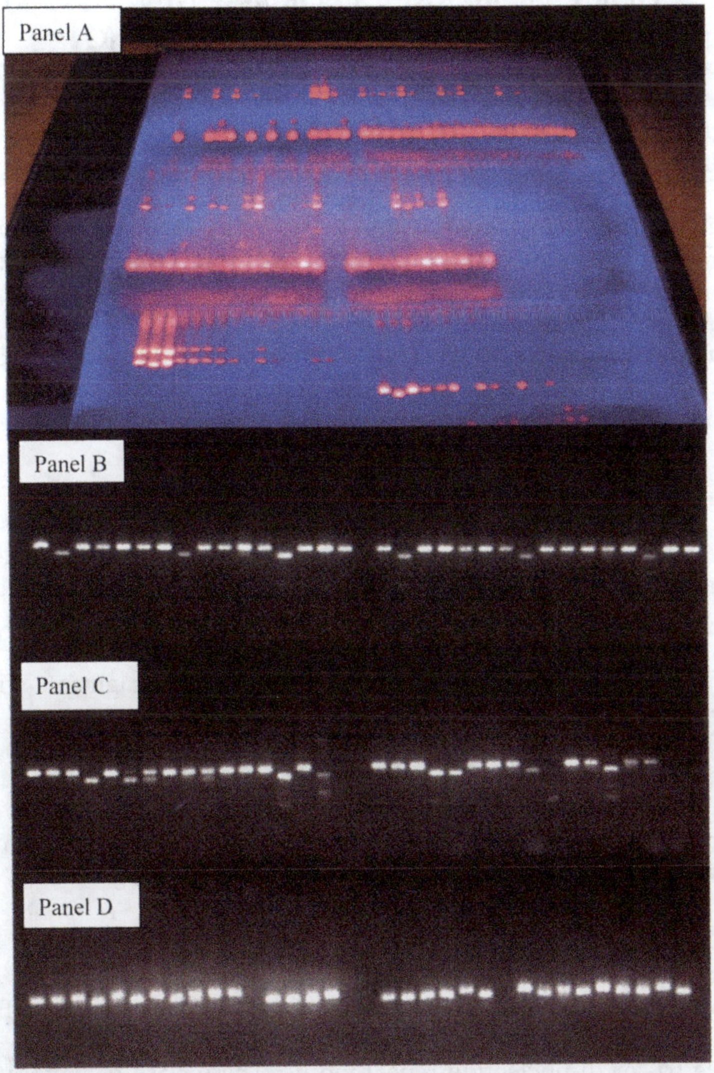

Fig. 3 Segregation of SSR locus, Satt in a population. Individuals can be clearly defined as carrying the alleles of either parent 1 (*P1*), parent 2 (*P2*) or heterozygous. The 10-μl PCR reaction was electrophoresed on a 4% (w/v) metaphore agarose gel and stained with ethidium bromide. DNA was isolated from radical tissue in a 96-well plate. PCR was carried out in a 96-well plate and samples were loaded on gel with an 8-channel pipette. **Panel A:** A typical gel. **Panel B-D**. SattB54L22 scored among 92 individuals segregating for resistance to biotic stress. First two bands in **A** and last 2 in **C** were the parental DNAs.

2 CHOICE OF MARKERS FOR MAP DEVELOPMENT AND MARKER-ASSISTED BREEDING

The choice of marker(s) depends upon the objective of the selection program, tools available to the individual/group researcher, markers available for a particular region or trait and the information content associated with a particular marker system. Intellectual property must be considered too, many modern marker systems are owned. Ironically, older technology that is no longer protected, like oligomer hybridization, is widely used. Linkage disequilibrium (LD) must be considered. Linkage blocks in soybean (350 Kbp) are about two-fold larger than in *Arabidopsis thaliana* on average (Zhu et al. 2003) but a hundred-fold greater than in outbreeding crops like *Zea mays*. Hence, comparatively distant markers work well in some crops but not others.

2.1 Restriction Fragment Length Polymorphism

The first marker system to be used was restriction fragment length polymorphism (RFLP). RFLPs can be adapted to high-throughput markers by determination of the SNP or IWDLL detected (Marek et al. 2001; Ashfield et al. 2003). However, the polyploid origin of most plant genomes can lead to the detection of multiple DNA fragments with all RFLP probes. Single RFLP probes have been used to detect up to 19 independent loci (Mansur et al. 1996). The multiplicity of RFLP loci can make the locus identity ambiguous. Other factors that prevent the use of RFLP in mapping and marker-assisted breeding are the low levels of polymorphisms observed (Shoemaker and Specht 1995) and expense of automating the procedure for high-throughput screening.

2.2 Microsatellites

Microsatellite markers, also known as simple sequence repeats (SSRs), use DNA sequence data and PCR to show polymorphism within 10-20 simple tri- or di-nucleotide repeat motifs. These class II microsatellite markers have been shown to be highly polymorphic in soybean and many other crops (Fig. 2; Akkaya et al. 1992; Morgante and Olivieri 1993). Some highly polymorphic microsatellite loci can have as many as 26 alleles (Rongwen et al. 1995; Powell et al. 1996). Such a high level of allelic diversity increases the possibility of detecting polymorphism between parents of populations derived from the hybridization of adopted soybean genotypes.

Class II SSRs are comparatively rare in the genome (one per 50 Kbp) and seem not to be distributed randomly throughout the soybean genome (Marek et al. 2001). As the map become more dense further evidence of clustering has emerged (Akkaya et al. 1995; Kassem et al. 2006). Analysis of genomic DNA has shown shorter class I microsatellites that contain less than ten simple tri- or di-nucleotide repeats (small simple sequence repeats or SSSRs) are more common than SSRs and frequently polymorphic in soybean and other crops (Fig. 3; McCouch et al. 2002; Rota et al. 2005; Shultz et al. 2007). There are about ten such DNA markers (as defined) per 100 Kbp of soybean sequence. Class I markers can overcome the problem of marker clustering because of their abundance in the genome.

In addition, complex microsatellites that contain two different tri- or di-nucleotide repeats are very common and frequently polymorphic in soybean (Witsenboer et al. 1998; Shultz et al. 2007). Selection for class I and II SSRs, SSSRs, and SAMPLs (selectively amplified microsatellite polymorphic loci) in soybean can be made with standard gel or capillary electrophoresis (Cregan et al. 1999; Prabhu et al. 1999), or electrophoresis-free assays like MALDI-TOF MS (Chen et al. 2003).

Therefore, microsatellite markers are currently the method of choice in most of the public laboratories based on their distribution in the genome, the polymorphism frequency and information content attached to each identified locus. The relative ease of automation for high-throughput screening programs is another bonus.

2.3 Single Nucleotide Polymorphisms

Single nucleotide polymorphisms (SNPs) are single base pair changes (or 1-3 bp indels) between individuals or cultivars. They also occur between homeologs in duplicated genomes. They occur about twice as often in non-coding compared to coding DNA. In coding DNA, about one quarter to one half of SNPs alter amino acid sequence depending on the genes examined (Meksem et al. 2001b; Zhu et al. 2003; Ruben et al. 2006). The observed SNP frequency in soybean (0.053 and 0.1%) is about two-fold lower than *A. thaliana* considering known genes and cDNAs (Zhu et al. 2003). However, the SNP frequency is four-fold higher than expected within AFLP (amplified fragment length polymorphism) bands as the sequence dataset (Meksem et al. 2001a, b). However, genes at loci under intense selection like the major resistance genes *rhg1*, *Rhg4* (resistance to *Heterodera glycines*), *Rfs1* and *Rfs2* (resistance to *Fusarium solani* f. sp. *glycines*) are encoded by 6-8 alleles (Meksem et al. 1999; Hauge et al. 2001; Triwitayakorn et al. 2005; Ruben et al. 2006) likely reflecting the extra diversity imported to cultivated soybean by gene introgression. Selection for SNPs in soybean can be made with standard gel or capillary electrophoresis, or electrophoresis-free assays like TaqMan (Fig. 1; Meksem et al. 2001a, b), HPLC and MALDI-TOF (Zhu et al. 2003).

2.4 Indels

Deletions and insertions (indels) are multi-base pair changes between soybean cultivars. They occur in non-coding DNA more often than coding DNA. In coding DNA they alter amino acid sequence (Meksem et al. 2001b). The observed frequency in soybean is about five-fold lower than for SNPs even considering AFLP bands as the sequence dataset (Meksem et al. 2001a, b). Some genes under intense selection like the major resistance gene *Rhg1*, bear indels in introns for two of nine alleles (Meksem et al. 1999; Hauge et al. 2001; Ruben et al. 2006) likely reflecting the extra diversity imported to cultivated soybean by gene introgression. Selection for indels can be made by non-PCR methods (Landegren et al. 1998) and isothermic assays like Invader (Mein et al. 2000). The standard PCR-based gel or capillary electrophoresis, or electrophoresis-free assays like TaqMan (Meksem et al. 2001a, b) or MALDI-TOF MS (Chen et al. 2003) can also be used (Fig. 1). The ease of selection makes indels the most versatile DNA marker available.

2.5 Identification of Polymorphism

The first step in applying marker-assisted selection (MAS) to a breeding program is the identification of polymorphic loci between the parents that can be traced in the segregating population. The frequency of polymorphism is related to genetic distance between the parents and the type of marker used (Zhu et al. 2003; Shultz et al. 2007). When low diversity or a paucity of markers is a problem, physical map-based marker identification is an efficient approach (Meksem et al. 1998; Cregan et al. 1999b; McCouch et al. 2000; Rota et al. 2005; Shultz et al. 2006, 2007).

2.6 Genetic Map Development

There are a wide variety of map building and QTL analysis programs available and many work well (Lander et al. 1987; Basten et al. 2001; see also Chapters 4 and 11 in Volume I). At a minimum the map should be built at a high LOD (4-5) then expanded at lower LODs to reduce transitive map errors. QTL should be detected by ANOVA, an interval mapping (IM) program and a composite interval mapping (CIM) program. Map errors must be removed before IM and CIM so that double recombinants must be looked at closely.

3 MARKER-ASSISTED RECOVERY OF RECURRENT PARENT GENOME

In plant breeding for resistance to biotic stress, the backcross procedure is often used to transfer favorable alleles from donor genotypes into a recipient elite genotype (Anand 1993; Frisch et al. 1999). The donors often have many poor agronomic properties requiring several backcrosses to eliminate. The goal of this 'introgression' is to recover transgressive segregants that yield more than the elite parent by 1-2% per year for each year necessary to develop the line. Only genes of large effect on biotic stress resistance can be selected in this way. In backcross gene introgression programs, marker-assisted selection has a great impact on the genome recovery of recurrent parent genotype (RPG; Tanksley et al. 1989). Individuals that are homozygous for the alleles of the recurrent parent at a large number of marker loci covering the entire genome are selected. Population size and marker density required in a background selection program can be determined (Frisch et al. 1999). The minimum recommended is the use of four markers per chromosome (of 200 cM-length) and a selection strategy for proximal recombinants of the target allele. In the case of soybean, having 20 linkage groups and an estimated map length of 2,478 cM, 80 uniformly distributed markers have been effective in recovering a sufficiently high proportion of RPG in three generations. The number of markers used for RPG recovery varies, about a 20 cM-marker density represented an optimal trade-off between percentage recurrent parent recovery and management of data-point throughput requirements (Frisch et al. 1999). Uniformly distributed markers are more effective than equal numbers per chromosome because smaller population sizes and fewer data points are required with equally spaced markers (Frisch et al. 1999). Therefore, selection of markers from a physical map rather than a less precise genetic map can improve the efficiency of RPG recovery.

4 MARKER-ASSISTED SELECTION (MAS) IN RECURRENT CROSS POPULATIONS

The majority of plant breeding programs make advances by re-assorting the entire genomes of two elite lines with favorable and complementary characteristics, when possible advances in biotic stress should be made this way (Iqbal et al. 2001). The goal of this 'shuffling of the deck' is to recover transgressive segregants that yield more than either parent by 1-2% per year for each year necessary to develop the line. Traditional recurrent selection cycles last from five to seven years from cross to release. The average cultivar is sold for two-three years before becoming obsolete. Therefore, marker selection that reduces development time by stacking selections into a shorter time frame will be very valuable. Further, since released cultivars appear to contain more recombination events than the population mean, markers can be used to find well shuffled genomes for phenotypic analysis (Stefaniak et al. 2005; Karangula et al. 2008).

MAS for biotic stress resistance often competes with field scores, greenhouse assays and even low cost laboratory assays of phenotype for price and effectiveness (Njiti et al. 2001). This competition is not always market driven, as investments in infrastructure made in the past determine economics in the present. Over time though, we expect MAS to be widely adopted by all effective soybean breeders. MAS, as employed today, usually focuses on two to four genes per population with 25 to 6.25% of lines 'passing the test'. Samples are selected after preliminary visual selection for yield potential (at the F_3-F_5), rarely are F_2 selections made. Hence, about 10% of a program's recombinants will be tested in the MAS lab. Since a single lab will serve dozens of breeders each with a hundred thousand lines, the task can quickly reach millions of selections per year. Automation is very important in this situation.

Selections are made by a variety of methods often driven more by due diligence of patent holders' rights than technical efficiency. Patent holders have the right to exclude competitors from the use of patented genes, loci and wider genomic regions. However, with the exception of a few markers, fair non-excluding licensing deals can be negotiated for most trait-marker associations.

5 EXAMPLES OF MARKER-ASSISTED SELECTION FOR TARGETED BIOTIC STRESS RESISTANCE GENES/ TRAITS

Most crop genomes have thousands of QTLs controlling hundreds of agronomic traits. Databases also contain gene class, locus, alleles, phenotypes, and two-point data (Shultz et al. 2006). The availability of all this information and commercial drive to release new and improved varieties assists the use of MAS for new and improved variety development. Two examples from soybean marker-assisted breeding are outlined below.

5.1 SCN

One example of MAS in soybean is selection for resistance to soybean cyst nematode (SCN). Soybean cyst nematode, *Heterodera glycines,* is a small plant-parasitic roundworm that attacks the roots of soybeans and causes significant crop losses in the infected fields (Wrather et al. 1996; 2003). SCN resistance was introgressed from weedy types (Anand 1992; Arrelli 1994) by backcrossing and is associated with both a linkage drag on yield and a genome load affecting yield (Fig. 4). Two QTLs

Fig. 4 The effects of gene stacking on resistance to SDS. To the left of the white line is a cultivar with a stack of six resistance genes. To the right of the line is a cultivar with few or no resistance genes showing the chlorosis and defoliation symptomatic of SDS.

significantly contributing to soybean resistance to *H. glycines, rhg*1 and *Rhg*4 were mapped on linkage groups G and A2 and the candidate genes isolated (Ruben et al. 2006). Cultivars selected for *rhg1* provide good resistance to *H. glycines* Hgtype 0 and 7 (race 3). SSR marker Satt309 was mapped at a 0.5-2 cM distance from *rhg*1 (Cregan et al. 1999). It was shown to be 42 Kbp in physical distance from *rhg1* (Ruben et al. 2006). Satt 309 was widely used for MAS for resistance to SCN race 3 but only for research purposes because of patent positions held privately.

With the sequencing of a large region of linkage group G of soybean genome by Monsanto (St. Louis, Mo), and the competing Monsanto and SIUC patents for perfect markers for SCN (Meksem et al. 1999; Hauge et al. 2001; Jamai et al. 2002; Ruben et al. 2006), it is now possible to select all nine *Rhg1/ rhg1* alleles for resistance with 100% certainty. There are satellite markers for the promoter region, an indel in the intron and SNPs adapted to Taqman in the leucine-rich repeat region of the *rhg1* receptor like kinase. In total 26 SNP markers are directly located on the gene associated with resistance to SCN race 3, six alter protein allotype. There are five susceptibility alleles and four alleles that confer resistance. Association analysis across germplasm shows this gene is part of the phenotype defining multigenic locus (Ruben et al. 2006). Confirmation by reverse genetics is dogged by the functional homeologs elsewhere in the genome.

5.2 SDS

A different example of MAS in soybean is selection for resistance to sudden death syndrome (SDS). SDS in the US is caused by a member of the *Fusarium* sp. disease complex, *F. virguliforme* (ex. *solani* f. sp. *glycines*) that attacks the roots of soybeans and causes significant crop losses in the infected fields (Njiti et al. 1997, 2001; Aoki et al. 2003). SDS resistance was found among elite adapted types and was exploited by gene stacking during inter-crossing (Fig. 4). Resistance to SDS was associated with no linkage drag on yield (Fig. 5), indeed it is a disease of high yielding genotypes. Among several resistance genes associated with resistance to SDS of soybean (Table 1), three are clustered with SCN resistance alleles at *rhg*1 (Meksem et al. 1999; Iqbal et al. 2001). One of the clustered genes, *Rfs1*, was responsible for part of a bi-genic root resistance to *F. virguliforme* and can be selected for by a set of bacterial artificial chromosome (BAC)-derived SSRs (Meksem et al. 1999; Shultz et al. 2007), a RAPD-derived SCAR (Triwitayakorn et al. 2005), or a candidate gene sequence (Lightfoot, unpub). The second recently

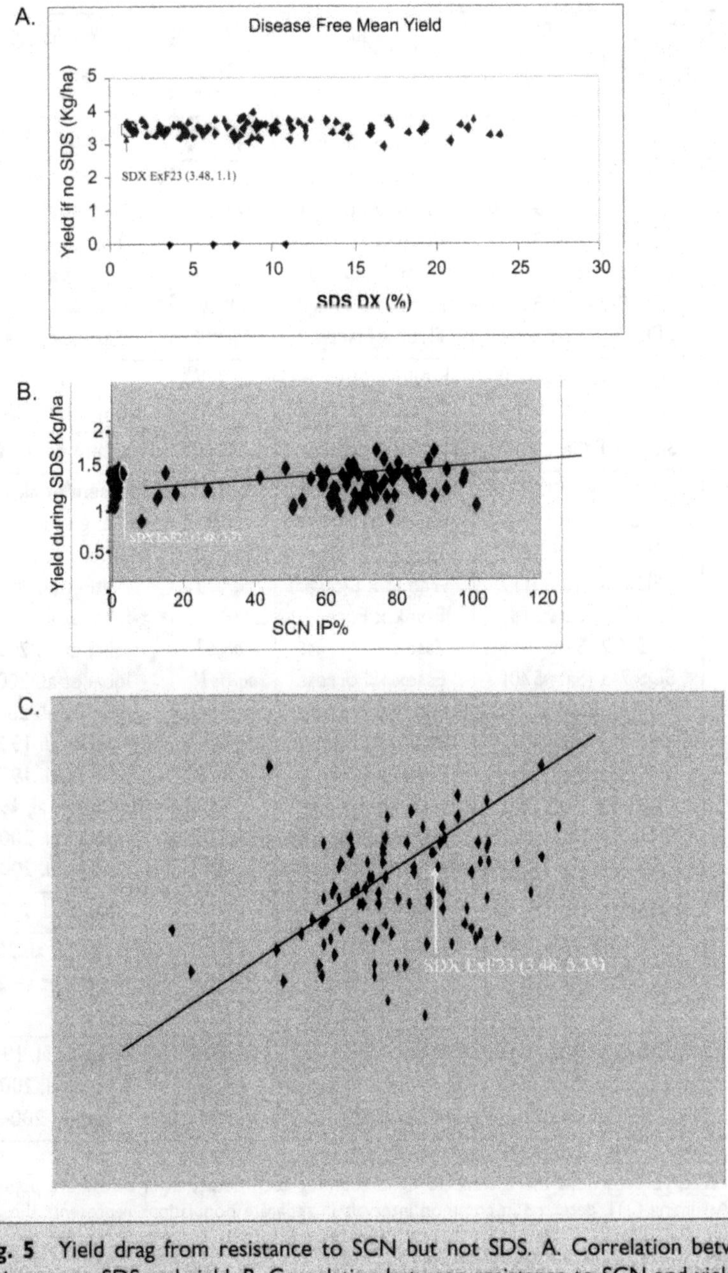

Fig. 5 Yield drag from resistance to SCN but not SDS. A. Correlation between resistance to SDS and yield. B. Correlation between resistance to SCN and yield. C. Correlation between yield during SDS infestation and yield in disease free fields.

Table 1 SDS QTL found in adapted germplasm and different RIL populations.

Linkage Group	*Soybase QTL Name (Linked Markers)*[a]	*Populations*	*Probable Locus Names*[b]	*References*
A2	SDS11-1 (Satt 187)	Ripley x Spencer	*qRfs7*	Hashmi, 2004
C2	SDS 1-1 (0005-250)	Essex x Forrest	*cqRfs4*	Hnetkovsky et al. 1996
	SDS 4-2 (K455_1)	Essex x Forrest	*cqRfs4*	Njiti et al. 1998
	SDS 7-5 (Satt 371)	Essex x Forrest	*cqRfs4*	Iqbal et al. 2001
	SDS 8-2 (Satt 307)	Pyramid x Douglas	*cqRfs4*	Njiti et al. 2002
	SDS 8-3 (Satt 316)	Pyramid x Douglas	*qRfs9*	Njiti et al. 2002
	SDS 9-2 (Satt 277)	Flyer x Hartwig	*cqRfs4*	Kazi et al. 2008
D2	SDS11-2 (Satt 528)	Ripley x Spencer	*cqRfs11*[c]	Hashmi 2004
	SDS 9-1 (Satt574)	Flyer x Hartwig	*cqRfs11*[c]	Kazi et al. 2008
F	SDS10-1 (Satt 160)	Essex x Forrest	*qRfs10*	Kassem et al. 2006
G	SDS 6-1 (Bng 122_1)	Essex x Forrest	*cqRfs1*[c]	Meksem et al. 1999
	SDS 3-1 (0G13_490)	Essex x Forrest	*cqRfs3*	Chang et al. 1996
	SDS 6-2 (Satt 309)	Essex x Forrest	*cqRfs2*	Meksem et al. 1999
	SDS 8-4 (0G01)	Pyramid x Douglas	*cqRfs3*	Njiti et al. 2002
	SDS 7-1 (Satt 214)	Essex x Forrest	*cqRfs2*	Iqbal et al. 2001
	SDS 7-2 (Satt 309)	Essex x Forrest	*cqRfs2*	Iqbal et al. 2001
	SDS 7-3 (Satt 570)	Essex x Forrest	*cqRfs1*[c]	Iqbal et al. 2001
	SDS 7-4 (OEO2_1000)	Essex x Forrest	*cqRfs3*	Iqbal et al. 2001
	SDS 4-1 (0103-512)	Essex x Forrest	*cqRfs1*[c]	Njiti et al. 1998
	SDS 4-3 (Bng122_1)	Essex x Forrest	*cqRfs1*[c]	Njiti et al. 1998
	SDS 5-1 (Satt 038)	Flyer x Hartwig	*cqRfs1*[c]	Prabhu et al. 1999
	SDS 8-1 (Satt 163)	Pyramid x Douglas	*cqRfs2*	Njiti et al. 2002
	SDS9-3 (Satt427)	Flyer x Hartwig	*cqRfs3*	Kazi et al. 2008
I	SDS 7-6 (Satt 354)	Essex x Forrest	*qRfs5*	Iqbal et al. 2001
J	SDS10-2 (Satt 285)	Essex x Forrest	*cqRfs8*	Kassem et al. 2006
	SDS(Satt183)	GC87018-12-2B-1 x GC89045-13-1	*cqRfs8*	Sanitchon et al. 2004
N	SDS 2-7 (0C01-650)	Essex x Forrest	*cqRfs6*	Chang et al. 1996
	SDS 8-3 (Satt 080)	Pyramid x Douglas	*cqRfs6*	Njiti et al. 2002
	SDS (Satt 631)	Ripley x Spencer	*cqRfs6*[d]	Hashmi 2004

[a]QTL numbers are from Soybase where assigned.

[b]From 27 QTL detections there were 15 QTL counting each detection in a separate population once. Assuming QTL detected in common intervals in separate populations represents the alleles there are 11- 12 loci.

[c]QTL associated with resistance to root infection.

[c]QTL detected in common intervals in separate populations or derived NILs are considered confirmed and suffixed with c under Soybean Genetics Committee recommendations (*http://soybase.agron.iastate.edu/nomenclature/QTL.html*).

discovered root resistance gene on linkage group (LG) D2 can be selected for by Sat_001 (Kazi et al. 2007).

Each of the several resistance genes discovered for SDS resistance has associated a SSR marker (Table 1). Each gene has been detected in multiple populations. Therefore, gene pyramids have been developed incorporating up to six genes (Iqbal et al. 2001) and gene pyramids of 10-12 genes are envisioned (Fig. 5). Gene pyramids have been shown effective in eliminating SDS symptoms from fields (Fig. 4). In recent works, an expression fingerprint for resistance to SDS has been identified (Iqbal et al. 2005). The expression profile or associated protein changes will be used to develop a very high-throughput assay for resistance to SDS (Shopinski et al. 2006). However, the map positions of the responsive genes did not correspond with the location of resistance QTL.

The SDS success story in Illinois has lead to some complacency among breeders that may be misplaced. The fungus is still present in the soil and is under strong selective pressure to adapt to the resistant germplasm. Further, a second pathogen that causes SDS in South America, *F. tucumanes* is likely to infest US fields (Aoki et al. 2003). Resistance to SDS in Argentina is not controlled by the same resistance gene set as in Illinois (Bashir, Lightfoot and Salas unpub).

6 METHODS FOR MARKER-ASSISTED SELECTION

Screening for markers linked to a trait or recovery of RPG can be carried out with leaf or seed samples. A high throughput method of MAS is described (Fig. 2; Triwitayakorn et al. 2005; Ruben et al. 2006). Briefly, DNA can be isolated from small leaf discs collected from fields as well as from the seeds. For sample, collection from plants, a 1-cm^2-leaf disc from a young leaf was removed and stored in small note pads and kept on ice or dry ice. Alternately, seeds are placed on small filter paper discs soaked with sterile distilled water for 48 h for germination. Radicals are carefully removed with sharp scalpel blades. Leaf discs or cotyledons are placed in 96-well plates. Samples can be ground either by a matrix mill in the presence of 100 μl of 0.5 N sodium hydroxide or by 96-pin crusher in the presence of liquid N_2 (Fig. 2).

Once the DNA is isolated (Xin et al. 2003 for an alternate method for DNA isolation), PCR with the selected primers is carried out in 96- or 384-well plates. Electrophoresis is in Metaphore agarose gel/regular agarose gel (Fig. 3), standard acrylamide gel or capillary (www.egeneinc.com) depending on the size difference of the target alleles. For a higher-throughput, SSR or SCAR primers can be labeled with fluorescence tags and multiplexed at the electrophoresis stage. The samples can then be electrophoresed using capillary-based DNA sequencers and analyzed by fragment analysis software as provided by the fragment analysis equipment used. The shortcomings of gel electrophoresis can be avoided by developing TaqMan-TM allelic discrimination (Fig. 1; Meksem et al. 2001a, b), or MALDI-TOF base extension methods (not shown). Also to consider for use are allele specific hybridizations, PCR-OLA, molecular beacons, padlock probes and well fluorescence assay (Landegren et al. 1998). Cost, throughput and technical capability will determine which method each laboratory adopts for SNP analysis.

7 CONCLUSIONS

The main aim of MAS for resistance to biotic stress is the selection of highly desirable lines among thousands of genotypes. Selection can only be achieved by high-throughput, rapid, automated procedures for the detection of DNA polymorphisms attached with the desirable traits. With the availability of capillary electrophoresis-based genetic analyzers, thousands of samples can be screened per day per machine. However, the availability of new non-electrophoresis-based tools such as TaqMan (Meksem et al. 2001, 2001a), padlock probes (Landegren et al. 1998), the invader assay (Mein et al. 2000) and MALDI-TOF for the detection of polymorphisms further simplifies the screening procedures for large-scale MAS procedures. Moreover, the availability of gene markers or perfect markers such as SNPs directly on the genes of interest makes the MAS procedures more desirable. With the availability of over 350,000 soybean EST reads, an emerging genomic sequence of 7.4 million reads and a physical map (Shultz et al. 2006) crops like soybean will have (www.phytozone.net) sufficient genomic resources for the rapid development of new markers for resistance to new biotic stresses. The application in marker-assisted breeding of new and improved varieties will become a widely accepted breeding tool.

References

Akkaya MS, Bhagwat AA, Cregan PB (1992) Length polymorphism of simple sequence repeat DNA in soybean. Genetics 132: 1131-1139

Akkaya MS, Shoemaker RC, Specht JE, Bhagwat AA, Cregan PB (1995) Integration of simple sequence repeat DNA markers into a soybean linkage map. Crop Sci 35: 1439-1445

Anand SC (1992) Registration of 'Hartwig' soybean. Crop Sci 32: 1060-1070

Aoki T, O'Donnell K, Homma Y, Lattanzi AR (2003) Sudden-death syndrome of soybean is caused by two morphologically and phylogenetically distinct species within the *Fusarium solani* species complex-*F. virguliforme* in North America and *F. tucumaniae* in South America. Mycologia 95: 660-684

Arelli AP (1994) Inheritance of resistance to *Heterodera glycines* race 3 in soybean accessions. Plant Dis 78: 898-900

Ashfield T, Bocian A, Held D, Henk A, Marek L, Danesh D, Penuela S, Meksem K, Lightfoot DA, Young N, Shoemaker R, Innes R (2003) Genetic and physical mapping of the soybean *Rpg1-b* disease resistance gene reveals a complex locus containing several tightly linked families of NBS/LRR genes. Mol Plant-Micr Interact 16: 817-826

Basten CJ, Weir BS, Zeng Z (2001) QTL Cartographer Version 2.0. Dept of Statistics, North Carolina State Univ, Raleigh, NC, USA

Chen CH, Potter NT, Taranenko NT (2003) Detection of trinucleotide repeat containing genes by matrix-assisted laser desorption/ionization (MALDI) mass spectrometry. Meth Mol Biol 217: 91-100

Cregan PB, Mudge J, Fickus EW, Danesh D, Denny R, Young ND (1999) Two simple sequence repeat markers to select for soybean cyst nematode resistance conditioned by the *rhg1* locus. Theor Appl Genet 99: 811-818

Frisch M, Bohn M, Melchinger AE (1999) Comparison of selection strategies for marker-assisted backcrossing of a gene. Crop Sci 39: 1295-1301

Hashmi R (2004) Analysis of SDS resistance in Ripley by Spencer. PhD Dissertation, SIUC, Carbondale, Illinois, USA

Hauge BM, Wang ML, Parsons JD, Parnell LD (2001) Nucleic acid molecules and other molecules associated with soybean cyst nematode resistance. US Patent: WO 0151627-A 19-JUL-2001

Hnetkovsky N, Chang SJC, Doubler TW, Gibson PT, Lightfoot DA (1996) Genetic mapping of loci underlying field resistance to soybean sudden death syndrome (SDS). Crop Sci 36: 393-400

Iqbal MJ, Meksem K, Njiti VN, Kassem MyA, Lightfoot DA (2001) Microsatellite markers identify three additional quantitative trait loci for resistance to soybean sudden death syndrome (SDS) in Essex × Forrest RILs. Theor Appl Genet 102: 187-192

Iqbal MJ, Yaegashi S, Njiti VN, Ahsan R, Cryder KL, Lightfoot DA (2002) Resistance locus pyramids alter transcript abundance in soybean roots inoculated with *Fusarium solani* f.sp. *glycines*. Mol Genet Genom 268: 407-417

Iqbal MJ, Afzal J, Yaegashi S, Ruben E, Triwitayakorn K, Njiti VN, Ahsan R, Wood AJ, Lightfoot DA (2002) A pyramid of loci for partial resistance to *Fusarium solani* f. sp. *glycines* maintains myo-inositol-1-phoshate synthase expression in soybean roots to Theor Appl Genet 105: 1115-1123

Iqbal MJ, Lightfoot DA (2004) Application of DNA markers: Soybean improvement. In: Lörz H, Wenzel G (eds) Molecular Marker Systems in Plant Breeding and Crop Improvement. Springer, New York, USA

Iqbal MJ, Yaegashi S, Ahsan R, Shopinski KL, Lightfoot DA (2005) Root response to *Fusarium solani* f. sp. *glycines*: Temporal accumulation of transcripts in partially resistant and susceptible soybean. Theor Appl Genet 110: 1429-1438

Jamai A, Meksem K, Ishihara H, Bensmail I, Arelli P, Lightfoot DA (2002) *Glycine max* receptor-like kinase RHG1 and RHG4, mRNA, gene and genomic DNA, complete cds. GenBank AF506516-26; AY163905; AF526257-61

Karangula U, Kassem AM, Gupta L, Lightfoot DA (2008). Genome architecture differs in soybean lines with high seed yield and disease resistance. Theor Appl Genet (in review)

Kassem MA, Shultz J, Meksem K, Cho Y, Wood AJ, Iqbal MJ, Lightfoot DA (2006) An updated 'Essex' by 'Forrest' linkage map and first composite interval map of QTL underlying six soybean traits. Theor Appl Genet 113: 1015-1026

Kazi S, Shultz J, Bashir R, Afzal J, Njiti V, Lightfoot DA (2008) Separate loci underlie resistance to soybean sudden death syndrome in 'Hartwig' by 'Flyer'. Theor Appl Genet 116: 967-977

Kazi S, Njiti VN, Doubler TW, Yuan J, Iqbal MJ, Cianzio S, Lightfoot DA (2007) Registration of the Flyer by Hartwig recombinant inbred line mapping population. J Plant Reg 1: 175-178

Landegren U, Nilsson M, Kwok PW (1998) Reading bits of genetic information: methods for single nucleotide polymorphism analysis. Genom Res 8: 769-776

Lander E, Green P, Abrahamson J, Barlow A, Daley M, Lincoln S, Newburg L (1987) MAPMAKER: An interactive computer package for constructing primary genetic linkage maps of experimental and natural populations. Genomics 1: 174-181

Lightfoot DA (2001) Soybean sudden death syndrome resistant soybeans, soybean cyst nematode resistant soybeans and methods of breeding and identifying resistant plants. United States Patent 6,300,541

Lightfoot DA, Njiti VN, Gibson PT, Kassem MA, Iqbal JM, Meksem K (2005) Registration of Essex x Forrest recombinant inbred line (RIL) mapping population. Crop Sci 45(4): 1678-1681

Mansur LM, Orf JH, Chase K, Jarvick T, Cregan PB, Lark KG (1996) Genetic mapping of agronomic traits using recombinant inbred lines of soybean. Crop Sci 36: 1327-1336

Marek LF, Mudge J, Darnielle L, Grant D, Hanson N, Paz M, Huihuang Y, Denny R, Larson K, Foster-Hartnett D, Cooper A, Danesh D, Larsen D, Schmidt T, Staggs R, Crow JA, Retzel E, Young ND, Shoemaker RC (2001) Soybean genomic survey: BAC-end sequences near RFLP and SSR markers. Genome 44: 572-581

McCouch SR, Teytelman L, Xu Y, Lobos KB, Clare K, Walton M, Fu B, Maghirang R, Li Z, Xing Y, Zhang Q, Kono I, Yano M, Fjellstrom RJ, DeClerck G, Schneider D, Cartinhour S, Ware D, Stein L (2002) Development and mapping of 2240 new SSR markers for rice (*Oryza sativa* L.). DNA Res 9: 199-207

Mein CA, Barratt BJ, Dunn MG, Siegmund T, Smith AN, Esposito L, Nutland S, Stevens HE, Wilson AJ, Philips MS, Jarvis N, Law S, de Arruda M, Todd JA (2000) Evaluation of single nucleotide polymorphism typing with invader on PCR amplification and its automation. Genom Res 3: 330-343

Meksem K, Doubler TW, Chang SJC, Chancharoenchai K, Suttner R, Cregan P, Rao Arelli P, Gibson PT, Lightfoot DA (1999) Clustering among genes underlying QTL for field resistance to Sudden Death Syndrome and cyst nematode race 3. Theor Appl Genet 99: 1131-1142

Meksem K, Hyten D, Ruben E, Lightfoot DA (2001a) High-throughput genotyping for a polymorphism linked to soybean cyst nematode resistance gene *Rhg*4 by using Taqman probes. Mol Breed 77: 63-71

Meksem K, Ruben E, Hyten D, Triwitayakorn K, Lightfoot DA (2001b) Conversion of AFLP bands into high-throughput DNA markers. Mol Genet Genom 265: 207-214

Morgante M, Olivieri AM (1993) PCR-amplified microsatellites as markers in plant genetics. Plant J 3: 175-182

Njiti V, Gray L, Lightfoot DA (1997) Rate-reducing resistance to *Fusarium solani* f.sp. *phaseoli* [*nee: glycines*] underlies field resistance to soybean sudden-death syndrome (SDS). Crop Sci 37: 1-12

Njiti VN, Doubler TW, Suttner RJ, Gray LE, Gibson PT, Lightfoot DA (1998) Resistance to soybean sudden death syndrome and root colonization by *Fusarium solani* f. sp. *glycines* in near-isogeneic lines. Crop Sci 38: 472-477

Njiti V, Johnson JE, Torto TA, Gray LE, Lightfoot DA (2001) Inoculum rate influences selection for field resistance to soybean sudden death syndrome in the greenhouse. Crop Sci 41: 1726-1731

Njiti VN, Meksem K, Iqbal MJ, Johnson JE, Kassem MA, Zobrist KF, Kilo VY, Lightfoot DA (2002) Common loci underlie field resistance to soybean sudden death syndrome in Forrest, Pyramid, Essex, and Douglas. Theor Appl Genet 104: 294-300

Njiti V, Myers O, Schroeder D, Lightfoot DA (2003) Glyphosate on Roundup Ready soybean: Effects on root infection by *Fusarium solani* f. sp. *glycines* and sudden death syndrome. Agron J 95: 1140-1145

Njiti VN, Lightfoot DA (2006) Genetic analysis infers Dt loci underlie resistance to SDS caused by *Fusarium virguliforme* in indeterminate soybeans. Can J Plant Sci 41: 83-89

Powell W, Morgante M, Andre C, Hanafey M, Vogel J, Tingey S, Rafalski A (1996) The comparison of RFLP, RAPD, AFLP, and SSR (microsatellite) markers for germplasm analysis. Mol Breed 2: 225-238

Prabhu RR, Njiti VN, Bell-Johnson B, Johnson JE, Schmidt ME, Klein JH, Lightfoot DA (1999) Selecting soybean cultivars for dual resistance to soybean cyst nematode and sudden death syndrome using two DNA markers. Crop Sci 39: 982-987

Ruben E, Aziz J, Afzal J, Njiti VN, Triwitayakorn K, Iqbal MJ, Yaegashi S, Arelli P, Town C, Meksem K, Lightfoot DA (2006) Genomic analysis of the 'Peking' *rhg1* locus: candidate genes that underlie soybean resistance to the cyst nematode. Mol Genet Genom 276: 320-330

Rongwen J, Akkaya MS, Bhagwat AA, Lavi U, Cregan PB (1995) The use of microsatellite DNA markers for soybean genotype identification. Theor Appl Genet 90: 43-48

Rota ML, Kantety RV, Yu JK, Sorrells ME (2005) Nonrandom distribution and frequencies of genomic and EST-derived microsatellite markers in rice, wheat, and barley. BMC Genomics 6: 23-32

Shoemaker RC, Specht JE (1995) Integration of the soybean molecular and classical genetic linkage groups. Crop Sci 35: 436-446

Shopinski K, Iqbal J, Afzal J, Shultz J, Jayaraman D, Lightfoot DA (2006) Development of a pooled probe method for locating small gene families in a physical map of soybean using stress related paralogues and a BAC minimum tile path. Plant Meth 2: 20-28.

Shultz JL, Jayaraman D, Shopinski KL, Iqbal MJ, Kazi S, Zobrist K, Bashir R, Yaegashi S, Lavu N, Afzal AJ, Yesudas CR, Kassem MA, Wu C, Zhang HB, Town CD, Meksem K, Lightfoot DA (2006) The Soybean Genome Database (SoyGD): A browser for display of duplicated, polyploid regions and sequence tagged sites on the integrated physical and genetic maps of *Glycine max*. Nucl Acids Res 34: D758-765

Shultz J, Kazi S, Bashir, R, Afzal J, Lightfoot DA (2007). The development of BAC-end sequence-based microsatellite markers and placement in the physical and genetic maps of soybean Theor Appl Genet 114: 1081–1090

Stefaniak TR, Hyten DL, Pantalone VR, Klarer A Pfeiffer TW (2006) Soybean cultivars resulted from more recombination events than unselected lines in the same population. Crop Sci 46: 43-51

Song QJ, Marek LF, Shoemaker RC, Lark KG, Concibido VC, Delannay X, Specht JE, Cregan PB (2004) A new integrated genetic linkage map of the soybean. Theor Appl Genet 109: 122-128

Tanksley SD, Young ND, Paterson AH, Bonierbale MW (1989) RFLP mapping in plant breeding: new tools for an old science. Bio/Technology 7: 257-264

Triwitayakorn K, Njiti VN, Iqbal MJ, Yaegashi S, Town C, Lightfoot DA (2005) Genomic analysis of a region encompassing *QRfs1* and *QRfs2*: Genes that underlie soybean resistance to sudden death syndrome. Genome 48: 125-138

Webb DM, Baltazar BM, Rao-Arelli AP, Schupp J, Keim P, Clayton K, Ferreira AR, Owens T, Beavis WD (1995) QTLs affecting soybean cyst-nematode resistance. Theor Appl Genet 91: 574-581

Webb DM (1996) Soybean cyst nematode resistant soybeans and methods of breeding and identifying resistant plants. United States Patent 5,491,081

Webb DM (2000) Positional cloning of soybean cyst nematode resistance genes. United States Patent 6,162,967

Webb DM (2003) Quantitative trait loci associated with soybean cyst nematode resistance and uses thereof. United States Patent 6,538,175

Witsenboer H, Vogel J, Michelmore RW (1998) Identification, genetic localization and allelic diversity of selectively amplified microsatellite polymorphic loci (SAMPL) in lettuce and wild relatives (*Lactuca* spp.). Genome 40: 923-936

Wrather JA, Anderson TR, Arsyad DM, Gai J, Ploper DL, Portapuglia A, Ram HH, Yorinori JT (1996) Soybean disease loss estimates for the top ten producing countries during. Plant Dis 79: 107-110

Wrather JA, Koenning SR, Anderson TR (2003) Effect of diseases on soybean yields in the United States and Ontario (1999 to 2002). Plant Health Progr (online Doi 10. 1049)

Wu C, Sun S, Nimmakayala P, Santos FA, Springman R, Ding K, Meksem K, Lightfoot DA, Zhang H-B (2004) A BAC and BIBAC-based physical map of the soybean genome. Genom Res 14: 319-326

Xin Z, Velten JP, Oliver MJ, Burke JJ (2003) High-throughput DNA extraction method suitable for PCR. Bio/Techniques 34: 820-826

Yuan, Z, Njiti V, Meksem K, Iqbal MJ, Triwitayakorn K, Kassem MA, Davis GT, Schmidt ME, Lightfoot DA (2002) Identification of yield loci in soybean populations that segregate for disease resistance. Crop Sci 42(1): 271-277

Zhu YL, Song QJ, Hyten DL, van Tassell CP, Matukumalli LK, Grimm DR, Hyatt SM, Fickus EW, Young ND, Cregan PB (2003) Single-nucleotide polymorphisms in soybean. Genetics 163: 1123-1134

6 Molecular Breeding for Abiotic Stress Resistance

Christopher J. Lambrides[1*], Gregory J. Rebetzke[2], Hunter K.C. Laidlaw[1] and Ian D. Godwin[1]

[1]School of Land, Crop and Food Sciences, The University of Queensland, Brisbane 4072, Australia

[2]CSIRO Plant Industry, ACT, Australia

*Corresponding author: chris.lambrides@uq.edu.au

1 INTRODUCTION

Agricultural plant species are grown in environments where production is limited by biotic (pests and diseases) and/or abiotic stresses. Abiotic stress is defined as any non-living variable that adversely affects plant growth. Abiotic stresses have been estimated to result in about 70% of yield reduction caused by all stresses in crop production (Acquaah 2007). There is now evidence that climate change has exacerbated this effect with significant yield reductions in some major crops, e.g. wheat, maize, and barley (Lobell and Field 2007). The major abiotic stresses discussed in this chapter include those caused by drought, salinity, nutrient toxicity/deficiency, and low temperature, although it is recognized that there are many others, e.g, those caused by high temperature, waterlogging, and oxidative stress to name a few. We have concentrated on drought, salinity, nutrient stress and low temperature because these have been mostly targeted for molecular breeding.

Our world is now confronted with a series of factors including (i) climate change, (ii) land degradation, (iii) population growth and poverty, and (iv) scarcity of water resources all of which will impact heavily on global agriculture and highlight the importance of abiotic stress research. These factors discussed briefly below, will collectively bring great pressure on the existing resources we use to produce food and will emphasize the need for a greater investment into the development of plant varieties with abiotic stress resistance.

1.1 Important World Factors – A Case for More Intense and Efficient Agriculture

1.1.1 Climate Change

In February 2007, the Inter-government Panel on Climate Change (IPCC) reported unequivocally that climate change is anthropogenic and that the observed climate trends will continue (IPCC 2007). The IPCC was established by WMO (World Meteorological Organisation and UNEP (United Nations Environment Program) to "assess scientific, technical and socio-economic information relevant for the understanding of climate change, its potential impacts and options for adaptation and mitigation". The implications of climate change for plant production are significant with clear ramifications for researchers studying abiotic stress resistance in plant systems. The report made the following statements about climate change.

Since 1750, greenhouse gases (CO_2, methane and nitrous oxide) have increased markedly and far exceed pre-industrial values as determined from ice-cores representing thousands of years. CO_2 concentrations have risen from 280 ppm to about 379 ppm in 2005 primarily from burning fossil fuels and land-use changes. The annual increase has been about 1.9 ppm over the last decade.

Radiative forcing (W/m^2) is a measure of a factor in terms of its effect on the earth-atmosphere energy balance. Negative forcing tends to cool the earth's surface while positive forcing heats it. There is now 'very high confidence' that the averaged, net, global, anthropogenic effect of greenhouse gases on the earth has been one of warming with a radiative forcing of 1.6 W/m^2 in the period from 1750 to 2005. It should be noted that some man-made aerosols, such as sulphate, organic carbon, black carbon, nitrate, and dust have had a combined cooling effect (global dimming).

Global temperatures are rising with 11 of the last 12 years to 2006 ranking among the 12 warmest since 1850. Average air temperature has risen by 0.76°C since 1850. Higher ocean temperatures have been associated with melting of snow and ice and higher global mean sea levels. It is predicted that global temperatures will rise on average by 0.2°C per decade for the next two decades.

On a continental and regional scale large changes in climate have been observed including widespread changes in precipitation amounts, wind patterns, aspects of extreme weather including droughts, heavy precipitation and heat waves and the intensity of tropical cyclones. More intense droughts have been observed in wider areas since 1970 particularly in the tropics and subtropics. For example, in 2006 parts of Australia suffered their worst droughts on record, a result of a prolonged *El-Niño* weather pattern and Europe experienced their warmest winter on record. Generally cold days, cold nights, and frost have become less frequent while hot days, hot nights, and heat waves have been more frequent.

The likely future impacts of climate change on agriculture have been summarized (FAO 2002). Higher global temperatures will mean higher rates of evapo-transpiration and plants will be exposed to heat stress particularly with higher night-time temperatures. Dryland agricultural systems will be exposed to drier conditions as serious reductions in global precipitation particularly in the middle latitudes are expected. Parts of Central America, South Asia, northern and southern Africa, and Europe may suffer large falls in available water resources. There will be an increase in the frequency of extreme weather events including *El Niño* droughts, floods, tropical cyclones, and hailstorms.

Large changes will occur in the availability of water from rivers and aquifers because of reductions in runoff and groundwater recharge. Substantial reductions are projected for Australia, India, southern Africa, the Near East/North Africa much of Latin America, and parts of Europe. The main decrease in water availability is likely to occur after 2030.

Higher atmospheric concentrations of CO_2 will impact plant production through a 'CO_2 fertilization effect' which may improve plant water use efficiency. By how much the CO_2 fertilization effect reverses the effects of global warming and erratic rainfall is unknown.

1.1.2 Land Degradation

On a world scale, FAO figures suggest salt-affected soils exceed 800 million hectares extending over all the continents and in some semi-arid countries 10-50% of irrigated land is affected, for example, India has lost seven million hectares (Martinez-Beltran and Manzur 2005). World-wide, Umali (1993) estimates a 1-1.5 million hectare increase in saline land per annum. Naturally occurring salt-affected lands, e.g. coastal salt marshes and inland deserts, are not agriculturally significant compared to those affected by secondary salinization brought about by irrigation and deforestation. Of the 275 million hectares of irrigated land 20% are salt-affected (Flowers and Flowers 2005). An increasing problem particularly in Australia is dry-land salinity (Munns et al. 2002) which has the potential to affect 8 million hectares of land. Dryland salinity is brought about by rising water tables after the replacement of deep-rooted perennials with shallow-rooted annual crops.

Saline soils effect plant growth adversely for two reasons (Munns et al. 2006). Firstly, plants have a reduced ability to take up water because the osmotic potential of the water in the soil around the root zone is too high, leading to water deficits and slower plant growth. Secondly, the excessive amounts of salts entering the plant will eventually injure transpiring leaves causing premature leaf senescence and loss of photosynthetic activity that further reduces growth. Clearly, these effects can lead to lower economic yields particular grain yield in field crops (e.g., see Husain et al. 2003).

Plant productivity can be severely affected on acid soils ($pH_{water} < 5.5$) due to aluminium and/or manganese toxicities, calcium and/or molybdenum deficiencies or a lack of root penetration (Sumner 2005). Importantly many developing countries of South-East Asia, sub-Saharan Africa, and Latin America have numerous agricultural regions where soils are predominantly acidic. On a world scale these soils occupy 1.75 billion hectares of arable land and occur in the soil orders: Oxisols, Utisols, Alfisols, Entisol, and Inceptisols (Sumner 2005).

1.1.3 Population Growth and Poverty

While the world growth rate of population peaked in the 1960s at 2.04% per annum, the world population is still predicted to reach 9.4 billion with a growth rate of 0.5% per annum by 2050. Importantly, most of the increase in world population will occur in developing countries. By 2030, world cereal production will need to increase by almost 1 billion tons from the current level of 1.9 billion tons. Fifty percent of the cereal increase will be for stock feed and 42% for human consumption (FAO 2002).

By 2030, crop production in the developing world is projected to increase by 67% over the base year 1997/99. Eighty percent of this increase is likely to come from an increase in intensification of cropping via higher yields, multiple cropping and fewer fallow periods and the remainder from expansion into new arable land. Globally the new arable land is not evenly distributed. Of the 1.8 billion hectare available for expansion, most traces to South America and sub-Saharan Africa yet South Asia, Near East/North Africa have little opportunity for expansion. Much of the new arable land is under forest, in protected areas or occupied by human settlements.

An assessment of food insecurity (FAO 2001) estimated that 17% of the people in developing countries (777 million persons) were malnourished in 1997/99. Consequently, the Millenium Project Taskforce (commissioned by the UN Secretary General and supported by the UN Support group) on hunger aims to halve world hunger by the year 2015 and has called for the implementation of seven recommendations, one of which is to increase agricultural productivity for food-insecure farmers.

Professor Per Pinstrup-Anderson from the Cornell University, New York, says that improving agriculture is the key. "When you put money in the hands of farmers, that money is spent on creating employment and reducing poverty elsewhere....We have found in our research that for every dollar you invest in agricultural research you generate about $6 of additional income among the farmers and about $15 of additional economic growth in the society as a whole. Much of that will help poor people in those countries."

1.1.4 Scarcity of Water Resources

For the developing countries, the increase in crop production in many cases will be associated with an increased dependence on irrigation (FAO 2002). Consequently, water use for agriculture as a percentage of total renewable water resources will increase for many countries although this will vary considerably across the developing world. For example, countries within North East/North Africa use about 50%, 36% in South Asia yet only 1-2% in Latin America and sub-Saharan Africa. Countries using more than 40% of available water are considered to be in a critical situation. By 2030, 20 developing countries will be confronted by water scarcity issues. Within a country also, differences can be significant. For example, North China is facing substantial constraints compared to the abundant water supplies of South China. To compound the situation, there is intense competition in these regions from other sectors.

2 OVERVIEW: MOLECULAR BREEDING FOR ABIOTIC STRESS RESISTANCE

For several decades plant breeders have made efficient use of simply-inherited traits in breeding programs; for example, many resistance genes to biotic stresses caused by plant pathogens are controlled by single dominant genes. However, many traits of interest to plant breeders including abiotic stress resistance have quantitative inheritance. Quantitative traits are those that show a continuous distribution in segregating populations and are usually controlled by many genes, each with a small effect (Falconer and Mckay 1996) although for some quantitative traits the inheritance may involve only a few, environmentally sensitive genes. Modern genetic linkage mapping techniques have given geneticists an excellent means for dissecting complex traits. Genetic linkage mapping has now reached a level of sophistication where many of the genes underlying a quantitative trait can be located on genetic maps: these locations are referred to as quantitative trait loci (QTL). Therefore, the QTL approach enables the genetic dissection of a complex trait into a series of resolvable components which individually have simpler inheritance and can be used in breeding. The QTL approach is one of many that can be used when developing abiotic stress resistant varieties. Ultimately it will be a multidisciplinary approach that includes genetics, breeding, statistics, physiology, biochemistry, molecular biology, simulation modeling, and the -omics (i.e. genomics, transcriptomics, proteomics, metabolomics) (Borrell et al. 2006). While this chapter will focus on the QTL approach it does not attempt to cover all QTL studies on abiotic stress resistance; the shear volume of literature in this area prevents us from doing so.

Trait-based research programs have played a major role in the study of abiotic stresses. To illustrate the complexities involved in QTL studies, we have included detailed physiological discussions on the target traits that have been associated with improved abiotic stress resistance in field conditions.

2.1 The Need for QTL Studies

The rate of adoption of transgenic crops has increased dramatically in the past decade; however, to date this technology is limited to only a few species. For these species, the combination of marker-assisted selection (MAS) and transgene technology offer researchers the best chance of overcoming the most problematic abiotic stresses. However, for those species where transformation protocols are difficult or do not exist, the genetic dissection of traits employing QTL analysis and the introgression of these loci into elite germplasm using MAS will be of great benefit.

As we pass through the '-omics' phase, the need for QTL studies will not diminish. Genetic solutions to abiotic stress are never likely to be simple and here the role of MAS will be important, particularly if there is a requirement to pyramid traits into single genotypes. The key for researchers, now and in the future, will be to identify those physiological traits that confer a yield improvement in stress environments without penalty in more favorable environments. Once these traits are identified, it will be necessary to understand the range of genetic variation that may exist for a trait within a particular species, the inheritance and the penetrance of the trait in a range of genetic backgrounds and environments all of which can be better studied with a QTL approach. In addition, the QTL approach can sometimes be used to decipher the underlying nature of genetic correlations between important traits. Genetic correlations can result from a number of mechanisms including genetic linkage, pleiotropy or gametic phase linkage disequilibrium (LD).

Target traits will need to be incorporated into plant breeding programs using MAS strategies and these will be enhanced by the use of 'candidate genes', an approach that is likely to contribute significantly to future mapping studies as sequence information of additional plant genomes becomes available. Despite the clear role for the QTL approach and MAS strategies these should be underpinned by sound plant breeding and physiological principles.

2.2 Genetic Considerations While Identifying QTLs for Abiotic Stress Resistance

2.2.1 Background

Abiotic stresses are among the greatest challenges confronting plant breeders in improving adaptation. Yields and quality are reduced to the grower both directly in terms of reduced profitability, and indirectly, in the use of varieties with reduced potential in overcoming these constraints. Selection varies between that of broad adaptation encompassing both favorable and unfavorable environments to specific adaptation where separate breeding programs target adaptation to the recognized constraints. For both, genetic gain is slowed, particularly when improvement in genotypic response is addressing adaptation to both stress and non-stress conditions (Rosielle and Hamblin 1981).

A case in point is the slowing of genetic progress for grain yield in wheat. Seasonal differences in the temporal distribution and severity of water stress contributes to large genotype × environment interaction for grain yield (Cooper et al. 1997). Large within-environment (spatial) variability is also of concern where plants are experiencing water stress. The presence of this interaction and spatial variability lower heritability and reduce the effectiveness of empirical selection and genetic gain for yield (Calhoun et al. 1994). As described earlier, an alternative to selection for yield directly is the use of a biological framework where yield reflects the culmination of components of yield, which in themselves, are heritable and amenable to selection in early generations. Sometimes these components may be under simple genetic control (e.g., aluminium tolerance to improve water uptake; Delhaize et al. 2004), or polygenic control (e.g., stem carbohydrates for grain-filling; Rebetzke et al. 2007a).

2.2.2 *Heritability*

Breeding traditionally employs direct selection for genetic gain of target traits. However, the success of direct selection is contingent on populations containing adequate additive genetic variance and high narrow-sense heritability. The trait should also be simple and inexpensive to measure particularly in mass selection where many families may be evaluated. For a trait measured on a target population in a replicated study evaluated across multiple environments, heritability (H) can be calculated in its simplest form on a single-plot basis as:

$$H_{\text{Single-plot}} = \sigma^2_G/\sigma^2_P$$
$$= \sigma^2_G/(\sigma^2_G + \sigma^2_{G.E} + \sigma^2_{Residual})$$

......or on an entry-mean basis as:

$$H_{\text{Entry-mean}} = \sigma^2_G/(\sigma^2_P/re)$$
$$= \sigma^2_G/(\sigma^2_G + \sigma^2_{G.E}/e + \sigma^2_{Residual}/re)$$

where σ^2_G, σ^2_P, $\sigma^2_{G.E}$, and $\sigma^2_{Residual}$ are estimates of the genotypic, phenotypic, genotype × environment, and residual variances, respectively, and e and r are the number of environments and replications per environment, respectively (Nyquist 1991).

Failure to account for large genotype × environment interaction and residual (or 'error') variation will reduce heritability and reduce the correlation of phenotype with genotype. Replicating entries across environments and replicating studies within environments, reduces the contribution of non-genetic effects to the phenotypic variance to increase heritability. This will be the case even when heritability on a single-plot basis is small and the detection of differences between entries is difficult. Importantly, entry-mean heritability (or line-mean heritability when assessing related lines from a cross) increases our confidence in detecting among-genotype differences to increase the correlation of phenotype with genotype.

2.2.3 *QTL Mapping*

The issue of heritability is as relevant to QTL mapping of traits related to improved productivity under abiotic stress as it is to phenotypic selection in conventional plant breeding programs. Simply, QTL mapping relies on a strong correlation of phenotype to genotype. Data quality will impact on successful marker detection particularly where the underlying genetic basis of target traits is polygenic.

2.2.3.1 Detection of QTL

Heritability gives an indication of repeatability for a trait and confidence in subsequent QTL mapping. The stronger the correlation of phenotype with genotype for a trait, the greater is the confidence in detecting genomic regions and their estimated effects for this trait (Schon et al. 2004). Furthermore, some QTL mapping software (e.g., PLABQTL and QTLNetwork) allow one to model the underlying genotype × environment interaction structure to increase heritability and assess repeatability of QTL effects across environments (i.e. QTL × environment interaction). These interactions are sometimes interpretable as physiological or response to a pathogen, and so contribute to our understanding of adaptation.

As an example of the complicating effects of genotype × environment interaction, we describe the mapping of leaf conductance in wheat. Stomata on the leaf surface open and close to affect transpiration. Slow transpiration results in lower water loss to increase transpiration efficiency in wheat. Genotypic differences exist for leaf conductance but these differences are difficult to detect owing to large genotype × environment interaction and time-of-day effects to reduce repeatability within an environment (Rebetzke et al. 2003). Phenotyping of leaf conductance was undertaken on 161 wheat doubled haploid lines in a Cranbrook/Halberd population. A total of 24 environments were sampled for leaf conductance and QTL analysis undertaken. Heritability varied from a low 12% on a single-plot basis to 87% on a line-mean basis. Analysis of individual environments detected, at most, one QTL coincident with variation at the *Rht-B1* dwarfing gene locus. Extension to all 24 environments in a one-step, multi-environment analysis identified 10 QTLs including *Rht-B1* on 4DS (Table 1). All the QTLs interacted significantly with environment, and this interaction appeared to be associated with differences in soil water in each environment.

Of the QTLs identified for leaf conductance, one was putatively associated with genotypic variation in plant height and three with genotypic differences in time to flowering (Table 1). The recognition of leaf conductance QTLs co-locating with alleles for plant height and flowering highlights an additional benefit of precise QTL analysis. That is, existence of any genetic correlation between characterized traits may be dissected further through coincident QTLs. Here high leaf conductance appears somewhat related to genotypic reductions in plant

Table 1 Quantitative trait loci for leaf conductance in the wheat doubled-haploid population Cranbrook × Halberd (n = 161 lines) evaluated near flowering across 24 environments in southern Australia. High additive genetic effects reflect alleles at QTL for greater leaf conductance.

Chromosome	*Nearest marker*	*Marker interval (cM)*	*Additive effect (sec)*	*% Additive variance*	*Add × Env interaction*
1BS	*Glu-B3*	0 – 9	0.05*	2	**
2BS	*XwPT-3983*	37 – 51	0.17**,a	3	**
2BL	*Xgwm388*	97 – 102	0.10**	4	**
2DS	*Xwmc112*	1 – 10	-0.09**,a	2	**
2DL	*Xs551tgag*	123 – 134	-0.10**,a	3	**
4AS	*Xpsr115*	76 – 97	0.07*	2	**
4BS	*Rht-B1*	46 – 54	0.14**,b	5	**
5AL	*Xpsr426*	106 – 130	0.19**	4	**
7AL	*Xabg652*	150 – 161	-0.02ns	2	**
7BS	*Xcdo949*	62 – 74	0.06*	3	**

[a]co-location of conductance QTL with QTL for altered flowering date
[b]co-location of conductance QTL with QTL for altered plant height

height (r_g = –0.31**) consistent with relatively greater leaf conductance previously reported for semidwarf wheat near-isogenic lines (LeCain et al. 1989). Similarly, progeny containing the earlier-flowering alleles from Cranbrook (chromosome 2D) produced smaller sink (fewer grains) to reduce leaf conductance.

In addition to providing evidence to the strength of the correlation of phenotype with genotype, heritability also defines the maximum proportion of total phenotypic variance that can be accounted for by QTL. If heritability is 75% then the sum of effects for all QTLs cannot account for more than 75% of the phenotypic variance. However, occasionally QTLs together may account for more of the phenotypic variance than predicted from heritability. Many of the procedures used in QTL-mapping are regression-based (e.g., interval mapping), hence colinearity where QTLs are not statistically independent may contribute to over-fitting. This is likely to be of greater concern in small populations where estimated effects at QTLs will be biased upward (Beavis 1998; Holland 2004; Vales et al. 2005).

In some cases, adaptation to abiotic stress is under simple genetic control (e.g., increased water use through tolerance to low soil pH: Delhaize et al. 2004) and so MAS is justified. However, the situation of many QTLs with small individual effects as demonstrated above for leaf conductance has been commonplace in our experience. Many of the traits associated with adaptation to abiotic stress are polygenic and of small genetic effect (e.g., carbon isotope discrimination: Teulat et al. 2002; and stem carbohydrates: Rebetzke et al. 2007a) and so are difficult for utilization in MAS. Furthermore, high heritability for some of these traits makes phenotypic selection a worthwhile consideration when screening large populations. In such cases, Holland (2004) justifiably questions the benefit of MAS for polygenic traits in forward selection programs. Markers must be tightly linked to alleles under selection, and must be informative to other potential crosses.

Markers may be useful in selecting QTL of relatively large genetic effect (e.g., 10-20% phenotypic variance) in tandem with QTL for expensive or difficult to measure traits (e.g., tolerance to soil pH and salinity, quality, disease, etc.). However, such opportunity cost still suffers with the issue of handling too many markers in the smaller population sizes common to early generations of a breeding program. For example, 3,100,000 F_2 individuals need to be assayed to be 95% confident of recovering a single F_2 line homozygous at 10-unlinked loci. Selection of carriers (either heterozygotes and/or homozygotes) at each locus would require a much smaller population size (minimum of around 50 F_2 individuals at $P = 0.05$) necessary to recover all 10 markers (Wang et al. 2007). This genetic enrichment strategy should reduce the critical population size thereby enabling larger populations to be carried through for inbreeding and selection of genes at other important loci that might otherwise be lost through genetic drift.

2.2.3.2 Reliability of QTL Estimates and Locations

Efforts toward the use of molecular markers in MAS and map-based cloning are reliant on good estimates of the location and size of QTL. A number of factors can limit the quality of locating QTL for genes associated with adaptation to abiotic factors. Firstly, gene order on molecular maps may be problematic owing to small population sizes and the correspondingly high standard errors associated with recombination frequency estimates (Mather 1951). Consensus maps only guide marker ordering as the consensus maps are themselves usually summaries of gene ordering in other small populations. Secondly, inadequate phenotyping in the presence of large genotype × environment interaction and/or within-environment sampling variation contribute to lowering repeatability and heritability as indicated earlier. Reduction in heritability risks misclassification and misidentification of QTL and regions, while confidence intervals may be so large to reduce confidence in markers targeted in selection.

Time to flowering and plant height are both important traits in conferring adaptation to drought in wheat. Both are relatively simple traits to screen on large populations and are associated with variation at both major and minor loci. Figure 1 shows the locations of single major QTLs on chromosome 2DS (*Ppd-D1*) and 4BS (*Rht-B1*) for time to flowering and plant height, respectively, in a Cranbrook/Halberd population. Despite both traits being simple to measure, heritability on a single-plot basis ranged from 20 ± 6% to 54 ± 5% for time to flowering and plant height, respectively. Of the 11 environments sampled for both traits, composite interval mapping identified significant ($P < 0.05$) QTLs

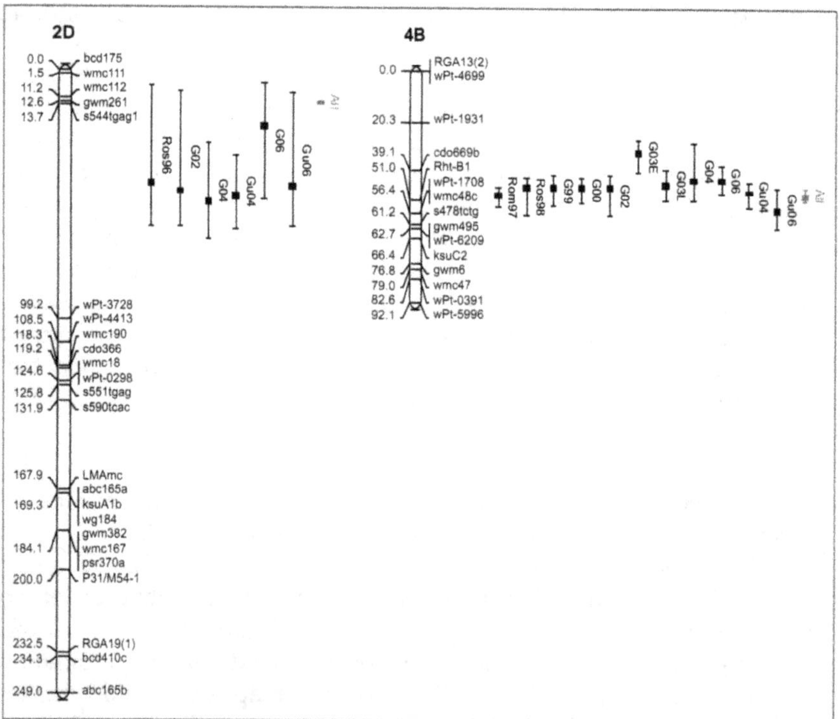

Fig. 1 Location and 95% confidence intervals for time to flowering (chromosome 2D) and plant height (4B) QTL in a wheat mapping population evaluated across 11 environments. Where detected, the environment is provided. The location of QTL (with 95% confidence interval) for the cross-environment analysis is given as 'All'. Both the 2DS and 4BS regions contain known major genes for time to flowering (*Ppd-D1*) and plant height (*Rht-B1*), respectively.

for time to flowering in only 6 of the 11 environments (Fig. 1). Cross-environment QTL analysis increased heritability for time to flowering to 73 ± 3% to locate a very large and significant ($P < 0.01$) QTL adjacent to *Xgwm261*. Previous studies have identified the day-length-insensitive flowering locus *Ppd-D1* to be located close to *Xgwm261* (Worland et al. 1998). The 95% confidence interval for the 2DS flowering QTL was very small (12.2 to 13.6 cM) contrasting with the large confidence intervals for the single environment analysis.

The QTL analysis for plant height was successful in locating a significant ($P < 0.05$) plant height QTL on chromosome 4BS in each environment (Fig. 1). These QTLs were often located adjacent to the SNP marker for *Rht-B1* developed from the dwarfing gene sequence itself (Ellis et al. 2002). The identification of height QTLs in each environment presumably reflected the higher heritability for this trait in comparison to time to flowering. Nonetheless, extension to multi-environment analysis of plant height increased line-mean heritability to 94 ± 2%, and co-located the plant height QTL proximal to the *Rht-B1* SNP marker (Fig. 1). As for time to flowering, the 95% confidence interval from the multi-environment analysis was considerably smaller than for the single environment analyses.

There is benefit in replicating across- and within-environments to increase heritability. However, replication of environments should be considered in a way that environments are meaningful to sampling in the target population of environments. That is, environments should be relevant. However, there are some situations where genotypic performance for target traits is correlated across both stress and non-stress environments. Genetic variance can sometimes be greater under favourable conditions for traits associated with improved adaptation to abiotic stress. For example, stomatal conductance and CID show greater genetic variance under irrigated conditions despite their value in altering water use efficiency (Rebetzke et al. 2003; Condon et al. 2004). Similarly, line performance for stem carbohydrate concentration is strongly correlated ($r_p = 0.71^{**}$) across droughted and irrigated conditions (Rebetzke et al. 2007a). The benefit in evaluation under favourable conditions lies in the improved repeatability when constraints are removed. Clearly, this will not always be the case as some traits (e.g., aluminium and salt tolerance) rely on conditions appropriate to eliciting differential genotypic responses. In our experience, a minimum of two replicates are needed when assessing genotypes for QTL analysis. Augmented, single and partially-replicated experimental designs are less expensive when testing large numbers of genotypes. However, it is difficult to confirm unusual genotypic responses when only a single replicate is present. Further, any unusual response is confounded with the potential for genotype × environment interaction. Additional replication also contributes to improved detection and removal of any spatial correlation that may exist.

2.2.4 *Linked Markers Should Provide Correlated Genetic Response*

Linked molecular markers provide a surrogate for alleles controlling traits that are costly or difficult to measure performance under abiotic stress (e.g., stomatal conductance). Considerations critical to the successful implementation of markers in a breeding program are: (1) strong association between marker and the target gene; and (2) informativeness of both the marker and target genes across different populations (Holland 2004). Molecular markers are not always diagnostic although in some cases (e.g., for *Rht-B1* and *Rht-D1* dwarfing genes in wheat: Ellis et al. 2002) SNP markers have been developed using part of the gene or promoter sequence. The issue of linkage and potential for recombination must not be ignored, and in some cases, markers previously considered as diagnostic for use in tolerance to abiotic stress (e.g., *Xgwm261* for *Rht8* in wheat: Worland et al. 1998) have been demonstrated to be informative in only some populations. Ellis et al. (2007) described how the *Xgwm261* 192 bp allele is commonly ineffective in selection for the *Rht8* dwarfing gene despite the 192 bp sequence being conserved. In contrast, the linked marker for tolerance to soybean cyst nematode in soybean is a large distance from the gene (*ca.* 5 cM) yet is effective in MAS. This is because all SCN genes are derived from outside the elite US germplasm pool and founder effects maintain linkage disequilibrium (LD) in US populations (Holland 2004). Selection within this gene pool should maintain LD to facilitate implementation of markers for selection in forward crossing.

2.2.5 *Design of QTL Experiments*

Properly designed and managed experiments that allow for precise phenotypes to be measured are crucial. For controlled environment (including glasshouses) and field experiments the use of designs that will allow for spatial variation (Cullis and Gleeson 1989) to be accounted for so that experimental error can be minimized are strongly recommended. At the very least row-column and incomplete block (e.g., α-lattice) or augmented designs to name a few, should be considered. Lowering the experimental error (including environmental variance) will improve estimates of heritability/repeatability and provide more precise phenotypic information. In Fig. 2 we provide a breeding flow-chart that details our recommended selection strategies. For successful QTL studies trait heritability is only one component. Other factors that influence the detection of QTL (see Chapter 6 in Volume 1) include, population size, and marker-informativeness, -density, and -distribution.

3 QTL FOR ABIOTIC STRESS RESISTANCE

3.1 Drought Resistance

Developing drought resistant plant varieties remains one of the most difficult breeding objectives in many plant improvement programs. In dryland environments, dry periods can occur during many, if not all, phases of plant growth and the severity of dryness can vary considerably. The problem then becomes which phase of growth and which traits should bear the focus of a breeding program? This problem has, in part, been resolved by the use of rain-free managed environments where dry periods can be imposed at any phase of growth (Campos et al. 2004).

Plant breeders have been highly successful in matching plant varieties to specific environments by manipulating phenology, effectively providing a mechanism of drought avoidance. Empirical selection for yield in dry environments has been successful, however, breeding progress is somewhat slow because of the 'yo-yo' effect associated with selection in different years where year-year variation can be large for timing and intensity of stress.

Wild-species have evolved many mechanisms for tolerating dry conditions where survival is imperative. However, for plant species of economic importance these mechanisms will only be useful, if they fit two important criteria; higher economic yields when challenged by drought and no yield reduction when conditions are favorable. To date, very few traits that fit these criteria have been identified. In this section, we will discuss three extensively studied traits that fit both these criteria i.e. anthesis silking interval (ASI), transpiration efficiency (TE), and stay green. The section will end with a brief discussion on root characteristics, an attractive yet unsuccessful area of drought resistance research.

3.1.1 Anthesis Silking Interval – Maize

On a world scale and in the developing world, maize is the third most important crop. The world's largest producer of maize, the USA is facing unprecedented demand driven by its thirst for biofuels. Maize production in the developing world is forecast to almost double by 2020.

The majority of the maize crop is rainfed and the extra demand may push maize to marginal lands with less favorable conditions of water and nutrients.

Generally maize is considered a drought-sensitive crop despite being a C_4 species (Ribaut and Ragot 2007). Under dry conditions and during anthesis, maize tassels and silks flower asynchronously causing the kernel set to fail. The time interval between tasselling and silking referred to as the anthesis silking interval (ASI) is correlated to kernel set and final yield (for review see Edmeades et al. 2000). Those maize genotypes that are able to keep ASI as short as possible in dry environments outperform those with a long ASI.

Using a recurrent selection scheme CIMMYT has made a concerted effort to develop drought tolerant, tropical maize (see review by Bruce et al. 2002). After several cycles of selection the performance under drought improved by 126 kg/ha/cycle associated with reduced barrenness (more ears/plant), slightly earlier anthesis date, reduced root growth in the top 50 cm and a reduction in ASI. ASI was also an indicator of ear growth rate under stress.

Long term studies of Pioneer Hi-Breed's maize breeding program (see overview by Campos et al. 2004) have shown that conventional breeding has improved stress tolerance, in part, due to the association with a comprehensive testing program that included extensive multi-environment evaluations. The ERA hybrid set consisting of three open-pollinated varieties and 50 Pioneer hybrids released from the 1930s to the present were studied to observe the variation in primary (e.g., yield) and secondary traits (Duvick 1997, 2004). Campos et al. (2004) evaluated in a rain-free environment, a subset of 18 ERA hybrids representing different eras and studied performance when water was withheld during different phases of growth. The greatest yield reduction was associated with dry conditions during the 25 days after flowering. In terms of yield advances across the 18 ERA hybrids in dry conditions there was 146 kg/ha/year improvement in these when the stress was applied during flowering. Improvements to performance under dry conditions at flowering were associated with a significant reduction in barrenness and in ASI. Grain yield and ASI were closely correlated (r = 0.88**) when compared across treatments. These authors suggest that a more directed approach to future breeding efforts for drought tolerance should involve the use of rain free 'managed stress environments' (MSE). They also caution that when using MSE there is a need to ensure that hybrid performance in such environments has high genetic correlation with performance in the target population of environments (i.e. those environments where commercial crops are grown).

3.1.1.1 Anthesis Silking Interval QTL

The importance of ASI to the development of drought tolerant maize has spawned many QTL mapping studies. Ribaut et al. (1996) evaluated 234 $F_{2:3}$ families under several field water regimes to map ASI. The maximum genetic variance for ASI was observed in the driest environment with broad sense heritability estimated to be 0.78. Six QTLs for ASI under drought were located to chromosomes 1, 2, 5, 6, 8, and 10 collectively explaining 47% of the phenotypic variance. Transgressive segregation was observed and the drought sensitive parent contributed alleles at two loci for reduced ASI. The alleles contributed by the drought resistant parent were stable over years and stress levels in this study and others (Ribaut et al. 1997).

Vargas et al. (2006) measured yield and ASI in a study of QTL x environment interactions (QEI) and genotype x environment interactions (GEI) for $F_{2:3}$ families of maize grown in eight environments differing in drought stress and N status. Using factorial regression models and partial least squares estimation procedures, the QEI and GEI were modelled where co-variables (contrasts) based on the levels of genotypic and/or environmental factors are used to describe main effects and interactions. Co-variables for genotypes and environments were based on marker information and environmental variables such as temperature and precipitation, respectively. These methods identified six ASI QTL main effects on chromosomes 1, 2, 6, 8, and 10, and showed a close alignment to the locations (other than for chromosome 5) reported earlier by Ribaut et al. (1996). Each locus accounted for 3.45 to 6.37% of the phenotypic variation for ASI and all loci had an additive effect except for a second locus on chromosome 1 which expressed dominance. The sign of the additive component of each of the QTL was the same across the eight environments suggesting a clear indication of the stability of ASI. QEIs were observed for the locus on chromosome 8 and this was explained by the environmental variable precipitation during flowering.

A study of the genetic linkage between source and sink strengths in maize involved experiments where the responses of leaf growth and ASI to water deficit were measured (Welcker et al. 2007). Leaf growth is a determinant of light interception and ASI is a determinant of yield particularly in dry environments. Consequently, the experiment was designed to test if the genes controlling leaf growth and ASI were linked. A recombinant inbred (RI) population grown in controlled environments and in the field was used to measure ASI and maximum leaf elongation rate (MLER) per unit thermal time in response to evaporative demand and soil water status. In well watered conditions MLER was found to be repeatable and was accounted for by five QTLs, three of which co-localized with ASI QTLs. The responses of MLER to evaporative demand and soil water status were linear allowing each RI lines to be characterized by the slopes of these relationships. Under dry conditions, these slopes had three QTLs that co-localized with ASI QTLs. In all cases the allele for leaf growth rate maintenance was associated with shorter ASI. Some ASI QTLs had no association with MLER. The authors concluded that some of the genetic determinants of MLER and ASI under well watered and dry conditions may be the same.

ASI is now a target trait in marker-assisted backcross (MABC) programs that aim to improve the grain yield of tropical maize in dry environments (Ribaut and Ragot 2007). It was possible to introgress ASI into drought susceptible genotypes using a MABC strategy where 85% of the recurrent parent genotype was recovered in four generations by selecting within large segregating populations (2,200 individuals). MABC-derived BC_2F_3 families were then crossed to two testers and evaluated in a range of dry environments. Their yield was compared to the yield of control hybrids made by crossing the recurrent parent with the same two testers. Under a severe water stress treatment the yield of the top five MABC-derived hybrids was at least 50% greater than the control hybrids.

3.1.1.2 Summary ASI Inheritance and QTL

ASI is a highly heritable, quantitative trait and up to six QTLs have been identified with many of these repeatable over years, locations and genetic backgrounds. These QTLs are largely independent and collectively are able to account for large proportions of the phenotypic variation in segregating populations. Gene action is predominantly additive and in some pedigrees, drought sensitive genotypes contribute positive alleles (i.e. alleles that reduce ASI) explaining the observation of transgressive segregation. MAS for ASI has been demonstrated and is likely to contribute to the development of drought resistance in future maize breeding programs.

3.1.2 Transpiration Efficiency

Plants incorporate isotopes of carbon into their tissue at different rates because of discrimination against ^{13}C relative to ^{12}C. Carbon isotope discrimination or delta ^{13}C (Δ) measures the ratio $^{13}C/^{12}C$ in plant material relative to the $^{13}C/^{12}C$ ratio in the atmosphere that the plants were growing in. For example, if 2 genotypes have a 1 per mil difference in carbon isotope discrimination this corresponds to 1.0×10^{-3} times the difference in the $^{13}C/^{12}C$ ratio in the source atmospheric CO_2.

Experimentally, Δ is negatively correlated with TE in many C_3 species (see review, Hall et al. 1995) and so these discrimination techniques have been proposed as a potential tool for selecting genotypes with improved performance in dry environments (Farquhar and Richards 1984; Hubick et al. 1988; Ehleringer et al. 1991; Condon and Richards 1992; Rebetzke et al. 2002; Lambrides et al. 2004). Measures of Δ provide a time integrated measure of TE. Two factors are potentially associated with improved TE in crop plants, improved photosynthetic capacity (the amount of photosynthetic machinery per unit leaf area) and/or differences in stomatal conductance (the regulation of water and CO_2 through stomata) (Farquhar et al. 1982). There is a classic trade-off between CO_2 fixation and water losses through transpiration because both these processes occur primarily through stomata. Therefore, the challenge is to select genotypes with high productivity as well has high TE.

Rebetzke et al. (2002) demonstrated that early generation divergent selection for Δ affected grain yield among high and low Δ, $BC_2F_{4:6}$ progeny of the wheat cross Quarrion (low Δ) x Hartog (high Δ). The low Δ lines had a yield advantage compared to the high Δ lines of up to 11% in low rainfall Australian environments. A direct result of this research was the commercial release of the world's first high TE bread wheat cultivars, Drysdale and Rees.

The influence of selection for grain yield on agronomic characteristics was examined with retrospective selection of the highest-yielding eight lines (ca. 25% selection intensity) in both low and high Δ groups (Rebetzke et al. 2002). Selection for increased grain yield was associated with significantly higher yield than for the recurrent parent Hartog (Fig. 2). Plant height and anthesis date were not different from Hartog while selection for increased grain yield was associated with significantly smaller Δ in both groups. Increases in grain yield were associated with significantly greater aerial biomass, harvest index, kernel number, and size for selected lines in the low Δ group. High-yielding selections in the high Δ group produced significantly greater harvest index and kernel number than Hartog while aerial biomass and kernel size were not different from Hartog.

The ultimate aim of selecting genotypes with high TE is to improve water use efficiency on a whole crop basis. Since the pioneering work of Farquhar and Richards (1984), many researchers have attempted to select for high TE however, this research is not without pitfalls. Condon et al. (2004) provide an excellent review of how breeding for high TE and water use efficiency should be structured. They make the following

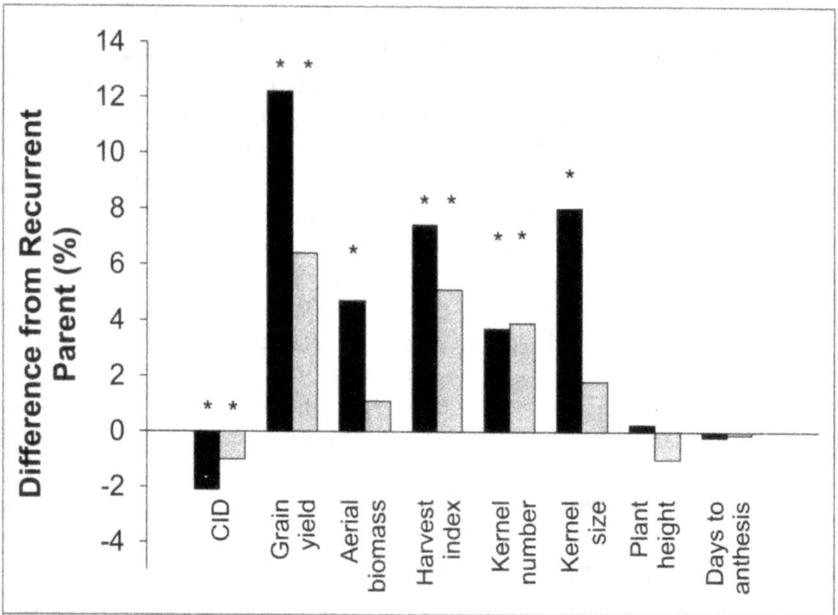

Fig. 2 Differences between recurrent parent Hartog and the means of the highest-yielding 25% of lines from low- (dark bars) and high- (light bars) carbon isotope discrimination (Δ) groups. Differences from Hartog (%) were estimated as: $[(\bar{x}_{\text{Selected group}} - \bar{x}_{\text{Hartog}})/\bar{x}_{\text{Hartog}}]*100$.

*Indicates selected-Δ group and Hartog means are significantly different at $P = 0.05$.

important points (among others) for cereals but these can equally be applied to many other C3 species.

Low Δ cereals are typically associated with more conservative crop growth which ensures that soil available water is used sparingly over the life of the crop. This strategy of development is appropriate where crops are grown on stored moisture and there is little soil evaporation during early growth. However, they conclude that breeding for low Δ (high TE) may not be appropriate for other dry environments. For example, they demonstrate that for the Mediterranean environments of SW Australia selection for high Δ wheat may be more desirable. The shallow, sandy soils of this region have a more favorable water profile early in the season and because high Δ wheat provides greater coverage of the soil surface, the high evaporation rates from the soil surface that occur early in the season can be reduced. This reduction in water loss from the soil surface early in growth ultimately leads to higher yields.

Studies of Δ versus grain yield have been variable because other traits such as flowering date and plant height can confound the interpretation of results particularly when the varieties are unrelated. Segregating populations should be used where it is possible to reduce genetic background affects. Plants should be sampled for Δ early in growth and under well watered conditions. An understanding of what mechanisms (i.e. photosynthetic capacity and/or stomatal conductance) drive variation in Δ for any given species will greatly enhance the ability to design plant varieties for target production environments.

They advise that studies of TE should not be done in isolation of other drought trait studies and where possible the use of crop simulation using long-term weather data should be used to best match Δ and other drought traits to target environments. They provide a good example for wheat grown in a range of Australian dryland environments. The simulations show that selection for improved drought resistance can be enhanced substantially by combining TE with early vigor (defined as rapid leaf area development in the early stages of growth).

QTL studies of Δ are potentially very important in plant improvement programs that aim to improve productivity in dry environments. To date relatively few studies have used appropriate strategies to optimize the use of this trait. Those researchers embarking on QTL studies of Δ should bear the review of Condon et al. (2004) and its recommendations closely in mind.

3.1.2.1 Δ QTL

We have restricted our focus to post-2000 studies that evaluated traits over multiple environments/or years unless landmark studies appeared before this time. In some studies described below the variable for carbon-isotope-discrimination calculated was $\delta^{13}C$ which is a different but related variable to Δ by the following equation:

$$\Delta = \frac{\delta_a - \delta_p}{1 + \delta_p}$$

where δ_a (source air) and δ_p (plant material) measure the deviation of the isotope composition from the molar abundance ratio of a standard.

In this review we describe all studies in terms of Δ even if not done so in the original manuscripts.

3.1.2.1.1 *Arabidopsis*

The small and sequenced genome of *Arabidopsis thaliana* and the large natural variation in Δ among ecotypes (McKay et al. 2003) have made this species attractive for studying TE. A study of phenotypic plasticity

(Hausmann et al. 2005), mapped six QTLs for Δ in two RI populations, Landsberg *erecta* (L*er*) x Columbia (Col) and Cape Verde Islands (Cvi) x L*er* grown under two watering regimes, i.e. water limited and well-watered. Plant material for Δ was sampled from each RI line by pooling leaf material across replications from within each water treatment. By pooling across replications they were unable to estimate heritability of delta in these materials. Two Δ QTLs from the L*er* x Col population and four Δ QTLs from the Cvi x L*er* population were located. In most cases these QTLs were identified using Δ data from either water regime however a greater proportion of the phenotypic variance was genetic when the data from the well-watered treatment was used. QTLs appeared on chromosomes II, III, and V and two of the Δ QTLs were also associated with shoot biomass. The QTLs detected on chromosome II were 12-22 cM away from the *Erecta* locus. One example of epistasis was detected between loci identified on chromosomes III and V.

In a sister study Juenger et al. (2005) used reciprocal crosses of Cvi x L*er* population grown under well-watered conditions to map flowering time, Δ, TE, and stomatal conductance. There was considerable transgressive variation for both Δ and flowering time but no cytoplasmic effects were detected. The range for Δ in the population was 4.65 per mil despite the small difference between the parental lines. There was also a strong negative correlation between Δ and flowering time. Five Δ QTLs were detected on chromosomes I, II, III, and V and alleles from Cvi and L*er* had both positive and negative effects. Two flowering QTLs mapped closely to two Δ QTLs suggesting the possibility of physical linkage or pleiotropy. The authors suggest that this may be a genetic trade-off between avoidance of and escape from drought. An epistatic interaction between loci on chromosomes I and V accounted for 2.5% of the genotypic variance for Δ. Further experiments were conducted using two near isogenic lines (NIL) designated NIL Δ 2.1 and NIL Δ 3.1 because they were derived by backcrossing the Δ QTLs at positions 2.1 and 3.1 into the L*er* nuclear and cytoplasmic background. There were no differences in the assimilation rate of these stocks, however, both NILs had lower TE, higher Δ, and higher stomatal conductance than the recurrent parent L*er* suggesting that Δ 2.1 and Δ 3.1 affect Δ through changes in stomatal conductance. The NILs had about 20% lower TE compared to L*er*.

Masle et al. (2005) mapped Δ in the RI population Col-4 x L*er*. One significant difference between this and the previously described studies in *Arabidopsis* is that L*er* was used as a high Δ parent. A single Δ QTL (TE1) that mapped at 50.64 cM on chromosome II explained between 21-64% of the variation depending on the indoor environmental conditions the plants were grown in. The most successful environment for phenotyping was a glasshouse with; radiation 150-500 μE m^{-2} s^{-1}, RH 50%, 20/22°C and 12 h day-length. Under these conditions, the additive effect of the Landsberg *er1* allele was to increase Δ by 0.293 per mil. TE1 mapped to a region of the *Arabidopsis* genome that included 37 genes including the *ERECTA* locus (At2g26330).

This observation led to the study of *erecta* mutants of *Arabidopsis* and their isogenic counter parts. The difference in Δ between the mutant and wild type pairs ranged from 0.5-1.1 per mil. The largest difference involved Coler105 mutant which carries a null allele for *ERECTA*. To study the effect of the *ER* locus on Δ and TE a series of transgenic plants were made by complementing the mutant *erecta* lines with the Col-0 *ERECTA* gene (*ER*). These transformation experiments confirmed that the *ER* locus was directly involved in the control of leaf Δ and TE. Further experiments were conducted to determine the mode of action of *ER* on TE. In general mutant ecotypes had much higher stomatal conductance and lower maximum rubisco carboxylation rate a measure of photosynthetic capacity. One exception was the *er1* mutation in Landsberg (i.e. *Ler*) in which there was no reduction in photosynthetic rate despite an increase in Δ.

Morphological characterization of the mutant and wild types suggested that *ER* was involved in leaf organogenesis and affected cell expansion, cell division and cell-cell interaction although some of these effects were dependant on genetic background. Generally, the *er* mutants were characterized by greater stomatal density and smaller epidermal cells. There were no clear differences in leaf blade thickness but the mutant lines typically had fewer, looser packed spongy mesophyll cells and therefore less photosynthetic machinery/unit leaf area. The large effects on the complemented mutants led to the conclusion that *ER* may act as a master gene. A phylogenetic analysis of the *ER* homologues from other plant taxa suggested that the *ER* locus most likely evolved during or before angiosperm evolution. Several complemented mutants had Δ values below the wild type without any detectable growth penalty leading the authors to conclude that *ER* manipulation has potential for the improvement in crop species.

3.1.2.1.2 Tree Crops

Maritime Pine (*Pinus pinaster*)

Brendel et al. (2002) used a diallel mating design and the pseudo-testcross mapping strategy to understand the inheritance and map several traits including Δ of main stem wood cellulose and mean ring width (MRW). Δ of the full-sib F_1 trees showed a continuous distribution and the range was large at 3.2 per mil. Four significant and four suggestive QTLs for Δ were identified, none of which co-located with QTLs for MRW. Collectively the Δ QTLs were able to explain 51.4% of the phenotypic variation for Δ with one QTL explaining 12.4%. Δ QTLs were located on both the female and male maps but not on the same chromosomes. The narrow sense heritability for Δ was moderate but significant at 0.17 ± 0.06. The authors suggest that the low estimate of heritability of Δ may have resulted from the repeated exposure to periods of water stress during the period of the experiment. No dominance effects were associated with Δ in this study. The phenotypic correlation between Δ and MRW based on four years of growth was largely driven by environmental influences.

European Chestnut (*Castanea sativa* Mill.) and Oak (*Quercus robur* L.)

European chestnut and oak are two forest trees species belonging to the *Fagaceae* family. Two diverse natural Turkish ecotypes of European chestnut were intercrossed and a pseudo-testcross mapping strategy was used to map Δ (Casasoli et al. 2004). The western ecotype adapted to a Mediterranean climate had higher stomatal conductance, photosynthetic capacity, and Δ compared to an eastern ecotype selected from a cool humid region. For the segregating population of full-sib trees Δ ranged by around six per mil in each of three years. The authors did not explain why there was such a large and unprecedented range of Δ values measured from leaf tissue harvested at the end of the growing season. While there were large year effects on Δ there was little genotype x environment interaction. Of eight unique QTLs detected for Δ only two were temporally stable across years. Significant negative phenotypic correlations were observed between Δ and plant height/diameter. Not surprisingly, several Δ QTLs were linked to QTLs associated with plant height/diameter.

A comparative QTL mapping study between European chestnut and oak (Casasoli et al. 2006) was designed to understand the genetic basis of adaptive traits. A consensus set of expressed sequence tags (ESTs) was used to map the 12 chromosomes of each species. Five and seven Δ QTL were mapped on the oak and chestnut maps, respectively, but none of the QTL locations were conserved between the two species.

3.1.2.1.3 Field Crops

Wheat

The most comprehensive study of Δ QTLs in any species has been conducted in wheat by Rebetzke et al. (2007b). They studied three populations of doubled haploids derived from the crosses of unrelated lines Cranbrook x Halberd (C/H), Sunco x Tasman (S/T) and CD87 x

Katepwa (CD/K). This material was grown over three years during 2002-2004 in the field under well-watered conditions using augmented designs that included 10 replicates of the parental genotypes and the low Δ check Quarrion. Leaf material for Δ was sampled from 10-20 plants at the peak tillering phase of growth (Z31; Zadoks et al. 1974) or from flag leaves at anthesis (Z60; Zadoks et al. 1974).

There were large environmental effects on Δ and a strong negative correlation was observed between Δ and vapor pressure deficit (VPD) suggesting that wheat plants grown in dry atmospheres on average have lower Δ values. Frequency distributions for Δ were normal or close to it for each population/year combination and transgressive segregation was common. Estimates of narrow sense heritability for Δ were moderate to high (0.54-0.91) and G x E interaction was low when leaf samples were collected at Z31, but heritabilities low (0.30-0.43) and G x E high when leaf samples were sampled later in growth at Z60. The estimated number of genes segregating for Δ in these populations ranged from 6-9 for the C/H and S/T populations and 4-5 for the CD/K population. These estimates were a slight underestimate of the number of genomic regions identified through QTL analysis. The mid-parent Δ was an excellent predictor of the mean progeny Δ in these populations suggesting largely additive genetic control.

QTL analysis was performed using a two stage process. Firstly, composite interval mapping (CIM) was used to identify significant QTLs associated with Δ, secondly, these significant QTLs were used as starting values for multiple interval mapping (MIM) of the QTL Network software (Yang et al. 2005). Nine, twelve, and seven Δ QTLs were detected for the C/H, S/T, and CD/K populations, respectively, many of which were common to two or more populations and repeatable over years. The percent additive genetic variance contributed by each QTL was quite small and ranged between 1-12%. Depending on the year and population, the amount of phenotypic variance accounted for by these QTLs ranged between 30-75%. The authors were not surprised by the large numbers of QTLs detected because Δ is an integrated measure of stomatal conductance and/or photosynthetic capacity both of which are polygenic traits. Interestingly, alleles for low or high Δ were contributed by both parents in each cross. The authors suggest that it should be possible to combine different alleles from the different populations to develop lines with further reduced or increased Δ. Some Δ QTLs were associated with maturity and plant height particularly with the dwarfing gene *Rht-B1b*. Generally, epistatic interactions accounted very little to the overall phenotypic variance for Δ in these populations.

Rice

An RI population made from the cross between an *indica* type Milyang 23 and a *japonica* type Akihikari was used to map Δ, days to heading and leaf nitrogen content over two years (2002, 2003) in a well watered environment (Takai et al. 2006). Leaf samples for Δ were collected at heading stage from the flag leaf and second leaf. Despite the small difference in the parents large transgressive segregation for Δ was observed for both years suggesting complementary gene action. For the 2002 trial five QTLs were detected on chromosomes 2, 6, 7, and 11 that explained 7.1 to 14.3% of the phenotypic variance. In 2003, seven QTLs that explained 5.9 to 10.6% of the phenotypic variation were detected including the same five QTLs detected in 2002. The QTLs identified in this study did not correlate with previous studies although Ishimaru et al. (2001) identified different QTLs on chromosomes 2, 6, and 11.

In a similar but unrelated experiment (Laza et al. 2006) Δ was mapped in a *japonica* (IR69093-41-3-2) x *indica* (IR72) RI population grown over four years 2001-2004 under fully irrigated lowland conditions. Different plant parts including grains, leaf and stem, were sampled for Δ. The variation between RI lines for Δ was least for grain and greatest for leaf samples taken at midgrowth. As there was a large difference between the parental lines for Δ (IR69093-41-3-2 high Δ and IR72 low Δ), little transgressive segregation was observed. The low Δ signature of the *indica* type was consistent with other studies that show *indica* to be generally lower than *japonica* types (Kondo et al. 2004). For grain Δ only one QTL on chromosome 9 was detected in three out of four years, this QTL near RM144 on chromosome 9 accounted for 8.9 to 19.0% of the phenotypic variation. In 2003 and 2004, Δ QTL from either leaf, stem, or shoot sampled at midgrowth or maturity mapped to chromosome 12 near the interval between the closely linked markers G148 and G2140. Interestingly, chromosome 12 had not been associated with Δ in previous mapping studies of rice (Ishimaru et al. 2001; Takai et al. 2006).

Cotton

An interspecific cross of cotton (*Gossypium hirsutum* x *G. barbadense*) was used as a deliberate strategy to exploit the genetic potential of elite germplasm from each species to improve productivity under dry conditions (Saranga et al. 2004). The progeny were evaluated using two watering regimes (wet and dry) over a two year period. The mapping experiment yielded 11 QTLs, 10 of which were from the D subgenome of cotton and favorable alleles were identified from both *hirsutum* (four alleles) and *barbadense* (five alleles). Overdominance was expressed at one locus and underdominance was expressed at another. This study showed that there was not a clear association between productivity characters and Δ. However, the plant material sampled for Δ was collected late in growth (between 30-50% bolling) a strategy we do not recommend. It is preferable to sample plant material early in growth from well watered plants. A refinement of the leaf sampling procedures in cotton may have led to other conclusions. An important finding of this study was the presence of complementary favorable alleles for Δ located on homeologous linkage groups suggesting that each species involved in the interspecific cross could be improved beyond which may have been possible within.

3.1.2.2 Summary on Δ Inheritance and QTL

Plant Δ is clearly a highly heritable, quantitative trait particularly when sampled from plant tissues early in growth. When sampled early in growth, genotype x environment interactions are typically low which makes Δ an attractive trait for manipulation in breeding programs. The correct time to sample plant tissue for Δ has not been resolved for some species, particularly tree species. Generally, studies of Δ in tree species lag behind those conducted in field crops and may reflect the lack of physiological studies of TE in tree species. All studies of delta should include an analysis of the associations of delta with plant height and maturity.

In general, QTL studies reveal a number of independent loci each of small additive effect. Many of these QTLs are stable across years, environments and genetic backgrounds. Transgressive segregation is suggestive of complementary gene action and generally occurs in segregating populations where there is little difference between the parental Δ values. The value of MAS for Δ may be questionable because of the large number of loci involved that account for small amounts of the phenotypic variation. Although one presumes the Δ QTLs detected are those that have the largest effect then these loci may have value in MAS and understanding any deleterious repulsion phase linkages.

The failure to detect greater levels of the phenotypic variation may be related to the Beavis effect, i.e. the use of small populations (< 200) used in the reported QTL experiments. However, phenotyping and genotyping large populations (> 200) can be resource rich and expensive. The trade-off between statistical resolution and resources to carry out QTL experiments will always be a concern to researchers of abiotic stress resistance. Future studies may need to better understand the inheritance of the component traits stomatal conductance and photosynthetic capacity that underlie variation for TE. The role of *Erecta* as the first cloned *TE* gene (Masle et al. 2005) will no doubt help unravel the complex inheritance of Δ.

3.1.3 Stay-green

3.1.3.1 Sorghum

Worldwide, the fifth most important cereal is sorghum, a staple food in parts of Africa and Asia and an important feed crop in many parts of the world including Australia and America. Sorghum is very well adapted to hot environments and possesses a remarkable tolerance to water stress, making it an important model for investigating drought resistance in cereals. Two responses to drought have been recognized in sorghum and are classified as either pre-flowering or post-flowering responses. Different genetic mechanisms have been thought to control the plant response in each of these environments (Rosenow et al. 1983). Sorghum germplasm with differing phenotypic responses to these drought types have been identified and widely used in breeding programs. Some estimates have suggested that of the 80% of the sorghum hybrids grown in the USA under dryland conditions, almost all have pre-flowering drought resistance, but very few have any significant resistance to post-flowering drought (Sanchez et al. 2002).

Pre-flowering drought stress reduces panicle size, grain number and yield, and may have the most significant impact, if present from panicle differentiation to flowering. In contrast, post-flowering drought responses in sorghum occur during the grain development stage and can lead to premature leaf senescence and susceptibility to lodging, charcoal rot and substantial yield losses (Rosenow and Clark 1995). Some sorghum lines are resistant to premature senescence when they encounter a water stress during the post-flowering growth stage. Such genotypes have become known as 'stay-green' types, as the plants retain their chlorophyll and the capacity for photosynthesis during the drought conditions, compared with other 'senescent' types. The stay-green trait has been identified in several sorghum lines (for genotypes see Harris et al. 2007), and other cereals such as maize (Tollenaar and Daynard 1978), oats (Helsel and Frey 1978), wheat (Verma et al. 2004), and rice (Cha et al. 2002).

Five classes of stay-green have been identified (Thomas and Smart 1993; Thomas and Howarth 2000). These include two classes of functional stay-green, which are obviously of great interest to the plant breeders. Plants in two other classes remain green under post-flowering drought conditions, but lack photosynthetic activity and thus exhibit only a 'cosmetic' stay-green phenotype. The final class describes stay-green plants that contain initially high chlorophyll content, resulting in greater leaf greenness due to the slower reduction in chlorophyll content during senescence.

Several contributing components to the stay-green phenotype have been identified. There is considerable continuing research to clarify the physiological basis of stay-green, particularly relating to the rate and onset of leaf senescence, transpiration and transpiration efficiency, leaf longevity and nitrogen dynamics in the sorghum plant. For example, stay-green types have been shown to contain higher levels of basal stem sugars and cytokinins than senescent types. Stay-green hybrids also assimilate more nitrogen and have higher specific leaf nitrogen (Borrell and Hammer 2000). Borrell et al. (2001) succinctly reviewed the nitrogen dynamics of stay-green at the cell, leaf and whole plant levels. However, it remains to be seen whether these traits are directly caused by the stay-green trait or if they are a consequence of it. Nevertheless, several studies have shown the yield benefits of the stay-green genotype under post-flowering drought conditions (Borrell et al. 2000a, b; Jordan et al. 2003).

Several studies have identified QTLs that influence drought tolerance in grain sorghum (Tuinstra et al. 1997; Tao et al. 2000; Kebede et al. 2001; Haussmann et al. 2002). Four major QTLs (*Stg1, Stg2, Stg3,* and *Stg4*) have been identified that are consistent in different genetic backgrounds (Subudhi et al. 2000). The level of phenotypic variation explained by each of these QTLs is approximate and has been difficult to quantify, particularly because of the relatively small population sizes that have been used to map the stay-green trait (Harris et al. 2007). Epistatic interactions have also been identified among these loci and between these loci and other genes (Subudhi et al. 2000), further demonstrating the complexity of stay-green and adding to the difficulties involved in determining the underlying basis of stay-green.

In order to overcome some of these complexities, NILs have been developed to contain one or more of the stay-green loci (Xu et al. 2000). This was achieved by crossing BTx642 (a stay-green type) with RTx7000 (a senescent type), followed by repeated backcrossing of the F_1 plants to the senescent parent. Following MAS for the stay-green loci, RTx7000 NILs were developed that contained either complete or partial stay-green loci derived from BTx642, or combinations of the same (Harris et al. 2007). Analysis of these lines showed that each of the loci contributed to the stay-green phenotype, and interestingly, under post-flowering drought conditions the *Stg2* NILs retained more green leaf area at maturity than the original stay-green parent RTx7000. Other trends relating to delayed onset and rate of leaf senescence were also observed for particular NILs. The lines generated through this research continue to be used for the thorough analysis of stay-green physiology and map-based cloning of the *Stg* genes (Harris et al. 2007).

3.1.3.2 Other Cereals

The genetic basis of stay-green in other cereals has been examined less thoroughly compared to sorghum. Initial work by Cha et al. (2002) identified a single recessive stay-green gene in rice that mapped to chromosome 9. QTLs for leaf senescence have also been identified on chromosomes 6 and 9 by Abdelkhalik et al. (2005). Additional work by Jiang et al. (2004) examined 190 doubled haploid rice lines from a cross between a senescent *indica* and stay-green *japonica* parent. By defining six traits to evaluate the stay-green phenotype, a total of 46 main-effect QTLs were detected which all had small effects. The epistatic effects and the QTL x year interactions were found to account for a large proportion of the variation observed in the population. Most of the stay-green traits were also negatively correlated with yield; however, the field trials were conducted in the absence of post-flowering drought. Ideally, the influence of stay-green loci on the physiological response of rice will be investigated under water-limiting conditions.

Few reports have examined the stay-green trait in wheat, although Verma et al. (2004) used a small population of 48 doubled haploid lines derived from a cross between parents with contrasting flag-leaf senescence to investigate the trait. The parental lines also differed in their photoperiod-sensitivity. The stay-green trait was shown to correlate positively with yield under optimal and drought-stressed conditions. Interestingly, QTLs were found for senescence on chromosomes 2B and 2D under drought-stressed and non-stressed conditions, respectively. These findings further emphasize the complex genetic mechanisms underlying the stay-green trait. Further characterization of the effects of stay-green loci on the phenotype of model plants such as sorghum, and the establishment of clearly associated QTLs and markers with the trait in other cereals, will remove or reduce the dependence on drought conditions during the selection phase in breeding programs.

3.1.4 Roots

This review has deliberately avoided detailed discussions on QTL studies involving root traits, despite the large number of studies in this area. In no way is this a comment on the quality of research undertaken nor do we suggest that this research area discontinue, rather, we were unable to find conclusive studies that showed any advantage in field performance of the root trait(s) being evaluated.

In a detailed review of drought resistance mechanisms in rice Price et al. (2002) indicated that for populations made from crosses between drought tolerant and drought sensitive parents (e.g., Bala x Azucena) there was poor co-incidence of QTLs for drought avoidance and root morphological traits. They questioned the notion that improving the roots of rice will improve drought resistance. Root studies in maize show that high root density in the upper layer of the soil profile may be negatively correlated to yield (Bolaños et al. 1993; Bruce et al. 2002; Campos et al. 2004).

However, some promising research on root characteristics is beginning to emerge. The role of abscisic acid (ABA) has long been known to play a role in the adaptive response of plants to drought. A set of NILs in maize for a leaf ABA (L-ABA) QTL has been field tested in well watered and dry environments (Landi et al. 2005) and in hybrid combinations (Giuliani et al. 2005) which revealed an effect of this QTL on root traits. The hypothesis that the L-ABA QTL was associated with root architecture in maize and in turn controls leaf ABA levels was proposed (Giuliani et al. 2005; Landi et al. 2005). Consequently, the L-ABA QTL was named root-ABA1 (Giuliani et al. 2005).

A more comprehensive evaluation of the root-ABA1 QTL in maize was undertaken to test the effect of this QTL in a range of genetic backgrounds and over a range of different environments (Landi et al. 2007). The root-ABA1 NILs were testcrossed with testers of different origins and the hybrids grown in Italy and China in well watered and dry environments. These materials effectively gave a comparison between hybrids where the root-ABA1 QTL was either present or absent.

Unexpectedly, the hybrids that had positive alleles for the root-ABA1 QTL were lower yielding in both countries for the dry treatment and lower yielding in the well watered treatment in China (Landi et al. 2007) primarily because of fewer ears per plant. Conversely, the hybrids with positive alleles for the root-ABA1 QTL were shorter, had higher levels of L-ABA and were superior for root lodging (Landi et al. 2007). Superior root lodging was thought to be because the root-ABA1 lines had a higher number of roots per node, roots with wider angle or insertion on the stem and greater diameter with more branching in the shallow soil layers with greater dry weight (Guiliani et al. 2005).

An analysis of maize hybrids released over the past 70 years has shown a reduced capacity of modern releases to accumulate ABA when seedlings are screened in hydroponic systems with polyethylene glycol (PEG) induced water deficits (Sanguineti et al. 2006). These results suggest that selecting against the presence of the root-ABA1 QTL may improve the yield of maize in dry environments although there may be undesirable pleiotropic effects such as increased root lodging.

A marker-assisted backcross (MABC) program in rice that aimed to introgress root QTL from a *japonica* variety into an upland rice variety has shown some promise (Steele et al. 2006, 2007). Five chromosomal segments of variety Azucena, four QTLs for root characteristics and one QTL for aroma were introgressed into Kalinga III in three backcrosses and a further intercross between two BC_3 lines (Steele et al. 2006). In all, trait pyramids and a set of NILs were made using eight generations, six years and 3,000 marker assays. One target region on chromosome 9 was shown to significantly increase root length in both irrigated and dry environments (Steele et al. 2006) although this segment was not associated with any yield advantage in replicated field trials conducted over three years in India (Steele et al. 2007). However, some NILs that contained segments of non-target loci did have higher grain yield than Kalinga III but this study was unable to confirm that higher yield was due to improved root systems.

3.2 Salinity

Salinization is the accumulation of water-soluble salts in the upper soil profile (solum) or regolith to the extent that it impacts on agricultural production, environmental health, and economic viability (Rengasamy 2006). Definitions of saline soils are somewhat variable but generally soils are defined as saline if the electrical conductivity (EC) of its saturation extract exceeds a certain threshold usually 4-15 dS/m. (The FAO Land and Water Division, US Salinity Laboratory Staff 1954). Salt-affected soils broadly fall into two categories; saline and sodic (Flowers and Flowers 2005). Saline soils are usually dominated by sodium (Na), chloride (Cl), and sulfate (SO_4) ions, and lower pH and sodium absorption ratios (SAR) but higher EC than sodic soils. Sodic soils are also dominated by sodium ions and high carbonate/bicarbonate, high pH (8.5-10.8) and high SAR.

The development of salt tolerant varieties in the major crops has been limited. Estimates suggest that as of the year 2000, only 25 cultivars in 12 species have been released for their salt tolerance (Flowers et al. 2000). Flowers et al. (2000) argue that for plants to survive saline environments multiple adaptations are required. For example, they list a number of typical halophyte characteristics including a tight coupling of growth and ion accumulation associated with the early development of the endodermis, an ability to compartmentalize ions reaching the leaves which in turn requires adaptation of membranes, and an ability to synthesize a compatible solute.

Salt tolerant cereal varieties have at least one of the following characteristics as listed by Munns et al. (2006): Na^+ exclusion, K^+/Na^+

discrimination (i.e. preferential transfer of K^+ to the shoots), retention of ions (e.g., Na^+) in the leaf sheath, tissue tolerance of high ion concentration, ion partitioning into different aged leaves, osmotic adjustment, TE, early vigor, and early flowering leading to an early growing season. Generally two types of mechanism of salt tolerance exist (Munns et al. 2006); salt exclusion attempts to minimize the amount of salt entering the plant and tissue tolerance attempts to minimize the concentration of salts in the cytoplasm.

One of the major limitations to screening and selecting crop species for salt tolerance has been the lack of suitable assays. The development of a repeatable bioassay to measure salt-sensitivity will greatly facilitate the progress of selecting salt tolerant varieties. For example, Munns and James (2003) conducted a range of experiments in an attempt to develop an assay to screen durum wheats for salt tolerance. Their experience can be broadly applied to other species and a summary of the important findings of this and their other studies are briefly reported here. Assays based on specific physiological traits are most likely to be successful. Short-term experiments (one week) that study biomass or leaf elongation effectively measure the osmotic effect of salt but are not useful for identifying genetic variation. Long-term experiments (3-4 weeks) may be useful because it takes longer for salt-specific effects to show, but these can be resource rich and not ideal for screening large numbers of genotypes. For durum wheats, Na^+ exclusion correlated well with salinity tolerance as did K^+/Na^+ discrimination to a lesser extent. For other durum types tolerance of high Na^+ could be assessed directly by measuring chlorophyll retention and was a measure of compartmentation ability.

Leaf injury scores can be useful but symptoms can arise from a number of factors including; osmotic effects of salt, toxic effects of salt, and nutrient imbalances. Extrapolating scores of leaf injury from glasshouse or controlled environment experiments to the field can be problematic particularly where low Ca^{2+} and high pH may be factors in saline and sodic soils. Often saline soils are drying soils therefore nitrogen can be less available and be a cause of early senescence. Factors promoting leaf longevity, deep water extraction, and TE will enhance crop yields in saline environments. Germination tests in high saline solutions do not correlate with genotypic performance in later stages of growth and are not recommended. Stomatal conductance and assimilation can be reduced during early phases of growth in a saline environment, an effect most likely due to osmotic stress. Therefore screening for high stomatal conductance may be an effective way of screening for salt tolerance. Yeo and Flowers (1986) make the important point that no single trait is likely to give a plant variety complete salt tolerance and emphasize the need to pyramid a number of traits from a range of sources that can be used to develop varieties resistant to saline environments. Despite the polygenic nature of salt tolerance many of the salt QTL studies described below have identified loci that explain large proportions of the genotypic/phenotypic variance reported. This is because most researchers target specific salt resistance traits for which the inheritance may be quite simple, for example, the arie. GP locus in barley (Forster 2001). This observation lends support to the strategy of trait pyramids suggested by Yeo and Flowers (1986).

3.2.1 *Salt Resistance QTL*

Bread Wheat

The ranking of cereals for salt tolerance from best to worst is barley, bread wheat, durum wheat, rice. In saline environments, bread wheats accumulate less Na+ and more K+ than durum wheats, particularly in young and expanding leaves. This K^+/Na^+ discrimination correlates well with the higher salt tolerance of bread wheats compared to durum wheats (Colmer et al. 1995) and 50 to 60% of the difference between these species could be attributed to chromosome 4D of wheat (Dvorak and Gorham 1992). Homeologous recombination between 4B and 4D using the *ph1c* mutant of durum wheat resulted in the segregation of two classes of progeny that discriminated K^+/Na^+ in a manner similar to the parental types. Based on this evidence a single locus designated *Kna1* for K^+/Na^+ discrimination was proposed (Dvorak and Gorham 1992) and validated as contributing to salt tolerance in field screenings (Dvorak et al. 1994). The map position of *Kna1* was determined by using a population of recombinant 4D/4B families (Dubcovsky et al. 1996). *Kna1* was mapped as a qualitative trait using phenotypic data from a glasshouse experiment that assigned lines to either the *Kna1* or *kna1* class by measuring K^+/Na^+ discrimination in the youngest fully expanded leaves 14 days after exposure to 50 mM NaCl. K^+/Na^+ discrimination was also determined from the flag leaves of field grown plants of the same 4D/4B families exposed to various saline environments (Dvorak et al. 1994) which allowed *Kna1* to be mapped as a quantitative trait. When treated as a qualitative trait *Kna1* mapped (LOD >14) to the long arm of 4D and 14.1 cM from the centromere. When treated as a quantitative trait *Kna1* mapped near to this location although at a much lower LOD score (2-3).

Durum Wheat

A screen of a large number of durum wheats identified Line 149 (Munns et al. 1999) as having an unusually low concentration of sodium in leaf blades. Mapping in an F_2 population of the cross Line 149 x Tamaroi (salt susceptible) identified one region that explained 38% of the phenotypic variance (Lindsay et al. 2004). This QTL named *Nax1* mapped to the longarm of chromosome 2A. Microsatellite marker gwm312 closely linked to the trait has been used to backcross salt tolerance into commercial durum varieties. A second independent locus was also shown to contribute to the sodium exclusion trait which was then named *Nax2*. Later physiological characterization of *Nax1* and *Nax2* (James et al. 2006) showed that both these genes restrict the transport of Na^+ from roots to shoots which results in enhanced K^+/Na^+ discrimination in the leaf blade. *Nax1* differs from *Nax2* in removing Na^+ from the lower part of the leaf as well as in the root. *Nax1* and *Nax2* appear to trace to the wild wheat ancestor *Triticum monococcum* (James et al. 2006).

A later study (Huang et al. 2006) used the rice genome sequence and wheat ESTs mapped in deletion bins, to construct a high-resolution map of the *Nax1* locus. Genes on wheat chromosome 2AL had good colinearity with rice chromosome 4L although these regions differed by an inversion that included the *Nax1* locus. Two putative sodium transporter genes with homology to the *OsHKT7* from rice were mapped to the wheat region containing the *Nax1* locus. The *TMHKT7*-A1 gene co-segregated with the *Nax1* locus and was polymorphic between the salt-sensitive and -insensitive parents. A second putative transporter gene *TMHKT7*-A2 was located on the same BAC (bacterial artificial chromosome) contig as *TMHKT7*-A1. The expression of *TMHKT7*-A2 was consistent with the physiological role of *Nax1*.

Rice

Lin et al. (2004) studied $F_{2:3}$ lines of a cross between the highly salt tolerant *indica* variety Nona Bokra and a salt-susceptible *japonica* variety Koshihikari. Parental and the $F_{2:3}$ lines at eight days after sowing were transferred to a 140 mM solution of NaCl and monitored for survival. In a second experiment the lines were grown under similar conditions but 10 days after treatment with a 140 mM solution of NaCl the Na^+ and K^+ concentrations of shoots and roots were determined. The distribution of each trait was continuous suggesting polygenic inheritance and transgressive segregation was common. Three QTLs for survival were detected on chromosomes 1, 6, and 7 individually accounting for upto 18% of the phenotypic variance. Two QTLs, qSNC-7 for shoot Na^+

concentration and qSKC-1 for shoot K^+ accounted for 48.5% and 40.1% of the total phenotypic variance, respectively. In a later related study (Ren et al. 2005) qSKC-1 of chromosome 1 was fine-mapped and the *SKC1* gene was cloned and found to be a member of the HKT-type transporters. *SKC1* encoded a Na^+ transporter and was preferentially expressed in the parenchyma cells surrounding the xylem vessels with a role in K^+/Na^+ discrimination. It was hypothesized that *SKC1* functions in the recirculation of Na^+ by unloading Na^+ from the xylem. These studies together with the recent discovery that the SOS (salt overly sensitive pathway) of *Arabidopsis* is conserved in rice (Martinez-Atienza et al. 2007) give great promise to researchers wanting to develop salt tolerant rice germplasm.

Sunflower

An investigation into the effect of hybridization on the evolution of several diploid species within the genus *Helianthus* (Rieseberg 1997; Welch and Riesberg 2002; Lexer et al. 2003a, b; Lai et al. 2005) led to some interesting conclusions about salt tolerance in this genus. This research was borne from the observation that the hybrid diploid species *Helianthus paradoxus* is ecologically distinct from its progenitors *Helianthus annuus* and *Helianthus petiolaris*. Curiously, *H. paradoxus* is adapted to extremely saline marshes while both the progenitors are not (Welch and Riesberg 2002). In terms of biomass and survivorship *H. paradoxus* was more than five-times as fit than its progenitors when screened in saline conditions of up to 200 mM NaCl. These observations were later confirmed with field studies (Lexer et al. 2003a) that also showed fitness associated with increased uptake of Ca^{2+} and decreased uptake of Na^+. The Ca^{2+}/Na^+ discrimination contrasts the K^+/Na^+ discrimination typically observed in salt tolerant cereal genotypes. An interspecific BC_2 population between *H. annuus* and *H. petiolaris* grown on a salt marsh was used to study elemental uptake in plant tissues and survivorship. While the parental species *H. annuus* and *H. petiolaris* did not survive the salt marsh conditions, a considerable number of the BC_2 progeny did. The survival of BC_2 progeny was evidence of transgressive segregation mostly likely due to complementary gene action. A QTL study of this population revealed three QTLs for Ca^{2+} uptake and four QTLs for Na^+ uptake and these QTL accounted for 78% and 65% of the phenotypic variance, respectively. Some of these QTL also coincided with QTL for 'survivorship'. Interestingly, there were several examples of traits where multiple QTLs inherited from the same parent had opposing effects. For example, *H. annuus* contributed QTL alleles that increased survivorship and other QTL alleles that decreased it. A later study (Lexer et al. 2004) mapped the candidate gene *CDPK3* (calcium dependant protein kinase 3) to a linkage cluster on linkage group 4 of sunflower that included QTLs for survivorship, and discrimination of Na^+ and Mg^{2+}.

Soybean

A major QTL conditioning salt tolerance in soybean has now been identified (Lee et al. 2005). These authors screened $F_{2:5}$ lines from the cross of tolerant line S-100 and salt sensitive line Tokyo in a glasshouse and a saline field. Entry mean heritabilities for salt tolerance in this population were high 48-85%, with the highest heritability estimate coming from the saline field. The genetic correlation between field and glasshouse ratings was 0.55 suggesting that glasshouse studies may replace more resource intense field evaluations. A major QTL for salt tolerance was mapped to linkage group N accounting for as much as 79% of the genetic variation and is likely to be the *Ncl* locus reported over 40 years ago. An analysis of US pedigrees dating back to more than 60 years showed that the flanking markers of this QTL were diagnostic of the S-100 allele and salt tolerance suggesting that these markers could be used for MAS in commercial breeding.

3.2.2 Summary of Salt Tolerance Inheritance and QTL

Generally the inheritance of component traits of salt resistance is simple and individual QTL tends to account for large proportions of the phenotypic variation in segregating populations. Transgressive segregation, an indication of complementary gene action is common particularly when interspecific populations are investigated. There are excellent opportunities to incorporate genes for salt tolerance via MAS although it remains to be seen whether these individual loci will provide sufficient tolerance to make large differences to field performance. Strategies that attempt to pyramid a range of salt tolerance mechanisms are more likely to succeed.

3.3 Nutrient Stress

3.3.1 Aluminium Toxicity

As indicated earlier, acid soils are widespread particularly in the developing world. Consequently, there is an imperative need to develop plant varieties that are adapted to acid soil conditions. Of the nutrient stresses typical of acid soils, aluminium toxicity has gained considerable attention in recent years. Under low pH conditions Al^{3+} is released into the soil solution where it inhibits root growth and development (Kochian 1995). In this section, we will describe the progress of developing aluminium resistant plant varieties.

The physiological mechanisms of Al resistance in higher plants has been reviewed by Ma (2005) which we briefly summarize here. Aluminium has no known function in plant nutrition and can be highly toxic to growth when in excess. However, plants have evolved several mechanisms for detoxifying Al. These fall into two broad categories; exclusion of Al from root tips and tolerance to the Al that enters the plant. These categories are generally referred to as the 'exclusion mechanism' and the 'internal tolerance mechanism', respectively. It now appears that organic acids (malate, citrate, oxalate) play an integral role in the detoxification of Al inside the plant and around the root zone. Not many plant species use the internal tolerance mechanism and most that do are woody species adapted to acid soils.

Two patterns of response have been described for the 'exclusion mechanism' based on how long the organic acids are released in response to excess Al. The rapid onset of organic acid secretion in response to Al suggests a pre-existing anion channel in the plasma membrane which has been activated and that gene expression is not involved. Conversely, gene expression may be involved when a delay in organic secretion is observed after exposure to excess Al.

3.3.1.1 Aluminium Tolerance QTL

The genetics of Al resistance in wheat have not been completely elucidated, with some studies suggesting monogenic inheritance and polygenic in others. Al resistance is typically evaluated using toxic levels of Al in nutrient solutions. Delhaize et al. (1993) identified a single dominant locus *Alt1* using a set of isogenic lines. This gene was subsequently cloned using subtractive hybridization of cDNAs isolated from the isogenic lines and called *ALMT1* (Sasaki et al. 2004). *ALMT1* (Al-activated malate transporter 1) encodes an Al-activated malate efflux transporter in roots. Riede and Anderson (1996) identified a single gene Alt_{BH} located on 4DL a location verified by others (Milla and Gustafson 2001). Luo and Dvorak (1996) identified the locus Alt_2 for Al resistance on the same chromosome arm using a set of aneuploid stocks.

Ma et al. (2006) studied Al resistance in a set of ditelosomic lines and observed lower Al resistance compared to the moderately resistant euploid Chinese Spring. This observation suggested that chromosome arms 5AS, 7AS, 2DL, and 4DL may be implicated in Al resistance. A QTL mapping project using Chinese Spring (Ma et al. 2006) identified a major

QTL on 4DL and two additional QTLs on 5AS and 2DL which were consistent with the locations determined from the aneuploid stocks. The mapping of an Al resistant locus to 4DL was consistent with previous studies outlined above.

Al resistant gene *ALMT1* (Sasaki et al. 2004) has now been used to clone a homolog from *Arabidopsis* referred to as *AtALMT1* and will no doubt be useful for identifying homologs from other species. Magalhaes et al. (2004) mapped Al resistance in sorghum and identified a single locus Alt_{SB} present in two highly resistant lines. Despite the significant macrosynteny between the Triticeae and sorghum for molecular markers closely linked to putatively orthologous Al resistance loci present in group 4 of wheat, barley, and rye, Alt_{SB} did not map to the homeologous region of sorghum. Rather, Alt_{SB} mapped to sorghum chromosome 3.

3.3.2 Hyperaccumulators

There is now interest in the hyperaccumulator *Thlaspi caerulescens* (Assuncao et al. 2003; Deniau et al. 2006) a member of the Brassicaceae family that shares 88% DNA identity in the coding regions with *Arabidopsis*. Hyeraccumulators are those plant species that are able to accumulate extremely high levels of heavy metals (e.g., zinc (Zn), cadmium (Cd) and have value in phytoremediation especially for reclaiming damaged and polluted land sites. QTLs for heavy metal (Zn and Cd) accumulation have now been mapped in *T. caerulescens* (Deniau et al. 2006) which will no doubt spawn many other studies in this area.

3.3.3 Summary of Nutrient Stress Resistance Inheritance and QTL

The inheritance of nutrient stress resistances appears to be simple with single loci expressing large phenotypic effects. It remains to be seen whether single loci will provide sufficient resistance to varieties grown in challenged environments. It is likely that MAS will play a major role in pyramiding genes that have different modes of action. The candidate gene approach has been successful and likely to have increased importance as more plant genome sequence information becomes available.

3.4 Low Temperature Stress

For many annual crop species, the duration of growth in any given season is largely determined by their response to combinations of temperature and photoperiod. The decision of when to plant is determined by the available soil moisture at sowing and when dry periods and unfavorable extremes of temperature can be avoided. However, un-seasonal extremes of temperature can often leave crops exposed to very high temperatures or low to freezing temperatures. Injury from freezing temperatures occurs largely from severe cellular dehydration associated with ice formation. The last decade has seen a vast amount of research aimed at understanding the molecular basis of cold acclimation in plants. Cold acclimation is the process by which a plant can increase its freezing tolerance after exposure to low temperature (Thomashow 2001).

The pathway of gene expression associated with cold acclimation in *Arabidopsis* has been well studied and described briefly below (for detailed review see Thomashow 2001). Cold acclimation is often associated with changes in gene expression associated with the 'CBF regulon'. The CBF regulon includes a family of transcription factors called the CBF (C repeat binding factor) or DREB1 (dehydration-responsive element binding protein 1) proteins which typically bind to a C-repeat (CRT) dehydration-responsive element (DRE) with a core sequence motif of CCGAC. CBF transcriptional activators in turn regulate *COR* (cold-regulated) genes. The net result is an increase in membrane stabilizing proteins and cryo-protectants such as sugar and proline. This pathway has provided researchers with some useful clues to help unravel the inheritance of freezing tolerance particularly in the crop and pasture species of the Poaceae.

One such member of Poaceae is cultivated wheat consisting of winter types and spring types that are differentiated by their response to cold temperatures. Winter wheats require a vernalization period of low but not freezing temperature that allows for the transition of the apical meristem from vegetative to reproductive phase when the photoperiod is optimal. Spring wheats do not require a vernalizing treatment to induce flowering. Several vernalization loci (*Vrn*) that differentiate winter and spring growth habits have been identified (for list of *Vrn* loci in cereals see Table 1 in Snape et al. 2001). For example, in hexaploid wheat the group 5 homeologous chromosomes contain several *Vrn* loci including *Vrn-A1*, *Vrn-B1*, and *Vrn-D1*. The winter growth habit is determined by recessive alleles at each of these loci and dominant alleles at one or more loci are required for spring growth habit.

In addition to vernalization genes winter growing species also benefit from tolerance to low or freezing temperatures (see for review Baga et al. 2007). Often cold acclimation occurs simultaneously with vernalization and tolerance to low temperatures is maximized when vernalization has reached saturation. Co-incidentally, several frost tolerance loci (*Fr*) have been mapped in the proximity of *Vrn* loci in winter cereals. In wheat *Fr-A1* mapped near *vrn-A1* on chromosome 5A (Galiba et al. 1995) and *Fr-D1* near *vrn-D1* on chromosome 5D (Snape et al. 2001). *Fr-A1* and *Fr-D1* appear homeologous to each other and to an *Fr* locus mapped in barley on chromosome 7 (Hayes et al. 1993). Recently, *Fr-B2* (was *Fr-B1*) has been mapped to the distal portion of the long arm of chromosome 5B a region syntenic to that of chromosomes 5A and 5D that also contain vernalization and frost tolerance loci (Tóth et al. 2003).

In *T. monococcum*, the 'A' genome donor of tetraploid wheat, *Fr-A^{m}2* has been mapped on chromosome 5A (Vágújfalvi et al. 2003). Interestingly, an expression QTL for the cold-regulated gene *COR14b* completely overlapped with the *Fr-A^{m}2* locus for frost survival at -13°C (Vágújfalvi et al. 2003). A similar association between *COR14b* and *Fr-H2* was also observed in barley (Francia et al. 2004). *COR14b* proteins may have a role in the protection of chloroplast membranes from freezing temperatures (Crosatti et al. 1999). Not surprisingly, *CBF* orthologs also co-segregated with *Fr-A^{m}2* (Vágújfalvi et al. 2003) in wheat and *Fr-H2* in barley (Francia et al. 2004). These observations suggested that *CBF* transcriptional activators which regulate *COR* genes may play a major role in frost tolerance in cereals and several studies have been initiated to test this hypothesis. For example, Miller et al. (2006) characterized 13 *TmCBF* genes from *T. monococcum* and found that 11 clustered around the *Fr-A^{m}2* locus within five different BAC clones. Using recombinant individuals segregating for the *Fr-A^{m}2* locus, studies will be made to determine which *CBF* gene(s) are responsible for differences in frost tolerance. In hexaploid wheat, a low temperature tolerance locus was also mapped to chromosome 5A in a region syntenic to *Fr-A^{m}2* (Baga et al. 2007). This hexaploid wheat locus was also associated with two *CBF* genes (Baga et al. 2007) with strong sequence homology to *CBF14* and *CBF15* that mapped with *Fr-A^{m}2* in the study of Miller et al. (2006). Importantly, Baga et al. (2007) found that the 5A low temperature tolerance locus did not explain all of the variation in the mapping population studied and suggested that there were alternative pathways for cold acclimation in wheat.

COR/CBF candidate genes may also have potential in mapping frost tolerance in other species e.g., Brassica (Kole et al. 2002) and perennial ryegrass (*Lolium perenne* L.) (Tamura and Yamada 2007). Tamura and Yamada (2007) mapped a *CBF* cluster in a region predicted by synteny between other Poaceae species such as wheat and barley.

4 CONCLUSION

As our world responds to climate change and other world factors there will be increased reliance on agriculture and the need to develop plant varieties with abiotic stress resistance. There are now clear examples where breeding for abiotic stress resistance using a QTL approach has been successful although these were facilitated by years of solid stress physiology, breeding, and genetics. Most of the success with QTL is in separating away poor alleles from unadapted donors. Avoiding or minimizing linkage drag is critical when much of the work toward genetic improvement of abiotic stress revolves around novel germplasm and its inclusion in breeding programs.

Some major genes that confer abiotic stress resistance have been identified and MAS has begun. Single genes are unlikely to confer substantial levels of abiotic resistance; rather trait pyramids that combine different mechanisms of resistance into one genotype are preferable. Many highly heritable traits that confer abiotic stress resistance were quantitatively inherited. For those traits, multiple QTLs were unable to account for much of the phenotypic variation making MAS somewhat questionable. This problem can, in part, be resolved by using larger populations (> 200) in QTL experiments. Transgressive segregation and complementary gene action can be an important source of stress resistance. In the future, a multi-disciplinary approach will no doubt contribute significantly to the QTL approach and further improvements in abiotic stress resistance in plants can be expected.

References

Abdelkhalik AF, Shishido R, Nomura K, Ikehashi H (2005) QTL-based analysis of leaf senescence in an *indica/japonica* hybrid in rice (*Oryza sativa* L.). Theor Appl Genet 110: 1226-1235

Acquaah G (2007) Principles of Plant Genetics and Breeding. Blackwell Publ, Oxford, UK

Assuncao AGL, Schat H, Aarts MGM (2003) *Thlaspi caerulescens*, an attractive model species to study heavy metal hyperaccumulation in plants. New Phytol 159: 351-360

Baga M, Chodaparambil SV, Limin AE, Pecar M, Fowler DB, Chibbar RN (2007) Identification of quantitative trait loci and associated candidate genes for low-temperature tolerance in cold-hardy winter wheat. Funct Integr Genom 7: 53-68

Beavis WD (1998) QTL analyses: power, precision, and accuracy. In: Paterson AH (ed) Molecular Dissection of Complex Traits. CRC Press, Boca Raton, FL, USA, pp 145-162

Bolaños J, Edmeades GO, Martinez L (1993) 8 cycles of selection for drought tolerance in lowland tropical maize. 3. Responses in drought-adaptive physiological and morphological traits. Field Crops Res 31: 269-286

Borrell AK, Hammer GL (2000) Nitrogen dynamics and the physiological basis of stay-green in sorghum. Crop Sci 40: 1295-1307

Borrell AK, Hammer GL, Douglas ACL (2000a) Does maintaining green leaf area in sorghum improve yield under drought? I. Leaf growth and senescence. Crop Sci 40: 1026-1037

Borrell AK, Hammer GL, Henzell RG (2000b) Does maintaining green leaf area in sorghum improve yield under drought? II. Dry matter production and yield. Crop Sci 40: 1037-1048

Borrell A, Hammer G, Van Oosterom E (2001) Stay-green: a consequence of the balance between supply and demand for nitrogen during grain filling? Ann Appl Biol 138: 91-95

Borrell A, Jordan D, Mullet J, Henzell R, Hammer G (2006) Drought adaptation in Sorghum. In: Ribaut JM (ed) Drought Adaptation in Cereals. Food Products Press, New York, USA, pp 335-399

Brendel O, Pot D, Plomion C, Rozenberg P, Guehl JM (2002) Genetic parameters and QTL analysis of delta C-13 and ring width in maritime pine. Plant Cell Environ 25: 945-953

Bruce WB, Edmeades GO, Barker TC (2002) Molecular and physiological approaches to maize improvement for drought tolerance. J Exp Bot 53: 13-25

Calhoun DS, Gebeyehu G, Miranda A, Rajaram S, van Ginkel M (1994) Choosing evaluation environments to increase wheat grain yield under drought conditions. Crop Sci 34: 673-678

Campos H, Cooper A, Habben JE, Edmeades GO, Schussler JR (2004) Improving drought tolerance in maize: a view from industry. Field Crops Res 90: 19-34

Casasoli M, Pot D, Plomion C, Monteverdi MC, Barreneche T, Lauteri M, Villani F (2004) Identification of QTLs affecting adaptive traits in *Castanea sativa* Mill. Plant Cell Environ 27: 1088-1101

Casasoli M, Derory J, Morera-Dutrey C, Brendel O, Porth I, Guehl JM, Villani F, Kremer A (2006) Comparison of quantitative trait loci for adaptive traits between oak and chestnut based on an expressed sequence tag consensus map. Genetics 172: 533-546

Cha K-W, Lee Y-J, Koh H-J, Lee B-M, Nam Y-W, Paek N-C (2002) Isolation, characterization, and mapping of the stay green mutant in rice. Theor Appl Genet 104: 526-532

Colmer TD, Epstein E, Dvorak J (1995) Differential solute regulation in leaf blades of various ages in salt-sensitive wheat and a salt-tolerant wheat x *Lophopyrum elongatum* (Host) A. Love amphiploid. Plant Physiol 108: 1715-1724

Condon AG, Richards RA (1992) Broad sense heritability and genotype x environment interaction for carbon isotope discrimination in field-grown wheat. Aust J Agri Res 43: 921-934

Condon AG, Richards RA, Rebetzke GJ, Farquhar GD (2004) Breeding for high water-use efficiency. J Exp Bot 55: 2447-2460

Cooper, M, Stucker RE, Delacy IH, Harch BD (1997) Wheat breeding nurseries, target environments, and indirect selection for grain yield. Crop Sci 37: 1168-1176

Cullis BR, Gleeson AC (1989) Efficiency of neighbor analysis for replicated variety trials in Australia. J Agri Sci 113: 233-239

Delhaize E, Craig S, Beaton CD, Bennet RJ, Jagadish VC, Randall PJ (1993) Aluminum tolerance in wheat (*Triticum aestivum* L). 1. Uptake and distribution of aluminum in root apices. Plant Physiol 103: 685-693

Delhaize E, Ryan PR, Hebb DM, Yamamoto Y, Sasaki T, Matsumoto H (2004) Engineering high level aluminium tolerance in barley with the ALTM1 gene. Proc Natl Acad Sci USA 10: 15249-15254

Deniau AX, Pieper B, Ten Bookum WM, Lindhout P, Aarts MGM, Schat H (2006) QTL analysis of cadmium and zinc accumulation in the heavy metal hyperaccumulator *Thlaspi caerulescens*. Theor Appl Genet 113: 907-920

Dubcovsky J, Maria GS, Epstein E, Luo MC, Dvorak J (1996) Mapping of the K+/Na+ discrimination locus *Kna1* in wheat. Theor Appl Genet 92: 448-454

Duvick DN (1997) What is yield? In: Edmeades GO, Banziger M, Mickelson HR, Pena Vadivia CB (eds) Developing Drought and Low-M Tolerant Maize. CIMMYT, El Batan, Mexico, pp 332-335

Duvick DN, Smith JSC, Cooper M (2004) Long-term selection in a commercial hybrid maize breeding program. Plant Breed Rev 24: 109-151

Dvorak J, Gorham J (1992) Methodology of gene-transfer by homoeologous recombination into *Triticum turgidum* - transfer of K+/Na+ discrimination from *Triticum aestivum*. Genome 35: 639-646

Dvorak J, Noaman MM, Goyal S, Gorham J (1994) Enhancement of the salt tolerance of *Triticum turgidum* L. by the *Kna1* locus transferred from the *Triticum aestivum* L. chromosome 4D by homoeologous recombination. Theor Appl Genet 87: 872-877

Edmeades GO, Bolaños J, Elings A, Ribaut JM, Bänziger M, Westgate ME (2000) The role and regulation of the anthesis-silking interval in maize. In: Westgate ME, Boote KJ (eds) Physiology and Modelling Kernel Set in Maize. CSSA Special Publication No 29. Madison, WI, USA, pp 43-73

Ehleringer JR, Klassen S, Clayton C, Sherrill D, Fullerholbrook M, Fu QN, Cooper TA (1991) Carbon isotope discrimination and transpiration efficiency in common bean. Crop Sci 31: 1611-1615

Ellis MH, Spielmeyer W, Gale K, Rebetzke GJ, Richards RA (2002) Perfect markers for the *Rht-B1b* and *Rht-D1b* dwarfing mutations in wheat (*Triticum aestivum* L.). Theor Appl Genet 105: 1038-1042

Ellis MH, Bonnett DG, Rebetzke GJ (2007) A 192bp allele at the *Xgwm261* locus is not always associated with the *Rht8* dwarfing gene in wheat (*Triticum aestivum* L.). Euphytica (in press)

Falconer DS and McKay T (1996) Introduction to quantitative genetics. 4th ed. Longman Burnt Mill, England

FAO (2001) The State of Food Insecurity in the World 2001. Rome, Italy: *http://www.fao.org/DOCREP/003/Y1500E/Y1500E00.HTM*

FAO (2002) World Agriculture 2015-2030: An FAO perspective (Harrison P ed) Earthscan, England *http://www.fao.org/docrep/005/y4252e/y4252e00.htm*

Farquhar GD, Richards RA (1984) Isotopic composition of plant carbon correlates with water-use efficiency of wheat genotypes. Aust J Plant Physiol 11: 539-552

Farquhar GD, Oleary MH, Berry JA (1982) On the relationship between carbon isotope discrimination and the inter-cellular carbon-dioxide concentration in leaves. Aust J Plant Physiol 9: 121-137

Flowers TJ, Flowers SA (2005) Why does salinity pose such a difficult problem for plant breeders? Agri Water Manag 78: 15-24

Flowers TJ, Koyama ML, Flowers SA, Sudhakar C, Singh KP, Yeo AR (2000) QTL: their place in engineering tolerance of rice to salinity. J Exp Bot 51: 99-106

Forster BP (2001) Mutation genetics of salt tolerance in barley: An assessment of Golden Promise and other semi-dwarf mutants. Euphytica 120: 317-328.

Francia E, Rizza F, Cattivelli L, Stanca AM, Galiba G, Toth B, Hayes PM, Skinner JS, Pecchioni N (2004) Two loci on chromosome 5H determine low-temperature tolerance in a 'Nure' (winter) x 'Tremois' (spring) barley map. Theor Appl Genet 108: 670-680

Galiba G, Quarrie SA, Sutka J, Morgounov A, Snape JW (1995) RFLP mapping of the vernalization (vrn 1) and frost-resistance (fr 1) genes on chromosome 5a of wheat. Theor Appl Genet 90: 1174-1179.

Giuliani S, Sanguineti MC, Tuberosa R, Bellotti M, Salvi S, Landi P (2005) Root-ABA1, a major constitutive QTL, affects maize root architecture and leaf ABA concentration at different water regimes. J Exp Bot 56: 3061-3070

Hall AE, Richards RA, Condon AG, Wright GC, and Farquhar GD (1995) Carbon isotope discrimination and plant breeding. Plant Breed Rev 12: 81-112

Harris K, Subudhi PK, Borrell A, Jordan D, Rosenow D, Nguyen H, Klein P, Klein R, Mullet J (2007) Sorghum stay-green QTL individually reduce post-flowering drought-induced leaf senescence. J Exp Bot 58: 327-338

Hausmann NJ, Juenger TE, Sen S, Stowe KA, Dawson TE, Simms EL (2005) Quantitative trait loci affecting delta C-13 and response to differential water availability in *Arabidopsis thaliana*. Evolution 59: 81-96

Haussmann BIG, Mahalakshmi V, Reddy BVS, Seetharama N, Hash CT, Geiger HH (2002) QTL mapping of stay-green in two sorghum recombinant inbred populations. Theor Appl Genet 106: 133-142

Hayes PM, Blake T, Chen THH, Tragoonrung S, Chen F, Pan A, Liu B (1993) Quantitative trait loci on barley (*Hordeum vulgare* L) chromosome-7 associated with components of winterhardiness. Genome 36: 66-71.

Helsel LDB, Frey KJ (1978) Grain yield variations in oats associated with differences in leaf area duration among oat lines. Crop Sci 18: 765-769

Holland J (2004) Implementation of molecular markers for quantitative traits in breeding programs - challenges and opportunities. In: New Directions for a Diverse Planet. Proc 4th Intl Crop Sci Congr, 26 Sept-1 Oct 2004, Brisbane, Queensland, Australia

Huang SB, Spielmeyer W, Lagudah ES, James RA, Platten JD, Dennis ES, Munns R (2006) A sodium transporter (*HKT7*) is a candidate for *Nax1*, a gene for salt tolerance in durum wheat. Plant Physiol 142: 1718-1727

Hubick KT, Shorter R, Farquhar GD (1988) Heritability and genotype x environment interactions of carbon isotope discrimination and transpiration efficiency in peanut (*Arachis hypogaea* L.). Aust J Plant Physiol 15: 799-813

Husain S, Munns R, Condon AG (2003) Effect of sodium exclusion trait on chlorophyll retention and growth of durum wheat in saline soil. Aust J Agri Res 54: 589-597

IPCC (2007) Climate Change 2007: The physical science basis. Summary for policy makers. Contribution of Working Group I to the fourth assessment report of the Intergovernmental Panel on Climate Change: *http://www.ipcc.ch/*

Ishimaru K, Yano M, Aoki N, Ono K, Hirose T, Lin SY, Monna L, Sasaki T, Ohsugi R (2001) Toward the mapping of physiological and agronomic characters on a rice function map: QTL analysis and comparison between QTLs and expressed sequence tags. Theor Appl Genet 102: 793-800

James RA, Munns R, Von Caemmerer S, Trejo C, Miller C, Condon T (2006) Photosynthetic capacity is related to the cellular and subcellular partitioning of Na^+, K^+ and Cl^- in salt-affected barley and durum wheat. Plant Cell Environ 29: 2185-2197

Jiang GH, He YQ, Xu CG, Li XH, Zhang Q (2004) The genetic basis of stay-green in rice analyzed in a population of doubled haploid lines derived from an *indica* by *japonica* cross. Theor Appl Genet 108: 688-698

Jordan DR, Tao Y, Godwin ID, Henzell RG, Cooper M, McIntyre CL (2003) Prediction of hybrid performance in grain sorghum using RFLP markers. Theor Appl Genet 106: 559-567

Juenger TE, McKay JK, Hausmann N, Keurentjes JJB, Sen S, Stowe KA, Dawson TE, Simms EL, Richards JH (2005) Identification and characterization of QTL underlying whole-plant physiology in *Arabidopsis thaliana*: delta C-13, stomatal conductance and transpiration efficiency. Plant Cell Environ 28: 697-708

Kebede H, Subudhi PK, Rosenow DT, Nguyen HT (2001) Quantitative trait loci influencing drought tolerance in grain sorghum (*Sorghum bicolor* L. Moench). Theor Appl Genet 103: 266-276

Kochian LV (1995) Cellular mechanisms of aluminum toxicity and resistance in plants. Annu Rev Plant Physiol Plant Mol Biol 46: 237-260

Kole C, Thormann CE, Karlsson BH, Palta JP, Gaffney P, Yandell B, Osborn TC (2002) Comparative mapping of loci controlling winter survival and related traits in oilseed *Brassica rapa* and *B. napus*. Mol Breed 9: 201-210

Kondo M, Pablico PP, Aragones DV, Agbisit R (2004) Genotypic variations in carbon isotope discrimination, transpiration efficiency, and biomass production in rice as affected by soil water conditions and N. Plant Soil 267: 165-177

Lai Z, Livingstone K, Zou Y, Church SA, Knapp SJ, Andrews J, Rieseberg LH (2005) Identification and mapping of SNPs from ESTs in sunflower. Theor Appl Genet 111: 1532-1544

Lambrides CJ, Chapman SC, Shorter R (2004) Genetic variation for carbon isotope discrimination in sunflower: Association with transpiration efficiency and evidence for cytoplasmic inheritance. Crop Sci 44: 1642-1653

Landi P, Sanguineti MC, Salvi S, Giuliani S, Bellotti M, Maccaferri M, Conti S, Tuberosa R (2005) Validation and characterization of a major QTL affecting leaf ABA concentration in maize. Mol Breed 15: 291-303

Landi P, Sanguineti MC, Liu C, Li Y, Wang TY, Giuliani S, Bellotti M, Salvi S, Tuberosa R (2007) Root-ABA1 QTL affects root lodging, grain yield, and other agronomic traits in maize grown under well-watered and water-stressed conditions. J Exp Bot 58: 319-326

Laza MR, Kondo M, Ideta O, Barlaan E, Imbe T (2006) Identification of quantitative trait loci for delta C-13 and productivity in irrigated lowland rice. Crop Sci 46: 763-773

LeCain DR, Morgan JA, Zerbi G (1989) Leaf anatomy and gas exchange in nearly isogenic semi dwarf and tall winter wheat. Crop Sci 29: 1246-1251

Lee GJ, Carter TE, Villagarcia MR, Li Z, Zhou X, Gibbs MO, Boerma HR (2005) A major QTL conditioning salt tolerance in S-100 soybean and descendent cultivars. Theor Appl Genet 110: 787-787

Lexer C, Lai Z, Rieseberg LH (2004) Candidate gene polymorphisms associated with salt tolerance in wild sunflower hybrids: implications for the origin of *Helianthus paradoxus*, a diploid hybrid species. New Phytol 161: 225-233

Lexer C, Welch ME, Durphy JL, Rieseberg LH (2003a) Natural selection for salt tolerance quantitative trait loci (QTLs) in wild sunflower hybrids: Implications for the origin of *Helianthus paradoxus*, a diploid hybrid species. Mol Ecol 12: 1225-1235

Lexer C, Welch ME, Raymond O, Rieseberg LH (2003b) The origin of ecological divergence in *Helianthus paradoxus* (Asteraceae): Selection on transgressive characters in a novel hybrid habitat. Evolution 57: 1989-2000

Lin HX, Zhu MZ, Yano M, Gao JP, Liang ZW, Su WA, Hu XH, Ren ZH, Chao DY (2004) QTLs for Na+ and K+ uptake of the shoots and roots controlling rice salt tolerance. Theor Appl Genet 108: 253-260

Lindsay MP, Lagudah ES, Hare RA, Munns R (2004) A locus for sodium exclusion (Nax1), a trait for salt tolerance, mapped in durum wheat. Funct Plant Biol 31: 1105-1114

Lobell DB, Field CB (2007) Global scale climate-crop yield relationships and the impacts of recent warming. Environ Res Lett 2: 1-7

Luo MC, Dvorak J (1996) Molecular mapping of an aluminum tolerance locus on chromosome 4D of Chinese Spring wheat. Euphytica 91: 31-35

Ma HX, Bai GH, Lu WZ (2006) Quantitative trait loci for Aluminum resistance in wheat cultivar Chinese Spring. Plant Soil 283: 239-249

Ma JF (2005) Physiological mechanisms of Al resistance in higher plants. Soil Sci Plant Nutr 51: 609-612

Magalhaes JV, Garvin DF, Wang YH, Sorrells ME, Klein PE, Schaffert RE, Li L, Kochian LV (2004) Comparative mapping of a major aluminum tolerance gene in sorghum and other species in the Poaceae. Genetics 167: 1905-1914

Martinez-Beltran J, Manzur CL (2005) Overview of salinity problems in the world and FAO strategies to address the problem. In: Proc Intl Salinity Forum, April 2005, Riverside, California, USA, pp 311-313

Masle J, Gilmore SR, Farquhar GD (2005) The ERECTA gene regulates plant transpiration efficiency in Arabidopsis. Nature 436: 866-870

Mather K (1951) The Measurement of Linkage in Heredity. 2nd edn. John Wiley, New York, USA

McKay JK, Richards JH, Mitchell-Olds T (2003) Genetics of drought adaptation in *Arabidopsis thaliana*: I. Pleiotropy contributes to genetic correlations among ecological traits. Mol Ecol 12: 1137-1151

Milla MAR, Gustafson JP (2001) Genetic and physical characterization of chromosome 4DL in wheat. Genome 44: 883-892

Miller AK, Galiba G, Dubcovsky J (2006) A cluster of 11 CBF transcription factors is located at the frost tolerance locus *Fr-A^{m}2* in *Triticum monococcum*. Mol Genet Genom 275: 193-203

Munns R, Hare RA, James RA, Rebetzke GJ (1999) Genetic variation for improving the salt tolerance of durum wheat. Aust J Agri Res 51: 69-74

Munns R, Husain S, Rivelli AR, James RA, Condon AG, Lindsay MP, Lagudah ES, Schachtman DP, Hare RA (2002) Avenues for increasing salt tolerance of crops, and the role of physiologically based selection traits. Plant Soil 247: 93-105

Munns R, James RA (2003) Screening methods for salinity tolerance: a case study with tetraploid wheat. Plant Soil 253: 201-218

Munns R, James RA, Lauchli A (2006) Approaches to increasing the salt tolerance of wheat and other cereals. J Exp Bot 57: 1025-1043

Nyquist WE (1991) Estimation of heritability and prediction of selection response in plant populations. Crit Rev Plant Sci 10: 235-322

Price AH, Steele KA, Moore BJ, Jones RGW (2002) Upland rice grown in soil-filled chambers and exposed to contrasting water-deficit regimes II. Mapping quantitative trait loci for root morphology and distribution. Field Crops Res 76: 25-43

Rebetzke GJ, Condon AG, Richards RA, Farquhar GD (2002) Selection for reduced carbon isotope discrimination increases aerial biomass and grain yield of rainfed bread wheat. Crop Sci 42: 739-745

Rebetzke GJ, Condon AG, Richards RA, Farquhar GD (2003) Genetic control of leaf conductance in three wheat crosses. Aust J Agri Res 54: 381-387

Rebetzke GJ, Condon AG, Appels R, Farquhar GD, Richards RA (2007b) Location of quantitative trait loci for carbon isotope discrimination in multiple wheat (*Triticum aestivum* L.) mapping populations. Crop Sci (submitted)

Rebetzke GJ, van Herwaarden AF, Jenkins C, Ruuska S, Tabe L, Fettell N, Lewis D, Weiss M, Richards RA (2007a) Genetic control of water-soluble carbohydrate reserves in bread wheat. In: Buck H, Nisi J, Rajaram S (eds) Proc 7th Intl Wheat Conf, 27 Nov - 2 Dec 2005, Mar Del Plata, Argentina (in press)

Ren ZH, Gao JP, Li LG, Cai XL, Huang W, Chao DY, Zhu MZ, Wang ZY, Luan S, Lin HX (2005) A rice quantitative trait locus for salt tolerance encodes a sodium transporter. Nat Genet 37: 1141-1146

Rengasamy P (2006) World salinization with emphasis on Australia. J Exp Bot 57: 1017-1023

Ribaut JM, Ragot M (2007) Marker-assisted selection to improve drought adaptation in maize: the backcross approach, perspectives, limitations, and alternatives. J Exp Bot 58: 351-360

Ribaut JM, Hoisington DA, Deutsch JA, Jiang C, GonzalezdeLeon D (1996) Identification of quantitative trait loci under drought conditions in tropical maize. 11. Flowering parameters and the anthesis-silking interval. Theor Appl Genet 92: 905-914

Ribaut JM, Jiang C, GonzalezdeLeon D, Edmeades GO, Hoisington DA (1997) Identification of quantitative trait loci under drought conditions in tropical maize. 2. Yield components and marker-assisted selection strategies. Theor Appl Genet 94: 887-896

Riede CR, Anderson JA (1996) Linkage of RFLP markers to an aluminum tolerance gene in wheat. Crop Sci 36: 905-909

Rieseberg LH (1997) Hybrid origins of plant species. Annu Rev Ecol Syst 28: 359-389

Rosenow DT, Clark LE (1995) Drought and lodging research for a quality sorghum crop. In: Proc 5th Annu Corn and Sorghum Industry Res Conf, Dec 6-7 1995, Am Seed Trade Assoc, Illinois, USA, pp 82-97

Rosenow D, Quisenberry JE, Wendt CE, Clark LE (1983) Drought tolerant sorghum and cotton germplasm. Agri Water Manag 7: 207-222

Rosielle AA, Hamblin J (1981) Theoretical aspects of selection for yield in stress and non-stress environments. Crop Sci 21: 943-946

Sanchez AC, Subudhi PK, Rosenow DT, Nguyen HT (2002) Mapping QTLs associated with drought resistance in sorghum (*Sorghum bicolor* L. Moench). Plant Mol Biol 48: 713-726

Sanguineti MC, Duvick DN, Smith S, Landi P, Tuberosa R (2006) Effects of long-term selection on seedling traits and ABA accumulation in commercial maize hybrids. Maydica 51: 329-338

Saranga Y, Jiang CX, Wright RJ, Yakir D, Paterson AH (2004) Genetic dissection of cotton physiological responses to arid conditions and their inter-relationships with productivity. Plant Cell Environ 27: 263-277

Sasaki T, Yamamoto Y, Ezaki B, Katsuhara M, Ahn SJ, Ryan PR, Delhaize E, Matsumoto H (2004) A wheat gene encoding an aluminum-activated malate transporter. Plant J 37: 645-653

Schon CC, Utz HF, Groh S, Truberg B, Openshaw S, Melchinger AE (2004) Quantitative trait locus mapping based on resampling in a vast maize testcross experiment and its relevance to quantitative genetics for complex traits. Genetics 167: 485-498

Snape JW, Sarma R, Quarrie SA, Fish L, Galiba G, Sutka J (2001) Mapping genes for flowering time and frost tolerance in cereals using precise genetic stocks. Euphytica 120: 309-315

Steele KA, Price AH, Shashidhar HE, Witcombe JR (2006) Marker-assisted selection to introgress rice QTLs controlling root traits into an Indian upland rice variety. Theor Appl Genet 112: 208-221

Steele KA, Virk DS, Kumar R, Prasad SC, Witcombe JR (2007) Field evaluation of upland rice lines selected for QTLs controlling root traits. Field Crops Res 101: 180-186

Subudhi PK, Rosenow DT, Nguyen HT (2000) Quantitative trait loci for the stay green trait in sorghum (*Sorghum bicolor* L. Moench): consistency across genetic backgrounds and environments. Theor Appl Genet 101: 733-741

Sumner ME (2005) Food production on acid soils in the developing world: Problems and solutions. Soil Sci Plant Nutr 51: 621-624

Takai T, Fukuta Y, Sugimoto A, Shiraiwa T, Horie T (2006) Mapping of QTLs controlling carbon isotope discrimination in the photosynthetic system using recombinant inbred lines derived from a cross between two different rice (*Oryza sativa* L.) cultivars. Plant Prod Sci 9: 271-280

Tamura K, Yamada T (2007) A perennial ryegrass *CBF* gene cluster is located in a region predicted by conserved synteny between Poaceae species. Theor Appl Genet 114: 273-283

Tao YZ, Henzell RG, Jordan DR, Butler DG, Kelly AM, McIntyre CL (2000) Identification of genomic regions associated with stay green in sorghum by testing RILs in multiple environments. Theor Appl Genet 100: 1225-1232

Teulat B, Merah O, Sirault X, Borries C, Waugh R, This D (2002) QTLs for grain carbon isotope discrimination in field-grown barley. Theor Appl Genet 106: 118-126

Thomas H, Smart CM (1993) Crops that stay green. Ann Appl Biol 123: 193-219

Thomas H, Howarth CJ (2000) Five ways to stay green. J Exp Bot 51: 329-337

Tollenaar M, Daynard TB (1978) Leaf senescence in short season maize hybrids. Can J Plant Sci 58: 869-874

Thomashow MF (2001) So what's new in the field of plant cold acclimation? Lots! Plant Physiol 125: 89-93.

Toth B, Galiba G, Feher E, Sutka J, Snape JW (2003) Mapping genes affecting flowering time and frost resistance on chromosome 5B of wheat. Theor Appl Genet 107: 509-514

Tuinstra MR, Grote EM, Goldsbrough PB, Ejeta G (1997) Genetic analysis of post-flowering drought tolerance and components of grain development in *Sorghum bicolor* (L.) Moench. Mol Breed 3: 439-448

Umali D (1993) Irrigation induced salinity: A growing problem for development and the environment. World Bank, Washington DC, USA

United States Salinity Laboratory Staff (1954) Diagnosis and improvement of saline and alkali soils. US Department of Agriculture, Agricultural Handbook No 60. Washington DC, USA

Vágújfalvi A, Galiba G, Cattivelli L, Dubcovsky J (2003) The cold-regulated transcriptional activator *Cbf3* is linked to the frost-tolerance locus *Fr-A2* on wheat chromosome 5A. Mol Genet Genom 269: 60-67

Vales M, Schön C, Capettini F, Chen X, Corey A, Mather D, Mundt C, Richardson K, Sandoval-Islas J, Utz H, Hayes P (2005) Effect of population size on the estimation of barley stripe rust QTL. Theor Appl Genet 111: 1260-1270

Vargas M, van Eeuwijk FA, Crossa J, Ribaut JM (2006) Mapping QTLs and QTL x environment interaction for CIMMYT maize drought stress program using factorial regression and partial least squares methods. Theor Appl Genet 112: 1009-1023

Verma V, Foulkes MJ, Worland AJ, Sylvester-Bradley R, Caligari PDS, Snape JW (2004) Mapping quantitative trait loci for flag leaf senescence as a yield determinant in winter wheat under optimal and drought-stressed environments. Euphytica 135: 255-263

Wang J, Bonnett DG, Chapman SC, Rebetzke GJ (2007) Efficient use of marker-based selection in plant breeding. Crop Sci 47: 580-588

Welch ME, Rieseberg LH (2002) Habitat divergence between a homoploid hybrid sunflower species, *Helianthus paradoxus* (Asteraceae), and its progenitors. Am J Bot 89: 472-478

Welcker C, Boussuge B, Bencivenni C, Ribaut JM, Tardieu F (2007) Are source and sink strengths genetically linked in maize plants subjected to water deficit? A QTL study of the responses of leaf growth and of Anthesis-Silking Interval to water deficit. J Exp Bot 58: 339-349

Worland AJ, Korzun V, Ganal MW, Röder M, Law CN (1998) Genetic analysis of dwarfing gene *Rht8* in wheat. Part II. The distribution and adaptive significance of allelic variants at the *Rht8* locus of wheat as revealed by microsatellite screening. Theor Appl Genet 96: 1110-1120

Xu W, Subudhi PK, Crasta OR, Rosenow DT, Mullet JE, Nguyen HT (2000) Molecular mapping of QTLs conferring stay-green in grain sorghum (*Sorghum bicolor* L. Moench). Genome 43: 461-469

Yang J, Zhu J (2005) Methods for predicting superior genotypes under multiple environments based on QTL effects. Theor Appl Genet 110: 1268-1274

Yeo AR, Flowers TJ (1986) Salinity resistance in rice (*Oryza sativa* L.) and a pyramiding approach to breeding varieties for saline soils. Aust J Plant Physiol 13: 161-173

Zadoks JC, Chang TT, Konzak CF (1974) Decimal code for growth stages of cereals. Weed Res 14: 415-421

7 Molecular Mapping and Breeding of Physiological Traits

Julie Graham[1*], Milind B. Ratnaparkhe[2] and Wayne Powell[3]

[1]Scottish Crop Research Institute, Invergowrie, Dundee, DD2 5DA, UK

[2]210 Waters Hall, Division of Plant Sciences, University of Missouri-Columbia, MO 65211, USA

*Corresponding author: jgraha@scri.sari.ac.uk

[3]NIAB, Huntingdon Road, Cambridge, CB3 OLE, UK

1 INTRODUCTION

1.1 Genetics and Plant Breeding

An understanding of genetics has been the key to improving our knowledge of many aspects of biology. Plant breeding has advanced greatly through an understanding of the principles of heredity, with Mendelian genetics forming the basis of plant breeding. This together with an understanding of biometrical genetics which applies to traits showing continuous variation and controlled by more than one gene, has allowed the manipulation of quantitative traits. Plant breeding aims to develop cultivars which fit specific environment and production practices and high yielding products whether for food, feed or processing into another product. Progress in any breeding program is based on the amount of genetic variability available and how effective is the selection and evaluation of the trait in question. Selecting for improved phenotype began early in the domestication of plants and has been the primary means ever since. Figure 1 shows the diversity of different Peruvian landraces of potato through selection for desirable traits. Phenotypic selection has limitations especially when interest is

Fig. 1 Phenotypic variation caused by genetic diversity of potato tubers from Peruvian landraces. Courtesy of SCRI.

focused on more complex physiological traits. A more accurate way of selection would be at the genetic level where markers linked to the gene(s) or quantitative trait loci (QTLs) underlying the trait can be screened for. A prerequisite for genotypic selection is the establishment of associations between traits of interest and genetic markers.

1.2 Nature of Physiological Traits

Plant physiology is the study of the functions and processes occurring in plants, which include the various aspects of plant life such as yield, photosynthetic capacity, cold tolerance, water relations, mineral nutrition, growth, developmental stages, and response to environmental stimuli as well as their interactions *inter se*.

Understanding the genetic control of physiological traits and the linkage of these physiological characteristics to molecular markers on chromosomes, and ultimately the gene(s) underlying the trait is the future of plant breeding. Mapping and cloning of genes involved in various physiological traits is a rather complicated process though due to the complexity of plant functions. In contrast, the accelerated pace of gene sequence discovery has surpassed our capability to understand the biological function of many genes. Sequencing technology has provided a wealth of genomic (total DNA) and expressed sequence (DNA that is transcribed and translated into protein) information, that has been accompanied by projects to locate, annotate, and assign biological function. Despite these efforts only a fraction of these annotated genes are associated with a phenotype that provides a predictive framework for understanding and manipulation. Complex traits such as pest and disease resistance mechanisms and quality traits such as flavor, texture, and appearance are poorly understood at the molecular/biochemical level. A renewed focus on linking genes to phenotypes is required to ensure our understanding of commercially acceptable traits.

2 MOLECULAR MARKERS FOR PHYSIOLOGICAL TRAITS

Molecular markers are DNA sequences (both known and unknown) that are located near genes and inherited characteristics of interest, allowing selective breeding and identification of progeny with desired characteristics. Molecular markers have been rapidly adopted by researchers globally as an effective and appropriate tool for basic and applied studies addressing physiological traits. Markers are most informative when integrated into the genetic linkage maps. Molecular markers, their history, type, and applications are described in detail in Chapter 2 in Volume 1.

These molecular markers are used as tools that identify DNA polymorphisms between DNA samples of different individuals. These polymorphisms can be of many different types from single nucleotide changes, large or small insertions and deletions or length variation in repeat sequences. All, however, provide information on that particular locus in the genome and importantly when that locus is known to be associated with a particular plant phenotype. An important way of linking marker loci to a particular plant phenotype is through the use of genetic linkage maps (Fig. 2). These maps when coupled with field trails and glasshouse or laboratory experiments to measure traits of interest on the population of individuals used for map development can then be used to relate phenotypic data to marker data on linkage maps (Graham et al. 2006).

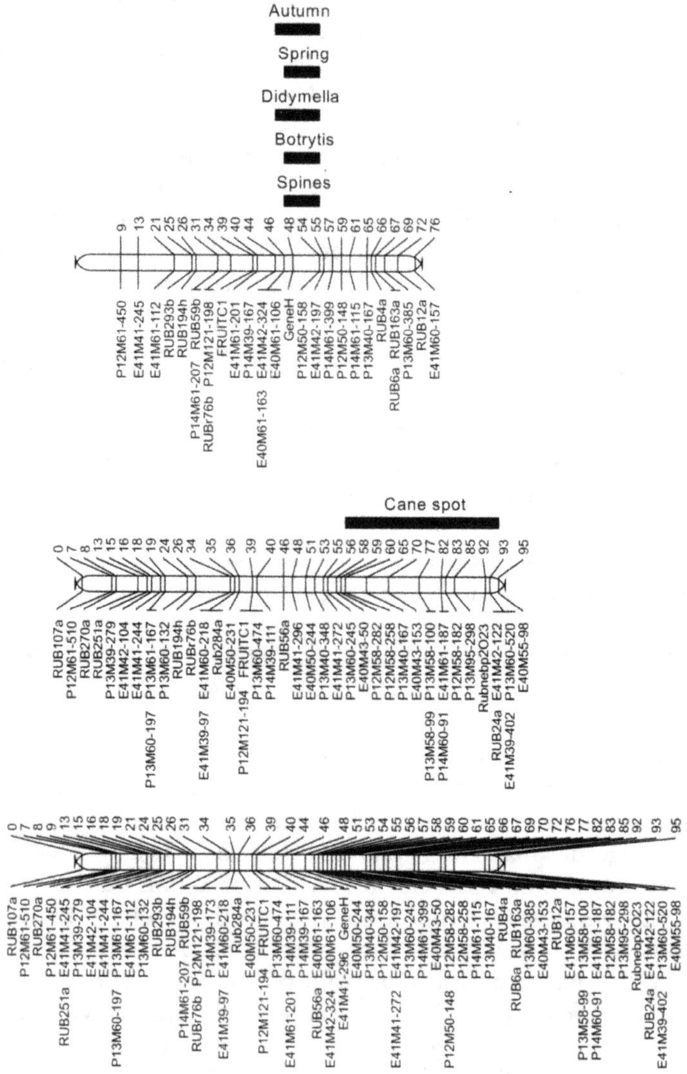

Fig. 2 Linkage group 2 of red raspberry with phenotypic data for disease and spines.

For map construction, individual marker loci are genetically characterized in a segregating population and the recombination rate of alleles at each pair of loci can be determined using classical linkage analysis. Loci can then be ordered into a linkage map and distance between loci can be expressed as recombination units given in centiMorgans (cM) where one cM is equal to 1% recombination. Once a sufficient number of markers have been mapped, the number of linkage groups should equal the haploid number of chromosomes. Several computer programs are available to quickly generate a map once markers have been applied to a segregating population. A detailed description of map construction is given in Chapter 4 in Volume I.

In the initial phase of map creation, genetically diverse parents are chosen which are known to segregate for the trait(s) of interest and depending on the biology of the crop an F_1, F_2, or backcross used for map construction. Once a map and segregating population have been developed attempts can be made to identify map locations of traits of interest. The speed and precision of crop enhancement can be improved by the development of genetic linkage maps which allow us to develop diagnostic markers for polygenic traits and in the future, aid the identification of the genes behind the traits (Fig. 3).

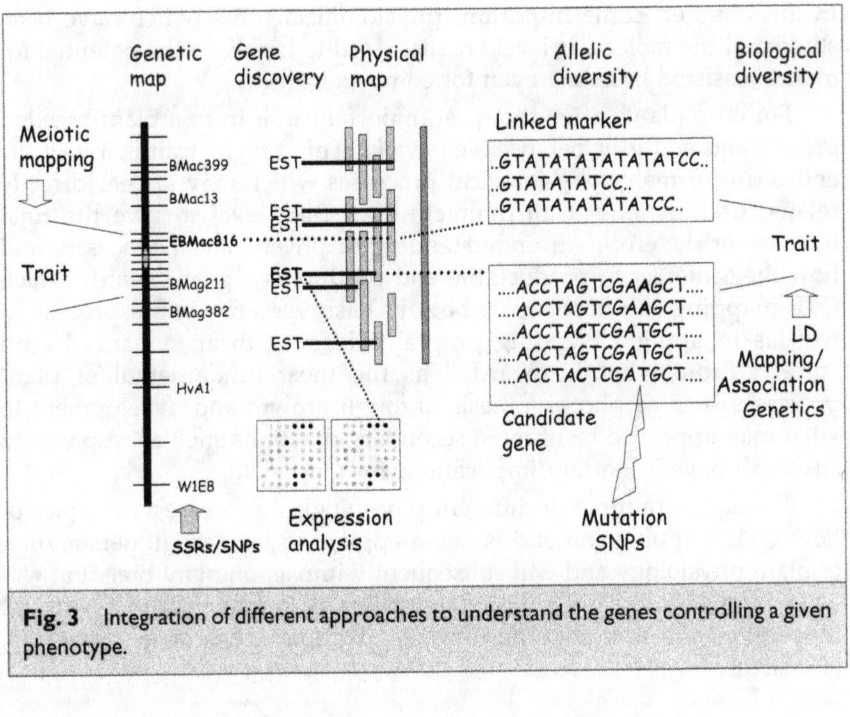

Fig. 3 Integration of different approaches to understand the genes controlling a given phenotype.

3 QTL MAPPING OF PHYSIOLOGICAL TRAITS

QTL mapping of physiological traits will provide crop breeders with a better understanding of the basis for the genetic correlation between economically important traits. This has potential to facilitate a more efficient incremental improvement of specific individual target traits.

Physiological traits may not be routinely screened for as the resulting phenotypes are time consuming or expensive to determine. Once experiments have been carried out however and DNA markers have been identified closely linked to the QTL involved in the expression of a physiological trait, they can be used efficiently in future marker-assisted breeding experiments. Identification at the same genomic location of QTLs related to physiological and morphological traits should be expected, given that changes in physiological pathways have an impact on the plant phenotype. Mapping single traits and QTLs are described in detail in Chapters 5 and 6, respectively in Volume 1. Once markers are generated, they can be used swiftly on various populations using marker-assisted selection (MAS). QTL mapping has become a standard procedure in dissecting the genetic controls of a variety of traits (see Table 1, for examples). The next step is the identification of the genes, alleles and physiological processes that are biologically important. In this chapter, some important physiological traits which have been studied at the molecular level are discussed to highlight the potential for marker-assisted breeding even for complex traits.

For crop plants often the most important trait from a plant breeder, grower and end user perspective is yield. This 'single' trait is in fact the end result of many physiological processes which may appear directly related or remote, but all interacting at some level to give the final biomass of the crop. An understanding of physiological processes and how they interact to produce the end product is a goal towards which QTL mapping can play a part both in basic research into a process as well as in applied breeding programs leading to an enhanced crop variety. From what is regarded as the most fundamental of plant processes such as photosynthesis, through growth and development to what may appear to be of more secondary functions such as response to stress, all have important implications for crop yield.

Through a number of different physiological processes, examples of how QTL mapping can and is being applied to increase understanding of plant physiology and will subsequently impact on plant breeding will be demonstrated. This will cover aspects of photosynthesis, nutrient absorption and assimilation, nitrogen fixation, senescence, flowering, and stress.

Table 1 Selected examples of some of the important traits for which genetic marker techniques have been used to map loci of physiological traits.

Crop	*Trait*	*Reference*
Rice	Ion transport	Koyama et al. 2001
Barley	Photoperiod response	Stracke 1998
	Photoperiod, vernalization	Karsai et al. 1997
Wheat	Flowering time	Tóth et al. 2003
	Flowering time	Sarma et al. 1998
	Vernalization	Sutka et al. 1999
	Flowering, physio-morphological traits	Börner et al. 2002
	Nitrogen absorption	Kato et al. 1999
	Metabolism, physiological traits	Gao et al. 2004
Pearl millet	Flowering, ear emergence time	Yadav et al. 2003
	Flowering, ear emergence time	Kato et al. 1999
Ryegrass	Heading date	Armstead et al. 2004
Soybean	Flowering, oil, protein	Zhang et al. 2004
Brassica oleracia	Flowering	Sebastian et al. 2002
Brassica rapa, Brassica napus	Flowering time, vernalization, winter survival, freezing tolerance	Kole et al. 2001, 2002
Sunflower	Flowering	Bert et al. 2003
Lotus, pea	Nitrogen fixation	Stracke and Borner 1998
Solanaceae	Flowering	Frary et al. 2003
Potato	Carbohydrate metabolism	Chen et al. 2001
Sugar beet	Sucrose content	Schneider et al. 2002
Arabidopsis	Flowering	Loudet et al. 2002
	Carbohydrate and nitrogen metabolism	Rauh et al. 2002
Sugi	Flowering, juvenile growth	Yoshimaru et al. 1998

3.1 Photosynthesis

Photosynthesis is the process by which plants, algae, and some bacteria use the energy from sunlight to produce sugar, which cellular respiration converts into ATP, the 'fuel' used by all living things. So, for example in rice, photosynthesis is the primary source of dry matter production and grain yield. The contribution of leaf photosynthesis to biomass has been estimated at 30%.

The conversion of unusable sunlight into usable chemical energy is associated with the actions of the green pigment chlorophyll. Most of the time, the photosynthetic process uses water, releasing oxygen essential for life. The raw materials of photosynthesis, water and carbon dioxide, enter the cells of the leaf, and the products of photosynthesis, sugar and oxygen, leave the leaf. The process is complex with many steps and many factors contributing to efficiency. Leaf photosynthesis cannot be measured in the field; hence the trait does not lend itself to large scale screening of segregants in a breeding program. If the process can be broken down, however, into the various processes and factors of importance, attempts can then be made to map these factors. Thus, a number of phenotypes and processes that impact on the photosynthetic potential of a plant can be measured. Factors such as chlorophyll content, gas diffusion, water availability, leaf size, root properties, and temperature, to name a few, will all have a role in determining the plants achievable photosynthetic potential. Research aimed at mapping QTLs relevant to photosynthetic potential has been initiated.

Recently a group working in rice (Teng et al. 2004) measured photosynthetic rate, chlorophyll content, stomatal resistance, and transpiration rate from the parents and a doubled haploid (DH) population containing 127 lines. A map was established from this population using restriction fragment length polymorphism (RFLP) and simple sequence repeat (SSR) markers and from this, two putative QTLs for net photosynthetic rate, three QTLs for chlorophyll content, one for stomatal resistance and two for transpiration rate were identified.

The interrelationship between photosynthesis and other physiological traits has been highlighted in works on both low temperature and water stress (discussed in more detail below). Low growth temperature parameters are considered to be important on the photosynthetic apparatus. In particular the combination of high light intensity and low temperature can cause photosynthetic inhibition of photosynthesis. This could be due to a number of factors including the ability to develop a functional photosynthetic apparatus at low temperature, as well as the susceptibility of enzymes in the C_4-cycle. In a study of maize grown under chilling stress, photosynthetic performance as well as shoot biomass was affected under low temperature and QTLs were mapped for a number of parameters (Jompur et al. 2005). The photosynthetic traits net CO_2 uptake, stomatal conductance, and abscisic acid (known to have a role in regulating stomatal opening) have also been measured in relation to water stress. Here the authors looked at normally irrigated and water-deprived young maize plants and found that QTL distribution was strongly influenced by water stress (Pelleschi et al. 2006). These examples begin to illustrate the relationships between traits and their final manifestation as yield potential.

3.2 Nutrient Absorption and Assimilation

Green plants are by far the earth's most important autotrophic organisms which can subsist in an inorganic environment by manufacturing their own organic compounds from raw materials small enough and soluble enough to pass through cell membranes. As seen in the last section, the raw materials most needed for photosynthesis are carbon dioxide and water which supply carbon, oxygen, and hydrogen, the predominant components in organic molecules. Carbon dioxide is obtained from the air through the leaves and water from the soil through the roots. There follows the synthesis of carbohydrate making up the bulk of dry weight of the plant. Other elements enter into the composition of the plant. Nitrogen is present in amino acids and two important amino acids also contain sulfur. ATP contains phosphorous and chlorophyll contains magnesium to name a few other elements. These other elements must therefore come from the soil, hence the development of fertilizer regimes in agriculture. Modern fertilizers are often designated by their ratios of N-P-K, these three elements are most rapidly removed from the soil and need to be replenished if crops are to continue to flourish. Functions of leaves and roots essential to nutrient absorption must therefore impact greatly on plant biomass. Thus for roots the capacity for water uptake is fundamental for crop productivity. The capacity of the root for uptake depends on the degree to which the root extends its absorption area, which is determined by complex root morphology. The lateral roots which occupy more than 90% of the total length of the root system play a major role in water uptake. To date studies have been carried out to identify QTLs related to root morphological features such as root mass and depth, root axis length and lateral branching (Yadav et al. 1997; Shen et al. 2001; Horii et al. 2006). QTLs for other factors, such as the effects on the plant of boron toxicity at high concentrations, have been identified (Jefferies et al. 2000). In studies on grain yield, differences in the concentration of boron have led to a 17% difference in grain yield. Other areas such as the ability to increase uptake of applied nitrogen fertilizer, whereby reducing losses into the environment have been studied (An et al. 2006). Understanding and optimization of plant nutrient uptake, therefore, have great potential for ultimate yield improvement not to mention other aspects such as tolerance to toxic compounds in the soil.

3.3 Nitrogen Fixation

Although the earth's atmosphere contains 78% nitrogen, free gaseous nitrogen cannot be utilized by animals or by higher plants. They depend instead on nitrogen that is present in the soil. To enter living systems, nitrogen must be 'fixed' (combined with oxygen or hydrogen) into compounds that plants can utilize, such as nitrates or ammonia. A certain amount of atmospheric nitrogen is fixed by lightning and by some cyanobacteria blue-green algae. But the great bulk of nitrogen fixation is performed by soil bacteria of two kinds: those that live free in the soil and those that live enclosed in nodules in the roots of certain leguminous plants (e.g., alfalfa, peas, beans, clover, soybeans, and peanuts). The symbiotic interaction between legumes and rhizobial bacteria accounts for a significant portion of biological nitrogen fixation worldwide. Nitrogen fixation is the process by which atmospheric nitrogen gas is converted into ammonia. The ammonia is subsequently available for many important biological molecules such as amino acids, proteins, vitamins, and nucleic acids. Legumes have the ability to use certain kinds of bacteria (genus *Rhizobium*) as a means of getting nitrogen through a process of nitrogen fixation. The site of fixation is the nodule, a unique plant organ located on the root, which functions to generate the aerobic environment essential for bacterial survival and nitrogenase activity. Nodule formation involves plant/bacterial signaling, with the bacterially generated signaling molecule, Nod factor, playing a critical role. The bacteria and the plant have a symbiotic relationship; the plant provides the bacteria with food, and the bacteria fixes nitrogen for the plant releasing much of the fixed nitrogen into the plants cytoplasm primarily in the form of amino acids. The bacteria in nodules generally produce a surplus of fixed nitrogen some of which is excreted from the roots into the soil. The whole nitrogen fixation process is very complicated because it involves both the bacterium and the plant. The process begins by host recognition followed by in most cases root hair infection. This leads to nodule development and to the reduction of nitrogen to ammonia catalyzed by an enzyme complex known as nitrogense. The plant genes determining nodulation and nitrogen fixation are not well known. Mapping out the genetic blueprints of the specific proteins and enzymes involved in the process allows us to understand how nitrogen fixation is regulated. Identification of the symbiotic genes controlling nodule formation and the efficiency of nitrogen fixation are one of the main goals in legume breeding. It involves localization of symbiotic genes on chromosomes with the use of genetic, biochemical, and ultrastructural analyses of symbiotic mutants.

Some nodulin genes and plant proteins with functions in oxygen transport, cell wall architecture, sugar, and N metabolism have been identified (Verma et al. 1992). A number of mapping studies have been carried out in this area. In common bean, QTLs affecting nodule number have been identified (Nodari et al. 1993). In soybean, QTLs controlling nodulation and shoot mass have been determined (Nicolas et al. 2005). *Sym* genes are essential for nodulation and/or symbiotic nitrogen fixation in legumes. Two novel *Sym* genes were isolated in *L. japonicus* by combining genetic and physical mapping with genome sequencing (Stracke et al. 2004). Approximately 40 *sym* mutants have been identified to date in various legume species, including pea, soybean, *M. truncatula*, and *L. japonicus*. Mapping QTLs and identification of regulatory genes are greatly increasing knowledge of the nitrogen fixation process.

3.4 Senescence

The biological importance and potential for improvement of crop characteristics, particularly plant productivity and post-harvest storage, have prompted extensive physiological, molecular, and genetic analyses of leaf senescence. Senescence is considered the final stage in leaf development. It is an active, ordered process that involves mobilization of nutrients from the senescing leaves to other parts of the plant, leading to the eventual death of the leaf. Plant hormones such as ethylene and cytokinins play vital roles in the regulation of senescence. Attention has focused on cytokinins that are key components of plant senescence. These compounds have been implicated in several aspects of plant development and are thought to be synthesized mainly in the roots and transported to the shoots via the xylem. Three main approaches have been used to study the effect of cytokinins in plant senescence, namely exogenous application of cytokinin solutions, measurement of endogenous cytokinins during senescence, and transgene-encoded cytokinin biosynthesis. Enzymes with varied functions have also been found to be involved in senescence. Senescence is a genetically programmed sequence of biochemical and physiological changes, a type of programmed cell death characterized by loss of chlorophyll, lipids, total protein, and RNA. The initiation of senescence is subject to regulation by various environmental and autonomous (internal) factors. Plants have evolved mechanisms by which leaf senescence can be induced by many stresses to reallocate nutrients to reproductive organs and to eliminate water consumption by older, less productive leaves. This regulation of leaf senescence has an obvious adaptive value as it allows the plant to complete its life cycle even under stressful condition. Senescence is a correlative event and is the cumulative effect of the formation of flowers, fruits, seeds, apices, and roots. Senescence can be induced by environmental stress, such as low light intensity, nutrient deficiency, pathogen attack, drought, waterlogging, and detachment from the plant. Endogenous factors, including leaf age and reproductive development, also trigger senescence. The identification of genetic mutants that control leaf senescence in *Arabidopsis thaliana* opened up new possibilities for genetically analyzing leaf senescence in a model system. ORE9, an F-box protein that regulates leaf senescence in *Arabidopsis* was identified by a map-based cloning approach. Genetic mapping, using cleaved amplified polymorphic sequence (CAPS) markers, placed *ORE9* locus on chromosome 2 (Woo et al. 2001).

Manipulation of leaf senescence may play an important role in crop production. For example, late leaf senescence has been shown to play an important role in increased photosynthetic activity in plants (Mae 1997). In rice, 60-90% of the total carbon found in rice panicles is produced by photosynthesis after heading. Genetic analysis of the trait has been hindered by the difficulty in measuring this trait. In a recent study, QTLs for senescence were identified by measuring leaf chlorophyll content, reduction in chlorophyll content and the number of late-discoloring leaves per panicle (Hu et al. 2005).

3.5 Flowering

Plant development is seasonal with flowering and other developmental processes occurring at particular times of the year. The obvious explanation for this timing is that these stages happen in response to environmental cues such as temperature or light intensity. These seasonal changes are in fact dependent on certain environmental characteristics acting as inductive signals to set in motion complex developmental processes. Photoperiodism, response to the length of day and vernalization, response through exposure to a period of cold, are two major mechanisms underlying such seasonal responses. Response to day-length and vernalization are two of the most important mechanisms controlling flowering, and therefore yield, in many of our crop plants. Flowering time is determined by genes that control the vernalization response and also by genes that control the photoperiodic response or affect flowering time independently of photoperiod and vernalization. Individual genes which have qualitative effects on these responses have been identified using molecular markers. The complex interactions among genes of these three classes often result in continuous variation in flowering time that is usually analyzed by quantitative trait loci (QTL) techniques (Laurie et al. 1995). Mapping of flowering time genes in relation to molecular markers is clearly a workable method and a consequence of colinearity is that markers mapped in other crosses or species become available for use.

The time of flowering–in some crops referred to as heading date–is a key target trait in breeding programs for many crop species. Flower development consists of several phases.

(1) The first step is the transition from vegetative to reproductive development, regulated by floral induction.

(2) Later steps include the initiation of individual flowers, the determination of organ identity, and organ-specific differentiation

In most plants flowering is triggered by either day-length, by exposure to a period of low temperature, or by both. The plant has parameters for the optimal, minimum and maximum temperatures for floral induction. Plants can be placed into three general groups based on their response to photoperiod (their requirement of light for initiation of flowering).

(1) *Short-day* plants such as corn require long night lengths to flower.

(2) *Long-day* plants such as lettuce require short night lengths to flower.

(3) *Day-neutral* plants such as tomatoes do not depend on photoperiod to flower, but high or low temperatures can retard flowering.

When the required photothermal units (the optimum amount of light and temperature) for floral induction are accumulated, the plant moves into the floral initiation phase. Floral initiation is the stage when the plant has decided to flower and is making the biochemical changes necessary to create inflorescences, flowers and fruits (Fig. 4). The length of the floral initiation period depends largely on temperature in many plants. Flowering time can be assessed in the same way as many other phenotypic variables by genetic mapping of segregating populations developed from parents with different attributes. In cases where genes of large effects are segregating, flowering time in a population may be resolvable into discreet Mendelian classes. In other cases where the trait behaves in a quantitative manner the underlying genes (QTL) must be located by statistical methods. Linked markers can be used for selection of specific flowering time alleles, but because their efficiency depends on the tightness of linkage they will be most efficient for genes that are accurately mapped. There are well established common principles that link crop model species and in several cases orthologous genes have been shown to regulate equivalent developmental process.

The most striking recent advances in our understanding of the genetic control of the timing of flowering have come from work on *Arabidopsis*. A large number of genes that control the timing of the transition to flowering have been identified in *Arabidopsis* by mutant

Fig. 4 Flowering in blackberry (*Rubus* spp) Courtesy SCRI.

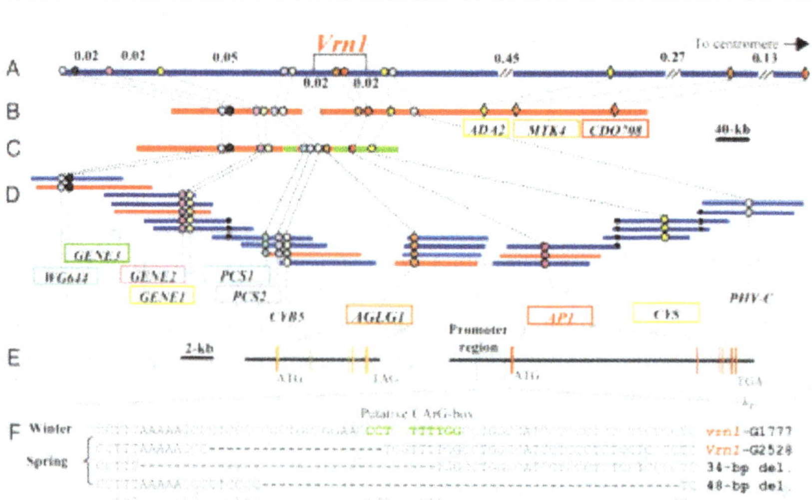

Fig. 5 Genetic and physical map of the *VRN1* region on chromosome $5A^m$ of *T. monococcum* (Yan et al. 2004).

analysis Genetic analysis of flowering time in pea, cereals, and *Arabidopsis* supports the hypothesis that the transition to flowering is under multifactorial control. Multiple genes that control flowering time have been identified in different species.

The understanding of the genetic control of flowering time is much more advanced in *Arabidopsis* than in crop species. Due to their close phylogenetic relationships, information about *Arabidopsis* flowering time can be readily exploited in *Brassica* species. In these models flowering is regulated by photoperiod, vernalization, and gibberellin pathways. Identification of the genes involved in flowering time has made it possible to determine the genetic control pathways for the response to photoperiod and vernalization in *Arabidopsis* (Levy and Dean 1998, Samach and Coupland 2000). In addition, homologs of *Arabidopsis* genes for flowering time also function in *Brassica napus* and *B. rapa* (Robert et al. 1998; Kole et al. 2001). Identification of the genes involved in flowering time has made it possible to determine the genetic control pathways for the response to photoperiod and vernalization in *Arabidopsis* Numerous studies have concentrated on mapping genome regions controlling flowering time and vernalization requirement (e.g., Ferreira et al. 1995, Kole et al. 2001, 2002)

In cereals, understanding the genes that control flowering time is best understood in rice (Yano et al. 2000; Ishimaru et al. 2001; Izawa et al. 2003; Hayama and Coupland 2003). Several flowering time genes have been isolated from rice, principally by positional cloning methods. Control of flowering in response to day-length has been analyzed extensively, and interestingly the genes cloned to date relate directly to known genes in *Arabidopsis*. Rice is very different from *Arabidopsis* in flowering behavior, being a tropical species that shows promotion of flowering in response to short-days and lacks a vernalization response. Rice is a short-day (SD) plant; its heading is promoted by short photoperiods. The response of the plant to length of day (referred to as photoperiod sensitivity) and its basic vegetative growth determine the heading date of rice. Heading date is a critical trait for adaptation to different cultivation areas and cropping seasons. Like many other important traits in plant breeding, heading date is a complex trait that shows continuous phenotypic variation among progeny and is controlled by multiple genes known as quantitative trait loci (QTLs). QTLs typically are difficult to identify because of the lack of discrete phenotypic segregation and because the phenotypic effects of each gene associated with a complex trait are relatively small. A major quantitative trait loci (QTL) controlling response to photoperiod, *Hd1*, was identified by means of a map-based cloning strategy (Yano et al. 2000). Many genes for heading in rice have been genetically identified (Yokoo et al. 1980; Yamagata et al. 1986; Poonyarit et al. 1989; Yokoo and Okuno 1993; Tsai 1995; Kinoshita 1998; Yamamoto 1998, 2000). The major genes or quantitative trait loci (QTLs) for heading date have been mapped by using DNA markers (Mackill et al. 1993; Xiao et al. 1996; Yano et al. 1997; Lin et al. 1998; Tamura et al. 1998).

Photoperiod sensitivity genes which control growth phase of rice plant from the vegetative stage to the reproductive stage are very important in breeding programs of rice. *Se-1* gene, one of the major genes for the photoperiod sensitivity of rice has been located on chromosome 6. In barley, two major loci regulating photoperiod response have been identified by genetic analyses (Laurie et al. 1995). Major loci controlling photoperiod response have also been mapped to the short arms of the group 2 chromosomes of wheat and comparative mapping using common RFLP markers shows that the wheat and barley genes fall into equivalent intervals.

Not all plants have a vernalization requirement, and the degree of vernalization required can vary within a species. In many plants, flowering can be accelerated or induced by exposure to a long period of near-freezing temperatures. This is a commonly employed reproductive strategy that allows for flowering and seed production in the environmentally favorably period following natural winter. Vernalization, has been studied for decades at the physiological level but only recently at the molecular level. The isolation of *FLOWERING LOCUS C*, in *Arabidopsis* has now provided an insight into the molecular mechanism involved, including the role of DNA methylation. The MADS-box protein encoded by *FLOWERING LOCUS C* (*FLC*) is a repressor of flowering. Vernalization, which promotes flowering in the late-flowering ecotypes and many late-flowering mutants, decreases the level of FLC transcript and protein in the plant. Many plants are responsive to vernalization, and use it as a cue to indicate that winter has passed ensuring flowering occurs in the favorable conditions of spring. Many late flowering *Arabidopsis* accessions and mutants exhibit a strong vernalization response. The roles of abscisic acid (ABA), gibberellins (GAs) and phytochrome B in the vernalization response have been investigated by combining mutations causing defects in their biosynthesis and response with *fca-1* (Chandler et al. 2000).

Winter wheat require several weeks at low temperature to flower. This process is controlled mainly by the *VRN1* gene (Yan et al. 2003). Genes controlling vernalization requirement in hexaploid wheat and other temperate cereals have been extensively characterized. Genes for vernalization response have been located in chromosomes of homeologous group 5 of wheat (Snape et al. 2001). Figure 5 shows the genetic and physical map of *VRN1* region of *T. monococcum*.

Three major loci controlling vernalization response have been identified in barley by genetic analysis. *Vrn-H1* maps to the middle of the long arm of barley chromosome 5H and comparative mapping shows that major vernalization response genes map to equivalent positions in wheat (*Vrn-A1, -B1,* and *-D1*) and rye (*Vrn-R1*; Laurie et al 1997; Börner et al. 1996).

Photoperiod and temperature are important for initiating pathways of development. As we have seen above in photosynthesis these parameters impact on photosynthetic potential.

3.6 Stress Tolerance

Plants are faced with a large number of stresses and have developed many physiological pathways to cope with them. Current breeding is hampered by a lack of understanding of the genetic mechanisms responsible. A large number of genes, transcripts, and proteins implicated in stress tolerance have been identified though their function remains unclear.

3.6.1 Salt Tolerance

It is estimated that saline soils cover between 400 and 950 million hectares of the Earth's surface. Accumulation of salt in the soil causes deleterious effects leading to a reduction in production from rice and other crops. Salt tolerance of a crop is the final manifestation of several components such as Na^+ and K^+ uptake and exclusion, ion balance, and ion compartmentalization. For example salt tolerance in rice is an important objective of rice breeding for coastal areas. Yeo et al. (1990) dissected the complex physiological trait of salt tolerance by measuring physiological components such as shoot sodium content, plant survival, and plant vigor. With the development of molecular marker technology the potential for genetic dissection of the mechanisms of salt tolerance are possible. QTL analysis of salt tolerance in rice has been carried out by several groups. These have identified QTLs for plant survival, Na^+ and K^+ uptake, Na^+ and K^+ concentration and Na+:K+ ratio in shoots (Zhang et al. 1995; Lin et al. 1998; Gong et al. 1999; Prasad et al. 2000; Koyama et al. 2001).

A QTL mapping approach concerning both shoots and roots has been carried out in rice and a number of QTLs identified. QTLs for Na^+ and K^+ concentration in shoots explaining over 40% of the phenotypic variation were detected (Lin et al. 2004). The QTLs however detected between the roots and shoots did not share the same map location suggesting different genes may be controlling the transport of Na+ and K+ between roots and shoots. A rice QTL, *SKC1* which maintains K+ homeostasis in a salt tolerant variety under salt stress has been mapped, and map based cloning carried out to isolate the gene which was shown to encode a sodium transporter (Ren et al. 2005).

3.6.2 Drought Tolerance

The development of drought tolerance in plants has been slow due to a lack of understanding of the physiological parameters that are diagnostic of genetic potential for improved productivity under water stress. Tolerance to drought involves a complex of mechanisms working in combination to avoid or to tolerate water stress (Tuberosa et al. 2002). Water deficit occurs when water potentials in the rhizosphere are sufficiently negative to reduce water availability to levels that are sub-optimal for plant growth and development. Plants under drought stress demonstrate a number of physiological responses. Osmotic adjustment (OA) has been found to occur in plants with resistance to water stress. Relative water content is a useful measurement of plant water status as affected by leaf water potential and OA. Other measures include water-use efficiency measured indirectly as carbon isotope ratio, abscisic acid levels, photosynthesis measurements (stomatal conductance and net CO_2 uptake) and ash content. QTLs have been identified for a range of water stress factors including morphological traits such as plant and leaf size, root properties and yield components to more specific evaluations such as leaf relative water content, leaf osmotic potential, osmotic potential at full turgor, water soluble carbohydrate concentration, osmotic adjustment, and carbon isotope discrimination as well as a number of crop productivity measures (Diab et al. 2004; Saranga et al. 2004). Candidate genes and ESTs have been identified underlying QTLs for drought tolerance (Diab et al. 2004).

3.6.3 Cold Tolerance

Exposure to low temperature early in development is essential for regulation of key processes as described above in 'Components of Flowering'. However this exposure can have a negative impact on the plant causing, for example poor photosynthetic performance. In maize, cold-induced decrease in photosynthesis has been associated to a number of changes such as damage to the PSII reaction centers (Haldimann et al. 1995), alteration of leaf pigment composition (Haldimann et al. 1995) and lower enzyme activity in the carbon cycle

(Kingston-Smith et al. 1997). The relationship between vernalization requirement and cold tolerance is unclear. Little is known about the genetic basis of cold tolerance and a better understanding would allow the identification of the complexity of different regulatory traits. Studies on different plant genotypes however have suggested there is considerable genetic variation for cold tolerance (Haldimann 1998). In a cross between a cold tolerant and a cold sensitive maize line, a number of significant differences in traits such as better tolerance to chronic photoinhibition, greater operating quantum efficiency and trapping efficiency of PSII, higher proportion of open reaction centers, higher carbon exchange rate and greater shoot dry mass were observed. QTLs for most of the measured traits were identified (Fracheboud et al. 2004).

4 CONCLUSION

Understanding physiological pathways and the many interactions between them has been one of the biggest challenges in plant improvement. An observed plant phenotype is the result of the expression of many diverse plant physiological and biochemical pathways. Given that most physiological tests cannot be applied routinely in a breeding program, because their protocols are generally too complex or time consuming, the genetic dissection of target traits and physiological parameters is of primary importance for plant breeding. In recent years, a number of practical examples have demonstrated the power of high-density genetic maps and candidate gene studies for the identification of genetic markers closely linked to important physiological traits. A recent review by Price (2006) has highlighted the accuracy of QTL mapping and where in fact, map-based cloning using the original data rather than fine-mapping is a realistic option for gene identification.

References

Abdelkhalik AF, Shishido R, Nomura K, Ikehashi H (2005) QTL-based anaylsis of leaf senescence in an indica/japonica hybrid in rice (*Oryza sativa* L.). Theor Appl Genet 110: 1226-1235

An D, Su J, Liu Q, Zhu Y, Tong Y, Li J, Jing R, Li B, Li Z (2006) Mapping QTLs for nitrogen uptake in relation to the early growth of wheat (*Triticum aestivum* L.). Plant Soil 284: 73-84

Armstead IP, Turner LB, Farrell M, Skøt L, Gomez P, Montoy T, Donnison IS, King IP, Humphreys MO (2004) Synteny between a major heading-date QTL in perennial ryegrass (*Lolium perenne* L.) and the *Hd3* heading-date locus in rice. Theor Appl Genet 108: 822-828

Bert PF, Jouan I, Tourvieille de Labrouhe D, Serre F, Philippon J, Nicolas P, Vear F (2003) Comparative genetic analysis of quantitative traits in sunflower (*Helianthus annuus* L.). 2. Characterization of QTL involved in developmental and agronomic traits. Theor Appl Genet 107: 181-189

Börner A, Korzun V, Worland A (1996) Comparative genetic mapping of loci affecting plant height and development in cereals. Wheat: Prospects for global improvement. Proc 5th Int Wheat Conf, June 10-14, 1996, Ankara, Turkey

Börner A, Schumann E, Fürste A, Cöster H, Leithold B, Röder M, Weber W (2002) Mapping of quantitative trait loci determining agronomic important characters in hexaploid wheat (*Triticum aestivum* L.). Theor Appl Genet 105: 921-936

Chandler J, Martinez-Zapater JM, Dean C (2000) Mutations causing defects in the biosynthesis and response to gibberellins, abscisic acid and phytochrome B do not inhibit vernalization in Arabidopsis fca-1. Planta 210: 677-82

Chen X, Salamini F, Gebhardt C (2001) A potato molecular-function map for carbohydrate metabolism and transport. Theor Appl Genet 102: 284-295

Diab AA, Teulat-Merah B, This D, Ozturk NZ, Benscher D, Sorrells ME (2004) Identification of drought-inducible genes and differentially expressed sequence tags in barley. Theor Appl Genet 109: 1417-1425

Ferreira ME, Satagopan J, Yandell BS, Williams PH, Osborn TC (1995) Mapping loci controlling verbalization requirement and flowering time in *Brassica napus*. Theor Appl Genet 90: 727-732

Fracheboud Y, Jompuk C, Ribaut JM, Stamp P, Leipner J (2004) Genetic analysis of cold tolerance of photosynthesis in maize. Plant Mol Biol 56: 241-253

Frary A, Doganlar S, Daunay MC, Tanksley SD (2003) QTL analysis of morphological traits in eggplant and implications for conservation of gene function during evolution of solanaceous species. Theor Appl Genet 107: 359-370

Gao LF, Jing RL, Huo NX, Li Y, Li XP, Zhou RH, Chang XP, Tang JF, Jia JZ (2004) One hundred and one new microsatellite loci derived from ESTs (EST-SSRs) in bread wheat. Theor Appl Genet 108: 1392-1400

Gong JM, He P, Qian Q, Shen LS, Zhu LH, Chen SY (1999) Identification of salt tolerance QTL in rice (*Oryza sativa* L.). China Sci Bull 44: 68-71

Graham J, Smith K, Tierney I, MacKenzie K, Hackett CA (2006) Mapping gene H and association with resistance to cane botrytis and spur blight, rust and cane spot. Theor Appl Genet 112: 818-831

Haldimann P (1998) Low growth temperature-induced changes to pigment composition and photosynthesis in *Zea mays* genotypes differing in chilling sensitivity. Plant Cell Environ 21: 200-208

Haldimann P, Fracheboud Y, Stamp P (1995) Carotenoid composition in *Zea mays* developed at sub-optimal temperature and different light intensities. Physiol Planta 115: 471-475

Hayama R, Coupland G (2003) Shedding light on the circadian clock and the photoperiodic control of flowering. Curr Opin Plant Biol 6: 13-19

Horii H, Nemoto K, Miyamoto N, Harada J (2006) Quantitative trait loci for adventitious and lateral roots in rice. Plant Breed 125: 198-200

Hu M-L, Wang C-M, Yang QH et al. (2005) QTL analysis for traits associated with photosynthetic functions in rice (*Oryza sativa L.*). Acta Genet Sin 32(8): 818-824

Ishimaru K, Yano M, Aoki N, Ono K, Hirose T, Lin SY, Monna L, Sasaki T, Ohsugi R (2001) Toward the mapping of physiological and agronomic characters on a rice function map: QTL analysis and comparison between QTLs and expressed sequence tags. Theor Appl Genet 102: 793-800

Izawa T, Takahashi Y, Yano M (2003) Comparative biology comes into bloom: Genomic and genetic comparison of flowering pathways in rice and Arabidopsis. Curr Opin Plant Biol 6: 113-120

Jefferies SP, Pallotta MA, Paull JG, Karakousis A, Kretschmer JM, Manning S, Islam AKMR, Langridge P, Chalmers KJ (2000) Mapping and validation of chromosome regions conferring boron toxicity tolerance in wheat (*Triticum aestivum*). Theor Appl Genet 101: 767-777

Jompur C, Fracheboud Y, Stamp P, Leipner J (2005) Mapping o quantitative trait loci associated with chilling tolerance in maize (*Zea mays* L.) seedlings grown under field conditions. J Exp Bot 56: 1153-1163

Karsai I, Mészáros K, Hayes PM, Bedő Z (1997) Effects of loci on chromosomes 2 (2H) and 7 (5H) on developmental patterns in barley (*Hordeum vulgare* L.) under different photoperiod regimes. Theor Appl Genet 94: 612-618

Kato K, Miura H, Sawada S (1999) QTL mapping of genes controlling ear emergence time and plant height on chromosome 5A of wheat. Theor Appl Genet 98: 472-477

Kingston-Smith AH, Harbinson J, Williams J, Foyer CH (1997) Effect of chilling on carbon assimilation, enzyme activation and photosynthetic electron transport in the absence of photoinhibition in maize leaves. Plant Physiol 114: 1039-1046

Kinoshita T (1998) Report of the committee on gene symbolization, nomenclature and linkage groups. II. Linkage mapping using mutant genes in rice. Rice Genet Newsl 15: 13-74

Kole C, Quijada P, Michaels SD, Amasino RM, Osborn TC (2001) Evidence for homology of flowering-time genes *VFR2* from *Brassica rapa* and *FLC* from *Arabidopsis thaliana*. Theor Appl Genet 102: 425-430

Kole C, Thormann CE, Karlsson BH, Palta JP, Gaffney P, Yandell B, Osborn TC (2002) Comparative mapping of loci controlling winter survival and related traits in oilseed *Brassica rapa* and *B. napus*. Mol Breed 9(3): 201-210

Koyama ML, Levesley A, Koebner RMD, Flowers TJ, Yeo AR (2001) Quantitative trait loci for component physiological traits determining salt tolerance in rice. Plant Physiol 125: 406-422

Laurie DA (1997) Comparative genetics of flowering time in cereals. Plant Mol Biol 35: 167-177

Laurie DA, Pratchett N, Bezant JH and Snape JW (1995) RFLP mapping of five major genes and eight quantitative trait loci controlling flowering time in a winter × spring barley (*Hordeum vulgare* L.) cross. Genome 38: 575-585

Levy YY, Dean C (1998) The transition to flowering. Plant Cell 10: 1973-1989

Lin HX, Yanagihara S, Zhuang JY, Lu J, Senboku T, Zheng KL, Yashima S (1998) Identification of a QTL for salt tolerance in rice via molecular markers. Chn J Rice Sci 12: 72-78

Lin HX, Zhu MZ, Yano M, Gao JP, LIang ZW, Su WA, Hu H, Ren ZH, Chao DY (2004). QTLs for Na+ and K+ uptake of the shoots and roots controlling rice salt tolerance. Theor Appl Genet 108: 253-260

Lin SY, Sasaki T, Yano M (1998) Mapping quantitative trait loci controlling seed dormancy and heading date in rice, *Oryza sativa* L., using backcross inbred lines. Theor Appl Genet 96: 997-1003

Loudet O, Chaillou S, Camilleri C, Bouchez D, Daniel-Vedele F (2002) Bay-0 × Shahdara recombinant inbred line population: a powerful tool for the genetic dissection of complex traits in Arabidopsis. Theor Appl Genet 104: 1173-1184

Mackill DJ, Salam MA, Wang ZY, Tanksley SD (1993) A major photoperiod-sensitivity gene tagged with RFLP and isozyme markers in rice. Theor Appl Genet 85: 536-540

Mae T (1997) Physiological nitrogen efficiency in rice: nitrogen utilization photosynthesis and yield potential. Plant Soil 196: 201-210

Nicolas MF, Hungria M, Arias CAA (2005) Identification of quantitative trait loci controlling nodulation and shoot mass in progenies from two Brazilian soybean cultivars. Field Crops Res 95: 355-366

Nodari RO, Tsai SM, Guzman P, Gilbertson RL, Gepts P (1993) Towards an integrated linkage map of common bean. III. Genetic factors controlling host-bacteria interactions. Genetics 134: 341-350

Pelleschi S, Leonardi A, Rocher JP, Cornic G, de Vienne D, Thevenot C, Prioul JP (2006) Analysis of the relationships between growth, photosynthesis and carbohydrate metabolism using quantitative trait loci (QTLs) in young maize plants subjected to water deprivation. Mol Breed 17: 21-39

Poonyarit M, Mackill DJ, Vergara BS (1989) Genetics of photoperiod sensitivity and critical day length in rice. Crop Sci 29: 647-652

Prasad SR, Bagali PG, Hittalmani S, Shashidhar HE (2000) Molecular mapping of quantitative trait loci associated with seedling tolerance to salt stress in rice. Curr Sci 78: 162-164

Price AH (2006) Believe it or not, QTLs are accurate. Trends Plant Sci 11: 213-216

Rauh B, Basten C, Buckler E (2002) Quantitative trait loci analysis of growth response to varying nitrogen sources in *Arabidopsis thaliana*. Theor Appl Genet 104: 743-750

Ren ZH, Gao JP, Li LG, Cai XL, Huang W, Chao DY, Zhu MZ, Wang ZY, Luan S, Lin HX (2005) A rice quantitative trait locus for salt tolerance encodes a sodium transporter. Nat Genet 37: 1141-1146

Robert LS, Robson F, Sharpe A, Lydiate D, Coupland G (1998) Conserved structure and function of the *Arabidopsis* flowering time gene *CONSTANS* in *Brassica napus*. Plant Mol Biol 37: 763-772

Samach A, Coupland G (2000) Time measurement and the control of flowering in plants. Bioessays 22: 38-47

Saranga Y, Jiang CX, Wright RJ et al. (2004) Genetic dissection of cotton physiological responses to arid conditions and their inter-relationships with productivity. Plant Cell Environ 27(3): 263-277

Sarma RN, Gill BS, Sasaki T, Galiba G, Sutka J, Laurie DA, Snape JW (1998) Comparative mapping of the wheat chromosome 5A Vrn- region with rice and its relationship to QTL for flowering time. Theor Appl Genet 97: 103-109

Schneider K, Schäfer-Pregl R, Borchardt D, Salamini F (2002) Mapping QTLs for sucrose content, yield and quality in a sugar beet population fingerprinted by EST-related markers. Theor Appl Genet 104: 1107-1113

Sebastian RL, Kearsey MJ, King GJ (2002) Identification of quantitative trait loci controlling developmental characteristics of *Brassica oleracea L.* Theor Appl Genet 104: 601-609

Shen L, Courtois B, McNally KL, Robin S, Li Z (2001) Evaluation of near isogenic lines of rice introgressed with QTLs for root depth through marker assisted selection. Theor Appl Genet 103: 75-83

Snape JW, Butterworth K, Whitechurch E et al. (2001) Waiting for fine times: Genetics of flowering time in wheat. Euphytica 119(1-2): 185-190

Stracke S, Börner A (1998) Molecular mapping of the photoperiod response gene ea_7 in barley. Theor Appl Genet 97: 797-800

Stracke S, Sato S, Sandal N, Koyam M, Kaneko T, Tabat S, Parniske M (2004) Exploitation of colinear relationships between the genomes of *Lotus japonicus, Pisum sativum* and *Arabidopsis thaliana*, for positional cloning of a legume symbiosis gene. Theor Appl Genet 108: 442-449

Sutka J, Galiba G, Vagujfalvi A, Gill BS, Snape JW (1999) Physical mapping of the *Vrn-* and *Fr1* genes on chromosome 5A of wheat using deletion lines. Theor Appl Genet 99: 199-202

Tamura K, Nomura K, Oshima I, Namai H, Yano M, Sasaki T, Kikuchi F (1998). Identification of restriction fragment length polymorphism markers tightly linked to a major photoperiod sensitivity gene, *Se-1*, and to a blast resistance gene, *Pi-z*^{t}, in rice. J Breed Genet 30: 61-67

Teng S, Qian Q, Zeng D, Kunihiro Y, Fujimoto K, Huang D, Zhu L (2004) QTL analysis of leaf photosynthetic rate and related physiological traits in rice (*Oryza sativa* L.). Euphytica 135: 1-7

Tóth B, Galiba G, Fehér E, Sutka J, Snape JW (2003) Mapping genes affecting flowering time and frost resistance on chromosome 5B of wheat. Theor Appl Genet 107: 509-514

Tsai KH (1995) Genetic analysis for heading time in wild rice strains. Jap J Genet 70: 555-562

Tuberosa R, Salvi S, Sanguineti MC, Landi P, Maccaferri M, Conti S (2002) Mapping QTLs regulating morphophysiological traits and yield: Case studies, short comings and perspectives in drought-stressed maize. Ann Bot 89: 941-963

Verma DPS, Hu CA, Zhang M (1992) Root nodule development: Origin, function and regulation of nodulin genes. Physiol Planta 85: 253-265

Woo HR, Chung KM, Park JH, Aeong S Oh, Ahn T, Hong SH, Jang SK, Nam HG (2001) ORE9, an F-Box protein that regulates leaf senescence in Arabidopsis. Plant Cell 13: 1779-1790

Xiao J, Li J, Yuan L, Tanksley, SD (1996) Identification of QTLs affecting traits of agronomic importance in a recombinant inbred population derived from a subspecific rice cross. Theor Appl Genet 92: 230-244

Yadav R, Courtois B, Huang N, McLaren G (1997) Mapping genes controlling root morphology and root distribution in a doubled-haploid population of rice. Theor Appl Genet 94: 619-632

Yadav R, Bidinger F, Hash C, Yadav Y, Yadav O, Bhatnagar S, Howar C (2003) Mapping and characterisation of QTL × E interactions for traits determining grain and stover yield in pearl millet. Theor Appl Genet 106: 512-520

Yamagata H, Okumoto Y, Tanisaka T (1986) Analysis of genes controlling heading time in Japanese rice. In: Rice Genetics. IRRI, Los Baños, The Philippines, pp 351-359

Yamamoto T, Kuboki Y, Lin SY, Sasaki T, Yano M (1998) Fine mapping of quantitative trait loci *Hd-1*, *Hd-2* and *Hd-3*, controlling heading date of rice, as single Mendelian factors. Theor Appl Genet 97: 37-44

Yamamoto T, Lin HX, Sasaki T, Yano M (2000). Identification of heading date quantitative trait locus *Hd6* and characterization of its epistatic interactions with *Hd2* in rice using advanced backcross progeny. Genetics 154: 885-891

Yan L, Loukoianov A, Tranquilli G, Helguera M, Fahima T, Dubcovsky J (2003) Positional cloning of the wheat vernalization gene VRN1. Proc Natl Acad Sci USA 100: 6263-6268

Yano M, Harushima Y, Nagamura Y, Kurata N, Minobe Y, Sasaki T (1997) Identification of quantitative trait loci controlling heading date in rice using a high-density linkage map. Theor Appl Genet 95: 1025-1032

Yano M, Katayose Y, Ashikari M, Yamanouchi U, Monna L, Fuse T, Baba T, Yamamoto K, Umehara Y, Nagamura Y, Sasaki T (2000) *Hd1*, a major photoperiod sensitivity quantitative trait locus in rice, is closely related to the Arabidopsis flowering time gene *CONSTANS*. Plant Cell 12: 2473-2484

Yeo AR, Yeo ME, Flowers SA, et al. (1990) Screening of rice (*Oryza-Sativa* L) genotypes for physiological characters contributing to salinity resistance and their relationship to overall performance. Theor Appl Genet 27(3): 263-277

Yokoo M, Kikuchi F, Nakane A, Fujimaki, H (1980) Genetical analysis of heading time by aid of close linkage with blast, *Pyricularia oryzae*, resistance in rice. Bull Natl Inst Agri Sci Ser D 31: 95-126

Yokoo M, Okuno K (1993) Genetic analysis of earliness mutations induced in the rice cultivar Norin 8. Jap J Breed 43: 1-11

Yoshimaru H, Ohba K, Tsurumi K, Tomaru N, Murai M, Mukai Y, Suyama Y, Tsumura Y, Kawahara T, Sakamaki Y (1998) Detection of quantitative trait loci for juvenile growth, flower bearing and rooting ability based on a linkage map of sugi (*Cryptomeria japonica* D. Don). Theor Appl Genet 97: 45-50

Zhang GY, Guo Y, Chen SL, Chen SY (1995) RFLP tagging of a salt tolerance gene in rice. Plant Sci 110: 227-234

Zhang WK, Wang YJ, Luo GZ, Zhang JS, He CY, Wu XL, Gaiand JY, Chen SY (2004) QTL mapping of ten agronomic traits on the soybean (*Glycine max* L. Merr.) genetic map and their association with EST markers. Theor Appl Genet 108: 1131-1139

8 Molecular Mapping and Breeding in Fruit Trees

Albert G. Abbott
Department of Genetics & Biochemistry, 116 Jordan Hall, Clemson University, Clemson, SC 29634, USA

1 INTRODUCTION

Fresh and processed products derived from the fruit species (almonds, apples, apricots, blackberries, peaches, pears, plums, sweet cherries, tart cherries, strawberries, raspberries, papaya, mango and others) make vital contributions to human health and well being, and collectively constitute the economic backbone of many rural economies across the world. Many fruiting crops contribute significantly to a healthy, diverse diet, and can help deter obesity and other health risks. Consumption of fruits is associated with reduced risk of chronic diseases and the ability to reverse age-related disorders. Fruits and nuts are rich sources of vitamins, minerals, dietary fiber, antioxidants (such as carotenoids, anthocyanins, phenolic compounds, and ascorbic acid), and other phytonutrients that can lower the risk of heart disease, cancer, and a host of other illnesses.

Some progress has been made in understanding the biological and non-biological factors that affect various sensory, nutritional, and processing qualities of fruits, nuts and flowers, and that affect various aspects of plant growth and flowering habits and plant architecture. Extensive databases of expressed genes for some fruit and nut species have been developed with the primary purpose of identifying the genetic determinants of such traits. Large genetic variation for almost every trait is available in fruit trees' germplasm underscoring the enormous potential for further genetic improvement. However, the overwhelming majority of useful genes that could be manipulated for varietal improvement in the fruit species remain uncharacterized or unidentified.

In the dicots, the model *Arabidopsis* genome is clearly the research standard for identification and characterization of important plant genes; however, as a useful comparative genomics model for fruiting species, it has limitations. There are traits and genes, primarily those for fruit, growth habit, and life history traits, that may not have recognizable homologs in the *Arabidopsis* genome. For this reason, it is now recognized that it is necessary to study and sequence model plants from other important families, such as the Rosaceae, Rutaceae and others, to develop a universally complete plant genome database.

As an example, the Rosaceae is typified by broad diversity in growth habit, fruit type, reproductive traits and other important characters; we have insufficiently capitalized on this unique resource for basic research. This is in part historical in that many Rosaceae species are, in general, not easily amenable to traditional basic genetic analysis (long generation time, significant requirements of space, and labor) and in part due to the strong emphasis of funding agencies on the plants with a perceived greater potential for economic return (e.g., maize, rice, soybean). However, with the introduction and application of molecular genetic technologies and the tools of genomics and transgenics, species, which in the past were not considered suitable for basic research, are now moving to the forefront because they present unique opportunities to extend and enhance our understanding of the evolution of specialized gene systems and their diverse functions in different plant families (e.g., the Floral Genome Project, *http://fgp.bio.psu.edu/*).

Despite the agricultural importance of the fruit crops, structural and functional genomic characterization of the member species lags well behind those of other crop species. Several factors have contributed to this lack of progress. First, although the fruit tree crops collectively are extremely important, individually each crop is considered a "minor" crop and, therefore, competitive funding has been limited and industry funding for genomics is not available. Additionally, linkage mapping in tree species (which represent the majority of fruit species) is difficult due to the field space required, time required to flowering and fruiting (at least two years), and the polyploid nature of many of these crops.

2 GENETIC MAPPING IN FRUITING TREES

The importance of high-quality fruit, and the intrinsic difficulties of breeding in perennial species, requires the development and application of structural and functional genomics databases for the sustained improvement of fruit tree crops. Identification and characterization of genes controlling the genetic basis of the traits, and their tagging with molecular markers, permit a more realistic estimate of the efforts needed to complete the introgression and to produce a new variety combining the best of traits formerly remaining isolated in different varieties. It also reduces effort and time, and improves the accuracy of marker-assisted selection. Thus, field evaluation is limited to trees containing the genes of interest, significantly reducing the costs associated with maintaining undesirable trees to maturity. The ability to preselect seedlings, using DNA-based markers, for traits such as sugar and acid levels, color, firmness, and fruit size while introgressing traits, such as biotic and abiotic stress resistance from exotic germplasm, speeds the development of commercially acceptable cultivars. For these reasons numerous fruit tree research programs worldwide have focused on the development of comprehensive genetic maps incorporating phenotypic characters both simple and complex and molecular markers. A fairly comprehensive compilation is presented in Table 1.

One of the main problems faced by these mapping programs is the lack of inbred lines to establish mapping population. Essentially the juvenility periods and limitations of time and space severely constrain the development of extensive homozygous inbred lines. Thus, most fruit tree mapping is done on F_1 intervarietal crosses or F_2 self-crosses (dependent on self-compatibility of the species). Taking advantage of the abundance of molecular markers and the knowledge of the parental genotypes at marker loci, two mapping strategies are utilized. One utilizes a pseudo-backcross strategy (Lodhi et al. 1995) in F_1 mapping thus generating two maps one for each parent; the other uses an F_2 mapping strategy in a progeny derived from an F_1 self-cross and produces a single map (Fig. 1).

A candidate gene map for one species of a family could serve as a substrate for comparative analysis of regions of interest in other related species and facilitate gene cloning investigations. With further integration of sequenced cDNA loci (EST loci), the map database immediately provides candidate genes located in the genetically marked intervals containing traits of interest. Having these associations has the potential to greatly speed up the process of gene discovery and characterization.

There is currently significant information available on the sequences of expressed genes from some key fruit crops. ESTs are a cost-effective way of identifying a large number of functional plant genes, many of

Table 1 Status of genetics maps in major fruiting trees.

Genera	*Genetic maps*	*Physical maps*	*Simple characters*	*QTL*	*EST*	*Genome sequence*	*Reference*
Prunus	22	1	+	+	+	Peach (in progress)	Arus et al. 2006
Malus	9	1	+	+	+	Apple (in progress)	Gardiner et al. 2006
Pyrus	2	-	+	Limited	-	-	Itai 2006
Olea	2	-	+	-	-	-	Doveri and Baldoni 2006
Vitis	9	?	+	+	+	Grape (completed)	Raiz et al. 2006
Pistacia	-	-	1 (sex determination)	-	-	-	Hormaza and Wunsch 2006
Vaccinium	4	-	+	+	+	-	Ratnaparkhe 2006
Rutaceae	15	In progress	+	+	+	In progress	Gmitter et al. 2006
Musa	1	-	+	+	-	-	Pillay and Tripathi 2006
Mangifera	1	-	-	In progress	-	-	Vasanthaiah et al. 2006
Persea	-	-	-	-	+ (Floral Genome Project)	-	Ashworth et al. 2006
Carica	1	In progress	+	+	+	Papaya (in progress)	Mishra et al. 2006
Diospyros	-	-	+	-	-	-	Kanzaki and Yonemori 2006

+ designates data available, - designates no significant data available

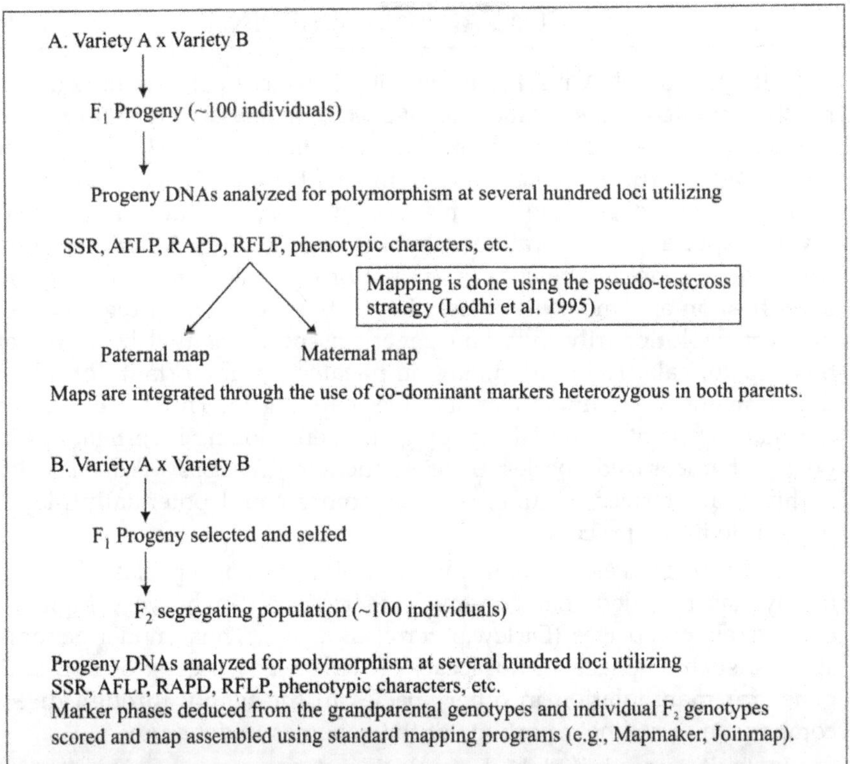

Fig. 1 Mapping strategies in fruit trees.

which may have novel functions. Generation of a large number of ESTs from the most economically important fruit producing crops will result in an annotated and publicly available fruiting species gene indices and EST databases. Equally importantly, the mapping to the unique EST set on the model fruiting species genetic maps will provide a rich candidate gene resource for identification and characterization of genes controlling or influencing traits of agronomic importance. The EST database will be a rich source of comparative sequence information for studies of evolution and functional genomics of all plant genomes. Furthermore, specific ESTs may be selected and arrayed to fabricate cDNA microarrays for genome-wide analyses in important processes such as fruit development and responses to biotic or abiotic stresses.

3 COMPARATIVE MAPPING

With the use of DNA marker technology it is now possible to generate highly saturated maps for most plant species in a relatively short period of time. One of the most important contributions of DNA marker technology to fundamental studies in plant biology is the ability to rapidly compare genome organization in closely related as well as diverse species. Comparative mapping studies can identify highly conserved genome blocks, and regions of lesser conservation. Identification and molecular dissection of these conserved regions may uncover evolutionarily invariant genetic associations that by virtue of their historical preservation, are implicated as important for plant development. In addition, comparative mapping information can serve as a starting point for initial mapping and gene cloning investigations in poorly characterized species. Species, such as *Arabidopsis thaliana*, with highly characterized, simple, small genomes could potentially play a major role in the process.

In fruiting trees, comparative mapping among species within a family has revealed that genome preservation can be very high for closely related species (Dirlewanger et al. 2004). Thus, model genome species, such as peach in Rosaceae, provide ready access to candidate genes for manipulation in other species in the family through direct comparison of collinear portions of the genomes under study.

In fruit trees, initial high-resolution genome sequencing studies among closely related species suggests significant preservation of the genome (Jung et al. 2008 manuscript in review). Comparisons across family boundaries such as those between peach and *Arabidopsis* suggest a fragmentary conservation of the genome (Georgi et al. 2003; Jung et al. 2006). Initial comparisons of papaya and other related species based on BAC-end sequences suggests some degree of conservation (Lai et al. 2006), however these fragmentary approaches may be misleading due to sampling and whole-genome sequence comparisons may reveal significantly more conservation than expected. In this regard, results by Jung et al. (2008 manuscript in review) suggest that there may be significant conservation between peach and poplar genomes with extensive colinearity at the high resolution level. Clearly much more work needs to be done to examine the degree of microsynteny among plants not only within families but also among families. Whole-genome sequences of key species in Rosaceae (peach and apple), Rutaceae (citrus), Caricaceae (papaya) are currently underway and this information should provide substrate for a much greater picture of genome evolution in fruiting trees.

4 MAPPING QUANTITATIVE TRAIT LOCI

Mapping of quantitative trait loci (QTL) is important in plant breeding because most traits of agricultural importance (e.g., yield, hardiness, fruit flavor, etc.) are controlled by polygenes. QTL analysis can be used to identify most of these loci, facilitating marker-assisted introgression, thereby enriching the genetic base and accelerating the rate of crop improvement (Tanksley and Nelson 1996; see Chapter 6 in Volume 1 for details).

Another benefit resulting from the molecular genetic localization of QTL is the opportunity to clone them using map-based approaches. In a few cases, for genes with relatively large effects, the map location has been precise enough to enable "chromosome landing" rather than "walking" (Tanksley et al. 1995; see Chapter 8 in Volume 1 for details).

As the EST and sequence databases are developed for model species in each of the major plant families, cross correlation of genomic regions among maps will facilitate candidate gene approaches to identify important genes controlling simple and complex traits in the other species of the families. This approach has the potential to deliver results in any species for which the requisite tools can be developed: tightly-linked flanking markers, a library of large genomic fragments, and a specific screen for expressed sequences in a defined genomic sequence.

In fruiting species, there exist some preliminary QTL maps for genes controlling fruit quality traits and disease resistance, (for review see, Genome Mapping and Molecular Breeding in Plants Volume 4, 2007, Springer), however, QTL mapping in fruit trees is very difficult due to the general restrictions on crossing and maintaining significant progeny numbers in orchards. In addition, many of the characters of interest are related to fruit development or resistance to abiotic and biotic stresses and these characters by necessity need to be scored during the fruiting cycle which in many fruit trees may not occur for at least three years or more, depending on the species juvenility requirements. This necessitates maintaining large trees for extended periods of time in orchard systems thereby limiting the number of trees that can be reasonably maintained in any one location much less in several locations. However, one advantage of QTL analysis in fruit trees is that the same tree can be analyzed over many years enabling significant year-to-year data to be obtained. Thus, one can estimate the environmental component of the QTL characters very effectively once populations are established. With these restrictions in mind, researchers have published QTL analyses for a number of fruiting trees species. A few key examples are outlined in Table 2.

Table 2 QTL mapping of important fruit tree characters.

Characters	*Species*	*QTL number*	*Reference*
Disease resistance:			
Fungal: Powdery Mildew	Grape	Three loci	Dalbo et al. 1997, Zyprian et al. 2002, Fischer et al. 2004
	Apple	Two loci	Kellerhals et al. 2000
	Peach	Three loci	Foulongne et al. (2003b)
Virus: Plum Pox Virus	Apricot	One major locus/ several minor loci	Lambert et al. 2007 Vilanova et al. 2006 Soriano et al. 2008 Lalli et al. 2008
Fruit quality:			
Sugars: Glucose, fructose, sucrose	Peach	Three loci	Abbott et al. 1998 Etienne et al. 2002
Acidity	Apple	Two Loci	Liebhard et al. (2003a)
Yield			
Fruit size/weight	Grape	Two to Four Loci	Riaz 2001, Doligez et al. 2002, Fischer et al. 2004
Endodormancy			
Bloom time	Peach	Three loci	Howad et al. 2006
Chilling requirement	Peach	Three loci	Fan et al. 2008

5 PHYSICAL MAPPING IN FRUITING TREES

Physical maps of several fruiting trees (peach, apple, citrus, papaya,) are currently in progress (Han et al. 2006, 2007a, b; Abbott et al. 2007; Zhebentyayeva et al. 2008; S. Korban and J. Beever, Pers. Comm.; A.H. Paterson, Pers. Comm.). To demonstrate the process for a typical fruit tree, we provide a summary of the efforts in peach.

Integrated physical/genetic maps are of critical importance for high-throughput EST mapping, QTL fine-mapping and effective positional cloning of genes (Zhang and Wing 1997; Green 2001; Zhang and Wu 2001). To construct the physical map for peach, we employed essentially the strategies utilized to develop the physical maps for the *Arabidopsis thaliana* and *Drosophila melanogaster* (Marra et al. 1999; Hoskins et al. 2000). The approach combines hybridization of the genetically mapped markers with BAC DNA-fingerprinting and in our case, hybridization of

EST sequences as well. Manual sequencing gel-based fingerprinting has been proven a reliable and cost-effective technique for BAC-fingerprinting and under certain circumstances performs better than other traditional fingerprinting methods (Xu et al. 2004). An initial acrylamide gel-based physical framework for peach was established and released (Zhebentyayeva et al. 2006). This framework based on random fingerprinting of 3x peach genome equivalents covered at least 50% of the genome and included hybridization data for 673 of 3,384 ESTs of the peach unigene set (PP_LE). On this framework, genetically anchored BAC contigs provided landmarks for a *Prunus-Arabidopsis* microsynteny study (Jung et al. 2006) and for further development of the *Prunus* transcript map by Horn et al. (2005). Since the first release, the peach physical framework underwent further enhancements. We have integrated global hybridization data that include additional ESTs, AFLPs (amplified fragment length polymorphisms), SSRs (microsatellites), gene-specific genomic probes and 'overgo' probes derived from the BAC-end sequences. The total number of markers incorporated into the physical framework was increased up to 2,636 markers. Selectively, we have fingerprinted all hybridization positive BACs omitted during random fingerprinting. As a result, we incorporated an additional 1x peach genome equivalent composed of marker-positive BACs. Finally, we took an advantage of a HICF (a high-information content fingerprinting) technique along with improved FPC v8.5.2 software to increase an average number of bands per BAC clone and improve accuracy in contig assembly (Nelson et al. 2007). The final HICF physical map for the peach consists of 2,138 contigs. Of these, 252 contigs are anchored to eight linkage groups of the *Prunus* reference map (Zhebentyayeva et al. 2008). The physical length of mapped contigs has been estimated at 303 Mb which is in excess of the estimated size of the peach genome (Baird et al. 1994). Due to the abundance of hybridization data the HICF physical map for peach is biased to the expressed genome regions and thus substantially covers the euchromatic portion of the peach genome.

6 ASSOCIATION MAPPING AND ITS APPLICATION IN FRUIT TREE BREEDING

Due to the physical restrictions (time, space, and labor) on molecular mapping in fruit trees, high resolution of QTLs through single or multiple family crosses is extremely difficult. An alternative to the use of individual segregating families for mapping QTL is association mapping (Fig. 2). Association mapping takes advantage of the phenomenon of linkage disequilibrium for markers and traits in diverse established

BOX 1 LINKAGE AND ASSOCIATION

Both linkage analysis and association studies rely on co-inheritance of neighbouring DNA variants. In both analyses, recombination is the main force eliminating linkage over generations. The difference is that in linkage analysis (Panel a), there have been only few opportunities for recombination to occur (one or two generations), resulting in comparatively low resolution of the map (i.e. markers may not be very closely linked with genes of interest). In association analysis with populations (Panel b), recombination events over many generations has eliminated linkage between a mutated gene (yellow diamond-shape) and molecular markers, except over very short distances. Only very close markers are in LD with the mutated gene, and form the basis for association mapping. LD measures the closeness of the genetic association between markers and a particular trait, and may be used to identify markers in close proximity to the gene(s) responsible for the trait.

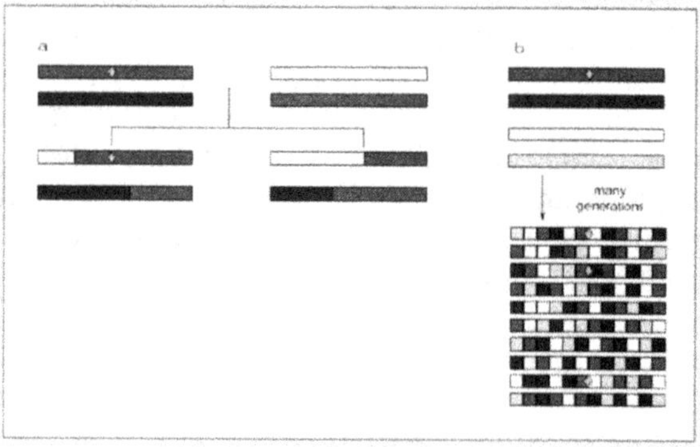

Fig. 2 From Leif Skot, Gene mapping in natural plant populations, guilt by association, Iger Innovations 2003

populations. This alternative QTL mapping method has been applied to analysis of human genetic characters and is currently being explored in plants (Skot 2003) as a way of resolving QTLs in particular species where requisite populations exist. It would seem that this approach would have great application for fruit tree mapping since many species of fruiting trees have broadly diverse existing populations maintained in germplasm centers around the world. Additionally perennial species provide opportunities to continually score different characters over multiple years in old populations to mark multiple phenotypes on a single population, a distinct advantage over traditional strategies that follow only a few polymorphic characters in a single population where most other phenotypic loci are fixed.

A basic strategy for linkage disequilibrium mapping in fruiting trees (modeled on that utilized for wheat in Breseghello and Sorrells 2006) could be as follows: 1. Establish and screen a genetically diverse population (population diversity established using programs such as Structure (Prichard et al. 2000a) for particular characters of interest. 2.

Individual trees are then scored for the traits of interest. 3. Utilizing highly polymorphic marker types (SNPs, or SSRs) score the trees across the genome for allelic composition at the various loci; 4. Utilizing statistical analysis packages such as Tassel (*www.maizegenetics.net*) to search for linkage associations of specific traits and specific alleles at marker loci. Markers obtained in such a manner can then be tested in field material and crosses for utilization in MAS. Of course this approach relies on adequate populations and significant numbers of polymorphic markers. In the near future for several fruit tree crops (citrus, peach, apple, papaya) whole-genome sequences will provide substantial resources for SNP development within important fruit tree families such as Rosaceae or Rutaceae. However, access to germplasm in centers of origin and significant infrastructure for scoring of phenotypic characters in these regions presents a significant challenge for the utilization of such and approach.

Recently an adaptation of association mapping called pedigree genotyping has been employed in apple QTL analysis (Van de Weg et al. 2004). In this approach, pedigree information is coupled with phenotypic information to follow the segregation of founder alleles for specific traits through breeding pedigrees. When alleles of molecular markers, such as SSRs, are found to associate with the phenotype through the pedigree, one can assign potential regions of the genome governing the trait. These regions are then targeted for further study. This system should, in theory work with any species that has:

1) a sufficiently saturated molecular genetic map; and 2) breeding materials of phenotyped, pedigreed individuals. In some fruiting crops however, extensive pedigree information may not be available thus limiting the application of this technique.

Finally, in the case of fruit trees the following strategy is currently under testing in a number of laboratories worldwide (Fig. 3).

7 TRANSGENIC REDUCTION OF JUVENILE PHASE

One of the major challenges to fruit tree breeding is the length of the juvenile phase (time to first flowering) in many species. Long juvenile phases (e.g., 5-10 years), which are common in many of these species, significantly hampers the mobilization of traits through standard introgressive crossing strategies. Recent studies utilizing transgenic technologies to knock out genes that control flowering time in *Arabidopsis* (Shannon and Meeks-Wagner 1991, 1993; Weigel et al. 1992;

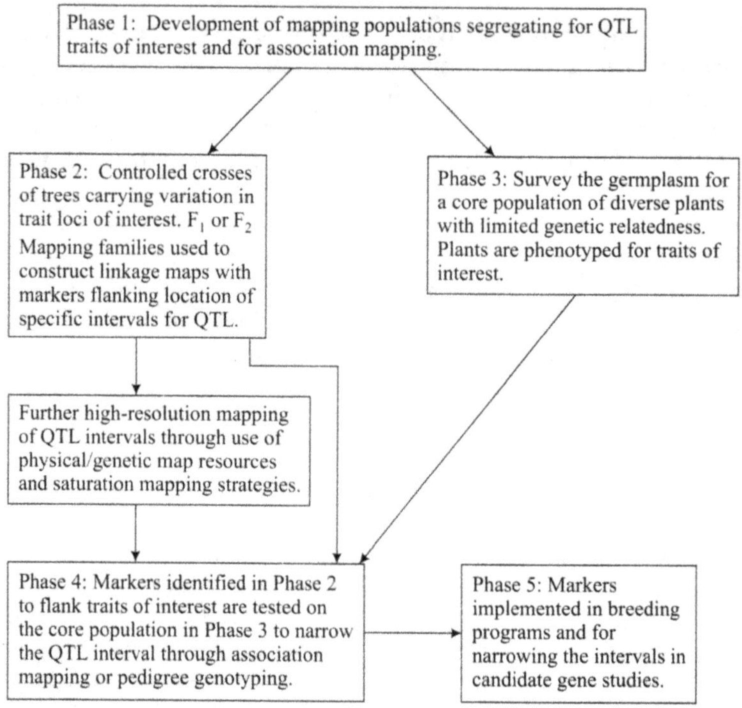

Fig. 3 Strategy for resolving QTL in fruit tree mapping.

Shultz and Haughn 1993) have provided some possible avenues to speed up the breeding process in fruiting trees (Endo et al. 2005; Kotoda et al. 2006). Broader application of these approaches to other fruit trees awaits development of efficient transgenic technologies in each species. This has proven relatively easy for some species (e.g., plum) but difficult for others (e.g., peach).

Of course, the use of genetically engineered materials is currently a subject of hot debate worldwide fueled by the lack of accurate public knowledge of the nature and risks of using such technologies. Development of educational outreach programs using regional crops of importance is an ideal way to educate the public: 1) in the use of these technologies; 2) in the risks associated with the application of the technologies; and 3) to the importance of these technologies for sustaining the production of safe local food products while minimizing the use of chemical control agents.

8 MARKER-ASSISTED SELECTIONS IN FRUIT TREES

One major goal for the development of molecular genetic maps in fruit tree research programs is the development of markers for marker-assisted selection (MAS) in breeding. In fruit tree research the identification and utilization of markers for such purposes is in its infancy. This is primarily because most of the mapping populations utilized in construction of molecular maps are fairly small (less than 100 trees) which makes identification of markers linked tightly enough for MAS very difficult. In spite of this limitation, several reports have suggested that MAS in fruit trees is feasible as exemplified by work in the Rosaceae (Gessler et al. 2000; Evans and James 2003; Dirlewanger et al. 2004; Lecouls et al. 2004). As the marker maps become more highly developed in these crops, breeding programs will progressively adopt MAS strategies, however, for the moment the state of MAS in fruit trees is in its infancy.

Concluding Remarks

Significant advances in the technologies for mapping and molecular breeding of important agronomic characters in fruit trees have been made in the past several years.

While the technology advances however, a significant effort needs to be made to develop appropriate mapping populations, to consistently phenotype these populations for important characters and to implement marker-assisted breeding strategies into ongoing traditional fruit tree breeding programs. This is no small task as the development and maintenance of breeding populations is labor limited, time intensive and costly and many breeding programs for fruit trees have disappeared. If these molecular mapping technologies are to pay off in the marketplace, the current funding emphasis must shift from supporting the basic research to a balanced support for both basic and applied translational research. With the emphasis shifting to translational strategies will come the successes in molecular breeding that have been forecast for many years. We are moving into this critical period at a time that is crucial for fruit tree agriculture and only time will tell if the industry can be sustained in its current form.

References

Abbott AG, Rajapakse S, Sosinski B, Lu ZX, Sossey-Alaoui K, Gannavarapu M, Scorza R, Reighard G, Ballard RE, Callahan A, Baird WV (1998) Construction of saturated linkage maps of peach crosses segregating for characters controlling fruit quality, tree archicture and pest resistance. Acta Hort 465: 41-50

Abbott AG, Georgi L, Inigo M, Sosinski B, Yvergniaux D, Wang Y, Blenda A, Reighard G (2002) Peach: The model genome for Rosaceae. Acta Hort 575: 145-155

Abbott AG, Arús P, Scorza R (2006) Peach. In: Kole C (ed) Genome Mapping and Molecular Breeding in Plants. Vol 4. Springer - Berlin, Heidelberg, New York, pp 137-156

Arús P, Yamamoto T, Dirlewanger E, Abbott AG (2006) Synteny in the Rosaceae. In: Janick J (ed) Plant Breed Rev 27: 175-211

Ashworth VETM, Chen H, Clegg T (2007) Avocado. In: Kole C (ed) Genome Mapping and Molecular Breeding in Plants. Vol 4. Springer - Berlin, Heidelberg, New York, pp 325-331

Baird WV, Estager AS, Wells J (1994) Estimating nuclear DNA content in peach and related diploid species using laser flow cytometry and DNA hybridization. J Am Soc Hort Sci 119: 1312-1316

Breseghello F, Sorrells ME (2006) Association mapping of kernel size and milling quality in wheat (*triticum aestivum* l.) cultivars. Genetics 172: 1165-1177

Dalbo MA, Guang NYe, Weeden NF, Reisch BI (1997) QTL analysis of powdery mildew resistance in grapes. In: Plant, Animal and Microbe Genome V Conf, San Diego, CA, USA

Dirlewanger E, Graziano E, Joobeur T, Garriga-Caldere F, Cosson P, Howad W, Arús P (2004) Comparative mapping and marker-assisted selection in Rosaceae fruit crops. Proc Natl Acad Sci USA 101: 9891-9896

Dirlewanger E, Cosson P, Boudehri K, Renaud C, Capdeville G, Tauzin Y, Laigret F, Moing A (2006) Development of a second-generation genetic linkage map for peach [*Prunus persica* (L.) Batsch] and characterization of morphological traits affecting flower and fruit. Tree Genet Genom 3: 1-13

Doligez A, Bouquet A, Danglot Y, Lahogue F, Riaz, S, Meredith CP, Edwards KJ, This P (2002) Genetic mapping of grapevine (*Vitis vinifera* L.) applied to the detection of QTLs for seedlessness and berry weight. Theor Appl Genet 105: 780-795

Doveri S, Baldoni L (2007) Olive. In: Kole C (ed) Genome Mapping and Molecular Breeding in Plants. Vol 4. Springer – Berlin, Heidelberg, New York, pp 253-265

Endo T, Shimada T, Fujii H, Kobayashi Y, Araki T, Omura M (2005) Ectopic expression of an FT homolog from Citrus confers an early flowering phenotype on trifoliate orange (*Poncirus trifoliata* L. Raf.). Transgenic Res 14(5): 703-712

Etienne C, Rothan C, Moing A, Plomion C, Bodenes C, Dumas LS, Cosson P, Pronier V, Monet R, Dirlewanger E (2002) Candidate genes and QTLs for sugar and organic acid content in peach (*Prunus persica* (L.) Batsch). Theor Appl Genet 105: 145-159

Evans KM, James CM (2003) Identification of SCAR markers linked to Pl-w mildew resistance in apple. Theor Appl Genet 106: 1178-1183

Fan S, Bielenberg D, Zhebentyayeva T, Reighard G, Abbott A (2008) The development of a genetic linkage map for chilling requirements in peach. In: Plant, Animal and Microbe Genome XV Conf, San Diego, CA

Fischer BM, Salakhutdinov I, Akkurt M, Eibach R, Edwards KJ, Topfer R, Zyprian EM (2004) Quantitative trait locus analysis of fungal disease resistance factors on a molecular map of grapevine. Theor Appl Genet 108: 501-515

Foulongne M, Pascal T, Pfeiffer F, Kervella J (2003b) QTL for powdery mildew resistance in peach x *Prunus Davidiana* crosses: Consistency across generations and environments. Mol Breed 12: 33-50

Gardner SE, Bus VGN, Rusholme RL, Chagne D, Rikkerink EHA (2007) Apple. In: Kole C (ed) Genome Mapping and Molecular Breeding in Plants. Vol 4. Springer - Berlin, Heidelberg, New York, pp 1-63

Georgi, LL, Wang Y, Reighard GL, Mao L, Wing RA, Abbott AG (2003) Comparison of peach and Arabidopsis genomic sequences: fragmentary conservation of gene neighborhoods. Genome 46:268-276

Gessler C, Patocchi A, Koller B, Kellerhals M (2000) DNA-technologies in the creation of disease resistant apple cultivars: Conventional breeding assisted by markers and gene-therapy. Phytopathology 90(6): S109

Gmitter FG, Chen C, Rao, MN, Soneji JR (2007) Citrus Fruits. In: Kole C (ed) Genome Mapping and Molecular Breeding in Plants. Vol 4. Springer - Berlin, Heidelberg, New York, pp 265-281

Green ED (2001) Strategies for the systematic sequencing of complex genomes. Nat Rev Genet 2: 573-583

Han J, Gasiæ K, Marron B, Beever JE, Korban SS (2007) A BAC-based physical map of the apple genome. Genomics 89: 630-637

Hormaza JL Wunsch A (2007) Pistachio. In: Kole C (ed) Genome Mapping and Molecular Breeding in Plants. Vol 4. Springer - Berlin, Heidelberg, New York, pp 243-253

Horn R, Lecouls A-C, Callahan A, Dandekar A, Garay L, McCord P, Howad W, Chan H, Verde I, Ramaswamy K, Main D, Jung S, Georgi L, Forrest S, Mook J, Zhebentyayeva TN, Yu Y, Kim HR, Jesudurai C, Sosinski BA, Arús P, Baird V, Parfitt D, Reighard G, Scorza R, Tomkins J, Wing R, Abbott AG (2005) Candidate gene database and transcript map for peach, a model species for fruit trees. Theor Appl Genet 110: 1419-1428

Hoskins RA, Nelson CR, Berman BP, Laverty TR, George L, Ciesiolka L, Naemuddin M, Arenson AD, Durbin J, David RG, Tabor PE, Bailey MR, DeShazo DR, Catanese J, Mammoser A, Osoegawa K, de Jong PE, Celniker

SE, Gibbs RA, Rubin GM, Scherer SE (2000) BAC-based physical map of the major autosomes of *Drosophila melanogaster*. Science 287: 2271-2274

Howad W, Arus P (2007) QTL analysis of blooming time in peach. In: Plant, Animal and Microbe Genome XV Conf, San Diego, CA, USA

Itai A (2007) Pear. In: Kole C (ed) Genome Mapping and Molecular Breeding in Plants. Vol 4. Springer - Berlin, Heidelberg, New York, pp 157-171

Jung S, Jiwan D, Cho I, Lee T, Abbott AG, Sosinski B, Main D (2008) Extensive conservation of gene order and content between *Prunus* and *Populus* but not between *Prunus* and *Arabidopsis* (in review)

Jung S, Main D, Staton M, Cho I, Zhebentyayeva T, Arus P, Abbott A (2006) Synteny conservation between the *Prunus* genome and both the present and ancestral Arabidopsis genomes. BMC Genom 7: 81

Kanzaki S, Yonemori K (2007) Persimmon. In: Kole C (ed) Genome Mapping and Molecular Breeding in Plants. Vol 4. Springer - Berlin, Heidelberg, New York, pp 353-359

Kellerhals M, Ganfranceschi L, Seglias N, Gessler C (2000) Marker-assisted selection in apple breeding. Acta Hort 521: 255-265

Kotoda N, Naozumi M (2006) Development of an apple capable of flowering in eight months. Agri Hort 81(9): 967-979

Lai CWJ, Yu Q, Hou S, Skelton RL, Jones MR, Lewis KLT, Murray J, Eustice M, Guan P, Agbayani R, Moore PH, Ming R, Presting GG (2006) Analysis of papaya BAC end sequences reveals first insights into the organization of a fruit tree genome. Mol Genet Genom 276(1): 1-12

Lalli DA, Krška B, Abbott, AG, Badenes ML, Damsteegt V, Polák J, Salava J (2008) A Genetic Linkage Map for an Apricot (*Prunus armeniaca* L.) BC1 Population Mapping *Plum Pox Virus* Resistance. Tree Genet Genom (in press)

Lambert P, Dicenta F, Rubio M, Audergon JM (2007) QTL analysis of resistance to sharka disease in the apricot (*Prunus armeniaca* L.) Polonais' x 'Stark Early Orange' F1 progeny. Tree Genet Genom 3(4): 299-309

Lecouls' A, Bergougnoux V, Cabetas M, Bosselut N, Voisin R, Poessel J, Faurobert M, Bonnet A, Salesses G, Dirlewanger E, Esmenjaud D (2004) Marker-assisted selection for the wide-spectrum resistance to root-knot nematodes conferred by the Ma gene from Myrobalan plum (*Prunus cerasifera*) in interspecific *Prunus* material. Mol Breed 13(2): 113-124

Liebhard R, Kellerhals M, Pfammatter W, Jertmini M, Gessler C (2003a) Mapping quantitative physiological traits in apple (*Malus x domestica* Borkh.). Plant Mol Biol 52: 511-526

Lodhi MA, Daly MJ, Ye GN, Weeden NF, Reisch BI (1995) A molecular marker based linkage map of *Vitis*. Genome 38: 786-794

Marra M, Kukaba T, Sekhon T, Hillier L, Martiensen R, Chinwalla A, Crokett J, Fedele J, Grover H, Gund G, McCombie WR, McDonald K, McPherson J, Mudd N, Parnell L, Schein J, Seim R, Shelby P, Waterson R, Wilson R (1999) A map for sequence analysis of the *Arabidopsis thaliana* genome. Nat Genet 22: 269-270

Mishra M, Chandra R, Saxena S (2007) Papaya. In: Kole C (ed) Genome Mapping and Molecular Breeding in Plants. Vol 4. Springer - Berlin Heidelberg, New York, pp 343-353

Nelson WM, Dvorak J, Luo MC, Messing J, Wing RA, Soderlund C (2007) Efficacy of clone fingerprinting methodologies. Genomics 89: 160-165

Pillay M, Tripathi L (2007) Banana. In: Kole C (ed) Genome Mapping and Molecular Breeding in Plants. Vol 4. Springer - Berlin, Heidelberg, New York, pp 281-303

Pritchard JK, Stephens M, Donnelly P (2000) Inference of population structure using multilocus genotype data. Genetics 155: 945-959

Ratnaparkhe MB (2007) Blueberry. In: Kole C (ed) Genome Mapping and Molecular Breeding in Plants. Vol 4. Springer - Berlin, Heidelberg, New York, pp 217-229

Riaz S (2001) A microsatellite markers based framework linkage map of *Vitis vinifera* L. and genetic dissection of fruit cluster morphology traits. PhD Dissertation, Univ of California, Davis, USA

Riaz S, Doligez A, Henry RJ, Walker MA (2007) Grape. In: Kole C (ed) Genome Mapping and Molecular Breeding in Plants. Vol 4. Springer - Berlin, Heidelberg, New York, pp 63-103

Schultz EA, Haughn GW (1993) Genetic analysis of the floral initiation process (FLIP) in Arabidopsis. Development (Cambridge) 119(3): 745-765

Shannon S, Meeks-Wagner DR (1991) A mutation in the Arabidopsis TFL1 gene affects inflorescence meristem development. Plant Cell 3(9): 877-892

Skot L (2003) Gene mapping in natural plant populations, guilt by association. Iger Innovations (*http://www.iger.bbsrc.ac.uk/Publications/Innovations/In2003/Ch2.pdf*)

Soriano JM, Vera-Ruiz EM, Vilanova S, Martínez-Calvo J, Llácer G, Badenes ML, Romero C (2008) Identification and mapping of a locus conferring *Plum Pox Virus* resistance in two apricot improved linkage maps. Tree Genet Genom (in press)

Tanksley SD, Nelson JC (1996) Advanced backcross QTL analysis: a method for the simultaneous discovery and transfer of valuable QTLs from unadapted germplasm into elite breeding lines. Theor Appl Genet 92(2): 191-203

Tanksley SD, Ganal MW, Martin GB (1995) Chromosome landing: a paradigm for map-based gene cloning in plants with large genomes. Trends Genet 11(2): 63-68

van de Weg WE, Voorrips RE, Finkers R, Kodde LP, Jansen J, Bink MCAM (2004) Pedigree genotyping: a new pedigree based approach of QTL identification and allele mining. Acta Hort 663: 45-50

Vasanthaiah HKN, Ravishankar KV, Mukunda GK (2007) Mango. In: Kole C (ed) Genome Mapping and Molecular Breeding in Plants. Vol 4. Springer - Berlin, Heidelberg New York, pp 303-325

Vilanova S, Soriano JM, Lalli DA, Romero C, Abbott AG, Llácer G, Badenes ML (2006) Development of SSR markers located in the G1 linkage group of apricot (*Prunus armeniaca* L.) using a bacterial artificial chromosome library. Mol Ecol Notes 6(3): 789-791

Weigel D, Alvarez J, Smyth DR, Yanofsky MF, Meyerowitz EM (1992) Leafy controls floral meristem identity in *Arabidopsis.* Cell 69(5) 843-859

Xu Z, Sun S, Covaleda L, Ding K. Zhang A, Wu C, Scheuring C, Zhang HB (2004) Genome physical mapping with large-insert bacterial clones by fingerprint analysis: methodologies, source clone genome coverage, and contig map quality. Genomics 84: 941-51

Zhang HB, Wing RA (1997) Physical mapping of the rice genome with BACs. Plant Mol Biol 35(1-2): 115-127

Zhang HB, Wu C (2001) BAC as tools for genome sequencing. Plant Physiol Biochem 39: 195-209

Zhebentyayeva TN, Horn R, Mook J, Lecouls A-C, Georgi L, Swire-Clark G, Reighard GL, Baird WV, Abbott AG (2006) A physical framework for the peach genome. Acta Hort 713: 83-88

Zhebentyayeva TN, Swire-Clark G, Georgi LL, Garay L, Jung S, Forrest S, Blenda AV, Blackmon B, Mook J , Horn R, Howad W, Arús P, Main D, Tomkins JP, Sosinski B, Baird WV, Reighard GL, Abbott AG (2008) A framework physical map for peach, a model Rosaceae species. Tree Genet Genom (in press)

Zyprian EM, Salakhutdinov I, Fischer B, Akkurt M, Topler R (2002) Molecular mapping of grapevine and QTL analysis of fungal disease resistances. In: Plant, Animal and Microbe Genomes X Conf, San Diego, CA, USA

9 Molecular Mapping and Breeding in Forest Trees

Preston R. Aldrich
Department of Biological Sciences, Benedictine University, Birck Hall 341, 5700 College Road, Lisle, IL, USA
E-mail: paldrich@ben.edu

1 PERSPECTIVES AND CHALLENGES

1.1 The Forest Tree

What is a forest tree? Clearly it is a tree which resides in a forest setting, surrounded by trees of few or several other species (e.g., temperate forest) or many different species (e.g., a tropical forest). But with respect to mapping and breeding of trees, the term can carry an added meaning centered on the utility of each species to humans. Some tree species provide fruit useful for human consumption (e.g., apple, *Pyrus malus*), some provide timber (e.g., oak, *Quercus* spp.), some provide both fruit and timber (e.g., Persian walnut, *Juglans regia*), while some species are useful in novel ways such as for their sap (e.g., maple, *Acer saccharum*) or bark (e.g., cork oak, *Quercus suber*). The forest tree is that species of tree commonly residing in a forest setting and whose primary use by humans is not as a fruit crop, with timber being the most common extractive use. For mapping and breeding of fruit crop trees, see Chapter 8 of this volume. Our ultimate interest in the forest tree species lies in altering and harnessing its vegetative rather than reproductive features, although reproductive modifications might serve a proximate goal such as reducing the age of flowering onset or limiting transgene spread. In addition to focusing our attention on which tree characters we wish to modify, the distinction between fruit crop versus forest tree species carries with it information about the probable history of domestication which can inform approaches to breeding.

1.2 Domestication History

1.2.1 Ancient History

Artificial selection has proven effective in altering the features of many plants and animals used by humans. Notable successes include the domesticated dog and many non-woody crop plants such as corn. Recent research on the dog (*Canis familiaris*) has revealed genome-level diversity underlying the numerous breeds that have arisen from a wolf-like ancestral stock over the last 100,000 years (Vila et al. 1997; Kirkness et al. 2003). The large phenotypic differences between modern corn (*Zea mays*) and its weedy progenitor teosinte are underlaid by rather modest changes in the genome initiated only 6,300 years ago (Doebley 1992; Jaenicke-Després et al. 2003).

As for fruit crop trees, there is evidence of human cultivation acitivities dating back between 5,000 and 10,000 years in the Middle East (Stubbe 1965). Species include the date palm (*Phoenix dactylifera*), the fig (*Ficus carica*), the olive (*Olea europaea*), and the pomegranate (*Punica granatum*). By 5000 BC, the Babylonians and Assyrians knew the date palm was dioecious, and there are reports of artificial pollinations by the time of Hammurabi (~2000 BC). Roberts (1929) noted over 400 varieties of date palms distributed amongst only four Saharan oases, each of these varieties differing in fruit traits involving size, shape, and flavor. Today we have considerable diversity in fruit products from trees (e.g., apples and olives). Why has there been such relatively rapid progress in the evolution of reproductive traits in plants (e.g., fruit and seed characteristics of corn and fruit crop trees), and somatic traits in animals (e.g., body traits of dogs), yet seemingly not in somatic traits of woody plants (e.g., vegetative characteristics of forest trees)?

It is generally thought that the greatest impediments to breeding forest trees have been (1) generation time and (2) breeding directly from natural populations rather than through the use of landraces (Libby et al. 1969; Grattapaglia et al. 1993; Lantz 2001). Otherwise the breeding of non-woody crops and the breeding of trees present many of the same problems that are addressed in similar fashion.

1.2.2 Generation Time

Probably the most critical limitation to tree breeding is the amount of time it takes for a tree to pass through a single life cycle (Lantz 2001). In non-woody crops the breeding cycle may take a little over a year whereas trees usually require between one and two decades for a breeding cycle. As a point of comparison, consider the domestication of corn from teosinte which began perhaps 6,300 years ago. This lineage has undergone ~6,300 generations of selection, whereas a tree species with a breeding cycle of 20 years will have gone through no more than 300 generations over the same interval.

Many other limitations are drawn from this protracted life cycle of trees (Fig. 1). One cannot make controlled crosses until individuals reach reproductive maturity, which is only a matter of months in crops but over a decade for most forest trees. Even if the tree can be induced to flower at an earlier age, many economically important traits are most readily appraised at an advanced age and large adult size. At times early selection on such traits is possible (Wu 1999) though often gains are slow due to weak correlations of some quantitative traits across size classes. In production forestry, one must wait for the adult traits to mature before harvest, regardless of successes with early crosses or early selection. Thus rotation lengths have been longer for trees (40+ years) compared with crops (< 1 year), and returns on investments are by necessity more tenuous. A severe storm at the production site may be unlikely in a single year, but much more likely over half a century.

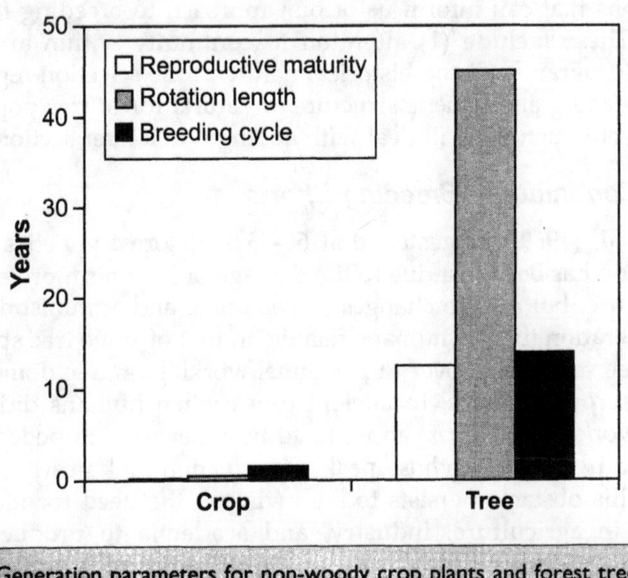

Fig. 1 Generation parameters for non-woody crop plants and forest trees. Values reported in Lantz (2001).

1.2.3 *Breeding from Natural Populations*

Humans have remained in the hunter-gatherer phase of their relationship with forest trees over most of history. Many cultures have treated the forest tree resource unsustainably; if more wood is needed, more trees are cut. The widespread planting and improvement of forest trees is largely a modern development of the twentieth and twenty-first centuries. The forest tree breeding that has occurred has mostly drawn directly from natural populations or from a few elite selections, whereas the breeding of most non-woody crop plants has involved extensive use of landraces, varieties, and inbred lines (Lantz 2001). For example, important progress has been made on black walnut (*Juglans nigra*) improvement in the past several decades in the midwestern US in part through the establishment of 350 phenotypically superior trees in a breeding program (Michler et al. 2005). About half of these superior trees were selected from wild populations and the other half from plantations and progeny tests. This approach represents an essential strategy, integrating wild germplasm and managed lines. The limitation is that many of the lines that were drawn from plantations and progeny tests had themselves been drawn from natural populations. Advanced generation selections simply are not available for most forest tree species.

There are three relevant subtexts to breeding from natural populations that can inform us of our approach to breeding forest tree species. These include (1) attention to continuity within and among breeding programs, (2) the historical nature of the selection regime, and (3) the breeding and genetic structure of natural forest tree populations, the latter of which we will deal with at length in a later section.

1.2.4 *Continuity in Breeding Efforts*

Libby et al. (1969) suggested that 50-75% of forest genetics research information has been lost due to the damage of long-term projects, as by storm or fire, but also to changes in personnel and administration. The short generation time of humans relative to that of some tree species has contributed to the turn over in personnel working on tree domestication projects across the ages. In ancient times when humans did not live much beyond 40, progress must have been severely impeded for any work on a tree species whose peak reproduction took nearly as long to accrue. This obstacle persists today owing to the need for individuals working in agriculture, industry, and academia to produce results within the 'lifetime' of the person's career (again, perhaps 30-40 years), be it for profit or publication. Cultural practices have undoubtedly exacerbated the situation as well. The agrarian culture has provided a certain stability in which people can invest themselves in a plot of land along with its plants and animals, yet wars, plagues, pestilence, depleted land, and simply boredom have encouraged a semi-nomadic life style in many human societies that would have favored domestication of annual plants over trees.

The net effect has been that the tree breeders have had to reinvent the wheel on numerous occasions. Setbacks assuredly befell the domestication of non-woody crop plants as well, though the history of progression is there to unravel, as relayed in the report by Jaenicke-Després et al. (2003) on archaeological specimens of maize from Mexico and the southewestern US. They examined three genes controlling plant architecture, storage protein synthesis, and starch production and found the alleles characteristic of modern maize were present by 4,400 years ago. However, allelic selection at one of these loci was incomplete by as recently as 2,000 years ago, showing that progress in the latter stages of domestication apparently had built on earlier work that had accrued for nearly 2,000 years (between 6,300 and 4,400 years ago).

1.2.5 The Historical Selection Regime

Prior to the twentieth century, the balance of humanity's selection pressures on forest tree gene pools may have inadvertently served to increase the representation of genes for slow-growth, small diameter, poor bole shape, and undesireable wood characteristics. For example, large portions of forest were cleared in the eastern US during the 1700s and 1800s as European settlers established farmsteds and cities. Most of the northern half of Indiana was cleared and many of the remnant forests persisted as woodlots on farmland where they were subject to repeated rounds of selective harvest (Den Uyl et al. 1938; Den Uyl 1954, 1961; Parker 1997). Culling of dead and dying trees would have had a negligible impact on the forest tree gene pools since these individuals were no longer reproducing. It was the practice of taking the biggest and the best healthy trees for their superior timber that may have made a significant and negative impact on forest tree domestication. This process, referred to as 'high-grading', would have removed the most desired trees from the gene pool, unless special care was taken to harvest seed from superior trees before they were taken for timber. Otherwise the net effect is the next generation of timber trees derives from the undesirable stock (left, Fig. 2) – altogether the opposite of what should occur during domestication.

By contrast, the harvesting of superior fruits of a fruit crop tree places the superior germplasm in the hands of the harvester. A portion of the select seed are easily set aside for planting and the surplus eaten,

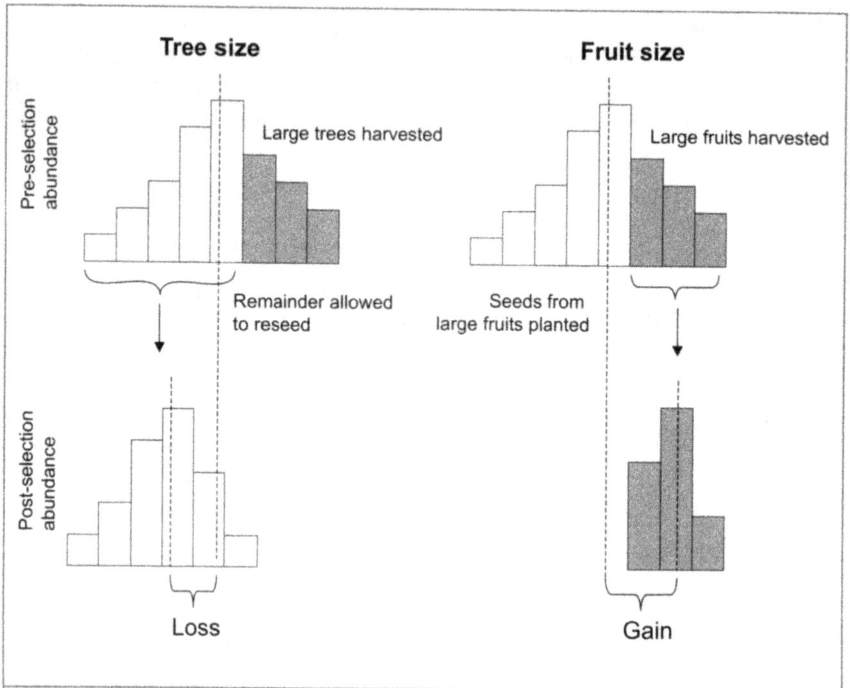

Fig. 2 The differential and response to selection on forest tree populations harvested for their timber by high-grading (left), compared to expectations for fruit crop trees (right).

or if it is the fruit that is consumed then the seed is processed from the fruit so the fruit tissue can be consumed without harming the seed, or the gene pool. In either event, the fruit crop trees of subsequent seasons derive from the best fruit producers (right, Fig. 2).

It remains unclear to what extent high-grading has eroded the gene pool of forest trees. This has much to do with the lack of records on forest tree attributes from past centuries that would provide a measure against which to appraise present-day stock. A notable exception to the lack of records is the Cedar of Lebanon (*Cedrus libani*), a conifer of the mountainous regions of the Mediterranean. Various ancient writings document the species had been favored for its exceptional size since the time of the Phoenicians, ancient Egyptians, Mesopotamians, and Romans who all used it for ship building and temple construction (reviewed in Talhouk et al. 2001). The Cedar of Lebanon now persists mostly at high elevations in fragmented and degraded populations with presumably few individuals that would have captured the interest of a Phoenician ship builder.

As a more recent example, consider black walnut in which rarity of large trees with good figure has long been attributed to the traditional practice of high-grading (Clark 1965; Elias 1980). McGuire et al. (1999) surveyed forest industry professionals about their perception of the scarcity of quality veneer logs in the upper Midwestern US. The consensus was that quality was indeed decreasing, as was the quantity of large to medium diameter logs, though hard data are lacking.

1.3 Advent of Modern Tree Breeding

1.3.1 Traditional Breeding

The technology applied to tree breeding has advanced far past the hunter-gatherer stage in its sophistication since the beginning of the twentieth century (see reviews by Namkoong et al. 1998; Eriksson et al. 2006). It is beyond the scope of this chapter to review all the various methods now employed in tree breeding programs, and indeed many of the approaches are the same as those used in other plant and animal breeding programs (Allard 1960; Falconer 1960). Traditional programs have concerned themselves with quantitative genetics and the tools and methods inherent to that proven technology, including breeding strategies, mating designs, controlled pollinations, selection strategies, clonal propagation, and the like. More recently, they have added molecular genetic and genomic methods of improvement.

1.3.2 Molecular Biology

The 1970s and 1980s brought an optimism that molecular biology would solve many of the aforementioned challenges faced by traditional tree breeders. Unfortunately, the pendulum probably swung too far for a period as many turned toward biotechnology and discarded traditional methods and long-term studies based on these older methods. For example, many of the black walnut plantations and provenance trials that had been painstakingly established were set aside and let to grow over as tree breeders went molecular (Woeste and McKenna 2004). Although biotechnology is likely to solve many problems in forest tree breeding that presently appear insurmountable, the short-term view is gradually embracing a more integrative approach to tree breeding that includes modern molecular methods applied to traditional breeding opperations (see Chapter 1 by Tuberosa et al. of this volume for a discussion of the fundamentals of molecular breeding in plants).

Molecular genetic markers have revolutionized tree breeding by allowing visualization of genome content and structure with minimal interference by the phenotype or environment (see also Chapter 1 of Volume 1 by Kole and Abbott). With regard to genome mapping, molecular genetic markers are used in several capacities in tree breeding. Many of these applications appear in breeding programs of non-woody plants and of animals as well, though not necessarily with equal weight since some are more critical to the breeding of forest tree species given their unique biology and history of domestication. The following are several applications of molecular markers to tree breeding:

- *Determination of species boundaries.* For several reasons noted below, it is useful to know the boundaries of the focal species, and molecular markers are useful to this end.
- *Characterization of geographic genetic structure.* Since most forest tree species are still largely undomesticated, the gene pool which includes both managed and natural popuations provides the base from which crosses are made, maps built, and tree improvements made. As it is useful to know the composition and organization of a genome, it is useful to know the composition and structure of a species gene pool.
- *Discrimination of sibships.* Many of the crosses used in breeding forest trees draw not from controlled matings between inbred lines but from open-pollinated canopies. Rogue pollination events can be detected by molecular markers and removed to produce pure families of full-sibs.
- *Production of a genetic map.* One can array the molecular genetic markers in order to produce a genetic map.
- *Detection of loci responsible for phenotypes.* Joint variation in the phenotype and mapped molecular genetic markers can be used to ascertain which portions of the genome are responsible for the phenotypic variation.
- *Marker-assisted selection.* Selection for or against genes responsible for certain phenotypes can be hastened by detecting the relevant genotypes early in the life cycle, bypassing the principal limit to forest tree breeding, the long generation time.

Strauss et al. (1992) highlighted several important concerns regarding the efficacy of applying molecular markers to the breeding of forest tree species. They noted that outbred forest tree species maintain considerable linkage equilibrium that dissolves associations among markers and between markers and traits, making both mapping and marker-assisted breeding a challenge. Traits of interest are likely to exhibit genotype x environment interactions and derive from numerous genes each of small effect, rather than a few genes of large effect. There also exist unfavorable genetic correlations between characters that may impede progress during selection. Markers should aid selection only in certain cirucumstances such as in an individually mapped full-sib family in which the genetic base has been severely restricted. They noted that the traits that should be most amenable to marker-aided selection in non-hybrid tree populations are those which are strongly inherited with low heritabilities, but this also may bring problems discriminating between the phenotypes. Such traits include wood quality, pathogen resistance, and resistance to air pollution. The good news is that these are the very traits that interest the breeder (Grattapaglia et al. 1993).

1.3.3 Genomics

Many of these cautionary notes offered by Strauss et al. (1992) have proven prescient though not fatal to the field of molecular genetic-based tree breeding as it has moved into the genomics era. Barriers exist but are dissolving, partly through innovations in the laboratory and in statistical analysis, and simply through the brute force afforded by high-throughput genomics technologies. Much of modern genomics research is based on procedures that have been done for years under the guise of molecular biology (e.g., DNA sequencing, surveys of gene expression), though now these assays are done over the weekend for thousands of genes instead of over several years for one or a few genes. PCR (polymerase chain reaction) was a significant technical innovation that helped usher in the genomics era and advance tree molecular genetics. PCR-based markers such as RAPDs made a significant contribution to tree genome mapping by presenting a large number of markers (beyond that offered by isozymes) with a reasonable throughput not possible with RFLPs (Grattapaglia et al. 1993).

Another technical innovation that changed the course of tree breeding was the analytical development of the pseudo-testcross mapping strategy (Grattapaglia and Sederoff 1994). While many were rightly concerned about the extensive linkage equilibrium present in outbred tree populations, Grattapaglia and Sederoff turned the coin over and noted that high individual heterozgosity is a product of outbreeding and that this diversity could be harnessed to find a suitable density of segregating markers along a chromosome. In combination with RAPD technology which provides a nearly inexhaustible number of markers from which to choose, researchers have completed many genetic maps of individual forest trees.

Genomics is beginning to provide basic information on the composition of forest tree genomes that will advance breeding efforts. It is now possible to answer the fundamental question of how many genes control the variation in a trait. For so long the only voices were the theoretical extremes of a single gene (population genetics perspective) or many genes each of equal and small effects (quantitative genetics perspective). Yet as researchers explore genomes they find the answer is often somewhere in the middle. For example, QTLs (quantitative trait loci) for wood traits often involve a few major loci and some modifiers. For example, Groover et al. (1994) found five QTLs controlling wood specific gravity in loblolly pine (*Pinus taeda*). Expression profiling offers novel insights into how genes are regulated, as with the survey of genes expressed in heartwood tissue of *Robinia pseudoacacia* (Yang et al. 2004). These results show floral regulatory genes present in the heartwood which underscores the cautionary note from Strauss et al. (1992) on correlated traits. The first tree genome has been fully sequenced (*Populus trichocarpa*, Tuskan et al. 2006) and this opens new roads for comparative genomics and the transfer of information to non-model system tree species.

At this point we will begin to explore the genetic base of the forest tree gene pool. With an understanding of the limits presented to us by the biology of forest trees and by their domestication history, we can then explore the resources available for molecular mapping and breeding.

2 FOREST TREE GENETIC RESOURCES: *EX SITU* SYSTEMS

An important element of breeding involves making use of prior progress, and with forest trees much of the improved stock resides in *ex situ* (off-site) management systems (NRC 1991; Eriksson et al. 1993). These entities maintain plant germplasm that has been removed from its natural forest setting. Management may be national, local (such as the state-level), or private industry (such as paper and lumber companies). Types of systems include seed storage facilities, seed orchards, clone banks, progeny or provenance tests, and various plantations (NRC 1991; Lipow et al. 2002; Eriksson et al. 2006). The subject of maintenance may be a planted tree or tree part, seed, pollen, or even tissue culture. Stock usually is considered improved to one degree or another, though not necessarily. *Ex situ* management systems are useful not only as a source of superior genetic material for breeding but they carry the added benefit that the reproductively mature individuals might serve directly as parents in the production of a mapping population.

The level and balance of *ex situ* versus *in situ* (on-site, in the original environment) management for any given forest tree species depends on the individual species biology and on the resources available for the investment in management of that species. Monterey pine (*Pinus radiata*) with its restricted natural distribution has a significant proportion of its total gene pool under *ex situ* management whereas the widely distributed Douglas-fir (*Pseudotsuga menziesii*) is well protected both *in situ* and through various *ex situ* resources (Burdon 1997; Lipow et al. 2002, 2003). Black poplar (*Populus nigra*) is a model organism for tree genomics and breeding and has been brought under extensive *ex situ* management in Europe (Storme et al. 2004). Some species are widely distributed and have limited infrastructure devoted to their *ex situ* management, such as many of the species in the Central Hardwoods region of the US which carries the largest contiguous assemblage of deciduous tree species in the world (Hicks 1998). Native Central Hardwood species include black walnut (*Juglans nigra*), several important oaks (*Quercus* spp.), hickories (*Carya* spp.), and black cherry (*Prunus serotina*). As the seeds of many of the temperate timber species lose their viability after only a year or two (Schopmeyer 1974), living populations (*ex situ* or *in situ*) have become the management method of choice for many of these species.

Molecular genetic studies suggest that *ex situ* management systems often perform well in capturing diversity present in wild forest tree populations, though more study of this important topic is warranted (for an example, see the study by El-Kassaby and Ritland 2004). Generally speaking though, much of the genetic variation present in forest tree species resides within populations as opposed to between populations according to reviews of the isozyme literature (Hamrick and Godt 1990; Hamrick et al. 1992). Thus, one can sample from several representative populations and expect to capture many of the common alleles in the species. There are reasonable concerns, nevertheless, concerning the loss of rare and endemic alleles, some of which could be vital to the long-term survival of the species in the face of climate change or introduced pests and pathogens (Ledig 1992). For a more thorough review of the matter consult Young et al. (2000), Rajora and Mosseler (2001), and regarding molecular markers in germplasm conservation, Labate (Chapter 2, this volume).

The following are brief descriptions of several *ex situ* systems used in US tree breeding programs, as well as the terminology used in forestry to grade individual trees as to their quality and level of testing (following Snyder 1972; Maynard 1996; Lippow et al. 2002; USDA-FS 2002; Eriksson et al. 2006). An understanding of these terms is useful for appraising which resources are available for a species when it comes to establishing a mapping and breeding program.

2.1 Tree Improvement Program Intensity Levels

The US Forest Service recognizes five levels of tree improvement (USDA-FS 2002):

- *Level 1 - Source-identified plant materials.* Seed and plants derive from known geographic locations (see *Provenance, Seed zone*).
- *Level 2 - Seed collection stands.* Natural stands consisting of trees that have been identified as superior in some capacity and used for repeated harvest of seed.
- *Level 3 - Seed production areas.* Similar to *Seed collection stands* except these are managed more intensively to optimize seed production of trees with demonstrated superior qualities. Undesirable trees that might cross-pollinate are removed (see *Rogue*) and density is adjusted to maximize seed production.
- *Level 4 - Seed orchards.* A plantation of trees comprised of clones (via grafts or cuttings) or seedling progeny from select trees. Orchards are usually spatially isolated from natural populations to limit the amount of pollen contamination from potentially inferior outside sources. Management is intensive and trees are planted and tended to yield frequent, large, and early seed crops. Selected trees have not necessarily proven their genetic quality through progeny testing.
- *Level 5 - Selective breeding program.* Production of seed with known parentage (see *Pedigree breeding*) to establish a solid genetic base from which to produce advanced-generation selections and mass-produce improved trees. Parents selected for inclusion in the program are usually superior in some respect.

2.2 Other Types of *Ex situ* Management

- *Clone bank.* This unit of management propagates a species vegetatively or clonally. Clonal propagation preserves the genotypic composition of a line due to the absence of meiosis between generations. (see also *Clonal test*).
- *Clonal test.* Clones planted in a field study. Vegetative propagation allows complete replication of genotypes in order to assess the role of the environment in determining the phenotype and the relative performance of different genotypes. The breeding behavior remains ambiguous, however, due to lack of sexual reproduction (see also *Combining ability*).

- *Gene bank.* A germplasm collection that is made to maximize retention of genetic variation through accessions of seed, pollen, plants, plant parts, and DNA.
- *Nursery.* A public or private sector forestry operation that propagates trees through mass germination and growth through the early stages of development before transplanting. Often involves only seeds and seedlings on-site, with seed-producing trees unfortunately often of unknown origin as seed may be taken by donation with little background information on the source (see *Provenance, Seed production area*).
- *Pedigreed breeding population.* A controlled-pollination management system in which information is retained on all breeding individuals in the local population dating back to the original tree(s) selected from the natural population (i.e., the *Plus* tree (+)).
- *Plantation.* A general term used to describe any ordered and maintained planting of trees, though usually trees are arrayed in plots with related trees within each plot (see *Family, Plot, Provenance*). Purposes vary though often will include various crosses and *Progeny tests* made in a breeding program, or for production forestry may represent a very narrow genetic base (see *Clone*) for production of an *Elite* line. A plantation can represent a valuable breeding resource to the extent that it contains lines that already have been selected for their *Superior* attributes.
- *Progeny study or test.* Plantings of numerous progeny per mating event or focal tree in order to compare the performance of offspring and judge the value of the parental genotypes. Although the genotypes of the progeny vary within each family due to sexual reproduction, the environment can be controlled through the common garden planting to test parents that might be growing in very different environments and conditions. (See *Clonal test, Combining ability, Provenance*).
- *Provenance study or test.* These common garden plantings bring together individuals from distant geographic sources (e.g., different *Race, Seed source, Seed zone*). The actual term 'provenance' means population, referring specifically to the population from which the seed was taken and transferred to this particular planting site. Plantings may be via a seed or cuttings. Provenance studies can include collections from numerous locations, thereby minimizing the effect of the environment on the phenotype and allowing a clearer determination of genetic effects. Older

provenance studies may include reproductively mature individuals with differentiated genomes that may serve as a ready source for wide crosses, either controlled or natural pollinations (see *Open-pollination*), in the production of mapping populations.

- *Seed bank.* A repository of seeds preserved in conditions (typically cool and dry) that allow for future germination. *Ex situ* conservation of most non-woody crop species relies heavily on seed banks, whereas seed collections of forest tree species are costly to maintain and much more variable in their abundance and utility, largely due to the lack of dormancy in many species. These species must be continually replanted and then the seeds are harvested and stored only so long as they remain viable, when they are replanted again. The seed which can be stored may be bulked though more often they are stored with reference to origin such as a *Seed zone,* management region, etc.
- *Seed zone.* A region with fairly uniform ecological conditions within which a seed can be collected in one spot and planted out at another without reasonable concerns regarding a lack of adaptation. Trees within a zone exhibit similar traits and tolerances, ideally identified through *Provenance studies*.

2.3 Individual-based Grading and Source Typing

- *Breeding value.* The genetic value of a tree judged from the phenotypes of its progeny, either based on an individual trait or a *Selection index.*
- *Candidate tree.* A tree that has been identified as potentially worth including in a breeding program but is as of yet untested.
- *Clones.* A set of trees that are genetically identical, derived through asexual propagation.
- *Combining ability.* The ability of an individual tree to produce quality progeny with respect to a given trait. Good general combining ability allows the individual to pass the trait on even when crossed to a variety of individuals.
- *Controlled pollination.* Individual produced through a hand-pollination event.
- *Ecotype.* A subpopulation of a species growing in a specific environment and seemingly better adapted to growing there than at some other sites.
- *Elite.* A tree or stand that has been identified as highly superior for desirable traits, and tested to verify the genetic basis.

- *Exotic.* A non-native tree, introduced from another country or region. May be invasive if growth is damaging to native flora or fauna.
- *Family.* A group of closely related genotypes.
- *Fitness.* A term from classical genetics which means the capacity of an individual to survive and reproduce in a given environment.
- *Hybrid.* Result of a cross between unlikely genotypes, perhaps between different varieties within the same species (intraspecific) or different species (interspecific).
- *Ideotype.* The ideal example of what the tree species should look like once it is domesticated or modified. It is mainly a starting point to guide selection.
- *Improved.* A general term denoting seed collected from select trees, though often from open-pollinated canopies.
- *Inbred line.* A population resulting from repeated selfing events, and thus is genetically homogeneous and largely homozygous.
- *Landrace.* A population of trees that has experienced several generations of selection, be it in a specific natural environment or through artiticial selection.
- *Open-pollinated.* The result of natural pollination(s) such as from the wind or insect activity.
- *Ortet.* The original tree used to establish an asexual line of propagation.
- *Pedigreed.* With a recorded ancestry.
- *Plus (+).* A stand or tree that displays a clearly superior phenotype (or set of phenotypes for a stand). Typically the superior character is specified.
- *Polyploid.* Having more than twice the haploid number of chromosomes.
- *Provenance.* The geographic source of seed, pollen, or propagules.
- *Race.* A subpopulation of a species exhibiting some phenotypic (and presumably genetic) discontinuities with the rest of the species. Often interbreeds freely with other members of the race, though perhaps not as freely outside of the race. May be associated with ecotypic variation.
- *Ramet.* A single individual of a clonal set, sharing the same genotype.
- *Rogue.* An individual with an undesired phenotype or genotype, usually marked for removal.

- *Seed collection zone.* Region in which trees display fairly uniform composition and tolerances verified by progeny testing.
- *Seedlot.* A group of seeds sharing an attribute such as provenance, sibship, date of collection.
- *Seed source.* The site from which the seed was collected. May be same as *Provenance* unless referring to an exotic species.
- *Selection index.* A weighted scheme in which parents are selected based on their composite economic value and heritability of certain traits.
- *Sublined.* A breeding population that has been divided into separate subpopulations.
- *Superior.* A general term indicating a select tree that has proved itself as outstanding.
- *Variety.* From the taxonomic view, a subdivision of a species based on geography or minor phenotypic difference. From the forestry perspective, a group of improved trees that carry a useful and heritable trait, similar to a cultivar.

3 FOREST TREE GENETIC RESOURCES: *IN SITU* SYSTEMS

In situ forest tree resources are populations of free-living trees residing more or less in their native state and condition. Wild, unimproved stocks are still critical to the establishment of breeding programs because of the rarity of advanced generation selections and the early stage of domestication for most species. The native populations form the base of a multitiered gene pool pyramid (Fig. 3) with the level of genetic diversity varying horizontally and genetic gain varying vertically (see Johnson et al. 2001). The gene resource population forms the foundation on which the breeding efforts rest, and incorporates all available genetic variation including native stands as well as *ex situ* resources like plantations and seed orchards. The breeding population is improved beyond the gene resource population but must retain sufficient genetic variation in order to realize gains in breeding over numerous generations. The apex of the pyramid is occupied by the most improved, elite lines that are deployed into orchards and other production operations.

We are interested in the gene pool pyramid in order to wisely manage the transfer of genetic material from wild populations into production forestry. Goals are mainly twofold: (1) harness existing

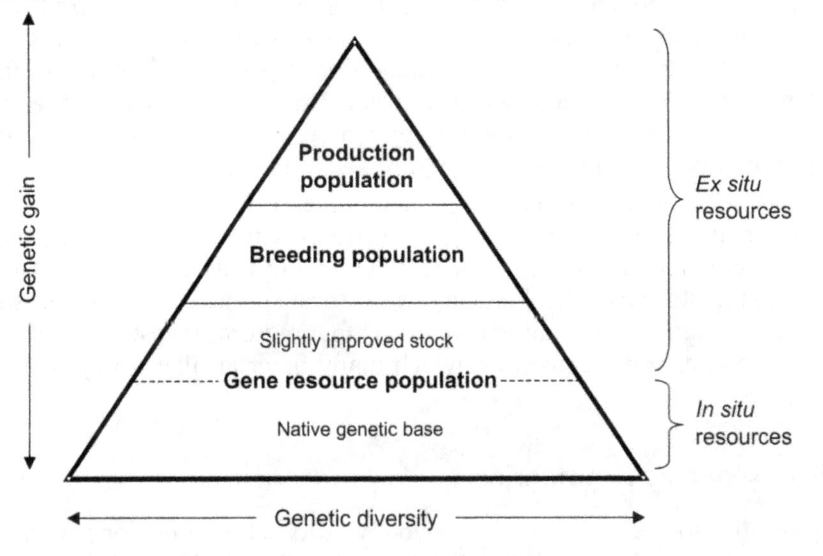

Fig. 3 Pyramid showing the tiers of the forest tree gene pool used in breeding. Modified from Figure 1, Johnson et al. (2001).

genetic variation for use as a tool with which to select adaptive or desirable genetic variation, and (2) select for that variation so as to improve the stock.

One of the genetic tools useful in domestication is the genetic map, and we need molecular genetic diversity for identifying linkage groups in the map. Lacking inbred lines and with controlled crosses difficult in some tree species, we often must rely on existing variation in the mapping parents to detect recombination events that disrupt linkage between alleles at different loci. Yet, often there is a high degree of linkage equilibrium in forest tree populations, leaving associatons between markers rather tenuous. There are several approaches that address this problem, some of which build on knowledge of the species' genetic structure. This can involve using high levels of individual heterozygosity in outbred trees or producing controlled crosses between genetically divergent, albeit historically outbred, lineages.

Selecting and improving the stock (goal 2) is the longer term and ultimate goal. Success depends on identifying and capturing variation for important marketable traits such as wood quality, growth rate, bole shape and the like. Also important are less obvious phenotypes that control the long-term success of a lineage through adaptation to climate, pests, and pathogens (Ledig 1992; Eriksson et al. 1993; Davis and Shaw 2001). As trees are very long-lived organisms, each individual must be resilient to a tremendous variety of conditions that will affect it over the course of its life. Domestication from too narrow a base may leave out important variation that presently is not adaptive but may be needed by an individual later in life, or by future generations.

Here we briefly consider some factors regarding the genetic structure of forest tree species and populations (primarily *in situ*) that are relevant to mapping and breeding. While it is possible to walk outside and select the first fruiting tree that looks like your focal species, there are benefits to a more rigorous beginning to a project that requires investment of time and expense, and a project on which many other ancillary projects will depend.

3.1 Species Boundaries

Domestication of trees, as of non-woody crops, should be done with an adequate knowledge of the focal species (Lantz 2001). Inadequate knowledge can lead to wasted effort and set limits to the gains possible in a breeding program, such as if a small subset of the available genetic base is included in the early stages of breeding (Fig. 4). As an example of

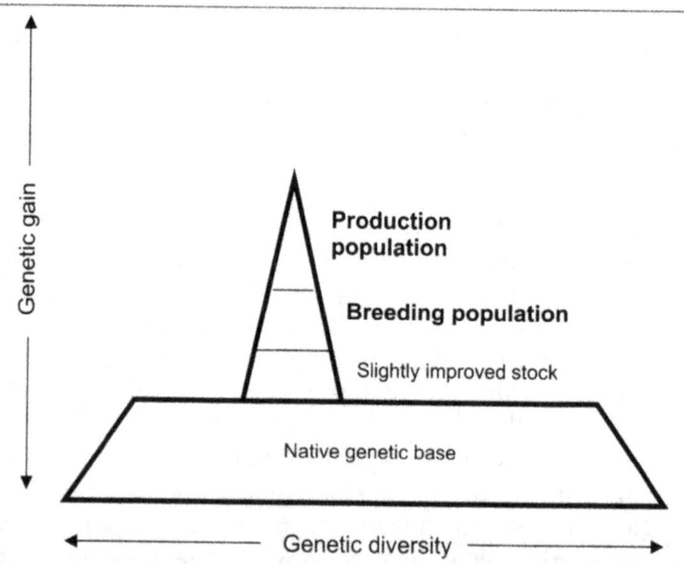

Fig. 4 The gene pool pyramid in which the breeding program is established from the outset using a very narrow genetic base.

the opposite practice, recall the black walnut breeding program (Michler et al. 2005) which includes roughly half improved and elite lines from plantations and progeny tests but the other half selections from wild populations.

3.1.1 Gene Pool Partitions

Maynard (1996) distinguished three partitions in the forest tree gene pool that are relevant. For a sexually reproducing species, the gene pool can be thought of as a field of sexual recombination through which gene exchange is possible between lineages. Free exchange of genetic material is possible within the primary gene pool of a species (species A triangle, Fig. 5), using here the commonly employed biological species concept (Mayr 1942). Maynard (1996) extends the primary gene pool to include closely related species that are freely able to form hybrids, though this definition extends beyond the simple definition of a biological species and has more likeness to Paterson's (1985) recognition species concept and to Templeton's (1989) cohesion species concept in some respects. Regardless, the secondary gene pool consists of a further set of species (species B triangle, Fig. 5) that also can engage in sexual reproduction

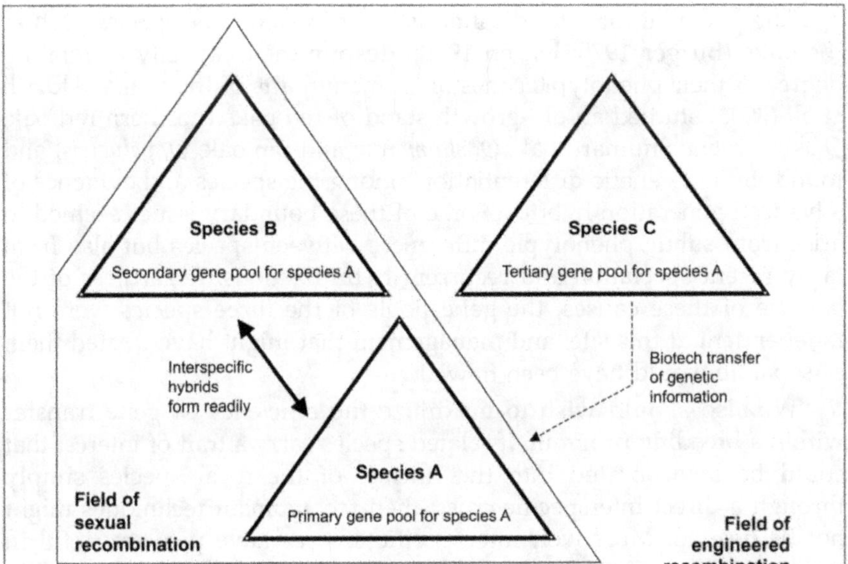

Fig. 5 Partitions of the forest tree gene pool that includes secondary components contributed by other species through hybridizations and tertiary components available via recombinant technologies.

with the focal species A, although Maynard (1996) suggests that these exchanges are not spontaneous and require intervention using such procedures as mentor pollen or embryo rescue, essentially biotech-assisted sexual exchange. Finally, the tertiary gene pool for species A consists of all other species (here, species C, Fig. 5) which cannot engage in sexual reproduction with species A (either independently or with assistance) but conceivably could contribute genes to species A by recombinant tecnologies.

3.1.2 Primary Gene Pool

We will take up the intraspecific portion of the primary gene pool in a later section. But will first consider that portion of the primary gene pool comprised of individuals of other species which are freely able to interbreed with the focal species. Subsections of a primary gene pool are not independent if there is recurrent gene exchange, and so management of one species without attention to the other species could be misguided. Pollen contamination in a seed orchard is a simple example. If one does not recognize and control species capable of crossing freely with the focal species, then there is little to stop these related taxa from diluting the breeding pool in an orchard.

The potential for hybridization is great in some tree species such as the oaks (Burger 1975; Jensen 1988), despite their capacity to retain a degree of their phenotypic cohesion in many parts of the range. Aldrich et al. (2003) studied an old-growth stand of red oaks (northern red oak, *Quercus rubra*; Shumard oak, *Q. shumardii*; and pin oak, *Q. palustris*) and found limited genetic differentiation among the species and evidence of advanced generation hybrids. Some of these boundary issues seemed to arise from subtle phenotypic differences between species but also from fairly recent speciation and recurrent hybridizations. Regardless of the balance of these causes, the gene pools of the three species were not independent at this site, and management that might have treated them as separate would have been flawed.

We also should wish to maximize the efficiency of gene transfer within a breeding program. If related species carry a trait of interest that could be incorporated into the lineage of the focal species simply through a direct interspecific cross, then recombinant techniques might not be needed. Moreover, interspecific crosses have proven useful in producing genetic maps in forest trees. The high linkage equilibrium in forest tree populations means that associations between markers are normally weak, even for loci that are fairly near each other on a chromosome. It is only the tightest linkage relationships that show through unambiguously. But two related species with somewhat divergent genomes can carry between them considerable linkage disequilibrium, as they often possess different alleles or at the very least are differentiated with respect to allele frequencies. Thus, a randomly selected pair would likely differ in their genotypic composition at many loci. These differences disappate rapidly with subsequent generations of backcrossing but can be harnessed in the short-term to track recombination events. For example, Grattapaglia and Sederoff (1994) used two species of *Eucalyptus* (*E. grandis* and *E. urophylla*) to produce a map of ~250 markers. Often the mapping population segregates not only for numerous genetic markers but also many phenotypic traits that can be used for QTL analysis (e.g., Frewen et al. 2000). We will discuss wide crosses in greater detail later in the chapter.

3.1.3 Secondary Gene Pool

The secondary gene pool represents a wealth of gene resources for improvement of a focal species, though infertility barriers could slow tree improvement efforts if seed crop yields are low because of poorly chosen matings. For example, Lopez et al. (2000) showed that F_1 hybrids of *Eucalyptus ovata* x *E. globules* were at a distinct disadvantage relative to open-pollinated half-sibs of one of the parental species. Some of the problems associated with certain crosses arise from incompatibility barriers. At times the mentor pollen method can be used to bypass the barrier. Mixtures of pollen from species B are mixed with pollen from species A, and then placed on the stigma of species A in the hope that the chemistry of the A pollen will provide a permissive environment in which the B pollen could germinate and cross-fertilize. This approach has been used successfully in poplar (Stettler 1968; Gaget et al. 2004) and several other tree species.

More severe barriers to interspecific hybridizations can arise from changes in genome architecture like inversions, translocations, and duplications. These mutations can produce abberant gametes during meiosis yielding hybrid sterility. Tuskan et al. (2006), in a comparison of *Arabidopsis* and poplar genomes, show that such mutations can arise repeatedly and have a marked result on genome architecture. This is not true of all groups of forest trees, though, as the pines (*Pinus*) have undergone rather limited structural genome differentiation despite large genome sizes.

3.1.4 Tertiary Gene Pool

Moving genes between species that cannot otherwise hybridize offer tremendous potential for the domestication of forest trees. Nearly any species is a candidate donor, though there are many points to consider and much to learn. But we cannot simply take a structural gene from one organism and stick it into another since genes do not express in a vacuum. One must consider the mode of gene regulation as well as unintended pleiotropic effects. We will take up this issue again briefly at the end of the chapter, though for a more thorough treatment of transgenics of plants in general (see Davey et al. 2008, Chapter 12 of this volume).

3.2 Genetic Structure of Forest Tree Species

The genetic structure of a species refers to the levels of diversity within its populations and the manner in which variation is partitioned among populations. Hamrick and Godt (1990) reviewed the allozyme literature and found that woody species are more diverse genetically on average than non-woody species. Woody species also carry much of their genetic variation within their populations, the balance accounting for genetic differences between populations (Hamrick et al. 1992). Within woody taxa, those with a large geographic range, predominantly outcrossed mating system, and seeds dispersed chiefly by animals or wind maintain the most variation within their populations. This latter group accounts for many of the important timber species in the temperate zone. Although general life history and ecological characters explained 34% of the variation in genetic diversity in the Hamrick et al. (1992) survey, the remainder was unaccounted for and must owe to species-specific and site-specific conditions. See also Krutovskii and Neale (2001) and FAO et al. (2004) for more recent reviews of forest tree genetic resources.

Several factors contribute to the genetic structure of the forest tree primary gene pool (Fig. 6), most notably mutation, gene flow, genetic drift, selection and the mating system (Hedrick 1985). Mutation and gene flow tend to introduce genetic variation into populations whereas genetic drift and selection often purge variation from a population. All four of these forces can act to change allele frequencies and genotypic frequencies whereas the mating system typically only influences genotypic frequencies, leaving the allele frequencies unchanged. The mating system simply repackages alleles into genotypes. However, large changes in genotypic frequencies can occur through a single mating event and so the effects of the mating system can be rapid and extensive. The mating system is especially interesting with respect to the crossing potential of a species and to the construction of mapping populations, so we will give it more than a passing consideration.

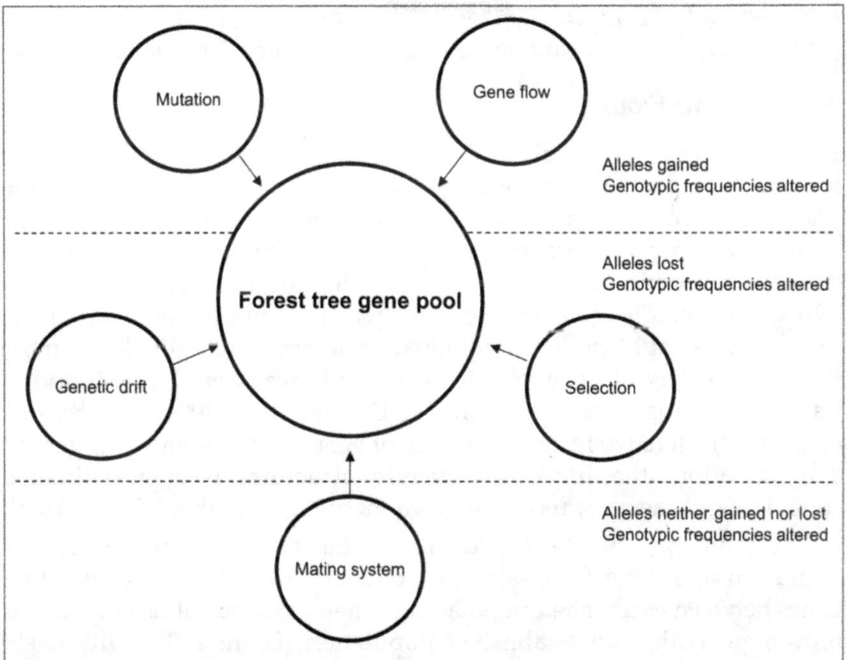

Fig. 6 Factors influencing the genetic structure of the forest tree species gene pool.

3.2.1 Mutation

Mutation provides the raw material for natural evolutionary change and for breeding improvement. Natural mutation rates are usually rather low although collectively a natural population can contain many mutant alleles (the term 'mutant' in this case being a purely relative label denoting a hereditary departure from the historically most abundant allele type in the population). Typically breeders make use of existing variation in populations although mutagenic experiments have been used in some tree breeding programs, and biotech applications are increasingly prominent. A standard mutagenic experiment is the induction of polyploidy, since often polyploids display more robust phenotypes than their diploid progenitors. As an example, Blakesley et al. (2002) used colchicine treatment to induce polyploidy in *Acacia dealbata* and *A. mangium* for inclusion in a breeding program. Although mutagenesis experiments using chemical agents and X-rays to produce point mutations have proven to be a powerful tool for dissecting genomes such as *Arabidopsis*, these fine-scale approaches have been used less successfully in forest trees. Nevertheless, poplar shows great promise for insertional-mutagenesis gene discovery (Groover et al. 2004).

3.2.2 Gene Flow

Libby et al. (1969) maintained that gene flow had been among the most intractable of problems in forest genetics. It usually was not possible to know with any certainty the pollen source of seed produced in an orchard or in any open-pollinated canopy until the application of molecular genetic markers to plant mating studies beginning in the 1970s. These studies soon showed that gene flow in tree species could be extensive, as could pollen contamination in seed orchards. For a more thorough review of gene flow in trees see Burczyk et al. (2004) and of transgene escape from plantations see DiFazio et al. (2004) and Brunner et al. (2007). Here we focus on the role of gene flow as a source of genetic contamination, the impact on family structure in open-pollinated canopies, and the relevance to linkage mapping in outbred forest trees.

Terminology can be confusing in this area, although normally migration and gene flow refer to movement of individuals and their genes between established populations whereas dispersal usually means movement within an established population (Levin 1981). Ultimately though, it is the extent to which pollen and seed move with respect to individuals in the landscape that determines the true biological dimensions of the population (Crawford 1984), even though humans may have a different view of the matter based on what they see in the landscape. Discrete patches of trees may be fully interconnected by high rates of gene exchange in which case they really are part of the same breeding unit, deme, or population even though they appear to be isolated by the intervening matrix habitat. This is good to remember when setting up a seed orchard or doing any management that is potentially concerned with pollen contamination.

Molecular marker studies confirm that gene flow into seed orchards can be considerable (reviewed in Adams and Burczyk 2000; Burczyk et al. 2004). Contamination estimates are commonly up to 40%, and are only loosely related to the size, age, and pollen production of the seed orchard itself (Friedman and Adams 1985; Pakkanen and Pulkkinen 1991; Wang et al. 1991; Adams et al. 1997; Pakkanen et al. 2000). Even orchards that are widely isolated from pollen sources are subject to contamination. Harju and Nikkanen (1996) found 48% pollen contamination in a Scots pine (*Pinus sylvestris*) seed orchard isolated over 2 km from any native conspecific populations, and Buiteveld et al. (2001) found 70% pollen contamination in a seed orchard of *Q. robur* despite separation of 400 m from a few free-standing oaks and another 10 km to the nearest large population. Many of the temperate timber species are wind-pollinated and there is some evidence that forest fragmentation may at times increase the rate of wind-mediated gene exchange between patches in a fragmented landscape, as noted in reports on temperate sugar maple (*Acer saccharum*, Fore et al. 1992) as well as dry tropical forest mahogany (*Swietenia humilis*, White et al. 2002).

If one had to generalize as to the shape of the pollen dispersal curve it is reasonable to assert that it is leptokurtic (e.g., Levin and Kerster 1974, Streiff et al. 1999). Relative to a normal distribution there is an excess of short pollen movements and an excess of large movements. When we estimate gene flow into a 'sink' population such as an orchard, we are measuring the cumulative effect of many heavy-tailed pollen dispersal curves emminating from different populations in the landscape, all of these tails having a collectively large effect on the net rate of immigration into the orchard. Pollen immigration into natural populations can be considerable even for stands surrounded by forest, where Strieff et al. (1999) found 65% of *Quercus petraea* and *Q. robur* pollen coming from outside a 5.8 ha plot, and Burczyk et al. (1996) found 56% of *Pinus attenuata* pollen immigrating into 11 m radius neighborhoods. Despite this inflated importance for long-distance gene movement, local breeding structure is still influential. For example, improved statistical estimates of pollen movement in white oak (*Quercus alba*, Smouse et al. 2001) revealed a pollination neighborhood comprised of roughly eight males covering an area less than 0.1 ha. This bipolar condition of dispersal (many near and far dispersal events) enables the production of full-sib families from an open-pollinated canopy, as one should expect to find numerous full-sib progeny in the canopies of two adjacent trees with overlapping phenologies despite plenty of pollination events from distant sources.

High gene flow between forest tree populations both hinders and promotes efforts to produce linkage maps. Consider a species with low rates of gene exchange between its populations. Population allele frequencies would diverge (see Genetic drift) and linkage disequilibrium would accrue between alleles at different loci. It then would be reasonable to expect that mapping populations made from inter-population crosses would segregate at several loci that would be informative to crossover events (not unlike crosses involving inbred lines). In contrast, high gene flow between populations yields a mixing of the overall gene pool, slowing or preventing the accrual of linkage disequilibrium between markers making it difficult to map using natural populations. On the other hand, woody plants maintain most of their variation within rather than between populations (Hamrick et al. 1992). This condition is likely to be the result of high rates of pollen movement, little self-fertilization (see Mating sytem), and delayed reproduction which may minimize the loss of genetic diversity during founding events from long-distance seed movement (Austerlitz et al. 2000). The high within-population diversity is manifest as high levels of individual heterozygosity, i.e. the average tree is heterozygous at many of its loci. Thus, there are many loci that are potentially informative regarding recombination events, though linkage phases may often be ambiguous, and we will consider this in greater detail later in the chapter.

3.2.3 Genetic Drift

Random genetic drift is the haphazard fluctuation in allele frequencies due simply to sampling properties of small populations (Hedrick 1985). Allele frequencies in a series of small populations are likely to differentiate with time unless the populations are linked by gene flow which acts to force the allele frequencies to converge. Effective population size (N_e) is directly related to genetic drift wherein small N_e entails substantial drift whereas large N_e implies a diminished role for drift. There are rather few estimates of N_e for tree populations. We do know that size variation within a population can translate into an unequal fecundity distribution which reduces the effective population size relative to the observed census, as seen in a tropical tree (*Symphonia globulifera*, Aldrich and Hamrick 1998) and in balsam fir (*Abies balsamea*, Dodd and Silvertown 2000). Jump and Penuelas (2006) report that populations of European birch (*Fagus sylvatica*) show evidence of genetic drift following six hundred years of forest fragmentation. On the other hand, Victory et al. (2006) studied the microsatellite genetic structure of black walnut (*Juglans nigra*) in 43 populations from 10 states of the Central Hardwoods regions of the US and found high levels of diversity and little differentiation among populations. The balance of evidence suggests that widespread pollen and seed movement, overlapping generations, and delayed reproduction all converge to at least buffer the populations of many forest tree species against extreme genetic drift in many instances.

3.2.4 Selection

Evolution by natural selection entails a change in allele frequencies through the differential survival and reproduction of individual genotypes, ultimately driven by the individual's phenotypic attributes which confer some advantage to garnering resources or mates. Evolution by artificial selection entails the same dynamics except that the extent to which a phenotype confers an advantage first passes through the filter of human perception and intent. In this way domestication can lead a species out of sync with its natural environment, becoming increasingly reliant on human intervention for survival. As an example, a key innovation in cereal crop domestication was retention of the seed on the plant (Paterson 2002; Doebley 2004). This failure to disarticulate allowed for the seed to be harvested rather than falling to the soil where it could germinate. As a result, corn and many other crops are fairly dependent on humans to complete their life cycle. Dependence of crops on the protections afforded by pesticides is another example.

This is not to say that all changes brought by domestication are abberations of perfection produced by nature since it is altogether unclear that natural selection leads to optimal phenotypic solutions (Gould and Lewontin 1979). This qualification applies especially to long-lived species such as trees. These organisms must be resilient to various diverse and conflicting environmental conditions that they experience not only in different habitats present in the landscape, but also over the course of their lifetime in a single habitat. Some oaks can live for three or four centuries that likely would include several rare environmental extremes in temperature or precipitation that normally come around only once a century. Thus, the tree phenotype cannot specialize in handling just one or a few sets of environmental conditions but must be a generalist, which typically requires tradeoffs and non-optimal solutions due to physical and physiological constraints (Mitchell-Olds 1996; Stearns 2004). The process of domestication, on the other hand, does seek an optimal phenotype adapted to a small subset of the possible growing conditions. Details regarding the methods of artificial selection used in tree breeding are beyond the scope of this chapter, however the following references can be reviewed (Falconer 1960; Namkoong et al. 1998; Eriksson et al. 2006).

In this chapter, we consider the role of the molecular genetic map in tree domestication. Before we leave natural selection though, it is worth noting two points relevant to molecular genetics and selection. As Strauss et al. (1992) warned, the gains resulting from artificial (and natural) selection will be countered at some point by the correlation among traits. This probably has less to do with the condition of being a tree as with the general condition that the genome and phenome of an organism are integrated with respect to structure, regulation, and development. As an example, traits may be controlled by genes that also control other traits (i.e. pleiotropy). Although this has been understood for many years, recent studies in systems biology reveal the tremendous density and complexity of the networks of DNA-protein and protein-protein interactions present in bacteria (e.g., Arifuzzaman et al. 2006), *Drosophila* (Giot et al. 2003), and humans (Ewing et al. 2007). In light of these findings, it is no surprise that Yang et al. (2004) found floral regulators epressing in heartwood tissue. The more we learn about such network architectures, the more likely we will be able to select in line with the constraints presented by an organism.

Secondly, important advances have been made using genome scans to detect the portions of a genome that have experienced natural selection in wild populations (reviewed in Storz 2005, Guevara et al. 2005). This owes much to the availability of molecular markers for mid- to high-density coverage of genomes allowing one to scan for outlier loci that show evidence of adaptive divergence relative to the state of the other markers in the genome. This is a variant of the association study which seeks to find correlations between molecular markers and variation in a phenotypic trait through surveys of natural populations. With genome scans, on the other hand, one might focus entirely on the molecular genetic variation to troll for any loci that show evidence of local adaptation. There are several methods being used though the general approach is to determine the neutral expectation of population divergence (F_{st}) and expected heterozygosity (H_e, Beaumont and Badling 2004), or shared ancestry among populations (F, Vitalis et al. 2001), to detect which loci depart from this state and are thus candidates for loci under selection. Ecotypic variants identified by such means could be valuable inclusions in a breeding program.

3.2.5 Mating System

There is a fifth factor, mating or breeding system that essentially speaks to how individuals mate in a population with respect to their genotype. Individuals can either self-fertilize, outcross (randomly or non-randomly), or display a mixed mating system with some combination of selfing (s) and outcrossing (t = 1-s). The mating system typically does not alter allele frequencies by itself, though it can greatly influence the frequency of genotypes by controlling how alleles are packaged together in organisms within a population. Whereas the genetic effects of mutation and natural selection may require a long time to become evident at the population level, the mating system can produce rapid changes in genotypic frequencies (reviewed in Charlesworth and Wright 2001). Random outcrossing produces the number of heterozygotes expected under Hardy-Weinberg equilibrium ($2pq$) whereas self-fertilization will bring a deficit of heterozygotes. Such inbreeding tends to reduce N_e and thereby increase the importance of genetic drift.

The propensity to outcross is high among temperate forest tree species. A number of these species are dioecious (reviewed in Burns and Honkala 1990) and so must outcross. Many of the monoecious taxa exhibit a mixed mating system or are mostly outcrossed (e.g., *Juglans nigra*, Rink et al. 1989; *Quercus rubra*, Schwarzmann and Gerhold 1991; *Carya illinoinensis*, Ruter et al. 2000) although changes in local pollen donor density and other factors can influence the amount of selfing (e.g., *Fraxinus lanuginosa*, Ishida and Hiura 2002). See Bawa et al. (1985b) for a consideration of tropical tree mating systems, and NRC (1991) for a general consideration of forest tree matings systems and their effect on gene conservation.

While a species with individuals that possess both gender functions may carry a certain propensity to outcross or self due to intrinsic incompatibilities and phenological hard-wiring, there are also environmental circumstances that intervene to influence the realized pattern of matings in a population. It is the combination of these factors that yields a particular rate of selfing in a canopy.

- *Separation of genders*. Most of the commercial timber species are monoecious (Lantz 2001), meaning 'one house' wherein both male and female gender functions reside within the same individual. Selfing can occur in monoecious species but not dioecious species which have separate male and female plants, as in *Fraxinus*, *Ilex*, *Juniperus*, *Populus*, *Salix*, and *Taxus*.
- *Phenological separation*. The male and female function may reach maturity at different times even when both gender functions are present in the same individual (monoecious species) or even in the same flower (bisexual flowers). Such temporal isolation prevents selfing. Phenological overlap between individuals within a population also determines potential mates during outcrossing.
- *Physical separation*. The spatial separation of male and female flowers or cones in a canopy can affect the mating system as well. For example, in conifers the female cones are higher in the canopy than the male cones, otherwise pollen would drift down onto the female cones and swamp out outcross pollen. Usually physical proximity also affects mating patterns during outcrossing.
- *Pollination syndrome*. Most of the temperate commercial timber species are pollinated by wind (Lantz 2001). Insect pollination does exist such as in *Acer*, *Liriodendron*, *Salix*, and *Magnolia*, as well as in many tropical tree species (Bawa et al. 1985a). The density-dependent foraging of animal pollen vectors can influence the rate of selfing (e.g., Aldrich and Hamrick 1998).

- *Genetic self-incompatibility*. The ultimate determinant of success of self versus outcross pollen can be the self-incompatibility system. Rather less is known about many of the species with bisexual flowers, though Libby et al. (1969) note that self-incompatibility is especially well-represented in the hardwoods.

Evaluation of *s*, the selfing rate, is a critical element used in some methods of genetic mapping in open-pollinated canopies, which we will consider later in the chapter. Estimation of *s* is fairly straightforward given a few variable molecular markers and a computer program (e.g., Ritland and Jain 1981; Cheliak et al. 2004). These largely address the parentage assignment problem of determining for how many seed in a canopy the father is the same as the mother (selfing) or different from the mother (outcrossing).

As with gene flow, the mating system contributes both a problem and solution to the challenge of mapping in outbred tree populations. Frequent outcrossing yields linkage equilibrium between loci, making it hard to detect crossover events when the parental linkage phase is unknown. Ingvarsson (2005) studied nucleotide polymorphism, haplotype structure, and population subdivision in five nuclear genes in European aspen (*Populus tremula*), an obligate outcrosser. They found high levels of genetic variation and low linkage disequilibrium, and what was found extended only a few hundred base pairs. This would seem to show that recombination rates are high and frequently disrupt linkage relationships, a condition that makes it hard to produce linkage maps and conduct QTL analysis (Grattapalgia et al. 1993). The solution presented by the forest tree mating system derives from the constraint that one can detect crossover events only between pairs of heterozygous loci. A predominantly outcrossed mating system leads to individual heterozygosities close to Hardy-Weinberg expectations, and expected heterozygosities (H_e) are high in part because high rates of gene flow maintain many alleles at each locus in each population. The net effect is high individual heterozygosities (H_o) that are useful for mapping.

It is difficult to construct inbred lines of trees for two main reasons, longevity and genetic load. As already noted, it is challenging to conduct repeated matings of any kind given their long generation time. But the highly outcrossed nature also leads to the accumulation of a genetic load comprised of recessive deleterious alleles at numerous loci (Ziehe and Roberds 1989), and inbreeding depression when self-fertilization does occur. Species that self-fertilize often expose these recessive alleles in the homozygous state and natural selection acts against them, purging them from the population. In predominantly outcrossed species such as most forest trees though, these recessive alleles can persist in the heterozygous state where they are not subject to selection. Abrupt selfing of a line that is historically outbred will expose recessive deleterious alleles that can cause low seed set and poor growth form. For example, black spruce (*Picea mariana*) is a predominantly outcrossed species (Perry and Bousquet 2001) in which selfing leads to considerable inbreeding depression in growth (Johnsen et al. 2003), whereas red pine (*Pinus resinosa*) readily self-fertilizes and shows no evidence of inbreeding depression (Fowler 1965).

3.3 Assessing Diversity in the Tree Gene Pool

We should want to know about the genetic structure of the focal species, and of related species, for two primary reasons when it comes to molecular mapping and breeding. With respect to breeding, we need to establish a breeding population that draws on sufficient variation at the characters of particular interest to allow progress toward our goal of reaching the ideotype, or ideal phenotype. For molecular mapping, we are looking for diversity for use as markers throughout the genome not only at the few loci that might interest us for breeding. A full genome scan of all individuals is not practical for either goal, and so we would like some economy in our methods. Here we consider several surrogate indicators of genetic diversity and genome composition (Fig. 7) that might correlate with molecular genetic variation, help guide acquisitions in a breeding program, and pare down a list of candidate mapping parents.

Our needs are to a certain extent driven by the sort of mapping we are doing, so it is useful to have the mapping design in mind from the outset. Most methods produce a mapping population directly from crosses between extant parents, be they *ex situ* or *in situ* parents, in which case parents must be highly heterozygous. Mapping using the megagametophyte of gymnosperms and in open-pollinated canopies require a single highly heterozygous parent, whereas other mapping approaches benefit from using two heterozygous outbred parents whose genomes are differentiated. So we might select these parents from different locations and move pollen between them, or use indicators of genome divergence to focus our sampling on adjacent parent pairs. Also, it is important to consider the multiple uses of a mapping population beyond production of the linkage map. For example, if the population is also to be used for mapping quantitative traits (see QTL mapping), then the parents must differ in phenotypic traits of interest that are readily scored.

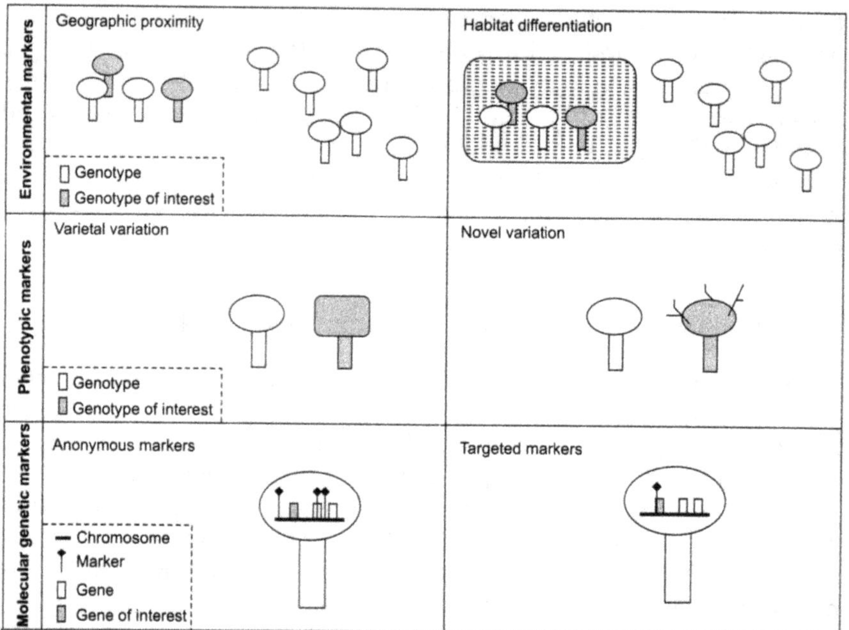

Fig. 7 Approaches to detecting and harnessing molecular genetic variation for mapping and breeding of forest trees.

3.3.1 *Environmental Markers*

We can use environmental structure as a surrogate for knowledge of genetic structure, and this may take the form of (1) geographic structure or (2) habitat-specific structure (Fig. 7). Geography may relate to genetic structure simply through an isolation by distance relationship that develops because of the limited scale of dispersal relative to the total geographic range of the species. Basically, adjacent populations are more alike one another genetically than are populations that are more spatially separated. Even in species that have minimal genetic structure this often is found to be the case, such as in black walnut (Victory et al. 2006). Habitat-specific genetic differentiation may occur in association with local environmental conditions, in discrete patches or in gradients and clines. This can arise by random genetic drift in which case associations with the environment are spurious. Other times it is driven by local adaptation in which case some of the genetic differentiation has been molded by natural selection, and the variants might be termed ecotypes. The major drawback for environmental markers is that, in the end, there may be no useful correlations between the environment and the tree gene pool. But lacking other information, it is a useful place to begin.

A common environmental surrogate used in tree breeding is the seed zone map. This is a habitat-specific representation of discrete regions or zones within which the environment is fairly uniform. One should therefore expect similar performance of seed extracted, transported, and planted within that zone. Variables considered usually include soils, temperatures (extremes, averages), precipitation, and elevation. Seed zones have been identified for the commercially important forest tree species in the US (e.g., Limstrom 1962; Deneke et al. 1981), and these guided many of the breeding efforts during the twentieth century. Molecular markers have since gained prominence for appraisal of genetic composition but the seed zone is still used as an important management tool.

3.3.2 Phenotypic Markers

A novel phenotype can indicate a novel genome (Fig. 7). Some intraspecific phenotypic variation is actually well-currated, such as variation associated with formal subspecific or varietal designations identified by taxonomic studies. So making a controlled cross between two known varieties of a species is one way to ensure at least some differentiation between the mapping parents. Phenotypic sports, or rare phenotypes that have not been raised to a formal taxonomic status, also may be useful indicators though usually their genetic basis has not been validated. Drawbacks of using phenotypic markers are that the environment may contribute a significant amount to the phenotypic variation (i.e. low heritability), or the phenotype being tracked may represent only a small portion of the genome.

One of the best ways to minimize environmental noise in the phenotype is the common garden experiment, which in forestry is termed the provenance (see above, Provenance). These experiments bring together at a communal site tree germplasm collected from different regions, or provenance. Resulting phenotypic variation should owe mostly to genetic factors. Note that a site containing reproductive-aged adults with diverse and divergent genomes situated adjacent to one another provides an ideal context for selection of genetically divergent parent pairs for mapping. Moreover, the minimal influence of the environment on the phenotype makes this a suitable site to appraise parental phenotypic variation for a QTL mapping study.

3.3.3 Molecular Genetic Markers

Molecular markers are ideal for surveying genetic variation (Fig. 7). See also Doveri et al. (Chapter 2 in Volume 1) for discussions of molecular genetic markers and plant breeding. This marker type is not subject to environmental noise and so provides a much more powerful means to directly assess genetic diversity. A drawback is that genotyping can be costly, at least more so than examining a phenotype in many instances, and rarely is it directly obvious how the molecular genetic variation translates (literally or figuratively) into phenotypic variation. Many of the markers used are anonymous, meaning it is unclear what portion of the genome is being queried. If there is sufficient genome coverage one can happen upon markers that are linked to genes that have experienced local selective pressures, this is achieved using genome scans (see Selection; Beaumont and Badling 2004; Vitalis et al. 2001).

Methods of targeted profiling use sequence-specific probes to query only genes of interest, though one must know or posses the sequence to use it as a probe. A gene of interest could be used as a probe in an RFLP application, or a PCR primer could be made that targets a specific gene. We can construct universal primers from the conserved domains of genes, and since these regions are retained across divergent taxa the markers are useful for applications in non-model tree species. For example, Aldrich et al. (2007) have used primers anchored to conserved domains of the cytochrome P450 gene superfamily to scan for potentially adaptive genome-level diversity in the invasive tree *Ailanthus altissima.* An increasingly prominent targeted marker is the ESTP (Expressed Sequence Tag Polymorphism) which is specific to a gene that was present as an RNA transcript in the study organism, and then sequenced (EST) so can we know its identity.

4 STRUCTURAL GENOMICS OF FOREST TREES

Much of what we concern ourselves with in this book is structural genomics as it has to do with the architecture of the plant genome. We will take only a cursory look at functional genomics, though not for lack of interesting material. Branching the divide between structural and functional genomics are such studies as QTL mapping, in which portions of the genome are ascribed the function of contributing to the development of a particular phenotype. As we are concerned here with the spatial organization of factors in the genome affecting the phenotype, this topic is also relevant to structural genomics.

Structural genomics has as its ultimate goal the elucidation of the complete DNA sequence for a species along with identification and localization of all the genes and their control regions (Krutovskii and Neale 2001). There are several paths toward this ideal goal, and some species are moving along this path more quickly than others and so serve as model systems for the forest tree breeder community. For example, the full genome sequence of the first tree, black poplar (*Populus trichocarpa*), was released recently (Tuskan et al. 2006). This rich resource will advance breeding efforts greatly. The linkage map is one of the primary keys to the world of genomics for most other forest tree species.

4.1 Mapping Genome Organiziation

Linkage maps provide information on the order of markers and the distance between them. This serves as a surrogate for the full DNA sequence in which it is the individual bases that are ordered along the chromosome. Much like a road map showing the order and distance between cities though leaving out much of the details of the trip, genome maps provide courser-scale information compared to DNA sequences. Maps nevertheless can be used for many applications in breeding, and can be obtained at a small fraction of the price of whole genome sequencing.

These maps come in two principal varieties, the physical map and the genetic or meiotic map (Fig. 8). Since genetic maps are produced more commonly for forest trees we will only briefly introduce the physical map. Physical mapping is often used in conjunction with genetic mapping, either in the clone-contig approach to whole-genome sequencing as used in the poplar genome project (Tuskan et al. 2006). Physical mapping also is used for chromosome walking in search of candidate genes controlling a trait of interest (see Trait mapping below). With both map types we are tracking the co-occurance of markers in a population of meiotic products (genetic map) or DNA fragments (physical map). The frequent co-occurrence of two markers suggests that they reside near one another on a chromosome whereas markers that are far apart will be detected together simply by chance in the mapping population.

4.2 Model Systems for the Forest Tree Genome

There are several plant genome initiatives, and Cullis (Chapter 10 in Volume 1) provides an overview of these. Here we explore the three primary forest tree genera, *Populus*, *Eucalyptus*, and *Pinus* that have received the most attention from genome researchers and molecular mapping efforts. We also consider some of the other non-model species that may prove important in the near future of forest tree breeding.

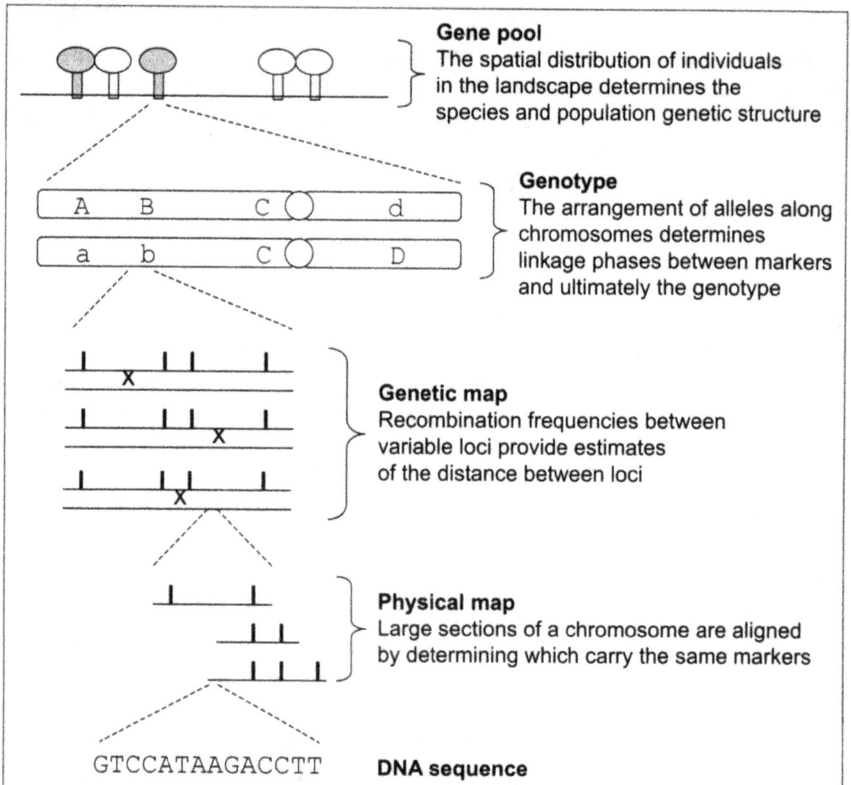

Fig. 8 The hierarchical organization of genetic diversity and maps in the forest tree genome.

4.2.1 Poplar

The angiosperm genus *Populus* from the family Salicaceae is considered the *Arabidopsis* of the tree world (see review by Taylor 2002). Poplar is a softwood tree with a predominantly northern hemisphere distribution, and its broad geographic range makes it readily available to many research groups. It is valued for timber, plywood, pulp, and paper, which helps advance its research. It has among the fastest growth rates of temperate trees. There are between 22 and 75 species depending on the method of classification (Eckenwalder 1996) and species hybridize readily, so wide crosses for mapping are possible, such as the cross by Cervera et al. (2001) that used *P. trichocarpa, P. nigra,* and *P. deltoides.* The species are dioecious, which is unfortunate from the perspective that selfing cannot be used to create inbred lines, though the prevention of selfing simplifies the logistics of controlled crosses. Clonal reproduction works well allowing replication of studies and immortalization of lines.

Along with the recent publication of a full genome sequence (Tuskan et al. 2006), *Populus* has numerous other resources that will advance its place as a model system for trees and tree improvement. Poplar has a fairly small genome, at 450-550 Mbp, which was an important criterion in its selection for whole-genome sequencing. The haploid chromosome number is 19 for the genus. There are many genetic maps available for different poplar species (e.g., Bradshaw et al. 1994; Wu et al. 2000, Cervera et al. 2001), including a high-density map consisting of 544 markers (Yin et al 2004b). QTL mapping has been done in numerous populations, such as for traits of stems, leaves, branches, as well as finer-scale traits like stomata (Taylor et al. 2001; Ferris et al. 2002). There also are freely distributed mapping pedigrees for QTL analysis. A variety of EST studies have been conducted as well including wood and root libraries (e.g., Sterky et al. 1998). Recently the new Poplar Genome Array by Affymetrix was released with over 56,000 transcripts and gene predictions, which will greatly improve work in this area. A good comparative genomic resource that includes annotated poplar sequences can be found online (*www.phytozome.net*). Genetic transformation is also fairly simple in poplar. Studies include control of flowering (Rottmann et al. 2000), genetic engineering of *Bt* and glyphosphate resistance (Meilan et al. 2000), and alteration of lignin properties (Chen et al. 2001).

4.2.2 *Eucalyptus*

Eucalyptus is a tropical hardwood angiosperm genus from the family Myrtaceae, native to Australia with over 700 species. Its genome is probably the second most-studied of tree genera. *Eucalyptus* is the most widely grown plantation tree in tropical and subtropical areas (Eldridge et al. 1993), valued for its paper, pulp, and timber. The species grow fast and adapt well to diverse and adverse environments.

There are a growing number of genomics resources for *Eucalyptus*. Genolyptus is a national research project in Brazil (Grattapaglia 2004) through which numerous public and private agencies are collaborating to advance eucalypt genomics. Poke (2006) summarizes the research on eucalypt genomics. The genus has a haploid chromosome number of 11, and there exist several genetic maps for different eucalypt species (e.g., Grattapaglia and Sederoff 1994; Byrne et al. 1995; Grattapaglia et al. 1995, 1996; Brondani et al. 1998, 2002), including a large microsatellite map of 230 markers (Brondani et al. 2006). Maps have been used to find QTLs for frost tolerance, seedling height, leaf area, wood density, pulp, yield, microfibril angle, fiber length, and cellulose; see Moran et al. (2002) for a summary of the genomics of eucalypt wood traits. BAC libraries and physical mapping are planned (reported in Poke 2006). There are various functional genomic studies of eucalypts as well. For example, Fullard and Moran (2003) studied microarray data to examine cold tolerance and exteme phenotypes. Cloning of genes has been done for the homolog of *Arabidopsis* floral homeotic genes (Southerton et al. 1993) and floral meristem identity genes (Southerton et al. 1998a, b). DDRT-PCR has been used to screen *E. microcorys* for genes down-regulated in high-salt environment (Chen and De Filippis 2001).

4.2.3 Pine

The genus *Pinus* is a temperate, boreal gymnosperm and softwood in the family Pinaceae, the most important of the eight families in the order Coniferales with 11 genera and 232 species (Frankis 1989). The genus has fewer resources available to it for genomics work, which has much to do with its large and unwieldy genome. However, it is an important source of timber, pulp, and resin products world-wide, and its genetics has been studied for some time with many of the early genetic maps produced in pine or a close ally. There are technical elements of mapping in gymnosperms (megagemetophyte tissue, Guries et al. 1978) that simplify the mapping in outbred individuals. For these reasons, *Pinus* has been and remains an important group for tree mapping and breeding. Full sequencing of the pine genome is unlikely in the near future, though. Pinaceae genomes are 100-fold larger than *A. thaliana* (Bennett and Smith 1991), and nuclear content varies a lot among pine species (Murray 1998; Leitch et al. 2001; Grotkopp et al. 2004; Bennett and Leitch 2005), with as much as 75% consisting of highly repetitive sequences (Kinlaw and Neale 1997; Schmidt et al. 2000; Friesen et al. 2001).

There are, however, many other genome-based resources for pines and other gymnosperms. Guevara et al. (2005) provide a review of the various pine genetic maps that have been produced with the cross types and marker systems (see also Chagne 2004). Several gymnosperms have been mapped, like *Larix* (Arcade et al. 2000) and *Picea* (Binelli and Bucci 1994), but pine is the most studied genus and loblolly pine (*Pinus taeda*) the most studied species. Linkage maps have been produced in several *Pinus* species (O'Malley et al. 1986; Devey et al. 1994, 1996, 1999; Nelson et al. 1994; Kubisiak et al. 1995; Echt and Nelson 1997; Remington and Plomion 2000; Chagne et al. 2002) to name only a few. Though not all pine maps are made using megagametophyte tissue, this method has proven useful for mapping. QTL maps also have been made, reporting loci controlling wood specific gravity (Groover et al. 1994), insect resistance (Hayashi et al. 2001), and early growth characters (Weng et al. 2002), among others. Several EST and microarray studies exist as well (Allona et al. 1998; Temesguen et al. 2001; Kirst et al. 2003; Komulainen et al. 2003). Gymnosperm research stands to make much progress using comparative genomic methods. Although the genomes are very large and complex, they have undergone limited structural rearrangements and pines retain a haploid number of 12 chromosomes (Krutovsky et al. 2004). Consequently, one can bypass full genome sequencing to a degree and transfer information between species using maps (e.g., Perry and Bousquet 1998a, b; Komulainen et al. 2003). The gymnosperms in general have a well-developed biointomratics infrastructure though the Dendrome project (*http://dendrome.ucdavis.edu/*). See also Edwards and Batley (Chapter 9 of Volume 1) for a review of bioinformatics for plant breeding.

4.2.4 Other Species

There are a growing number of linkage maps of forest tree species outside the three model genera. Chagne (2004) summarizes the mapping projects in gymnosperm taxa. Angiosperms are increasingly represented as well, including the rubber tree (*Hevea* spp., Lespinasse et al. 2000), Persian walnut (*Juglans regia,* Aradhaya et al. 2001), pedunculate oak (*Quercus robur,* Barreneche et al. 1998), European beech (*Fagus sylvatica,* Scalfi et al. 2004), birch (*Betula pendula,* Pekkinen et al. 2005), and pecan (*Carya illinoinensis,* Beedanagaril et al. 2005). Other model systems will without doubt arise since the three primary genera will not serve all needs. For example, *Juglans* is emerging as an important representative of the fine hardwoods, with black walnut (*J. nigra*) an important source of veneer and high value lumber, Persian walnut (*J. regia*) an important nut-producing tree, and the genomics and silvicultural infrastructure of the Hardwood Tree Improvement and Regeneration Center (Michler et al. 2005).

5 GENETIC MAPPING

A genetic map is an ordered set of markers with representation of their respective positions and spacing along the chromosomes. A genetic map is produced by tracking the inheritance, particularly the co-inheritance, of markers through the natural process of homologous recombination and segregation which occurs during meiosis. Thus a genetic map, as the term is typically used, is also a meiotic map. Mihovilovich et al. (Chapter 4 in Volume 1) provide a detailed discussion of the construction of genetic linkage maps and so we will focus much of our further attention to circumstances that arise when they are applied to forest tree species.

Genetic maps are used for several purposes in forest tree studies. (1) They are useful for the efficient tracking of diversity. At times tightly linked regions of the genome can be marked by a single or few markers, offering great economy for tracking segments of chromosomes through these haplotypes (e.g., Gwaze et al. 2003). (2) Genetic maps are used to organize structural genomics activities such as large-scale sequencing projects. For example, in the sequencing of the poplar genome, Tuskan et al. (2006) used a genetic map based on 356 microsatellite markers to anchor the 410 Mb DNA scaffolds that had been produced by shotgun sequencing. It is not uncommon that sequencing will leave gaps or reveal repetitive regions of the genome that are difficult to align. Anchoring the aligned regions onto a larger genetic map via common markers helps to resolve the positions of the scaffold in the genome. (3) Another common use of genetic maps is for QTL mapping (e.g., Groover et al. 1994) and other methods of associating a particular portion of the genome with a particular trait. Without a map saturated with markers it becomes difficult to find associations between the trait of interest and the relevant regions of the genome.

In proceeding let us keep in mind the central challenge to mapping in forest trees, the lack of disequilibrium in natural tree populations (Strauss et al. 1992; Grattapaglia et al. 1993). Recall that many forest tree species are widely distributed, long-lived, predominantly outcrossing, and their pollen can move large distances by the wind, all of which comes together to produce high levels of genetic variation within a population and little differentiation among populations (Hamrick et al. 1992). As a consequence there may be numerous alleles at any given locus and the associations between alleles along a chromosome are not strong due to past recombination events in heterozygous individuals. That is to say there is little linkage disequilibrium with which to map.

5.1 Typical Flow of Analysis

A common practice in mapping non-woody crop plants is to produce an F_2 generation from inbred parental lines (Fig. 9). Here, the parental lines (P_1, P_2) are differentiated for alleles at numerous loci, or said another way there is much linkage disequilibrium at these loci considering these two lines. Typically the parental phase is known and the F_2 population will segregate for markers based on meiosis in the highly heterozygous F_1, and crossover events are detectable between these heterozygous loci. As for the method of data analysis, the five-point outline presented by

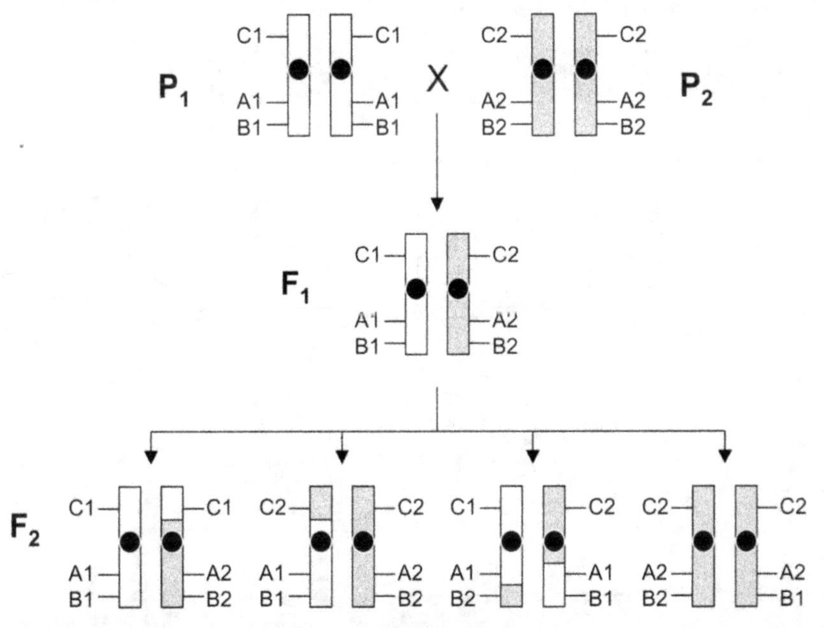

Fig. 9 Marker segregation in an F_2 population constructred from inbred lines. (Modified from Figure 1, Broman 2005).

Liu (1997) is enumerated here, which many computer programs do as part of the linkage analysis either in step or simultaneously depending on the approach. See Yandell and Bradbury (Chapter 11 in Volume 1) for a discussion of computing strategies and software in genetic mapping, and mapping software at URL *http://linkage.rockefeller.edusoft/list.html.*

Having introduced the general model, several issues that are relevant to mapping in natural populations of forest trees that require special attention or a departure from this standard model (Fig. 9) will be emphasized here. Chief among these differences is the heterogenous composition of the genomes of the parents used in the cross (F_1, Fig. 10). These ad hoc parents usually are not the result of crosses between inbred lines but are often drawn directly from the wild or indirectly via an *ex situ* management site. Indeed the P_1 parents are themselves a complex mosaic of ancestral haplotypes that have broken up by the repeated crossing over within heterozygous individuals. This high individual-level diversity owes to the extensive gene flow and limited self-fertilization characteristic of woody plants. The result is restricted haplotypic structure in the mapping parents, or put another way, low

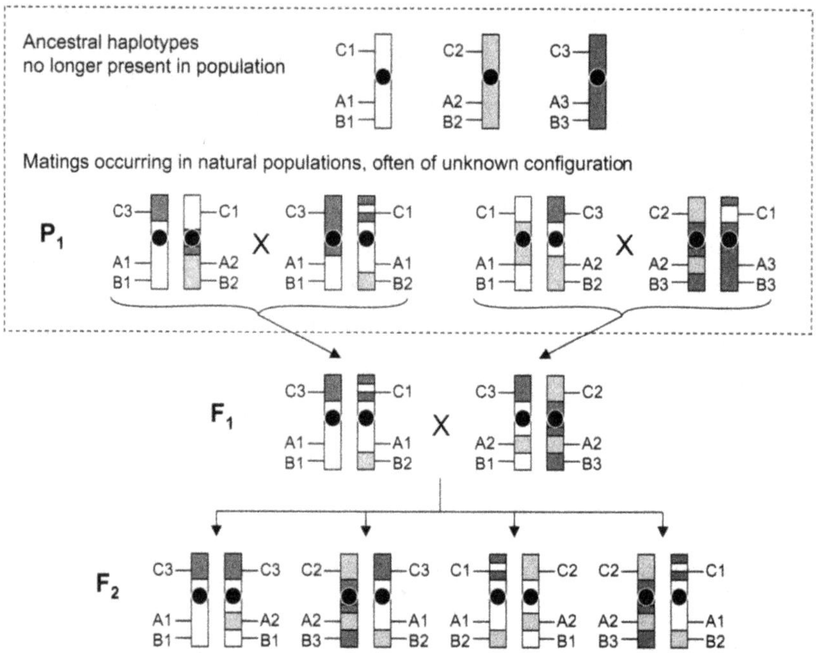

Fig. 10 Marker segregation in a mating between two outbred forest trees. Shading of chromosomal regions indicates haplotypic segments that are identical by descent.

linkage disequilibrium, which makes it difficult to map by traditional methods. At times a three-generation pedigree is used in which case we would know the genotypes and phase configuration of the grandparents (P_1) of the mapping population, though more often one must resort to other means. Increasingly software is available that accommodates the various complications presented by mapping from natural populations. At this point we review the five steps of mapping enumerated by Liu (1997).

5.1.1 Examine Single-Locus Segregation Ratios

This is the quality control step to determine that the data used to construct the linkage map are reliable. We expect certain genotypic frequencies in the progeny given a particular genetic model. For instance consider a backcross (BC) and codominant marker system, in which case segregation should follow a 1:1 ratio (i.e., $A_1A_2 \times A_2 A_2 \rightarrow A_1 A_2 : A_2 A_2$). Whereas a BC produced from inbred lines will exhibit the $A_1 A_2 \times A_2 A_2$ cross homogeneously across all variable loci, a mating between outbred forests trees will involve a variety of cross types that will be specific to a locus and individual in the population. Depending on the mechanism of inheritance (dominance or codominance), one locus could display a BC mating scheme (A_1 A_2 x A_1 A_1 or A_1 A_2 x A_2 A_2) with the expected 1:1 segregation while the next locus could display an F_2 mating scheme (A_1 A_2 x A_1 A_2) with genotypic expectations of 1:2:1 for co-dominant markers (3:1 for dominant).

It is therefore more complicated though tractable to test for distortions in segregation ratios from expected frequencies. Having broadened the scope of possible models that could occur in an outbred individual, one then should want to detect distortions from expected ratios that might arise from something other than the linkage-crossover dynamic, and this is done typically using a goodness of fit test. As Liu (1997) points out, a poor fit may draw from a bad model, bad data, or bad sampling. Here examples relevant to forest tree research are extrapolated:

- *An inappropriate genetic model.* A bad genetic model could arise if, for example, a mapping population were drawn from an open-pollinated canopy which resulted from mixed mating (partial selfing and outcrossing) whereas it was assumed erroneously that there had been no selfing. The assumption of random outcrossing could also be violated by heterogeneous patterns of pollen deposition and seed collecting (see also non-random sampling). It has been noted that canopies may be stratified, with different pollen donors better represented in the pollination events in some parts of the maternal canopy (Gibson and Hamrick 1991). Biased collection of seed in only a single, small part of the canopy could interact with this fine structure of pollination and violate the assumption, if so held by the program, that all male parents had an equal chance of contributing pollen to the progeny set.
- *Poor quality data.* Clearly this can arise in any study, but open-pollinated studies are especially prone to errors in genotyping. In a controlled cross the identity and genotypes of both parents usually are known and so we know exactly which alleles to expect in the progeny - each should inherit one allele from each parent. Any non-parental allele detected in the progeny signifies either a genotyping error or perhaps pollen contamination. In outbred populations of trees it is a bit more complicated in that we know the identity and genotype of one parent, the maternal one, but perhaps not the other (half-sib family), or perhaps the other parent has to be inferred by parentage analysis (inferred full-sib family, see later section). This condition makes it more

complicated and more likely that errors should arise in the data scoring. For a female that is heterozygous at a particular locus, and for n potential male donors, there will be between 2 and $2n$+2 alleles to score in the progeny set at that locus, depending on allele-sharing among trees and on the level of male heterozygosity.

- *Non-random sampling*. This could arise from artificial or natural causes interfering with the random sampling assumptions. For example, preferential collection of red versus green fruits could bias the segregation patterns at loci controlling fruit color along with adjacent linked marker alleles. Bradshaw and Stettler (1994) detected segregation distortion in a three-generation inbred pedigree involving interspefic crosses of *Populus*. They suggested the distortion arose from directional chromosome loss, a pollen lethal allele, genetic factors isolating the parent species, and inbreeding depression due to a genetic load. In this case, the progeny set was a biased sample due to underlying genetic factors that were particular to the species and hybrids under study.

5.1.2 Two-Point Linkage Analysis

The next step is to determine which loci are linked to one another. Linkage associations are explored by examining all possible two-locus combinations in the data set. Markers on separate chromosomes will be inherited together, 50% of the time simply by random processes thereby forming a ceiling on the map distance estimate. Markers that are linked on the same chromosome will travel together unless recombination distrupts their association, and recombination is more likely the farther apart the markers are along the chromosome. Map distance is estimated from the percentage of recombinants. Note that it is only the recombination events occurring between heterozygous loci within an individual that will provide linkage information. If that individual is heterozygous at only one of the two loci, then recombination will not alter the two-locus allelic associations and the crossover will go undetected.

In simple crosses it often is straightforward to determine which progeny carry the recombinant gametes as these usually occur at the lowest frequency in the set of genotypic classes in the progeny. This is because in some meiotic tetrads a crossover does not occur between the loci under consideration, especially for closely linked loci. Moreover, many crossover events in this interval will involve only two, sometimes three, of the four chromatids, generating an excess of parental configuration chromosomes. One also can directly infer the linkage phase of the parents by inspecting the genotypes of the progeny. For example, in Fig. 11 the two central gametic types A_1B_2 and A_2B_1 would occur at a lower frequency than the other two genotypes since the parents were in coupling phase, whereas the reverse would have held if the parents had been in a repulsion configuration (Fig. 12). The inference of gametic genotypes is especially simplified in a standard cross such as a BC in which the heterozygote is mated to a homozygote (or testcross with a homozygous recessive), in which case the gametic contribution of the homozygote is a constant that can be subtracted out to infer the genotype of the gametes contributed by the heterozygote parent.

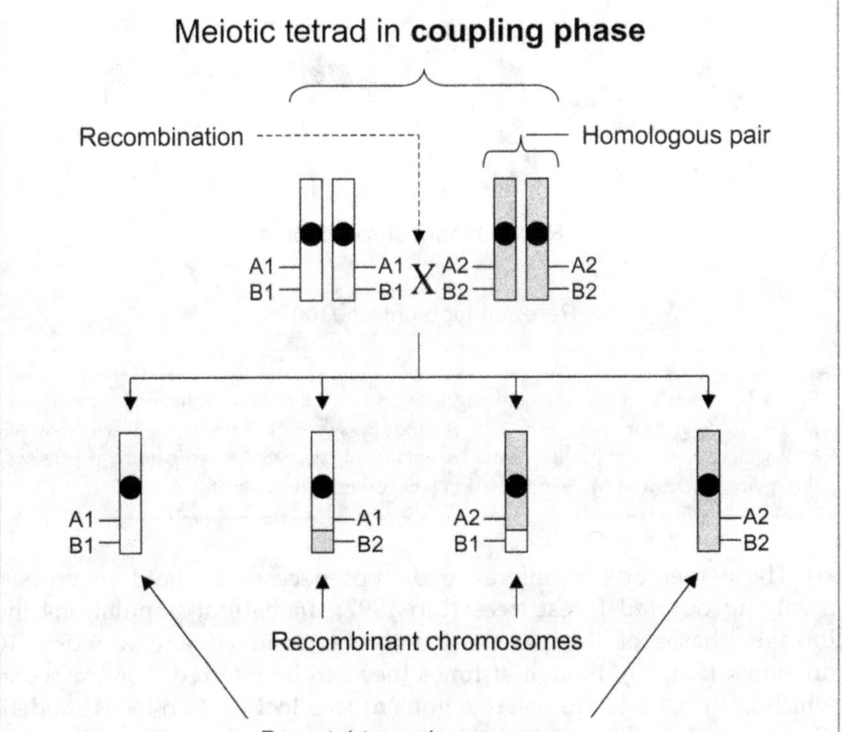

Fig. 11 Meiosis, crossover, and segregation in an individual exhibiting a coupling phase configuration between loci A and B. Here, chromosomes are shaded homogeneously per individual in the parental generation to unambiguously represent the region exchanged by recombination (represented by an 'x').

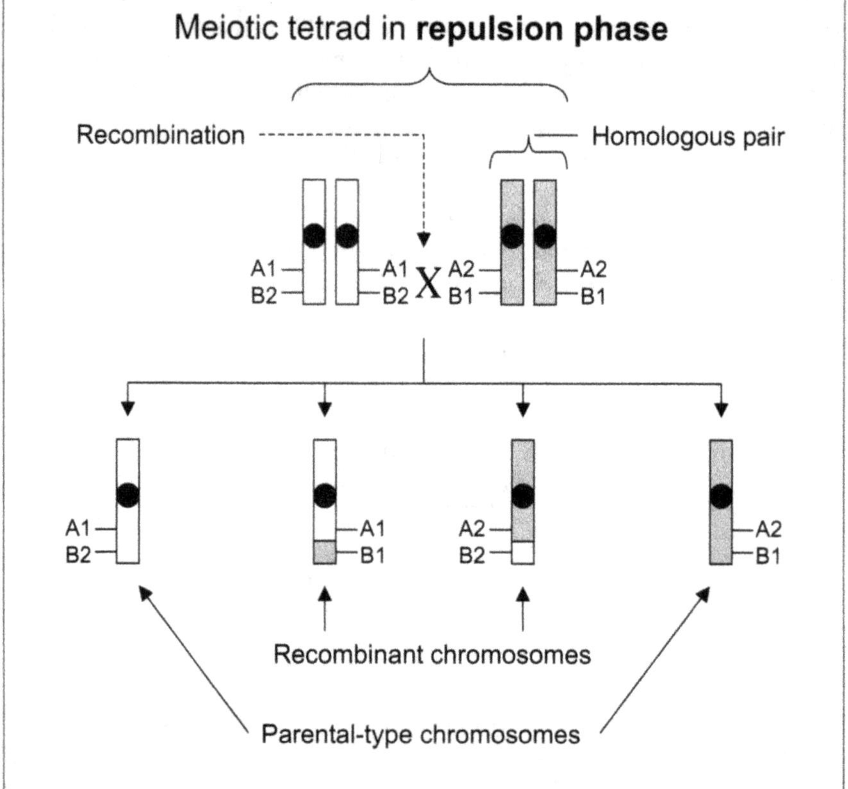

Fig. 12 Meiosis, crossover, and segregation in an individual exhibiting a repulsion phase configuration between loci A and B. Here, chromosomes are shaded homogeneously per individual in the parental generation to unambiguously represent the region exchanged by recombination (represented by an 'x').

These methods of inference do not necessarily hold in crosses involving outbred forest trees (Liu 1997). In natural populations the linkage phases of the parents cannot be assumed and are usually unknown (Fig. 10), though at times they can be inferred. For crosses in which both parents are heterozygous at two loci (F_2 two-locus model), the parents may differ in their phases across pairs of loci (Fig. 13) so that a single cross will entail a need to consider several different phase combinations. Clearly only one of these phases is the correct one, but it may not be obvious which is correct. Phases can be determined using three-generation pedigrees though these are not always available in forest tree mapping studies.

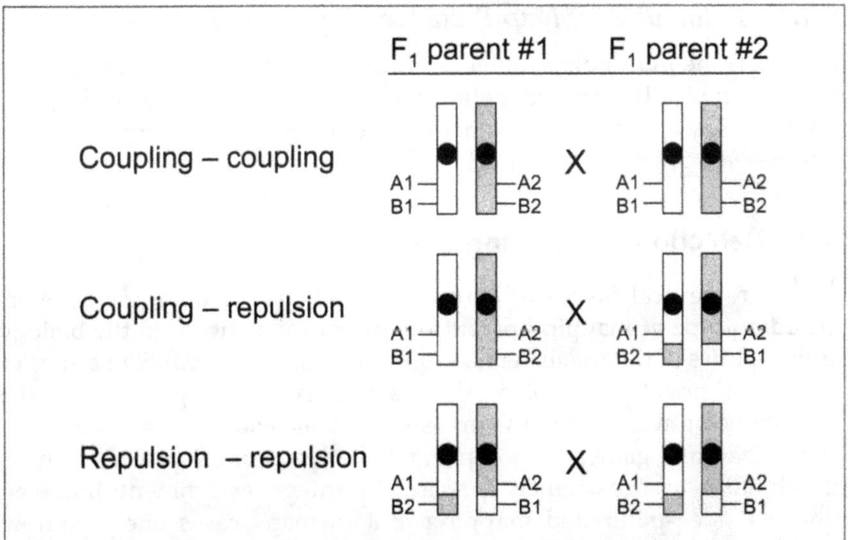

Fig. 13 Possible mating configurations with linkage phases between loci in an outbred tree species (F_2 two-locus model, after Table 7.1, Liu 1997).

Further complications can arise in forest tree mapping when one considers the mixed mating system, or combined selfing and random outcrossing. We consider this topic further in a later section on open-pollinated mapping. At this point several of the procedures and computer programs depart in their approaches to infer linkage phases. But once linkage phases are determined, the remaining steps follow as we are able to estimate the number of recombination events and map distances.

5.1.3 Determination of Linkage Groups

Markers are then organized into different linkage groups based largely on the recombination fraction, significance of the recombination fraction, and the number of chromosomes (Liu 1997). Given enough markers, the map should be saturated and the total number of linkage groups should match the known number of chromosomes in the species.

5.1.4 Ordering of Markers within Linkage Groups

This portion of the analysis determines the linear organization of markers within linkage groups. Prior to this step markers had simply been binned into clusters of linked loci. Ideally the order of markers on a genetic map should be co-linear with that of a physical map.

5.1.5 Estimation of Map Distances between Markers

Here, it is the multipoint, rather than two-point, recombination fractions that are typically used to determine map distances. The multipoint estimate may approach that achieved through physical mapping, though differences may remain (Liu 1997).

5.2 Selection of Mapping Scheme

There are several issues to consider in selecting a mapping scheme, including type of mapping parents, marker availability, and the biology of the species (Grattapaglia et al. 1993; Guevara et al. 2005). Some species present a novel mechanism that facilitates mapping, such as the megagametophyte tissue of gymnosperms which allows direct inference of the maternal gamete genotype, and the seeds of mimosoid legumes which allow for extremely precise parentage assignment that can simplify open-pollinated mapping. But in many cases one is simply working with a tree that has no obvious bells or whistles, and inbred lines are not available. In such a case crosses are made between trees selected for their genetic and phenotypic differences, or highly heterozygous trees are selected and used as ad hoc F_1 parents to generate a segregating progeny pool by artificial or natural pollinations. The selection of mapping parents draws on information developed in the prior sections regarding *ex situ* and *in situ* forest tree resources, and we will revisit the topic in a subsequent section. See also Hittalmani et al. (Chapter 3 in Volume 1) for discussions of the development of mapping populations for plants in general, and Doveri et al. (Chapter 2 in Volume 1) for discussions of markers used in mapping. Here the most commonly encountered types of mapping populations used in forest tree studies are presented:

- Haploid model
    - o Haploid megagametophyte
- Diploid models
    - o Pseudo-testcross
    - o Three-generation pedigree
        - Outbred model
        - Inbred model
    - o Open-pollinated families
        - Full-sib
        - Half-sib

5.2.1 *Haploid Megagametophyte*

The first genetic linkage studies of forest trees used isozymes to explore patterns of inheritance in the megagametophyte of gymnosperms (Guries et al. 1978; Rudin and Eckberg 1978; O'Malley et al. 1979; Conkle 1981; Cheliak et al. 1984). Since that time there have been many gymnosperm maps produced by this procedure, some directly from an open-pollinated canopy and some from an F_2 selfed family (reviewed in Guevara et al. 2005).

The megagametophyte is a haploid nutritive tissue that surrounds the developing gymnosperm embryo (Fig. 14). It derives from the same megaspore that produces the maternal gamete, and so the maternal meiotic contribution to the embryo can be detected unambiguously through genotyping the megagametophyte. As a consequence, the maternal (and paternal) linkage phases are obvious and need not be inferred from the diploid progeny genotype.

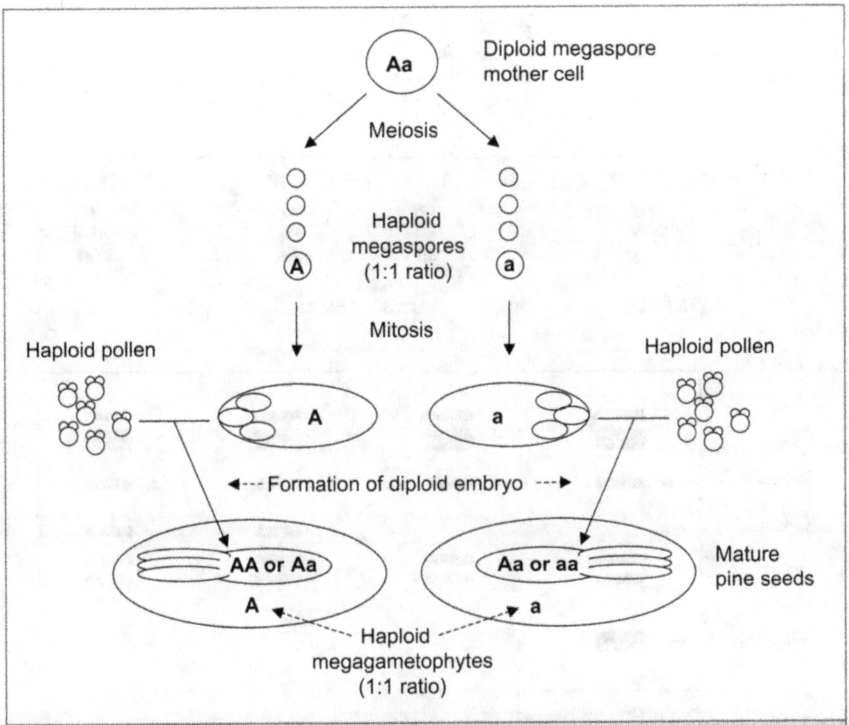

Fig. 14 Haploid megagametophyte mapping in gymnosperms. After Figure 3, Wilcox et al. (1996).

Loci that are heterozygous in the mother tree will behave as a BC with 1:1 segregatation. Nothing is lost here if dominant markers such as RAPDs are used since their segregation is the same so long as the mother tree is heterozygous. The inheritance of some RAPD markers is shown in Fig. 15, where alleles are represented in a pedigree (alleles producing a band: A_1, B_1, C_1; null alleles: A_0, B_0, C_0) and in a gel (below pedigree). Note that locus A is not heterozygous in the maternal parent and so does not segregate whereas loci B and C are variable and segregate. Consequently, locus A cannot be mapped in this tree whereas B and C are good candidates. Also note the male parent does not contribute to the megagametophyte tissue and so the paternal contribution of the mating event does not enter into the analysis, whether there is one sire (as shown here) or several sires. Since we are tracking recombination only in

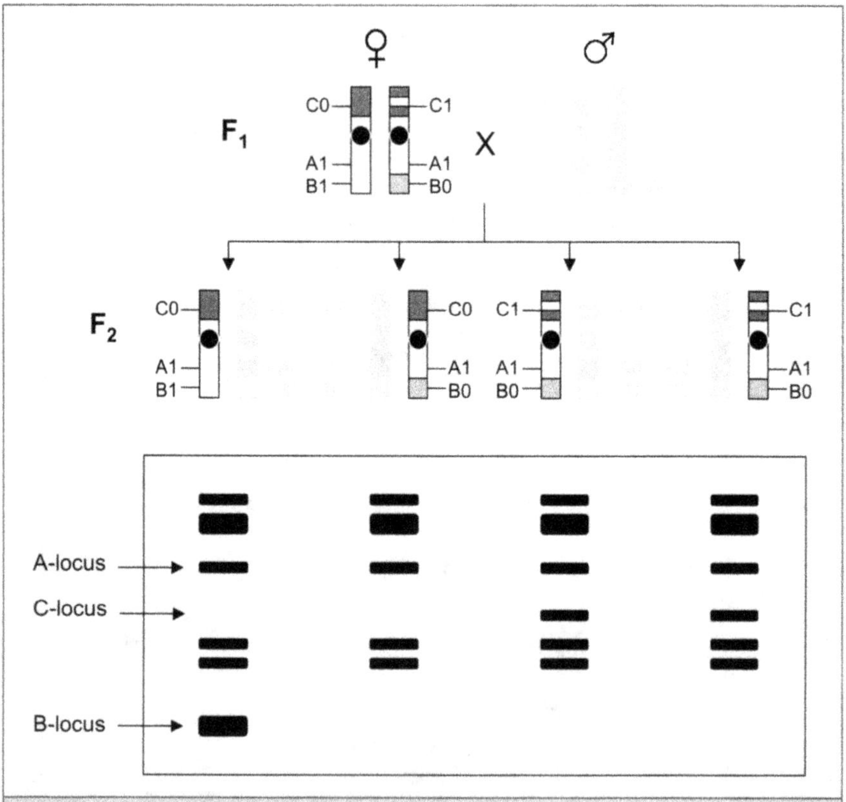

Fig. 15 Inheritance and scoring of RAPD genotypes from megagametophyte tissue in a gymnosperm. The male contributes to the diploid embryo but not to the haploid megagametophyte.

the maternal tree, we obtain a linkage map that pertains only to the maternal genome, i.e., a single-tree linkage map. Note that Gasbarra and Sillanpaa (2006) have proposed a method for inferring parental linkage phase and mapping from pooled-haploid DNA, allowing for rapid analysis of pollen rather than the megagametophyte, the latter of which constrains haploid analysis to gymnosperms.

5.2.2 Pseudo-Testcross

The next major innovation in forest tree mapping came with the development of the pseudo-testcross (Fig. 16). Grattipaglia and Sederoff (1994) developed the approach and applied it to a controlled cross of *Eucalyptus* species. The approach seems to work in most species and

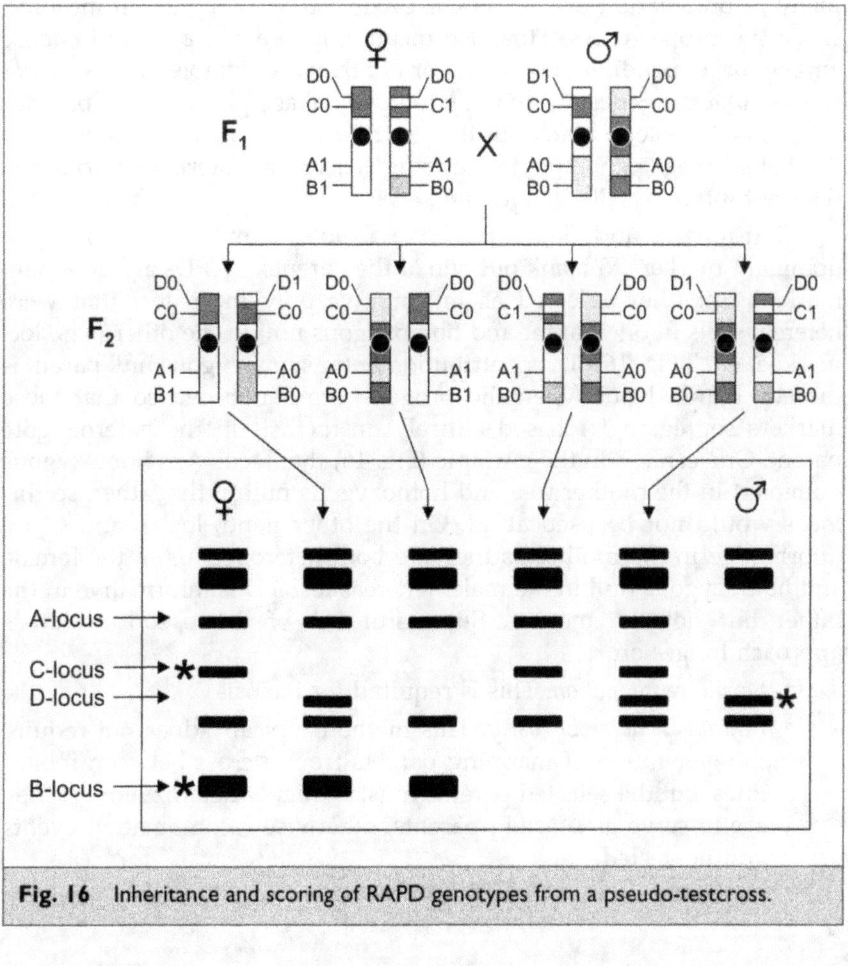

Fig. 16 Inheritance and scoring of RAPD genotypes from a pseudo-testcross.

without prior genetic knowledge other than the initial screens required of candidate loci. This flexibility has been important in transferring map-building beyond the gymnosperms and beyond the model organism. Maps of forest tree species produced using the pseudo-testcross include eucaltyps (*Eucalyptus grandis* and *Eucalyptus urophylla*, Grattipaglia and Sederoff 1994), the rubber tree (*Hevea* spp., Lespinasse et al. 2000), oak (*Quercus robur*, Barreneche et al. 1998), pecan (*Carya illinoinensis*, Beedanagari1 et al. 2005), and numerous others.

As with the megagametophyte mapping approach, the pseudo-testcross tackles the problem of high linkage equilibrium in tree populations by capitalizing on the fact that the forces which erode disequilibrium also increase levels of individual heterozygosity. A condition of high individual heterozygosity is a condition in which many recombination events should be detectable along a chromosome, given the proper cross. This also means that we can pick and choose among loci depending on whether or not they are suitable to a particular type of analysis. Since we do not know the linkage phases of the parents, it helps to be able to blank out the contribution of one of the parents, as in haploid megagametophyte mapping where the paternal contribution does not interfere with genotyping.

Grattipaglia and Sederoff (1994) used the null allele found in dominant markers to blank out one of the parents. RAPDs are dominant markers and they selected as informative only those loci that were heterozygous in one parent and homozygous null in the other (e.g., loci marked as '*', Fig. 16). The contribution of the homozygous null parent is thereby blanked out when the progeny are genotyped so that these markers segregate 1:1 based entirely on meiosis in the heterozygote parent. One can see in the example (Fig. 16) that locus A is homozygous dominant in the mother tree and homozygous null in the father, so this locus would not be used at all. On the other hand, loci B and C are informative in the mother as they are both heterozygous in the female and homozygous null in the male, whereas locus D is informative in the father but not the mother. Successful use of the pseudo-testcross approach hinges on:

- *Sexual reproduction*. This is required for meiosis.
- *Individual heterozygosity*. This method typically does not require the generation of mapping parents from a cross between inbred lines, but the selected parents must be highly heterozygous if they are to serve as *adhoc* F_1 parents, otherwise recombination events go undetected.

- *Genetic divergence between parents.* Greater genetic differentiation between parents improves the chances that a given locus will be heterozygous in one parent and homozygous null in the other.
- *Mating system.* Outcrossing is necessary as selfing does not allow for the condition that parents must differ with respect to their genotypes at a locus.
- *Mapping population.* Ideally a full-sib family is used, either a controlled cross or an open-pollinated canopy that is rogued to leave only full-sibs. Alternatively a half-sib family can be used but then only a maternal map can be made and a smaller fraction of the markers will be infomative.

The pseudo-testcross is a logical extension of both the testcross and haploid megagametophyte mapping. In a traditional testcross, it is known *a priori* that all variable loci will be in the testcross configuration (i.e., *Aa* x *aa*, where *A* is dominant to *a*). In a pseudo-testcross, mating configurations are known *a posteriori*, and it is based on genotypes and segregation information that candidate loci are selected. Although the rejection of some markers is unfortunate, there typically will be numerous loci that are heterozygous in one parent and not in the other when the cross is between two highly heterozygous individuals. Since PCR-based technologies such as RAPDs provide a wealth of markers, this does not pose a serious limitation. As for megagametophyte mapping, the main difference is that the pseudo-testcross does not require a special haploid tissue to infer the maternal contribution unambiguously. The pseudo-testcross does typically utilize a full-sib mapping family though half-sib maps are possible, while the sire is irrelevant in megagametophyte mapping.

The pseudo-testcross can yield two single-tree linkage maps, one for each parent, and so is sometimes refered to as a two-way pseudo-testcross. Grattapaglia and Sederoff (1994) used 62 progeny to produce a maternal map (240 markers, 11 linkage groups) and paternal map (251 markers, 14 linkage groups). The latter resolved into the expected 11 linkage groups with proper adjustments of the LOD score. Hacket et al. (2000) found notable segregation distortion in their mapping population of tea (*Camellia sinensis*), and through clustering and ordination techniques they concluded that three, not one, pollen donors had sired the progeny. Consequently, they only produced a maternal map, retaining just the loci in which the female carried the dominant allele and the progeny segregated with the expected ratio of 1:1. This pattern indicated the mother was heterozygous and the father homozygous null, regardless of which father it was. Typically both maternal and paternal maps are made where possible as this maximizes the number of markers that can be positioned on the map(s) per the amount of genotyping effort.

Some (e.g., Plomion and O'Malley 1996) have observed that recombination rates differ between pollen and seed parents, and so at times an average or consensus map is constructed through merging the separate individual parent maps. Merging requires markers that are common to both maps in order to anchor and align the maps properly. Co-dominant markers can be used though their segregation will be 1:1, 1:2:1, or 1:1:1:1 in the progeny. Maliepaard et al. (1998) cautioned that large differences in recombination rates between the maps might significantly alter the inferred marker order, though others (e.g., Lespinasse et al. 2000; Brondani et al. 2006) have found mostly modest effects on synteny.

5.2.3 *Three-Generation Pedigree*

A three-generation pedigree uses genetic information on grandparents, parents, and progeny in order to detect recombination events and determine linkage. The variety used in forest tree mapping is a derivation of the classic three-generation cross involving inbred lines used in non-woody crop plants. Typically, inbred lines are not used as grandparents for reasons already stated. Instead, trees that are genetically unrelated and have divergent genomes are often used (Neale et al. 2002). In the best circumstance, two controlled crosses are made between four genetically divergent trees that serve as the grandparent generation (same as P_1, P_2, P_3, P_4). Natural crosses are possible as well, provided there is reasonable evidence that the true pedigree is known, such as due to proximity or parentage analysis. However, natural crosses within wild populations are less likely to involve divergent genomes because of family structure within the population, though as already noted, heterozygosities may still be sufficiently high for mapping.

One variant on the three-generation pedigree is the F_2 inbred model (reviewed in Guevara et al. 2005) wherein grandparents are treated as inbred lines with genotypes A_1A_1 and A_2A_2. Consequently, there are three genotypes (A_1 A_1, A_1 A_2, and A_2 A_2) present at each of the loci segregating in proportions 1:2:1 in the progeny (e.g., Mukai et al. 1995, Tani et al. 2003). It is especially useful here to have divergent genomes that can readily be treated as containing non-overlapping allele sets.

Other times we are concerned with mapping from matings between outbred and highly heterozygous parents, and tracking the observed segregation patterns of each allele without recoding, which involves much greater complexity in the analysis. The F_1 intercross between two highly heterozygous parents should entail many loci that are informative with respect to recombination. Liu (1997) notes that there are 81 mating configurations for co-dominant markers, and 17 of these provide information on linkage. As for dominant markers, 7 of 9 of the mating configurations offer linkage information. Ritter et al. (1990) developed the segregational rules for crosses between heterozygous parents (see also Ritter and Salamini 1996; Maliepaard et al. 1997; Wu et al. 2002). Lu et al. (2004) provide an approach that uses an EM algorithm to produce simultaneous maximum likelihood estimates of linkage, parental diplotype, and gene order via a multi-point analysis. Jansen (2005) uses inheritance vectors to characterize the grandparental origin of alleles in an individual and combinatorial optimization to minimize the number of recombinations between nearby markers. He applied this approach to a cross between two apple cultivars, and his method will likely appear in the new version of JoinMap (H. Jansen, personal communication). Maliepaard et al. (1997) describe the complexities inherent to working with outbred crosses between heterozygous parents:

- For co-dominant markers there will be 2-4 alleles segregating in the F_2 per variable locus within a single outbred pedigree.
- At any variable locus, the numbers of heterozygous parents may be 1 or 2, and this may differ across loci.
- At loci in which both parents are heterozygous for the same alleles, we cannot be certain which parent contributed which allele.
- Some markers can carry null alleles along with codominant alleles.
- The linkage phase of some loci may be unknown in one or both parents prior to analysis. Sometimes this can be inferred in a three-generation pedigree, and sometimes inferred by differences in frequencies, though not always.

To date, the more common application is the three-generation outbred model which has been used to map numerous forest tree species (e.g., Devey et al. 1994; reviewed in Sewell and Neale 1995; Sewell et al. 1999; Guevara et al. 2005). The F_2 data on marker segregation contains the results of independent recombination events in both of the F_1 parents. Therefore, the data set is usually split by F_1 parent and two independent maps are constructed and later merged via markers that appear in both maps and can serve as anchors. Ignoring some of the complexities such as null alleles that Maliepaard et al. (1997) and Jansen (2005) consider, these co-dominant markers will display one of three principal patterns:

- *1:1.* In this testcross configuration, one parent is heterozygous and the other homozygous (see the F_1 parents, locus A, Fig. 17). The loci that are heterozygous in the female will be informative for the female map but not the male map, and vice versa.
- *1:2:1.* In this intercross configuration, both parents are heterozygous for the same alleles (see the F_1 parents, locus B, Fig. 17). Some segregation data at these loci is ambiguous with respect to the male and female maps since parental contributions to heterozygous progeny are unclear. These loci can be used to align maps.
- *1:1:1:1.* In this configuration, both parents are heterozygous but involving different alleles, and so this is fully informative (see the

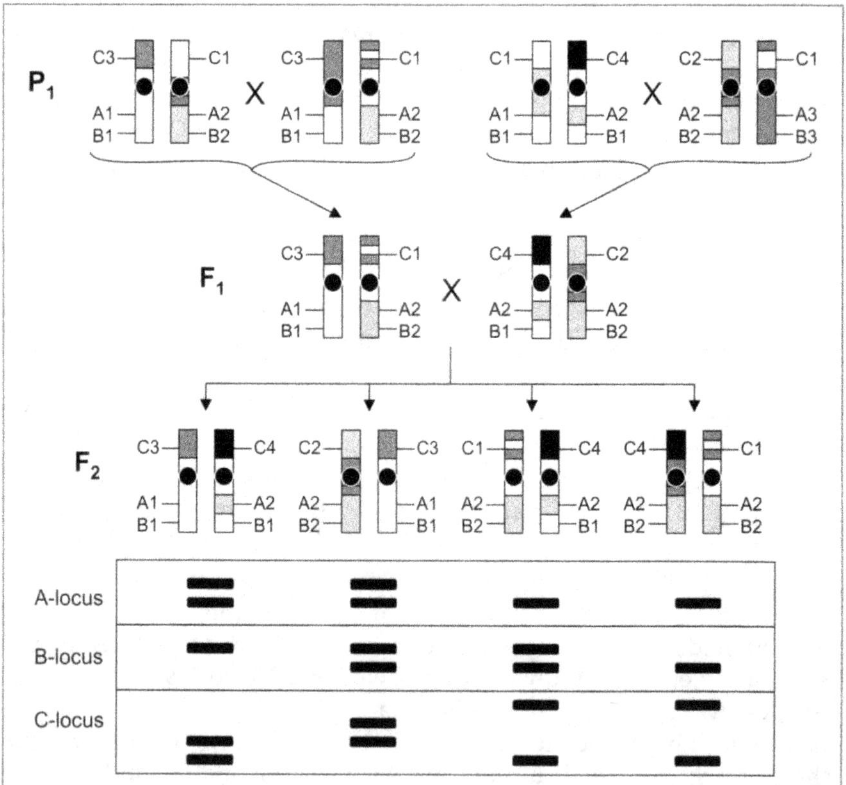

Fig. 17 Inheritance and scoring of codominant genotypes from a three-generation outbred pedigree.

F_1 parents, locus C, Fig. 17). Information can be partioned between the male and female maps, and these loci are useful as anchors for map alignment.

One advantage of the three-generation pedigree is that it can permit the inference of linkage phases. For example, we can recognize that the fourth F_2 progeny (far right) has a recombinant genotype. Its genotypes at the B and C loci are B_2B_2 C_1C_4. Clearly both F_1 parents contributed a B_2 allele while the C_1 allele had to come from the first F_1 parent and the C_4 from the second. In other words, the second F_1 parent contributed a B_2C_4 gamete to the fourth progeny, but, this haplotype did not exist in its parents (grandparents 3 and 4 with genotypes B_1B_1 C_1C_4 and B_2B_3 C_1C_2, respectively). Put in another way, the second F_1 parent must have been in cis configuration (B_1C_4/B_2C_2) based on the grandparental genotypes, and so the gamete it contributed to the fourth progeny (B_2C_4) must have been recombinant. The alternative explanation that the B and C loci are altogether unlinked is discredited by the evidence of association provided by the total progeny set, leaving recombination as the most parsimonious explanation. Not all determinations work out so cleanly of course.

5.2.4 *Open-pollinated Families*

A less explored though increasingly feasible approach for mapping forest trees utilizes open-pollinated families, both full-sib and half-sib. Some of the innovation involves extending mapping in full-sib families to open-pollinated canopies and half-sib mapping to angiosperms.

Whereas most full-sib families have been produced by controlled crosses, it is feasible to use an open-pollinated crown. Here the focus is to take the open-pollinated canopy, which may contain many different types of pollination events including selfing and outcrossing to several parents, and by parentage analysis remove or rogue all but the outcross events to a single male parent. Vogel et al. (1995) used such an approach to select against rogue progeny in a full-sib population of *Eucalyptus*. Once the full-sib family has been identified, many of the problems and strategies attendent to mapping in outbred heterozygous parents and outbred pedigrees described above apply.

There are innovations in open-pollinated, half-sib mapping as well. This general approach has been common in gymnosperm mapping where the megegametophyte makes the paternal contribution irrelevant. Open-pollinated, half-sib mapping also can be done in gymnosperms or angiosperms using the pseudo-testcross design though this does not necessarily optimize the use of resources, as noted above. At times this is a reasonable approach though, as Hacket et al. (2000) found in their linkage study of tea (*Camellia sinensis*). They had assumed their maternal canopy had been pollinated by a single male but after genotyping found segregation distortion that led them to discover evidence for three primary pollen donors. As a consequence, they produced only the maternal map from this half-sib progeny set. The recent innovation in open-pollinated, half-sib mapping comes in using, rather than throwing out, the paternal data in a multi-sired crown. This can be done using certain algorithms that incorporate estimates of outcrossing rate and allele frequencies in the pollen pool (Liu et al. 1993, Liu 1997).

Outcrossing rates and pollen pool allele frequencies are readily estimated with molecular markers (See Mating system). One can characterize the pollen pool by genotyping males in the vicinity of the maternal tree. Knowledge of phenological overlap and pollen dispersal distances (see Gene flow) improve the ability to describe the pollen gametes that are likely to affect fertilizations. Alternatively, outcrossing rates can be inferred by parentage analysis, the latter methodologies reviewed in Jones and Ardren (2003). Regardless, the outcrossing rate should be estimated independently of the linkage analysis since heterozygous parents are most informative to linkage analysis but least informative to the estimation of outcrossing (Liu 1997). Outcrossing rates can be obtained from standard programs such as that provided by Ritland and Jain (1981), and typically require an independent estimate of the genotypes present in the pollen donor pool as well. It also is good practice to sample seed from several places in the crown to minimize any effects of heterogeneity in the pollen pool, as noted by Gibson and Hamrick (1991).

The software package OutMap (Butcher et al. 2004) has been developed recently for mapping with codominant markers in outbred forest trees. It corrects some of the limitations inherent in other programs as it can order phase-ambiguous data and uses more than pair-wise information to order loci into linkage groups. The authors used it to generate linkage groups from two pedigrees of *Acacia mangium*. Other programs that can handle open-pollinated mapping are listed in Table 1. See also Yandell and Bradbury (Chapter 11 in Volume 1) for a discussion of computing strategies and software used for mapping plant genomes in general.

Table 1 Programs for mapping in open-pollinated tree canopies.

Program	*Contact*	*Email*
Gmendel	S Knapp	sknapp@helix.css.orst.edu
JoinMap	P Stam	Piet.Stam@wur.nl
OutMap	D Whitaker, ER Williams	Emlyn.Williams@csiro.au
PGRI	B Liu	benliu@unity.ncsu.edu

5.3 Further Considerations

Selection of parents and progeny for mapping should be informed by the goals, design, and resources of the study. See Hittalmani et al. (Chapter 3, Volume 1) for a general discussion on the construction of mapping populations in plants. Here we consider a few issues relevant to mapping in forest trees.

5.3.1 Selection of Mapping Parents

Forest tree mapping is often done using ad hoc, outbred parents in which case one must have highly heterozygous parents. But this will depend not only on the natural history of the species but also on the marker system as well (see also Doveri et al. Chapter 2, Volume 1). Some species exhibit limited differentiation using some marker types (e.g., isozymes) but high variation with others (e.g., AFLPs). Initial screens of candidate populations and secondarily of small sets of seed from canopies of adults proven to be genetically variable will provide useful information on the individual - marker system interaction. Polymorphism is gauged by the numbers (N_a) and frequencies of alleles per locus, and composite measures such as observed (H_o) and expected heterozygosities (H_e, Hedrick 1985). A reasonable proportion (> 70%) of the actual heterozygosity present in a population can be revealed by querying at least 15 individuals (Liu 1997). Also widely used is the PIC or polymorphic information content (Botstein et al. 1980). Liu (1997) provides details regarding surveying polymorphic markers in natural populations.

Wide crosses between divergent genomes have been used successfully to construct genetic maps for forest trees. For example, Grattapaglia and Sederoff (1994) used two species of *Eucalyptus* from the same subgenus and section: *E. grandis* from a provenance representing Zimbabwe and *E. urophylla* from a Brazilian landrace. The hybridization brings together considerable variation within a single line and produces a large amount of disequilibrium that is useful for mapping. However, interspecific crosses often can yield segregation distortion as well (reviewed by Rieseberg and Carney 1998; Burke and Arnold 2001). Yin et al. (2004b) revealed segregation distortion using a dense map of *Populus*. The mating involved a heterospecific cross between a hybrid (*Populus trichocarpa* x *P. deltoides*) and pure *P. deltoides* as the paternal tree. The markers that displayed segregation distortion were contributed by the donor parent, and most of these resided in large blocks on two separate linkage groups. Similarly, Myburg et al. (2004) found postzygotic barriers to introgression in a wide cross between *Eucalyptus grandis* and *E. globulus*. Strauss et al. (1992) raise several points of consideration for QTL mapping from wide crosses.

It is worth emphasizing, mapping projects that have as a goal the generation of both genetic and QTL maps should consider the phenotypic differences between mapping parents. Parents should differ with respect to traits of interest and these traits should be readily scored. In a three-generation pedigree, typically four divergent grandparents are selected and crossed. Two F_1 individuals are then selected that display intermediate phenotypes to ensure they are heterozygous at several of the relevant loci (the extreme phenotypes being more homozygous). These two individuals are crossed to produce the F_2 mapping population. General considerations are offered by Van der Beek et al. (1995), Wu (1996), Liu (1997), and Zeng et al. (Chapter 6 of Volume 1). It is feasible to produce a genetic map in one parent or pair and then transfer the map information to new parents and progeny for a QTL study, but there is added cost in that some markers will not transfer because they are not informative in the new population. So although it may be tempting to walk outside and select the first trees one runs into, forethought can be a good thing too.

Field considerations are relevant as well, including spatial and temporal proximity of the potential mapping parents and the global pollen pool. One should understand the natural history of the species as best as possible and consider how its traits would affect pollination dynamics in a natural or *ex situ* stand. Is the species dioecious, and if so, where are the males relative to females? Some forest tree species may be fully outcrossed and others exhibit a mixed mating system, which matters tremendously for mapping in half-sib, open-pollinated families but less so in some other designs. Recall that pollen dispersal is often leptokurtic, meaning that for any crown many of the seed will likely fall into one of a few large full-sib families and the remainder into a diverse assemblage of global pollen donorship classes. Are the putative mapping parents phenologically in sync with one another but out of sync with other nearby conspecifics, which would maximize the size of their full-sib family? Singleton or loner trees are useful for controlled crosses and half-sib families, and trees occurring in isolated pairs are useful for natural full-sib families, though rogueing of the progeny may nevertheless be needed.

5.3.2 Selection of Progeny

We have already addressed the importance of random sampling from a canopy so as not to interface with any heterogenous structure in the pollen pool. Let us now briefly address matters of sample size and rogueing.

The resolution of a genetic map is limited by sample size and constraints on meiotic recombination. Genotyping few progeny will lower the chances of detecting a crossover event between loci that are near one another on a chromosome, which will entrap markers and force them to coalesce to a common point on the map. The addition of more progeny in a study typically allows for the detection of small distances between adjacent loci and improves overall estimates of marker spacing (Liu 1997). Physical constraints on meiosis also can entrap markers to a point on a map such as when one chromosome contains an inversion that yields inviable crossover products within that region, thereby preventing their detection. Liu (Table 6.23) explores the samples sizes and statistical power needed to detect linkage, and a portion of this table (Table 2) is reproduced below. Natural mapping populations will contain some combination of BC and F_2 mating configurations. In general, more progeny are needed to detect a given level of recombination (θ) with a dominant marker as opposed to a codominant marker. Most mating configuration - marker combinations allow for the detection of actual recombination of q = ~0.25 with ~100 seeds. The worst circumstance arises in the F_2 configuration with dominant markers

Table 2 Minimum number of progeny needed in a mapping population in order to detect linkage (θ = recombination fraction) when α = 0.05 and statistical power is at γ = 0.90. Drawn from more complete Table 6.23, Liu (1997).

Actual θ	*BC*	*F2-CC*	*F2-CD*	*F2-DDc*	*F2-DDr*
0.1	29	22	38	40	144
0.2	55	47	77	85	209
0.3	128	118	186	223	397

BC, backcross; F_2, second filial; CC, both loci are codominant markers; CD, one codominant and one dominant marker; DDc, both dominant markers in coupling phase; DDr, both dominant markers in repulsion phase.

that are in repulsion phase, which requires three to four times as many progeny.

Often one may need more than 100 seeds for a study simply because not all the seeds initially collected are of the required sibship. For instance, mapping from an open-pollinated, full-sib family can require rogueing of unwanted genotypes sired by extraneous males (e.g., Vogel et al. 1995). This can be done by simple exclusion wherein the mapping parents and all the seeds are screened with a panel of markers, and seeds that do not carry the expected alleles (one from each of the mapping parents) are removed. Still, there are cases in which extraneous pollen donors might be capable of producing gametes that are genetically indistinguishable from those produced by the intended mapping parents. This hard-to-detect contribution is termed cryptic parentage, or if one were to circumscribe the mating pair as a mating population then cryptic gene flow. There are several procedures for parentage assignment in natural populations (reviewed in Jones and Ardren 2003) and some can correct for cryptic parentage though these often draw on a full population-level assessment of parentage in the local landscape and/or estimates of global allele frequencies. Methods of fractional assignment exist, in which a fraction of each seed is assigned to each parent pair in the study area based on that pair's genetic likelihood of having been able to produce the seed (e.g., Aldrich and Hamrick 1998). Although this approach can be good for estimating population-level parameters such as individual adult fecundities, these do not provide the judgments that are most useful for rogueing a crown which is a seed-by-seed assignment of a single pair of adults as the most likely parents of a seed. Popular parentage assignment software packages include CERVUS (Marshall et al. 1998) and FaMoz (Gerber et al. 2003).

Parentage studies can be fairly straightforward for a pair of spatially isolated mapping parents since the maternal crown should contain a large full-sib family resulting from matings between the mapping parents, provided their phenologies overlapped (see also Gene flow and Mating system). Most of the other seed should arise from long-distance pollination events from a diverse and genetically distinct stock that often can be removed with reasonable genotyping effort. Parentage assignment rapidly becomes troublesome in dense natural stands, though, because a full population census is required within a certain radius so as not to miss any donors, and a small radius in a temperate forest can include many adults of a focal species. Additionally, family structure within this radius means that cryptic parentage is more likely due to pollen donorship from relatives of the mapping parents. Parentage assignment within a provenance study site is simplified by the presence of diverse and distinct individuals from different geographic regions. However, the degree of replication of a provenance within a block could create an artificial family structure and a propensity for cryptic parentage not unlike a natural stand.

In certain cirumstances one can harness correlations within a dataset to increase the power of a parentage study. For example, haplotype analysis involves using the correlations between alleles at different linked loci to assess genome or gene pool structure. A parentage application seeks to better characterize individuals with respect to the gamete genotypes they are able to produce. An individual in the population that is heterozygous at several unlinked loci is capable of producing many different gamete genotypes. This broad gamete genotype space may overlap considerably with that of the intended sire, yielding instances of ambiguous or cryptic parentage. However, if several marker loci are tightly linked then the gametic genotype space is constrained by their linkage phase in the individual. Often the linkage phase can be inferred directly from marker data and can help to discriminate between individuals, aided by the collective power of the multiple loci present in the haplotype. This approach has been used in oaks (Lexer et al. 2000) where they determined the number of pollen donors to an open-pollinated oak canopy.

Another parentage application can be found in correlations between markers that arise from the packaging of pollen grains into aggregate dispersal units in some tree species, notably the figs (*Ficus*, Nason et al. 1996) and the mimosoid legumes (Apsit et al. 2001). The mimosoid legumes, including the timber genus *Acacia*, package their pollen in sticky conglomerates of multiple grains called a polyad. A single fruit pod is usually the result of fertilization by a polyad from a single male donor so they are all full-sibs. Since each seed in the pod represents a separate meiotic event from the same pollen donor, one can within reason reconstruct the pollen donor genotype even at heterozygous loci, which are usually a source of ambiguity. At a heterozygous locus, the donor is very likely to pass on at least one copy of both alleles into the eight or so seeds of a pod, failing to do so only < 1% of the time ($P = 2 \times (½)^8 = 0.008$).

5.3.3 Marker Multitasking

In most instances molecular markers can serve multiple purposes in a mapping study. A microsatellite locus might prove useful for rogueing a progeny set of unwanted genotypes, determining the outcrossing rate for the remaining progeny, and then including as a marker locus in the linkage map. Some marker types, such as certain STS markers used in physical mapping, are not variable in natural populations and so their applications would be constrained to the mapping exercise. But most markers have the potential for multitasking, and there are a few points to consider before we leave genetic mapping proper.

The rogueing of a canopy for the production of a full-sib family might generate distortion in a linkage map centered on those markers used in the rogueing, if there are systematic errors of inclusion or exclusion made based on the markers. Distortion would extend to linked markers as well but the distortion should appear greatest at the errant loci. The same bias could arise with phenotypic markers, as noted earlier, if preferential inclusion of red over green fruit were to influence the representation of loci controlling fruit color in the progeny set.

Secondly, should a full panel of mapping markers be used to rogue the progeny used in the mapping? Put another way, can we map and conduct parentage analysis at the same time? Aside from the fact that it is less economical to apply many markers to progeny that can be identified as rogues with far fewer markers, it also introduces a propagation of errors that can easily pare down a mapping population through repeated attrition. Let us say there is an error rate of $P = 0.01$ per locus per individual wherein a progeny is ascribed an allele that is not present in the mapping parents, which erroneously excludes the progeny from the study. If the progeny set is screened with 10 markers with equal chance of errors in genotyping, then each progeny stands a chance of $P = (0.99)^{10} = 0.90$ of being retained in the study, given that it truly was a member of the desired sibship. However, with 100 markers and the same error rate the retention likelihood drops to $P = 0.37$ and it becomes clear that while it is desirable to eliminate unwanted genotypes it can be wasteful to exclude desired genotypes erroneously. Consequently, some mechanism of accounting for this is useful. For example, bands can be weighted by their confidence such that a high quality band detected in the progeny but not the mapping parents might be considered seriously and perhaps re-genotyped for verification, whereas a low quality band might be dismissed as an artifact. Moreover, when data for the entire panel of markers is appraised, a progeny that matches the mapping parents at 99 out of 100 markers is probably worth looking at again before excluding whereas a progeny matching only 68 out of 100 markers is very likely an imposter.

6 PHYSICAL MAPPING

Physical mapping is the exercise of determining the order and distance between markers or genes on a chromosome based on their physical distance along a section of DNA or chromosome. This stands in contrast to genetic or meiotic mapping in which the frequency of recombination events is used as a surrogate for physical distance, and a mapping function is required to relate genetic to physical distances. Karyotypes and chromosome maps form the largest scale of physical map, the BAC-based map providing finer scale information. DNA sequence information provides the ultimate in physical map information as simply the position of particular bases along the DNA molecule. Each of these modes of physical mapping has played a role in the development of forest tree genomics. Although there are many chromosome counts of forest tree species available, there is limited physical map information on forest tree genomes compared to the extensive database of genetic maps developed over the past couple of decades.

6.1 Chromosome Mapping

The coarsest information on genome architecture is the chromosome count. The haploid chromosome number provides useful information for guiding a genetic mapping project since it sets an upper limit on the number of linkage groups to expect. Chromosome counts have been tallied for a variety of forest tree species and these have long been used to study speciation events involving euploidic and aneuploidic changes in chromosome number that can arise with race or species formation. Karyotypes are full representations of a genome with chromosomes arrayed by size and homology, and often stained to show architectural landmarks such as knobs, constrictions, and banding patterns. Modern cytogenetic techniques have greatly improved the specificity of stains and the FISH (fluorescent *in situ* hybridization) procedure allows specific targeting of DNA sequences, especially repetitive stretches such as telomeric and centromeric repeats, rDNA, and retrotransposons (e.g., Doudrick et al. 1995; Vischi et al. 2003) that allow for chromosome painting. Note that information on genome content as C-values are available online for a wide variety of plant taxa through the Royal Botanic Gardens, Kew (Bennett and Leitch 2005).

6.2 BAC Mapping

Physical mapping through a BAC (bacterial artificial chromosome) library provides greater fine-scale resolution than typically does genetic mapping. As such these are used to isolate genes from regions that have been identified as relevant in a QTL mapping study, or they can be used in the context of a larger genome sequencing project to organize the genome into smaller manageable units.

Here, the term physical map refers to an aligned array of cloned fragments of DNA. It is made by breaking the genome into fragments, often by restriction digests, and then tracking the co-occurance of genetic markers by fingerprinting the resultant fragments. Often the fragments are introduced into a large-insert vector such as a BAC to facilitate handling, isolation, and propagation. These vectors can handle inserts in excess of 300 kb. Unlike a genetic map, the physical map does not require genetic variation and can be made using non-variable STS markers and a single individual of the target species. The random breaks made during the shearing or digestion step in the production of a physical map serve the same purpose as do the recombination events in the mapping parents used for making a genetic map. Markers that tend to co-occur are likely to be spaced closer than markers whose co-occurance is frequently disrupted by the digestions (Fig. 18). Physical maps of this kind have been made of some forest trees species but usually on model systems such as poplar in the context of full genome sequencing (e.g., Tuskan et al. 2006) and often in conjunction with

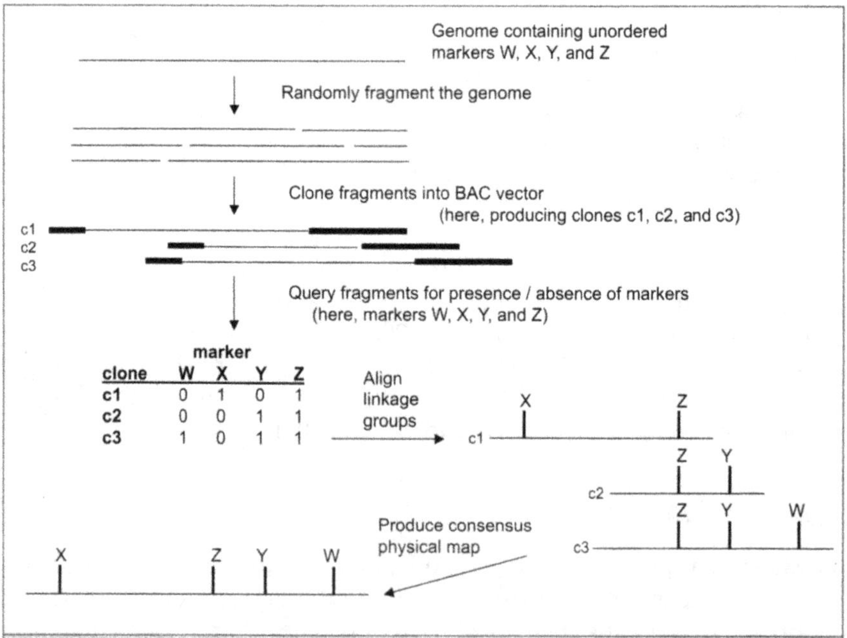

Fig. 18 Physical mapping.

genetic mapping. Yin et al. (2004a) used physical and genetic mapping to isolate *Melampsora* rust resistance genes in *Populus*.

7 TRAIT MAPPING

An important practical motivation for producing a genetic map is to identify locations in the genome where genes of interest reside that affect important traits such as growth rate and form, heartwood quality and content, disease resistance, etc. At this point structural genomics interfaces with functional genomics as we are interested in the locations and importance of genes that convey some critical function to the phenotype. Each trait may be under the control of a single gene or a complex of several genes of large effect and perhaps some of lesser importance. Perhaps several genes are controlled under a shared transcription factor which is in turn triggered by an environmental cue. The farther we move along this line of reasoning, the further we move away from structural genomics into the realm of functional genomics. Typically molecular markers are employed to lay bare the correlations present between the genotype and phenotype. We will briefly survey the field of trait mapping as it is currently applied in forest tree studies, providing examples along the way. Topics include QTL analysis, haplotype mapping, association mapping, bulked segregant analysis, functional genomic profiling, and candidate gene mapping. See also Paran and Levin (Chapter 5 in Volume 1) for a discussion on the mapping and tagging of simply inherited traits.

7.1 QTL Mapping

Quantitative Trait Locus mapping is one of the primary means by which phenotypic traits are associated with portions of the genome. Zeng et al. (Chapter 6 in Volume 1) discuss QTL detection in plants. Required here are (1) dense genetic maps with equally spaced markers, (2) statistical procedures, (3) and a mapping population that is segregating for both molecular markers and phenotypic traits of interest. Often oubred mapping populations including three-generation pedigrees are used in forest tree studies, even though these provide less precision compared with those constructed from inbred lines (Neale et al. 2002). Grandparent matings are made between divergent phenotypes, and then the intermediate phenotype from each cross is selected as it presumably is heterozygous at many loci, the extreme phenotypes being much more homozygous.

Forest tree QTLs have been identified for several adaptive and economically important traits. Wood quality has received much attention (Groover et al. 1994; Bradshaw and Stettler 1995; Grattapaglia et al. 1996; Kumar et al. 2000; Lerceteau et al. 2000; Arcade et al. 2002; Moran et al. 2002; Neale et al. 2002; Brown et al. 2003; Markussen et al. 2003). Other studies have found loci for tolerance to cold and drought, disease resistance, phenology, growth form and rhythm, and others (e.g., Groover et al. 1994; Bradshaw and Stettler 1995; Frewen et al. 2000; with reviews in Krutovskii and Neale 2001; Neale et al. 2002; Guevara et al. 2005). As a refinement to the standard approach, Gwaze et al. (2003) used haplotypic QTL mapping to trace inheritance from progeny back to individual founders in an outbred pedigree of *Pinus taeda,* thereby enhancing the clarity of the information extracted from the mapping study.

Krutovskii and Neale (2001) offer three important factors to consider in the selection of markers for QTL mapping in forest trees. (1) Due to their outbred history, parents may carry up to four alleles at a locus and so it is advantageous to use co-dominant markers that can detect all the allelic states. This allows the study of interactions between alleles (additive, dominant, over-dominant, under-dominant), as in Groover et al. (1994), (2) Comparative mapping is an important tool used to transfer the results of QTL mapping studies to other species, and so use of markers that can themselves be transferred between taxa is important. (3) Since the ultimate reason for attempting to associate a portion of the genome with a phenotype is to identify genes controlling the trait, it is useful to use markers with a known function such as ESTPs.

As an example of QTL mapping in trees consider the study of loblolly pine (*Pinus taeda*) by Groover et al. (1994). They used an outbred pedigree that had been characterized for variation in wood specific gravity and RFLP marker segregation in 177 progeny. They found five regions of the genome with at least one RFLP marker exhibiting a large effect, and allelic interactions in the QTL with the largest effect. There was evidence of QTL x environment interactions for two QTLs, but no evidence of digenic epistasis. The female map (Fig. 19) which shows four of the five QTLs is reproduced here, the fifth was detected only in the male map. The excess number of linkage groups found would likely coalesce into the expected haploid number (n = 12) if more markers were added to the map.

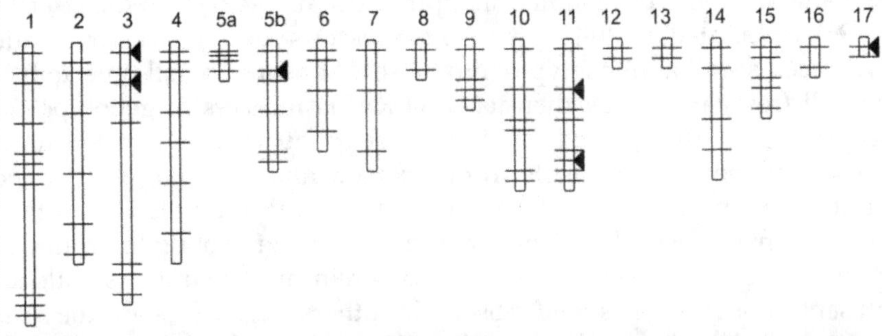

Fig. 19 Loblolly pine map of RFLPs with wood specific gravity QTLs. After a portion of Figure 3, Groover et al. (1994). Marker locations shown here are approximate.

7.2 Association Mapping

Whereas QTL analysis examines associations between markers and traits based on linkage disequilibrium that has accrued over a few generations of controlled crosses, association mapping draws on associations that have developed over generations of matings in natural populations (Jannink and Walsh 2002; Neale and Savolainen 2004; Guevara et al. 2005). Here, associations are a composite of not only physical linkage on the same chromosome but also historical contingency in the development of population structure as well as epistatic interactions between loci that may or may not be on the same chromosome. Since loose associations due to physical linkage or historical contingency are often broken up in one or a few generations of random mating, it is only the tight associations that persist. Thus, association mapping can provide for finely targeted mapping of traits to specific parts of the genome, given sufficient marker density. Spurious associations can accrue due to population substructure and the attendant differentiation in allele frequencies (historical contingency), and analysis of population substructuring is a useful tool to guard against such problems, as can be done with the program Structure (Pritchard et al. 2000).

7.3 Bulked Segregant Analysis

Bulked segregant analysis is a simple but useful variety of association mapping done in a controlled setting. Bulked samples that display a shared phenotype can be used to identify the markers that are closely linked to the gene controlling the phenotype. This permits the determination of markers flanking a specific, targeted portion of the genome without full genome mapping. As an example, Wilcox et al. (1996) inoculated loblolly pine (*Pinus taeda*) seedlings with rust and organized seedling phenotypes into those that formed a gall (infected) or no gall (resistant). They then used 60 RAPD markers to genotype the seedlings and then pooled all of the PCR products from infected lines, and all of the PCR products from resistant lines, keeping the pooled samples of the two phenotypes separate. Equal amounts of the PCR products from each of the lines were added in the pooling to ensure an unbiased representation of markers, wherein most markers should be present in some lines and absent in others regardless of the rust resistance phenotype. This manifests in the bulked samples as bands with only modest differences in intensity between infected and resistant lines. The exceptions are markers linked to genes controlling the rust resistance phenotype. These markers should report more intensely on the test gel in the PCR products from the resistant lines only, such as those from linkage group 9 (Fig. 20). As the RAPD markers had already been placed into a linkage map, the detection of the marker-phenotype association allowed immediate targeting of that portion of the genome for greater dissection and breeding.

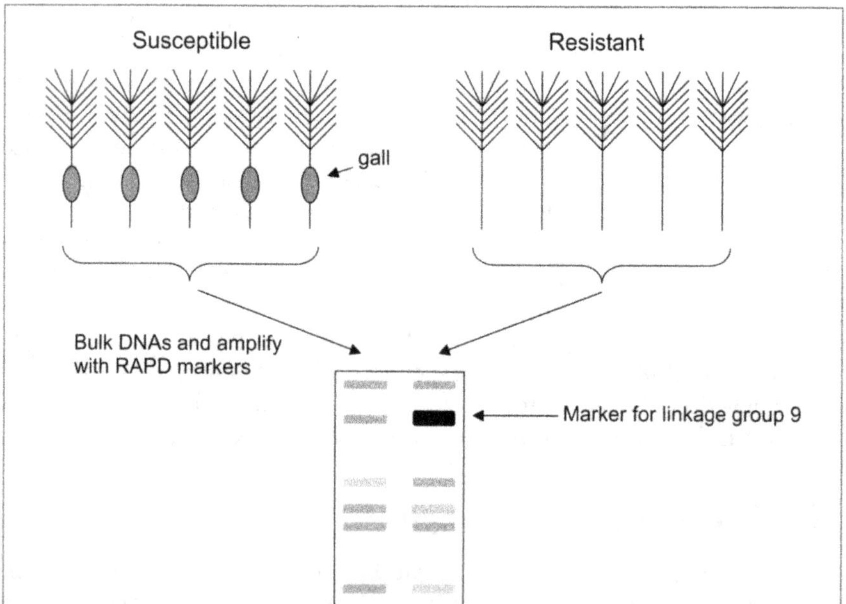

Fig. 20 Bulked segregant analysis of rust resistance in loblolly pine, from study by Wilcox et al. (1996).

7.4 Functional Genomic Profiling

High-throughput sequencing has allowed the detailed exploration of the transcriptome. Although we are clearly within the realm of functional genomics here, there are several applications that utilize various types of expression profiling for gene discovery, and then these findings feed directly into the mapping of critical genes.

7.4.1 ESTs and Gene Discovery

End sequencing of the full complement of RNAs yields a set of expressed sequence tags (ESTs) that indicate which genes are turned on in a given circumstance as well as the copy number of transcripts present. Such expression profiling is at its most powerful during comparisons of the transcriptome across varying conditions, such as with or without a particular nutrient or pathogen. The sequences obtained are typically compared to public database entries where sequences have been annotated for function. ESTs have become an important component of gene discovery especially for non-model organisms like most forest tree species. ESTs have been produced for latex from the rubber tree (*Hevea brasiliensis*, Han et al. 2000), floral pistils (*Prunus dulcis*, Jiang and Ma 2003), wood extractives (*Robinia pseudoacacia*, Yang et al. 2004) and different wood forming tissues (Allona et al. 1998; Sterky et al. 1998; Hertzburg et al. 2001; Kirst et al. 2003), among many others. Large EST databases are available for public evaluation at GenBank (NCBI) and through the Dendrome website (http://dendrome.ucdavis.edu/).

As an example, Sterky et al. (1998) produced 5,692 ESTs in wood-forming tissues of poplar. This pool entailed 3,719 unique transcripts for the two species studied (*P. tremula x tremuloides* and *P. trichocarpa*). Functions were ascribed for 2,245 of the unique sequences, and there were 820 associated protein functions. Roughly 4% of the identified genes were involved in cell wall formation, 5% were regulators and members of signal transduction paths, 2% were active in horomone biosynthesis, and 12% had no known function. Lack of function could indicate they were specific to the wood formation process since accessions from woody tissue were poorly represented in the database at the time. Slight differences were evident between the transcriptome of the cambium and xylem tissues (Fig. 21).

7.4.2 Candidate Gene Mapping with ESTPs

Candidate gene mapping combines (1) the power of QTL mapping to detect associations between phenotypes and specific parts of the genome with (2) the specificity provided by ESTs to signal which genes are

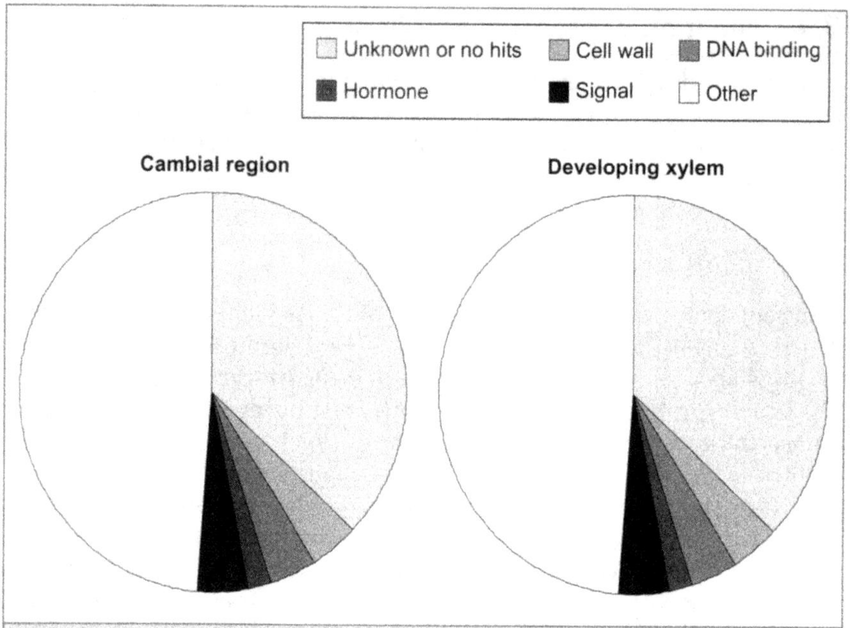

Fig. 21 ESTs from poplar cambium and developing xylem tissue, from study by Sterky et al. (1998).

present in a given circumstance. ESTs that carry a polymorphism can be used as polymorphic markers for that gene (ESTPs or expressed sequence tag polymorphisms), as has been shown in some tree species (e.g., Cato et al. 2001; Temesgen et al. 2004). If the ESTP represents a reasonable gene function, and the position of the gene maps to the same region as the QTL, then together these elements provide strong evidence that the gene represented by the EST is the gene controlling the QTL (Krutovskii and Neale 2001). Ultimately the QTL is merely a statistical association and so the ability to link the phenotype with something tangible such as a gene is an important additional step in the process of establishing causation between the genotype and phenotype, and in controlling the phenotype. See also Zhang (Chapter 8 in Volume 1) for additional considerations of map-based cloning of genes and QTLs.

Porth et al. (2005) mapped ESTs for osmotic stress induced genes to an oak map using an intraspecific cross of *Q. robur* and a cross between *Q. robur* (more drought sensitive) and *Q. petraea*. Using an EST set produced from osmotic stress in *Q. petraea* callus cell culture by Porth et al. (2005), they identified 25 cDNAs that they screened for polymorphism and produced ESTP markers to be placed on a pre-existing map of RAPD, AFLP, SSR, SCAR, and isozyme markers (Barreneche et al. 1998). Fourteen loci involving 34 SNP polymorphisms were used and thirteen of these were placed on the map. They were able to establish linkage, although precise placement required more than the 88 progeny they had used in the study.

Pot et al. (2006) studied QTLs and candidate genes for wood properties in maritime pine (*Pinus pinaster*). They used a three-generation outbred pedigree with four grandparents selected based on phenotypes and graded as + or - for breeding value relating to stem growth and straightness. Parents were selected from this cross and then backcrossed producing what is presently a 15-year full-sib family of 186 progeny, which eventually were cut and assayed for their chemical and physical wood properties. Parents and progeny were genotyped with AFLPs and a linkage map produced by the pseudo-testcross method (116 and 103 markers per respective map). A total of 54 QTLs were found with an average of 2.4 per trait. They then mapped 10 genes known to be involved in cell wall structure. ESTs were available for each of these genes and SNPs were detected via primer extension and the ESTPs profiled in the progeny set and placed on the map. Only one of the genes (Korrigan) co-localized to a QTL site, a cluster that had been identified as active in controlling hemicellulose content and fiber traits according to the QTL study (Fig. 22). Korrigan is a gene for beta-1-4 endo-glucanase known in *Arabidopsis* to participate in the biosynthesis of polysaccharides. Other genes such as PAL (phenylalanine ammonia-lyase) did not map to a region of the genome with a QTL, and so were unlikely candidate genes for the phenotypic traits under study.

7.4.3 Mapping with Microarrays

Microarrays provide a powerful means for transcriptional profiling that does not require extensive sequencing of the study species (Shena et al. 1995), though microarray data is most reliable when the transcriptome DNA and the arrayed DNA come from the same species. Unfortunately, there are a limited number of species with publicly available microarrays. Fortunately, one of these species is *Populus* given the recent release of the Affymetrix Poplar Genome Array. As arraying technology advances, more individual researchers are producing their own microarrays specific to their focal species and project.

As an example, Kirst et al. (2004) compared QTL analysis of growth and lignin content with cDNA microarray data in an interspecific backcross of *Eucalyptus*. They used three types of variation: phenotypic variation in diameter, genotypic variation, and transcript level variation from the microarray studies. The QTL work had been done previously

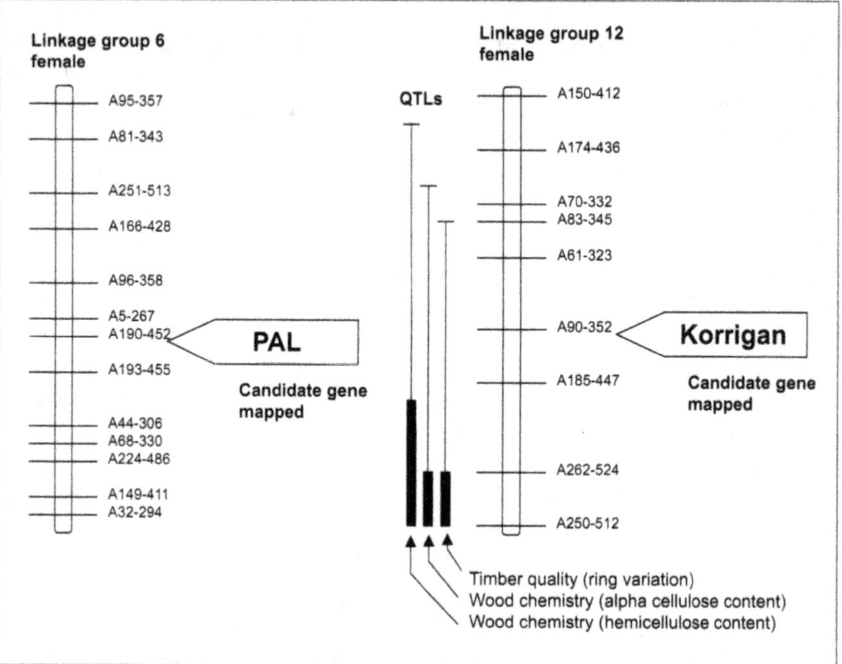

Fig. 22 Candidate gene mapping of ESTP markers for cell wall biosynthesis genes to a linkage map with identified QTLs for wood properties. After a portion of Figure 1, Pot et al. (2006). Marker positions are approximate.

and had produced a genetic map and QTLs for diameter growth (Myburg 2001; Myburg et al. 2003). Transcript levels were monitored on microarrays of 2,608 cDNAs from genes that were putative homologues thought to be involved in wood formation. Microarray data was collected from 91 progeny from an interspecific BC [(*E. grandis* x *E. globulus*) x *E. grandis*]. They used the signal intensity as a quantitative measure of variation in transcript level for each gene, and searched for associations between these transcript levels and variation in diameter growth in the mapping progeny, i.e. an eQTL or QTL for gene expression levels.

Some of the findings of Kirst et al. (2004) are reproduced in Fig. 23. Linkage groups (LG1-LG11) are represented along the bottom of the figure, with the top of the figure showing the positions of the two major QTLs for growth determined from the mapping population. At the bottom, black triangles indicate the positions of several lignin genes (CAD, CAld5H, DAHP, and SAMS) mapped using ESTPs. The positions

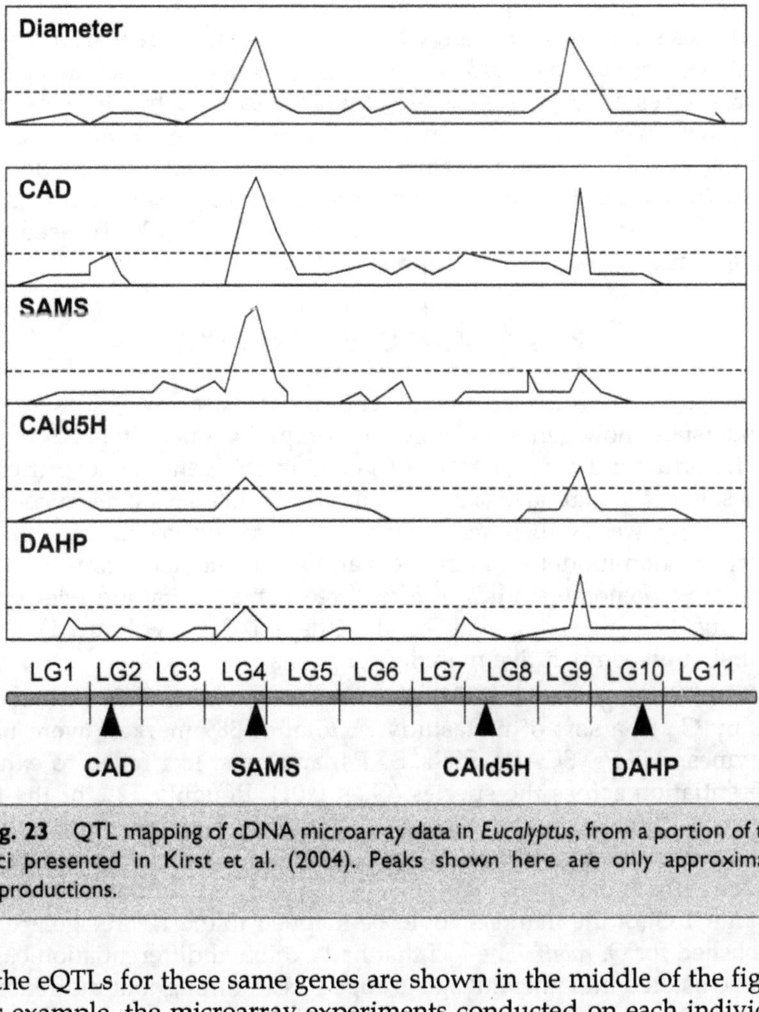

Fig. 23 QTL mapping of cDNA microarray data in *Eucalyptus*, from a portion of the loci presented in Kirst et al. (2004). Peaks shown here are only approximate reproductions.

of the eQTLs for these same genes are shown in the middle of the figure. For example, the microarray experiments conducted on each individual of the mapping population had shown that CAD gene expression was controlled by two primary linkage groups, though interestingly, not the linkage group known to contain the CAD gene. This discrepancy actually held for most of the lignin genes considered. The overall analysis showed that these eQTLs were controlled by two trans-acting regulatory loci that exerted coordinated control. These same two control loci co-localized to the two major QTLs for growth, offering strong evidence that these two loci were indeed responsible for controlling genes that influenced growth.

7.4.4 *Insertion-based Gene Discovery*

Mutagenesis has proven valuable for discovery of gene function in model systems such as *Arabidopsis* but has been less-successfully applied to forest trees. One possible alternative is presented by Groover et al. (2004) who use gene and enhancer trap tagging in *Populus*. Vectors carrying the GUS reporter gene were transformed into poplar via *Agrobacterium*, and the expression patterns of genes present at or near the point of insertion were revealed. Groover et al. (2004) used this mechanism to show vascular-expressed genes.

8 COMPARATIVE MAPPING

As the number of fully sequenced genomes increases, we are more able to understand how genomes evolve by comparing not only gene content but the structural arrangement of genes in the genomes of different species. For taxa that have experienced rather limited changes in genome organization, we can then transfer information between species, from the model to non-model system. A variety of methods are used in comparative genome studies of forest trees, and several are relevant to mapping. See also Doganlar et al. (Chapter 7 in Volume 1) for a discussion on comparative mapping in plants.

Scotti-Saintagne et al. (2004) examined interspecific differentiation of oaks by G_{st} in a sort of metastudy. A total of 389 markers were used (isozymes, AFLPs, SCARs, SSRs, SNPs) and most loci failed to exhibit differentiation across the species ($G_{st} < 0.01$). Roughly 12% of the loci displayed differences that exceeded neutral expectations with coding regions more divergent than noncoding regions, which could indicate genome differentiation via natural selection and microadaptation. About half (n = 158) of the markers could be mapped to the 12 linkage groups established for *Q. robur*. The loci that had exhibited differentiation based on G_{st} clustered into nine linkage groups spread throughout the genome.

Krutovsky et al. (2004) produced a comparative genetic map of loblolly pine (*Pinus taeda*) and Douglas fir (*Pseudotsuga menziesii*). These are two prominent members of the Pinaceae which have very large genomes and so are not readily studied by large-scale DNA sequencing. They used ESTP and RFLP (restriction fragment length polymorphism) markers, both of which can be used effectively as orthologous probes of divergent genomes. The map showed considerable synteny and colinearity, except for minor small-scale differences such as putative inversions (Fig. 24).

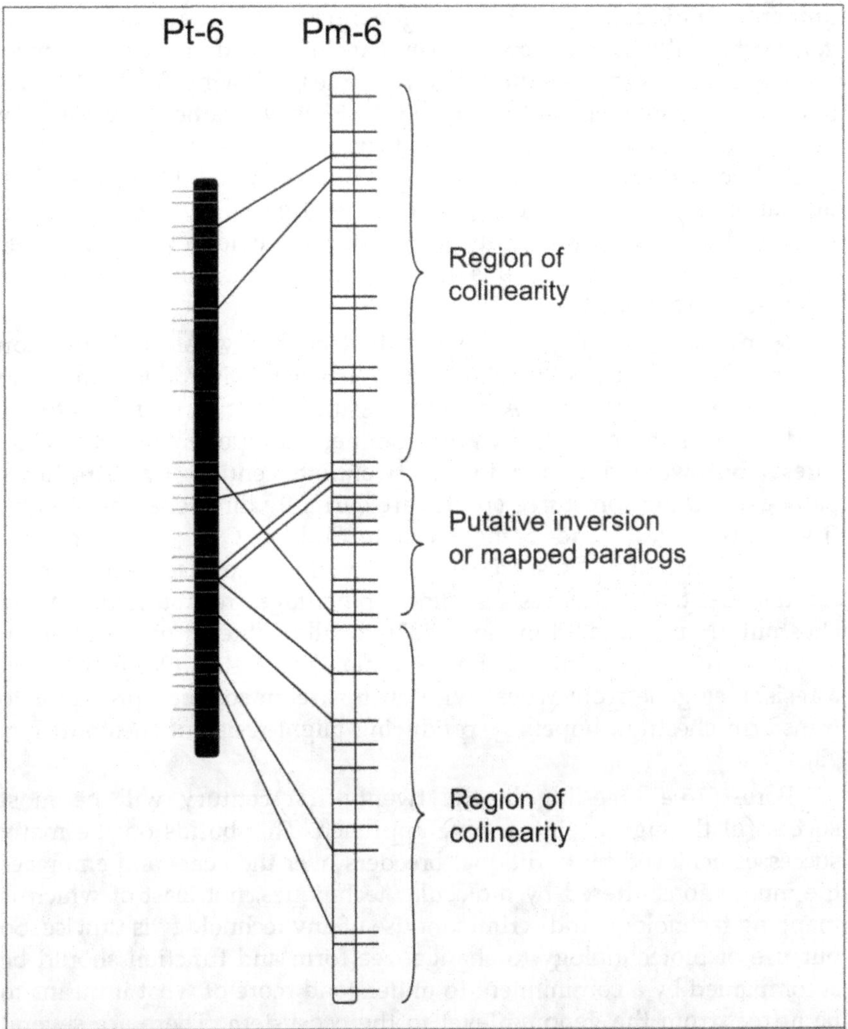

Fig. 24 Comparative mapping in linkage group 6 of loblolly pine (Pt, *Pinus taeda*) and Douglas fir (Pm, *Pseudotsuga menziesii*). After Figure 1, Krutovsky et al. (2004). Marker positions are approximate here.

9 FOREST TREE BREEDING IN THE TWENTY-FIRST CENTURY

Modern tree breeders are still playing catchup with their colleague breeders of crop plants. Nevertheless, the playing field has become much more even over the past several decades through the development of numerous molecular genetic and genomic applications, such as the genetic map. We have reviewed several ancillary methods that maximize the use of maps for breeding of forest trees, allowing for the several idiosyncrasies involved with being a tree. However rather little has been said about genetic engineering up to this point, because this field's greatest contributions lie in the near future in all likelihood. The utilization of the tertiary gene pool of forest trees is a largely untapped resource that, used wisely, could solve many problems in forestry. See Davey et al. (Chapter 11 of this volume) for a discussion on transgenics in plant improvement.

Many are watching the major biotechnology-based restoration projects that have been initiated to revitalize and reintroduce American elm (*Ulmus americana*) and American chestnut (*Castanea dentata*) into the wild. Both of these species were prominent components of many US forests but were devasted in the twentieth century by introduced pathogens. Other forest tree species are currently threatened by twenty-first century pests and pathogens like Sudden Oak Death and the emerald ash borer. Researchers are working to use recombinant technologies to engineer resistant lines of 'heritage' trees such as elm and chestnut (Mann and Plummer 2002) to allow these once dominant species to re-enter our forests. For example, Powell et al. (2005) are using a resistance gene from wheat with an upstream soybean promoter to transform chestnut, hopefully producing blight-resistant chestnut (Fig. 25).

Forest tree breeding in the twenty-first century will be most successful through an integrative approach. This builds on the many successes achieved by traditional breeders over the years and embraces the innovations offered by molecular techniques, not least of which is mapping technology. Indiscriminant use of any technology is unwise. So our use of biotechnology to change tree form and function should be accompanied by a commitment to understand more of what it means to be a tree, from the genome level to the ecosystem. There are several integrative programs in existence, including the Hardwood Tree Improvement and Regeneration Center (HTIRC) based at Purdue University, the Dendrome Project out of the University of California, Davis, the Pacific Northwest Tree Improvement Research Cooperative (PNWTIRC) out of the University of Oregon, the International *Populus* Genome Consortium (IPGC) organized through the DOE, and the Genolyptus Consortium based in Brazil, among others. These programs are playing key roles in the domestication of the forest tree.

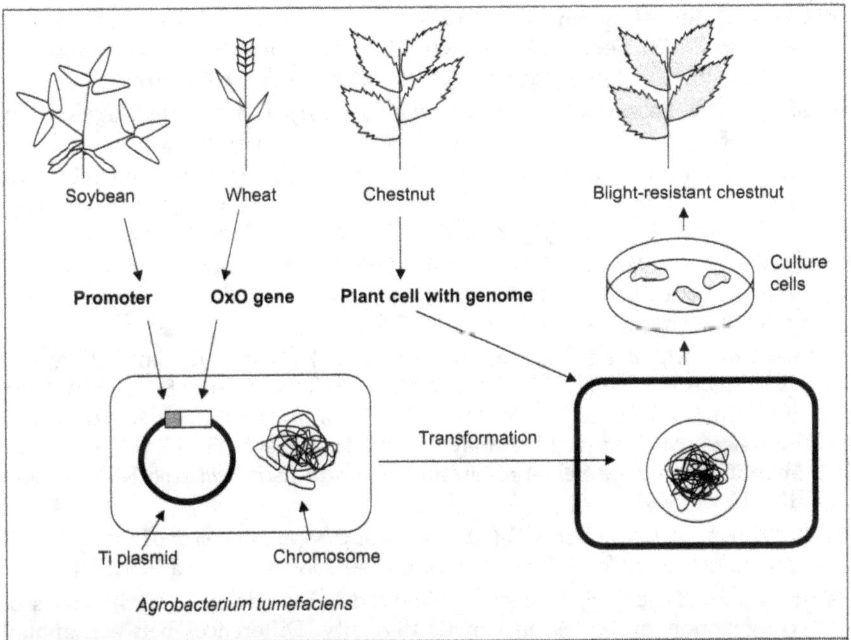

Fig. 25 Engineering blight resistance in American chestnut, from research by Powell et al. (2005).

References

Adams WT, Burczyk J (2000) Magnitude and implications of gene flow in gene conservation reserves. In: Young A, Boshier D, Boyle T (eds) Forest Conservation Genetics: Principles and Practice. CSIRO Publ, Collingwood, Australia, pp 215-224

Adams WT, Hipkins VD, Burczyk J, Randall WK (1997) Pollen contamination trends in a maturing Douglas-fir seed orchards. Can J For Res 27: 131-134

Aldrich P, Ahmed A, Ayub N, Briguglio J, Kapadia S, Morker M, Ondracek C, Rawal A, Salam A, Huebner C, Greer G (2007) Genome scans of adaptive diversity in US populations of the invasive tree *Ailanthus altissima*. Meet Bot Soc Am, Chicago, IL, USA

Aldrich PR, Hamrick JL (1998) Reproductive dominance of pasture trees in a fragmented tropical forest mosaic. Science 281: 103-105

Aldrich PR, Parker GR, Michler CH, Romero-Severson J (2003) Whole-tree silvic identifications and the microsatellite genetic structure of a red oak species complex in Indiana old-growth forest. Can J For Res 33: 2228-2237

Allard RW (1960) Principles of Plant Breeding. John Wiley, New York, USA

Allona I, Quinn M, Shoop E, Swope K, St. Cyr S, Carlis J, Riedl J, Retzel E, Campbell MM, Sederoff R, Whetten RW (1998) Analysis of xylem formation in pine by cDNA sequencing. Proc Natl Acad Sci USA 95: 9693-9698

Aradhaya MK, Potter D, Woeste K, Simon CJ (2001) An integrated high density AFLP-SSR map of Persian walnut, *Juglans regia* L. HortScience 36: 536

Arcade A, Anselin F, Faivre-Rampant P, Lesage MC, Laurans F, Paques LE, Prat D (2000) Application of AFLP, RAPD and ISSR markers to genetic mapping of European larch and Japanese larch. Theor Appl Genet 100: 299-307

Arcade A, Faivre-Rampant P, Pâques LE, Prat D (2002) Localisation of genomic regions controlling microdensitometric parameters of wood characteristics in hybrid larches. Ann For Sci 59: 607-615

Arifuzzaman M, Maeda M, Itoh A, Nishikata K, Takita C, Saito R, Ara T, Nakahigashi K, Huang HC, Hirai A, Tsuzuki K, Nakamura S, Altaf-Ul-Amin M, Oshima T, Baba T, Yamamoto N, Kawamura T, Ioka-Nakamichi T, Kitagawa M, Tomita M, Kanaya S, Wada C, Mori H (2006) Large-scale identification of protein–protein interaction of *Escherichia coli* K-12. Genom Res 16: 686–691

Apsit, VJ, Hamrick JL, Nason JD (2001) Breeding population size of a fragmented population of a Costa Rican dry forest tree species. J Hered 92: 415-420

Austerlitz F, Mariette S, Machon N, Gouyon PH, Godelle B (2000) Effects of colonization processes on genetic diversity: Differences between annual plants and tree species. Genetics 154: 1309-1321

Barreneche T, Bodenes C, Lexer C, Trontin J-F, Fluch S, Streiff R, Plomion C, Roussel G, Steinkellner H, Burg K, Favre J-M, Glössl J, Kremer A (1998) A genetic linkage map of *Quercus robur* L. (pedunculate oak) based on RAPD, SCAR, microsatellite, minisatellite, isozyme and 5S rDNA markers. Theor Appl Genet 97: 1090-1103

Bawa KS, Bullock SH, Perry DR, Coville RE, Grayum MH (1985a) Reproductive biology of tropical lowland rain forest trees. II. Pollination systems. Am J Bot 72: 346-356

Bawa KS, Perry DR, Beach JH (1985b) Reproductive biology of tropical lowland rain forest trees. I. Sexual systems and incompatibility mechanisms. Am J Bot 72: 331-345

Beaumont MA, Badling DJ (2004) Identifying adaptive genetic divergence among populations from genome scans. Mol Ecol 13: 969-980

Beedanagaril SR, Dove SK, Wood BW, Conner PJ (2005) A first linkage map of pecan cultivars based on RAPD and AFLP markers. Theor Appl Genet 110: 1127-1137

Bennett MD, Leitch IJ (2005) Plant DNA C-values database (release 4.0, Jan 2005). (*http://www.rbgkew.org.uk/cval/*)

Bennett MD, Smith JB (1991) Nuclear DNA amounts in angiosperms. Phil Trans Roy Soc B 334: 309-345

Binelli G, Bucci G (1994) A genetic linkage map of *Picea abies* Karts, based on RAPD markers, as a tool in population genetics. Theor Appl Genet 88: 283-288

Blakesley D, Allen A, Pellny TK, Roberts AV (2002) Natural and induced polyploidy in *Acacia dealbata* Link. and *Acacia mangium* Willd. Ann Bot London 90: 391-398

Botstein D, White RL, Skolnick M, Davis RW (1980) Construction of a genetic linkage map in man using restriction fragment length polymorphisms. Am J Hum Genet 32: 314-331

Bradshaw HD Jr, Stettler RF (1994) Molecular genetics of growth and development in *Populus*. II: Segregation distortion due to genetic load. Theor Appl Genet 89: 551-558

Bradshaw HD Jr, Stettler RF (1995) Molecular genetics of growth and development in *Populus*. IV. Mapping QTLs with large effects on growth, form, and phenology traits in a forest tree. Genetics 139: 963-973

Brondani RPV, Brondani C, Tarchini R, Grattapaglia D (1998) Development, characterization and mapping of microsatellite markers in *Eucalyptus grandis* and *E. urophylla*. Theo Appl Genet 97: 816-827

Brondani RPV, Brondani C, Grattapaglia D (2002) Towards a genus-wide reference linkage map for *Eucalyptus* based exclusively on highly informative microsatellite markers. Mol Genet Genom 267: 338-347

Brondani RPV, Williams ER, Brondani C, Grattapaglia D (2006) A microsatellite-based consensus linkage map for species of *Eucalyptus* and an novel set of 230 microsattelite markers for the genus. BMC Plant Biol 6: 1-16

Broman KW (2005) The genomes of recombinant inbred lines. Genetics 169: 1133-1146

Brown GR, Bassoni DL, Gill GP, Fontana JR, Wheeler NC, Megraw RA, Davis MF, Sewell MM, Tuskan GA, Neale DB (2003) Identification of quantitative trait loci influencing wood property traits in loblolly pine (*Pinus taeda* L). III. QTL verification and candidate gene mapping. Genetics 164: 1537-1546

Brunner A, Li J, DiFazio S, Shevchenko O, Mohamed R, Montgomery B, Elias A, Van Wormer K, DiFazio SP, Strauss SH (2007) Genetic containment of forest plantations. Tree Genet Genom 3: 75-100

Buiteveld J, Bakker EG, Bovenschen J, de Vries SM (2001) Paternity analysis in a seed orchard of *Quercus robur* L. and estimation of the amount of background pollination using microsatellite markers. For Genet 8: 331-337

Burczyk J, Adams WT, Shimizu JY (1996) Mating patterns and pollen dispersal in a natural knobcone pine (*Pinus attenuata* Lemmon) stand. Heredity 77: 251-260

Burczyk J, Di Fazio SP, Adams WT (2004) Gene flow in forest trees: How far do genes really travel? For Genet 11: 1-14

Burdon RD (1997) Genetic diversity for the future: Conservation or creation and capture? In: Burdon RD, Moore JM (eds) IUFRO '97 Genetics of Radiata Pine: Proc New Zealand FRI-IUFRO conf, Rotorua, New Zealand. New Zealand For Res Inst, Bull No 203: 237-246

Burger WC (1975) The species concept in *Quercus*. Taxon 24: 45-50

Burke JM, Arnold ML (2001) Genetics and the fitness of hybrids. Annu Rev Genet 35: 31-52

Burns RM, Honkala BH (tech coords) (1990) Silvics of North America: 1. Conifers 2. Hardwoods. Agriculture Handbook 654. USDA Forest Service, Washington, DC, USA, Vol 2

Butcher PA, Williams ER, Whitaker D, Ling S, Speed TP, Moran GF (2004) Improving linkage analysis in outcrossing forest trees - An example from *Acacia mangium*. Theor Appl Genet 104: 1185-1191

Byrne M, Murrell JC, Allen B, Moran GF (1995) An integrated genetic-linkage map for eucalypts using RFLP, RAPD and isozyme markers. Theor Appl Genet 91: 869-875

Cato SA, Gardner RC, Kent J, Richardson TE (2001) A rapid PCR-based method for genetically mapping ESTs. Theor Appl Genet 102: 296-306

Cervera MT, Storme V, Ivensa B, Gusmão G, Liu BH, Hostyna V, Van Slyckenc J, Van Montagua M, Boerjana W (2001) Dense genetic linkage maps of three *Populus* species (*Populus deltoides, P. nigra* and *P. trichocarpa*) based on AFLP and microsatellite markers. Genetics 158: 787-809

Chagné D (2004) Inheritance and mapping studies in gymnosperms. (*http://www.pierroton.inra.fr/genetics/labo/mapreview.html*)

Chagné D, Lalanne C, Madur D, Kumar S, Frigerio JM, Krier C, Decroocq S, Savouré A, Bou-Dagher-Kharrat M, Bertocchi E, Brach J, Plomion C (2002) A high density genetic map of Maritime pine based on AFLPs. Ann For Sci 59: 627-636

Charlesworth D, Wright SI (2001) Breeding systems and genome evolution. Curr Opin Genet Dev 11: 685-690

Cheliak WM, Morgan K, Dancik BP, Strobec KC, Yeh FC (1984) Segregation of allozymes in megagametophytes of viable seed from a natural population of jack pine, *Pinus banksiana* Lamb. Theor Appl Genet 69: 145-151

Cheliak WM, Morgan K, Strobeck C, Yeh FCH, Dancik BP (2004) Estimation of mating system parameters in plant populations using the EM algorithm. Theor Appl Genet 65: 157-161

Chen DM, De Filippis LF (2001) Differentially expressed genes identified during salt adaptation in *Eucalyptus* microarrays: Down-regulation of a cDNA sequence coding for alpha-tubulin. J Plant Physiol 158: 1195-1202

Clarke FB (1965) Forest service research in black walnut. Annu Rep Northern Nut Growers Assoc 56: 84-87

Conkle MT (1981) Isozyme variation and linkage in six conifer species. In: Conkle MT (ed) Proc Symp on Isozymes of North Am For Trees and For Insects. Pacific Southwest Forest and Range Exp Stn, Berkeley, CA, USDA Forest Service Gen Tech Rep PSW-48, pp 11-17

Crawford TJ (1984) What is a population? In: Shorrocks B (ed) Evolutionary Ecology. Brit Ecol Soc Symp no 23. Blackwell Sci Publ, Oxford, UK, pp 135-174

Davis MB, Shaw RG (2001) Range shifts and adaptive responses to quaternary climate change. Science 292: 673-679

Den Uyl DD, Diller OD, Day RK (1938) The development of natural reproduction in previously grazed farmwoods. Purdue Univ Agri Exp Stn Res Bull No 431, p 28

Den Uyl DD (1954) Indiana's old-growth forests. Proc Indiana Acad Sci 63: 73-79

Den Uyl DD (1961) Natural tree reproduction in mixed hardwood stand. Purdue Univ Resour Bull No 728, p 19

Deneke FJ, Funk DT, Bey C (1981) Preliminary seed collection zones for black walnut. Gen Tech Rep NA-FB/M-4, St Paul, MN, USDA For Serv, Northeastern Area State and Private Foresty, Broomall, PA, 5p

Devey ME, Fiddler TA, Liu B-H, Knapp SJ, Neale BD (1994) An RFLP linkage map for loblolly pine based on a three-generation outbred pedigree. Theor Appl Genet 88: 273-278

Devey ME, Bell JC, Smith DN, Neale DB, Moran GF (1996) A genetic map fpr *Pinus radiata* based on RFLP, RAPD and microsatellite markers. Theor Appl Genet 92: 673-679

Devey ME, Sewell MM, Uren TL, Neale DB (1999) Comparative mapping in loblolly and radiata pine using RFLP and microsatellite markers. Theor Appl Genet 99: 656-662

DiFazio SP, Slavov GT, Burczyk J, Leonardi S, Strauss SH (2004) Gene flow from tree plantations and implications for transgenic risk assessment. In: Walter C, Carson M (eds) Plantation Forest Biotechnology for the 21st Century. Res Signpost, Trivandrum, India, pp 405-422

Doebley J (1992) Mapping the genes that made maize. Trends Genet 8: 302-307

Doebley J (2004) The genetics of maize evolution. Annu Rev Genet 38: 37-59

Dodd ME, Silvertown J (2000) Size-specific fecundity and the influence of lifetime size variation upon effective population size in *Abies balsamea*. Heredity 85: 604-609

Doudrick RL, Heslop-Harrison JS, Nelson CD, Schmidt T, Nance WL, Schwarzacher T (1995) Karyotype of slash pine (*Pinus elliottii* var. *elliottii*) using patterns of fluorescence in situ hybridization and fluorochrome banding. J Hered 86: 289-296

Eckenwalder JE (1996) Systematics and evolution of *Populus*. In: Stettler RF, Bradshaw HD, Heilman PE, Hinckley TM (eds) Biology of *Populus* and Its Implications for Management and Conservation. NRC Res Press, Ottawa, Ontario, Canada, pp 7-32

Echt CS, Nelson CD (1997) Linkage mapping and genome length in eastern white pine (*Pinus strobus* L.). Theor Appl Genet 94: 1031-1037

El-Kassaby YA, Ritland K (2004) Impact of selection and breeding on the genetic diversity in Douglas-fir. Biodiver Conserv 5: 795-813

Eldridge K, Davidson J, Harwood C, Van Wye G (1993) Eucalypt Domestication and Breeding. Oxford Univ Press, Oxford, UK

Elias TS (1980) The walnut family. In: Elias TS (ed) Trees of North America: Field Guide and Natural History. Van Nostrand Reinhold, New York, USA, pp 265-275

Eriksson G, Namkoong G, Roberds JH (1993) Dynamic gene conservation for uncertain futures. For Ecol Manag 62: 15-37

Eriksson G, Ekberg I, Clapham D (2006) An Introduction to Forest Genetics. 3rd edn. (*http://www2.vbsg.slu.se/forgen/*)

Ewing RM, Chu P, Elisma F, Li H, Taylor P, Climie S, McBroom-Cerajewski L et al. (2007) Large-scale mapping of human protein–protein interactions by mass spectrometry. Mol Syst Biol 3: 1-17

Falconer DS (1960) Introduction to Quantitative Genetics. Ronald Press, New York, USA

FAO, FLD, IPGRI (2004) Forest genetic resources conservation and management. International Plant Genetic Resources Institute, Rome, Italy. (*http://www.bioversityinternational.org/publications/Pdf/1018.pdf*)

Ferris R, Long L, Bunn SM, Robinson KM, Bradshaw HD, Rae AM, Taylor G (2002) Leaf stomatal and epidermal cell development: identification of putative quantitative trait loci in relation to elevated carbon dioxide concentration in poplar. Tree Physiol 22: 633-40

Fore SA, Hickey RJ, Vankat JL, Guttman SI, Schaefer RL (1992) Genetic structure after forest fragmentation: A landscape ecology perspective on *Acer saccharum*. Can J Bot 70: 1659-1668

Fowler DP (1965) Effects of inbreeding in red pine, *Pinus resinosa*. II. Pollination studies. Silv Genet 14: 12-23

Frankis MP (1989) Generic inter-relationships in Pinaceae. Notes Roy Bot Gard Edinb 45: 527-548

Friedman ST, Adams WT (1985) Estimation of gene flow into two seed orchards of loblolly pine (*Pinus taeda* L.). Theor Appl Genet 69: 609-615

Friesen N, Brandes A, Heslop-Harrison JS (2001) Diversity, origin and distribution of retrotransposons (gypsy and copia) in conifers. Mol Biol Evol 18: 1176-1188

Frewen BE, Chen THH, Howe GT, Davis J, Rohde A, Boerjan W, Bradshaw HD Jr (2000) Quantitative trait loci and candidate gene mapping of bud set and bud flush in populus. Genetics 154: 837-845

Fullard KJ, Moran G (2003) Identification of frost tolerance genes in a *Eucalyptus* hybrid cross. IUFRO Tree Biotechnol Meet, June 7-12, 2003, Umea, Sweden

Gaget M, Villar M, Dumas C (2004) The mentor pollen phenomenon in poplars: A new concept. Theor Appl Genet 78: 129-135

Gasbarra D, Sillanpaa MJ (2006) Constructing the parental linkage phase and the genetic map over distances >1 cM using pooled haploid DNA. Genetics 172: 1325-1335

Gerber S, Chabrier P, Kremer A (2003) FaMoz: A software for parentage analysis using dominant, codominant and uniparentally inherited markers. Mol Ecol Notes 3: 479-481

Gibson JP, Hamrick JL (1991) Heterogeneity in pollen allele frequencies among cones, whorls, and trees of table mountain pine (*Pinus pungens*). Am J Bot 78: 1244-1251

Giot L, Bader JS, Brouwer C, Chaudhuri A, Kuang B, Li Y, Hao YL et al. (2003) A protein interaction map of *Drosophila melanogaster*. Science 302: 1727-1736

Gould SJ, Lewontin R (1979) The spandrels of San Marco and the Panglossion paradigm: A critique of the adaptationist programme. Proc Roy Soc Lond B Bio 205: 581-598

Grattapaglia D (2004) Integrating genomics into *Eucalyptus* breeding. Genet Mol Res 3: 369-379

Grattapaglia D, Sederoff R (1994) Genetic linkage maps of *Eucalyptas grandis* and *Eucalyptus urophylla* using a pseudo-testcross: Mapping strategy and RAPD markrs. Genetics 137: 1121-1137

Grattapaglia D, Chaparro JX, Wilcox PL, McCord S, Crane B, Amerson H, Werner D, Liu B-H, O'Malley D, Whetton R, McKeand S, Goldfarb B, Greenwood M, Kuhlman G, Bridgwater F, Sederoff R (1993) Application of genetic markers to tree breeding. Proc Southern For Tree Improv Conf, Atlanta, GA, USA, 22: 452-463

Grattapaglia D, Bertolucci FL, Sederoff RR (1995) Genetic-mapping of QTLs controlling vegetative propagation in *Eucalyptus grandis* and *E. urophylla* using a pseudo-testcross strategy and RAPD markers. Theor Appl Genet 90: 933-947

Grattapaglia D, Bertolucci FL, Penchel R, Sederoff RR (1996) Genetic mapping of quantitative trait loci controlling growth and wood quality traits in *Eucalyptus grandis* using a maternal half-sib family and RAPD markrs. Genetics 144: 1205-1214

Groover A, Devey M, Fiddler T, Lee J, Megraw R, Mitchel-Olds T, Sherman B, Vujcic S, Williams C, Neale D (1994) Identification of quantitative trait loci influencing wood specific gravity in an outbred pedigree of loblolly pine. Genetics 138: 1293-1300

Groover A, Fontana JR, Dupper G, Ma C, Martienssen R, Strauss S, Meilan R (2004) Gene and enhancer trap tagging of vascular-expressed genes in poplar trees. Plant Physiol 134: 1742-1751

Grotkopp E, Rejmanek M, Sanderson MJ, Rost TL (2004) Evolution of genome size in pines (*Pinus*), and its life-history correlates, supertree analysis. Evolution 58: 1705-1729

Guevara MA, Soto A, Collada C, Plomion C, Savolainen O, Neale DB, Gonzalez-Martinez SC, Cervera MT (2005) Genomics applied to the study of adaptation in pine species. Invest Agrar Sist Recur For 14: 292-306

Guries RP, Friedman ST, Ledig FT (1978) A megagametophyte analysis of genetic linkage in pitch pine (*Pinus rigida* Mill). Heredity 40: 309-314

Gwaze DP, Zhou Y, Reyes-Valdes MH, Al-Rababah MA, Williams CG (2003) Haplotypic QTL mapping in an outbred pedigree. Genet Res Camb 81: 43-50

Hackett CA, Wachira FN, Paul S, Powell W, Waugh R (2000) Construction of a genetic linkage map for *Camellia sinensis* (tea). Heredity 85: 346-355

Hamrick JL, Godt MJW (1990) Allozyme diversity in plant species. In: Brown AHD, Clegg MT, Kahler AL, Weir BS (eds) Plant Population Genetics,

Breeding, and Genetic Resources. Sinauer Assoc, Sunderland, MA, USA, pp 43-63

Hamrick JL, Godt MJW, Sherman-Broyles SL (1992) Factors influencing levels of genetic diversity in woody plant species. New For 6: 95-124

Han K-H, Shin DH, Yang J, Kim IJ, Oh SK, Chow KS (2000) Genes expressed in the latex of *Hevea brasiliensis*. Tree Physiol 20: 503-510

Harju AM, Nikkanen T (1996) Reproductive success of orchard and nonorchard pollens during different stages of pollen shedding in a Scots pine seed orchard. Can J For Res 26: 1096-1102

Hayashi E, Kondo T, Terada K, Kuramoto N, Goto Y, Okamura M, Kawasaki (2001) Linkage map of Japanese black pine based on AFLP and RAPD markers including markers linked to resistance against the pine needle gall midge. Theor Appl Genet 102: 871-875

Hedrick PW (1985) Genetics of Populations. Jones and Bartlett Publ, Boston, USA

Hertzburg M, Aspeborg H, Schrader J, Andersson A, Erlandsson R, Blomqvist K, Bhalerao R, Uhlen M, Teeri TT, Lundeberg J, Sundberg B, Nilsson P, Sandberg G (2001) A transcriptional roadmap to wood formation. Proc Natl Acad Sci USA 98: 14732-14737

Hicks RR Jr (1998) Ecology and Management of Central Hardwood Forests. John Wiley, New York, USA

Ingvarsson PK (2005) Nucleotide polymorphism and linkage disequilibrium within and among natural populations of European Aspen (*Populus tremula* L., Salicaceae). Genetics 169: 945-953

Ishida K, Hiura T (2002) Mating system and population genetic structure of an androdioecious tree, *Fraxinus lanuginosa* Koidz. (Oleaceae) in northern Japan. Heredity 88: 296-301

Jansen J (2005) Construction of linkage maps in full-sib families of diploid outbreeding species by minimizing the number of recombinations in hidden inheritance vectors. Genetics 170: 2013-2025

Jaenicke-Després V, Buckler ES, Smith BD, Gilbert MTP, Cooper A, Doebley J, Pääbo S (2003) Early allelic selection in maize as revealed by ancient DNA. Science 302: 1206-1208

Jannink JL, Walsh JB (2002) Association mapping in plant populations. In: Kang MS (ed) Quantitative Genetics, Genomics, and Plant Breeding. CABI, Wallingford, UK, pp 59-68

Jensen RJ (1988) Assessing patterns of morphological variation of *Quercus* spp. in mixed-oak communities. Am Midl Nat 120: 120-135

Jiang Y-Q, Ma R-C (2003) Generation and analysis of expressed sequence tags from almond (*Prunus dulcis* Mill.) pistils. Sex Plant Reprod 16: 197-207

Johnsen K, Major JE, Maier CA (2003) Selfing results in inbreeding depression of growth but not of gas exchange of surviving adult black spruce trees. Tree Physiol 23: 1005-1008

Johnson R, St. Clair B, Lipow S (2001) Genetic conservation in applied tree breeding programs. In: Kjaer ED, Graudal L, Nathan I (eds) Proc ITTO Conf on In Situ and Ex Situ Conservation of Comm Trop Trees, Yogyakarta, Indonesia, pp 215-230

Jones AG, Ardren WR (2003) Methods of percentage analysis in natural populations. Mole Ecol 12: 2511-2523

Jump AS, Penuelas J (2006) Genetic effects of chronic habitat fragmentation in a wind-pollinated tree. Proc Natl Acad Sci USA 103: 8096-8100

Kinlaw CS, Neale DB (1997) Complex gene families in pine genomes. Trends Plant Sci 2: 356-359

Kirkness EF, Bafna V, Halpern AL, Levy S, Remington K, Rusch DB, Delcher AL, Pop M, Wang W, Fraser CM, Venter JC (2003) The dog genome: Survey sequencing and comparative analysis. Science 301: 1898-1903

Kirst M, Johnson AF, Baucom C, Ulrich E, Hubbard K, Staggs R, Paule C, Retzel E, Whetten R, Sederoff R (2003) Apparent homology of expressed genes from wood-forming tissues of loblolly pine (*Pinus taeda* L.) with *Arabidopsis thaliana*. Proc Natl Acad Sci USA 100: 7383-7388

Kirst M, Myburg AA, De Leon JPG, Kirst ME, Scott J, Sederoff R (2004) Coordinated genetic regulation of growth and lignin revealed by quantitative trait locus analysis of cDNA microarray data in an interspecific backcross of Eucalyptus. Plant Physiol 135: 2368-2378

Komulainen P, Brown GR, Mikkonen M, Karhu A, Garcia-Gil MR, O'Malley D, Lee B, Neale DB, Savolainen O (2003) Comparing EST-based genetic maps between *Pinus sylvestris* and *Pinus taeda*. Theor Appl Genet 107: 667-678

Krutovskii KV, Neale DB (2001) Forest genomics for conserving adaptive genetic diversity. In: Palmberg-Lerche C, Iversen PA, Sigaud P (eds) Forest Genetic Resources. For Dept, For Genet Resour Working Pap 29, FAO, Rome, Italy (*http://www.fao.org/docrep/004/y2316e/y2316e03.htm#bm03*)

Krutovsky KV, Troggio M, Brown GR, Jermstad KD, Neale DB (2004) Comparative mapping in the Pinaceae. Genetics 168: 447-461

Kubisiak TL, Nelson CD, Nance WL, Stine M (1995) RAPD linkage mapping in a longleaf pine x slash pine F1 family. Theor Appl Genet 90: 1110-1127

Kumar S, Spelman RJ, Garrick DJ, Richardson TE, Lausberg M, Wilcox PL (2000) Multiple-marker mapping of wood density loci in an outbred pedigree of radiata pine. Theor Appl Genet 100: 926-933

Lantz CW (2001) Genetic improvement of forest trees. In: Nisley RG (manag ed) Bonner FT (tech coord) Woody Plant Seed Manual. USDA For Serv, Washington, DC, USA (*http://www.nsl.fs.fed.us/wpsm/*)

Ledig FT (1992) Human impacts on genetic diversity in forest ecosystems. Oikos 63: 87-108

Leitch IJ, Hanson L, Winfield M, Parker J, Bennett MD (2001) Nuclear DNA C-values complete familial representation in gymnosperms. Ann Bot Lond 88: 843-849

Lerceteau EC, Plomion C, Andersson B (2000) AFLP mapping and detection of quantitative trait loci (QTLs) for economically important traits in *Pinus sylvestris*: a preliminary study. Mol Breed 6: 451-458

Lespinasse D, Rodier-Goud M, Grivet L, Leconte A, Gegnate H, Seguin M (2000) A saturated genetic linkage map of rubber tree (*Hevea* spp.) based on RFLP, AFLP, microsatellite and isozyme markers. Theor Appl Genet 100: 127-138

Levin DA, Kerster HW (1974) Gene flow in seed plants. In: Dobzhansky TM, Hecht Mt, Steere WC (eds) Evolutionary Biology. Plenum Press, New York, USA, pp 139-220

Levin DA (1981) Dispersal versus gene flow in plants. Ann Mo Bot Gard 68: 233-253

Lexer C Heinze B, Gerber S, Macalka-Kampfer S, Steinkellner H, Kremer A, Glossl J (2000) Microsatellite analysis of maternal half-sib families of *Quercus robur*, pedunculate oak : II. Inferring the number of pollen donors from the offspring. Theor Appl Genet 100: 858-865

Libby WJ, Stettler RF, Seitz FW (1969) Forest genetics and forest tree-breeding. Annu Rev Genet 3: 469-494

Limstrom GA (1962) Proposed Seed Collection Zones for the Central States. Central States For Exp Stn, USDA For Serv, Columbus, OH, USA 3: 31-39

Lipow SR, St. Clair JB, Johnson GR (2002) Ex Situ Gene Conservation for Conifers in the Pacific Northwest. USDA For Serv, Pacific Northwest Res Stn, Portland, OR, USA, PNW-GTR 528, p 53

Lipow SR, Johnson GR, St Claiff JB, Jayawickrama KJ (2003) The role of tree improvement programs for ex situ gene conservation of coastal Douglas-fir in the Pacific Northwest. For Genet 10: 111-120

Liu BH (1997) Statistical Genomics: Linkage Mapping, and QTL Analysis. CRC Press, New York, USA

Liu B, Sederoff R, O'Malley D (1993) Linkage mapping using open-pollinated populations. Proc 23rd Southern For Tree Improv Conf, Atlanta, GA, USA

Lopez GA, Potts BM, Tilyard PA (2000) F1 hybrid inviability in *Eucalyptus*: The case of *E. ovata* x *E. globules*. Heredity 85: 242-250

Lu Q, Cui Y, Wu R (2004) A multilocus likelihood approach to joint modeling of linkage, parental diplotype and gene order in a full-sib family. BMC Genet 5: 1-14

Maliepaard C, Jansen J, van Ooijen JW (1997) Linkage analysis in a full-sib family of an outbreeding plant species: Overview and consequences for applications. Genet Res 70: 237-250

Maliepaard C, Alston FH, van Arkel G, Brown LM, Chevreau E, Dunemann F, Evans KM, Gardiner S, Guilford P, van Heusden AW, Janse J, Laurens F, Lynn JR, Manganaris AG, den Nijs APM, Periam N, Rikkerink E, Roche P, Ryder C, Sansavini S, Schmidt H, Tartarini S, Verhaegh JJ, Vrielink-van Ginkel M, King GJ (1998) Aligning male and female linkage maps of apple (*Malus pumila* Mill.) using multi-allelic markers. Theor Appl Genet 97: 60-73

Mann GC, Plummer ML (2002) Can genetic engineering help restore 'heritage' trees? Science 295: 1628

Markussen T, Fladung M, Achere V, Favre JM, Faivre-Rampant P, Aragones A, Da Silva Perez Harvengt L, Ritter E (2003) Identification of QTLs controlling growth, chemical and physical wood property traits in *Pinus pinaster* (Ait). Silv Genet 52: 8-15

Marshall TC, Slate J, Kruuk LEB, Pemberton JM (1998) Statistical confidence for likelihood-based paternity inference in natural populations. Mol Ecol 7: 639-655 (*http://www.fieldgenetics.com/pages/aboutCervus_Papers.jsp*)

Maynard C (1996) Forest Genetics Glossary. SUNY-ESF, Syracuse, NY, USA (*http://www.esf.edu/for/maynard/GENE_GLOSSERY.html*)

Mayr E (1942) Systematics and the Origin of Species. Columbia Univ Press, New York, USA

Meilan R, Han K-H, Ma C, James RR, Eaton JA, Stanton BJ, Hoien EA, Crockett RP, Strauss SH (2000) Development of glyphosate-tolerant hybrid cottonwoods. Tappi J 83: 164-166

McGuire M, Stevens J, Potter-Witter K (1999) Assessing scarcity of the north central veneer log resource. North J Appl For 16: 160-166

Michler CH, Meilan R, Woeste KE, Pijut PM, Jacobs D, Aldrich P, Glaubitz J (2005) Hardwood Genetics and Tree Improvement - A Midwest USA Perspective. The Thin Green Line In: Proc Symp State-of-the-art in Reforestation, 26-28 July, Thunder Bay, ON, USA, pp 26-28

Mitchell-Olds T (1996) Genetic constraints on life-history evolution: Quantitative-trait loci influencing growth and flowering in *Arabidopsis thaliana*. Evolution 50: 140-145

Moran GF, Thamarus KA, Raymond CA, Qiu D, Uren T, Southerton SG (2002) Genomics of *Eucalyptus* wood traits. Ann For Sci 59: 645-650

Mukai Y, Suyama Y, Tsumura Y, Kawahara T, Yoshimaru H, et al. (1995) A linkage map for sugi (*Cryptomeria japonica*) based on RFLP, RAPD, and isozyme loci. Theor Appl Genet 90: 835-840

Murray B (1998) Nuclear DNA amounts in gymnosperms. Ann Bot 82(Suppl A): 3-15

Myburg AA (2001) Genetic Architecture of Hybrid Fitness and Wood Quality Traits in a Wide Interspecific Cross of Eucalyptus Tree Species. PhD Thesis. North Carolina State Univ, Raleigh, NC, USA (*http://www.lib.ncsu.edu/theses/available/etd-20010723-175234/*)

Myburg AA, Griffin AR, Sederoff RR, Whetten RW (2003) Comparative genetic linkage maps of *Eucalyptus grandis, Eucalyptus globulus* and their F1 hybrid based on a double pseudo-backcross mapping approach. Theor Appl Genet 107: 1028-1042

Myburg AA, Vogl C, Griffin AR, Sederoff RR, Whetten RW (2004) Genetics of postzygotic isolation in *Eucalyptus*: Whole-genome analysis of barriers to introgression in a wide interspecific cross of *Eucalyptus grandis* and *E. globulus*. Genetics 166: 1405-1418

Namkoong G, Kang H, Brouard JS (1998) Tree Breeding: Principles and Strategies. Springer, New York, USA

Nason JD, Herre EA, Hamrick JL (1996) Paternity analysis of the breeding structure of strangler fig populations: Evidence for substanstial long-distance wasp dispersal. J Biogeogr 23: 501-512

Neale DB, Savolainen O (2004) Association genetics of complex traits in conifers. Trends Plant Sci 9: 325-330

Neale DB, Sewell MM, Brown GR (2002) Molecular dissection of the quantitative inheritance of wood property traits in loblolly pine. Ann For Sci 59: 595-605

Nelson CD, Kubisiak TL, Stine M, Nance WL (1994) A genetic linkage map of longleaf pine (*Pinus palustris* Mill) based on ramdom amplified polymorphicd DNAs. J Hered 85: 433-439

NRC: National Research Council (USA) (1991) Managing Global Genetic Resources: Forest Trees. Natl Acad Press, Washington DC, USA

O'Malley DM, Allendorf FM, Blake GM (1979) Inheritance of isozyme variation and heterozygosity in *Pinus ponderosa*. Biochem Genet 17: 233-250

O'Malley DM, Guries RP, Nordheim EV (1986) Linkage analysis for 18 enzyme loci in *Pinus rigida* Mill. Theor Appl Genet 72: 530-535

Parker GR (1997) The wave of settlement. In: Jackson M (ed) The Natural Heritage of Indiana. Indiana Univ Press, Bloomington, IN, USA, pp 369-381

Paterson AH (2002) What has QTL mapping taught us about plant domestication? Tanksley Rev no 137. New Phytol 154: 591-608

Paterson HEH (1985) The recognition concept of species. In: Vrba ES (ed) Species and Speciation. Trans Mus Monog 4, pp 21-29

Pakkanen A, Pulkkinen P (1991) Pollen production and background pollination levels in Scots pine seed orchards of northern Finnish origin. In: Lindgren D (ed) Pollen Contamination in Seed Orchards. Proc Meet Nordic Group Tree Breed 1991, Swed Univ Agri Sci, Dep For Genet Plant Physiol, Umea, Sweden, pp 14-21

Pakken A, Nikkanen T, Pulkkinen P (2000) Annual variation in pollen contamination and outcrossing in a Picea abies seed orchard. Scand J For Res 15: 399-404

Pekkinen M, Varvio S, Kulju KKM, Kärkkäinen H, Smolander S, Viherä-Aarnio A, Koski V, Sillanpää MJ (2005) Linkage map of birch, *Betula pendula* Roth, based on microsatellites and amplified fragment length polymorphisms. Genome 48: 619-625

Perry DJ, Bousquet J (1998a) Sequence-tagged-site (STS) markers of arbitrary genes: The utility of black spruce-derived STS primers in other conifers. Theor Appl Genet 97: 735-743

Perry DJ, Bousquet J (1998b) Sequence-tagged-site (STS) markers of arbitrary genes: Development, characterization and analysis of linkage in black spruce. Genetics 149: 1089-1098

Perry DJ, Bousquet J (2001) Genetic diversity and mating system of post-fire and post-harvest black spruce: an investigation using codominant sequence-tagged-site (STS) markers. Can J For Res 31: 32-40

Poke F (2006) Summary of World-wide Research in Eucalypt Genomics. Tech Rep, Coop Res Centre for Sustainable Prod For, Univ Tasmania, Australia (*http://www.ieugc.up.ac.za/Genomic_survey_page.htm*)

Porth I, Scotti-Saintagne C, Barreneche T, Kremer A, Burg K (2005) Linkage mapping of osmotic stress induced genes of oak. Tree Genet Genom 1: 31-40

Pot D, Rodrigues J-C, Rozenberg P, Chantre G, Tibbits J, Cahalan C, Pichavant F, Plomion C (2006) QTLs and candidate genes for wood properties in maritime pine (*Pinus pinaster* Ait.). Tree Genet Genom 2: 10-24

Powell WA, Merkle SA, Liang H, Maynard CA (2005) Blight resistance technology: Transgenic approaches. In: Steiner KC, Carlson JE (eds) Proc Conf Restoration of Am Chestnut to Forest Lands, May 4-6, 2004, North Carolina Arbor, Asheville, NC, USA, (*http://chestnut.cas.psu.edu/Meetings/NPS/proceedings/_2_09-Powell%20manuscript.pdf*)

Plomion C, O'Malley DM (1996) Recombination rate differences for pollen parents and seed parents in pine. Heredity 77: 341-350

Pritchard JK, Stephens M, Donnelly P (2000) Inference of population structure using multilocus genotype data. Genetics 155: 945-959

Rajoral OP, Mosseler A (2001) Challenges and opportunities for conservation of forest genetic resources. Euphytica 118: 197-212

Remington DL, Plomion C (2000) A genetic map of Maritime pine based on AFLP, RAPD and protein markers. Theor Appl Genet 100: 39-48

Rieseberg LH, Carney SE (1998) Plant hybridization. New Phytol 140: 599-624

Rink G, Carroll ER, Kung FH (1989) Estimation of *Juglans nigra* L. mating system parameters. For Sci 35: 623-627

Ritland K, Jain SK (1981) A model for the estimation of outcrossing rate and gene frequencies using *n* independent loci. Heredity 47: 35-52

Ritter E, Gebhardt C, Salamini F (1990) Estimation of recombination frequencies and construction of linkage maps from crosses between heterozygous parents. Genetics 125: 645-654

Ritter E, Salamini F (1996) The calculation of recombination frequencies in crosses of allogamous plant species with applications to linkage mapping. Genet Res 67: 55-65

Roberts HF (1929) Plant Hybridization before Mendel. Princeton Univ Press, Princeton, NJ, USA

Rottmann WH, Meilan R, Sheppard LA, Brunner AM, Skinner JS, Ma C, Cheng S, Jouanin L, Pilate G, Strauss SH (2000) Diverse effects of overexpression of LEAFY and PTLF, a poplar (*Populus*) homolog of LEAFY/FLORICAULA, in transgenic poplar and *Arabidopsis*. Plant J 22: 235-245

Rudin D, Ekberg I (1978) Linkage studies in *Pinus sylvestris* using macrogametophyte allozymes. Silv Genet 27: 1-12

Ruter B, Hamrick JL, Wood BW (2000) Outcrossing rates and relatedness estimates in pecan (*Carya illinoinensis*) populations. J Hered 91: 72-75

Scalfi M, Troggio M, Piovani P, Leonardi S, Magnaschi G, Vendramin GG, Menozzi P (2004) A RAPD, AFLP and SSR linkage map, and QTL analysis in European beech (*Fagus sylvatica* L.). Theor Appl Genet 108: 433-441

Schena M, Shalon D, Davis RW, Brown PO (1995) Quantitative monitoring of gene expression patterns with a complementary DNA microarray. Science 270: 467-470

Schmidt A, Doudrick RL, Heslop-Harrison JS, Schmidt T (2000) The contribution of short repeats of low sequence complexity to large conifer genomes. Theor Appl Genet 101: 7-14

Schopmeyer CS (tech coord) (1974) Seeds of Woody Plants in the United States. Agri Handb 450. US Dep Agri, For Serv, Washington, DC, USA

Schwarzmann JF, Gerhold HD (1991) Genetic structure and mating system of northern red oak (*Quercus rubra* L.) in Pennsylvania. For Sci 37: 1376-1389

Scotti-Saintagne C, Mariette S, Porth I, Goicoechea PG, Barrenche T, Bodenes C, Burg K, Kremer A (2004) Genome scanning for interspecific differentiation between two closely related oak species (*Quercus robur* L. and *Q. petraea* (Matt.) Liebl.]. Genetics 168: 1615-1626

Sewell MM, Neale DB (1995) Genetic mapping in loblolly pine. In: Proc 23rd Southern For Tree Improv Conf, June 19-22, Ashville, NC, USA

Sewell MM, Sherman BK, Neale DB (1999) A consensus map for loblolly pine (*Pinus taeda* L.). I. Construction and integration of individual linkage maps from two outbred three-generation pedigrees. Genetics 151: 321-330

Smouse PE, Dyer RJ, Westfall RD, Sork VL (2001) Two-generation analysis of pollen flow across a landscape. I. Male gamete heterogeneity among females. Evolution 55: 260-271

Snyder EB (1972) Glossary for Forest Tree Improvement Workers. Southern For Exp Stn, USDA For Serv (*http://www.sfws.auburn.edu/sfnmc/class/glossary.html*)

Southerton SG, Marshall H, Mouradov A, Teasdale RD (1998a) Eucalypt MADS-box genes expressed in developing flowers. Plant Physiol 118: 365-372

Southerton SG, Strauss SH, Olive MR, Harcourt RL, Decroocq V, Zhu XM, Llewellyn DJ, Peacock WJ, Dennis ES (1998b) *Eucalyptus* has a functional equivalent of the *Arabidopsis* floral meristem identity gene LEAFY. Plant Mol Biol 37: 897-910

Stearns SC (2004) The Evolution of Life Histories. Oxford Univ Press, USA

Sterky F, Regan S, Karlsson J, Hertzberg M, Rohde A, Holmberg A, Amini B, Bhalerao R, Larsson M, Villarroel R, van Montagu M, Sandberg G, Olsson O, Teeri TT, Boerjan W, Gustafsson P, Uhlen M, Sundberg B, Lundeberg J (1998) Gene discovery in the wood-forming tissues of poplar: Analysis of 5,692 expressed sequence tags. Proc Natl Acad Sci USA 95: 13330-13335

Stettler RF (1968) Irradiated mentor pollen: Its use in remote hybridization of black cottonwood. Nature 219: 746-747

Storme V, Vanden Broeck A, Ivens B, Halfmaerten D, Van Slycken J, Castiglione S, Grassi F, et al. (2004) Ex-situ conservation of Black poplar in Europe: Genetic diversity in nine gene bank collections and their value for nature development. Theor Appl Genet 108: 969-981

Storz JF (2005) Using genome scans of DNA polymorphism to infer adaptive population divergence. Mol Ecol 14: 671-688

Strauss SH, Lande R, Namkoong G (1992) Limitations of molecular-marker-aided selection in forest tree breeding. Can J For Res 22: 1050-1061

Streiff R, Ducousso A, Lexer C, Steinkelllner H, Gloessl J, Kremer K (1999). Pollen dispersal inferred from paternity analysis in a mixed oak stand of *Quercus robur* L. & *Q. petraea* (Matt.) Liebl. Mol Ecol 8: 831-841

Stubbe H (1965) History of Genetics: From Prehistoric Times to the Rediscovery of Mendel's Laws. Waters TRW (transl). MIT Press, Boston, MA, USA

Talhouk SN, Zurayk R, Khuri S (2001) Conservation of the coniferous forests of Lebanon: past, present and future prospects. Oryx 35: 206-215

Tani N, Takahashi T, Iwata H, Mukai Y, Ujino-Ihara T, Matsumoto A, Yoshimura K, Yoshimaru H, Murai M, Nagasaka K, Tsumura Y (2003) A consensus linkage map for sugi (*Cryptomeria japonica*) from two pedigrees, based on microsatellites and Expressed Sequence Tags. Genetics 165: 1551-1568

Taylor G (2002) *Populus*: *Arabidopsis* for forestry. Do we need a model tree? Ann Bot London 90: 681-689

Taylor G, Beckett KP, Bunn SM, Robinson KM, Rae AM (2001) Identifying QTL for yield in UK biomass poplar. Aspects Appl Biol 65: 173-182

Temesgen B, Brown GR, Harry DE, Kinlaw CS, Sewell MM, Neale DB (2004) Genetic mapping of expressed sequence tag polymorphism (ESTP) markers in loblolly pine (*Pinus taeda* L.). Theor Appl Genet 102: 664-675

Templeton A (1989) The meaning of species and speciation: A genetic perspective. In: Otte D, Endler J (eds) Speciation and its Consequences, Sinauer Assoc, Sunderland, MA, USA, pp 4-27

Tuskan GA, DiFazio S, Jansson S, Bohlmann J, Grigoriev I, Hellsten U, Putnam N, Ralph S, Rombauts S, et al. (2006) The genome of black cottonwood, *Populus trichocarpa* (Torr. & Gray). Science 313: 1596-1604

USDA-FS (2002) Forest Service Manual, FSM 3200, Rural and Urban Forestry Assistance, Chapter 3230, Cooperative Tree Improvement (*http://www.fs.fed.us/im/directives/fsm/3200/3230.doc*)

Van der Beek S, van Arendonk JAM, Groen AF (1995) Power of two and three-generation QTL mapping experiments in an outbred population containing full-sib or half-sib families. Theor Appl Genet 91: 1115-1124

Victory ER, Glaubitz JC, Rhodes OE Jr, Woeste KE (2006) Genetic homogeneity in *Juglans nigra* (Juglandaceae) at nuclear microsatellites. Am J Bot 93: 118-126

Vila C, Savolainen P, Maldonado JE, Amorim IR, Rice JE, Honeycutt RL, Crandall KA, Lundeberg J, Wayne RK (1997) Multiple and ancient origins of the domestic dog. Science 276: 1687-1689

Vischi M, Jurman I, Bianchi G, Morgante M (2003) Karyotype of Norway spruce by multicolor FISH. Theor Appl Genet 107: 591-597

Vitalis R, Dawson K, Boursot R (2001) Interpretation of variation across marker loci as evidence of selection. Genetics 158: 1811-1823

Vogel H, Gibbings M, Smail G, Chaparro J, Teasdale R (1995) Detection of rogue trees in eucalypt mapping populations. In: Plant Genome IV Conf, San Diego, CA, USA

Wang X, Lindgren D, Szmidt AE, Yazdani R (1991) Pollen migration into a seed orchard of *Pinus sylvestris* L. and the methods of its estimation using allozyme markers. Scand J For Res 6: 379-385

Weng C, Kubisiak T L, Nelson CD, Stine M (2002) Mapping quantitative trait loci controlling early growth in a (longleaf pine x slash pine) x slash pine BC1 family. Theor Appl Genet 104: 852-859

White GM, Boshier DH, Powell W (2002) Increased pollen flow counteracts fragmentation in a tropical dry forest: An example from *Swietenia humilis* Zuccarini. Proc Natl Acad Sci USA 99: 2038-2042

Wilcox PL, Amerson HV, Kuhlman EG, Liu BH, O'Malley DM, Sederoff RR (1996) Detection of a major gene for resistance to fusiform rust disease in loblolly pine by genomic mapping. Proc Natl Acad Sci USA 93: 3859-3864

Woeste KE, McKenna JR (2004) Walnut genetic improvement at the start of a new century. In: Michler CH, Pijut PM, Van Sambeek JW, Coggeshall MV, Seifert J, Woeste K, Overton R, Ponder F Jr (eds) Black walnut in a new century. In: Proc 6th Walnut Council Res Symp, July 25-28, Lafayette, IN. Gen Tech Rep NC-243, St Paul, MN, USA: US Dep Agri For Serv, North Central Res Stn, pp 9-17

Wu RL (1996) Quantitative genetic dissection of complex traits in a QTL-mapping pedigree. Theor Appl Genet 93: 447-457

Wu HX (1999) Study of early selection in tree breeding. 2. Advantage of early selection through shortening the breeding cycle. Silv Genet 48: 78-83

Wu RL, Han YF, Hu JJ, Fang JJ, Li L, Li ML, Zeng Z-B (2000) An integrated genetic map of *Populus deltoides* based on amplified fragment length polymorphisms. Theor Appl Genet 100: 1249-1256

Wu RL, Ma CM, Painter I, Zeng ZB (2002) Simultaneous maximum likelihood estimation of linkage and linkage phases in outcrossing populations. Theor Popul Biol 61: 349-363

Yang J, Park S, Kamden DP, Keathley DE, Retzel R, Paule C, Kapur V, Han K-H (2004) Novel gene expression profiles define the metabolic and physiological processes characteristic of wood and its extractive formation in a hardwood tree species, *Robinia pseudoacacia*. Plant Mol Biol 52: 935-956

Yin T-M, DiFazio SP, Gunter LE, Jawdy SS, Boerjan W, Tuskan GA (2004a) Genetic and physical mapping of *Melampsora* rust resistance genes in *Populus* and characterization of linkage disequilibrium and flanking genomic sequence. New Phytol 164: 95-105

Yin TM, DiFazio SP, Gunter LE, Riemenschneider D, Tuskan GA (2004b) Large-scale heterospecific segregation distortion in *Populus* revealed by a dense genetic map. Theor Appl Genet 109: 451-463

Young A, Boshier D, Boyle T (2000) Forest Conservation Genetics: Principles and Practice. CABI Publ, Australia

Ziehe M, Roberds JH (1989) Inbreeding depression due to overdominance in partially self-fertilizing plant populations. Genetics 121: 861-868

10 Molecular Mapping and Breeding in Polyploid Crop Plants

Dan Milbourne[1*], John E. Bradshaw[2] and Christine A. Hackett[3]

[1]Teagasc, Crops Research Centre, Oak Park, Carlow, Ireland

[2]Genetics Programme, Scottish Crop Research Institute, Invergowrie, Dundee DD2 5DA, UK

[3]Biomathematics and Statistics Scotland, SCRI, Invergowrie, Dundee DD2 5DA, UK

*Corresponding author: Dan.Milbourne@teagasc.ie

1 INTRODUCTION

1.1 The Evolutionary Significance of Polyploidy

Polyploidy is an incredibly important feature in the evolution of plant species. The prevalence of polyploidy in plants has been recognized for over 50 years (Stebbins 1950), and studies based on comparative analysis of chromosome numbers, and other methods, gave rise to estimations that between 70% and 80% of angiosperm species have experienced a polyploidization event in their ancestry (Grant 1963, 1981; Goldblatt 1980; Masterson 1994). Over the last 20 years, our ability to examine plant genomes in detail has progressed rapidly. Dense genetic linkage maps, extensive collections of expressed sequence tags (ESTs), and even the complete genome sequences of *Arabidopsis* and rice are now available. Analysis of these types of whole-genome data have brought about a revelation in our understanding of the prevalence, and perhaps more importantly, the frequency of polyploidization events in the evolution of plants, leading to the conclusion that polyploidy has occurred (sometimes repeatedly) in the lineages of all plant species.

Arabidopsis thaliana represents a striking example of the above point. With one of the smallest genome sizes in the plant kingdom (~150 Mb), *Arabidopsis* was once considered to be an excellent illustration of a diploid plant species. Surprisingly, however, analysis of the entire genome sequence of *Arabidopsis* has revealed extensive duplication of large blocks of genes on a scale that can only be accounted for by not just one, but possibly three rounds of whole-genome duplication during its evolution. These duplication events seem to have been followed by rapid and extensive loss of many duplicated genes, followed by a slower process of divergent evolution for the remaining duplicated segments, resulting in an eventual stabilization and a return to the diploid status. The three diploid *Brassica* species, *B. nigra* ($n = 8$), *B. oleracea* ($n = 9$), and *B. rapa* ($n = 10$), have genomes containing triplicated copies of an ancestral genome which is structurally similar to the genome of *Arabidopsis* (Lagercrantz and Lydiate 1996; Lagercrantz 1998). Furthermore, these diploid species are the progenitors of allotetraploids that display disomic inheritance, e.g., *B. napus* ($n = 19$) is the allotetraploid of *B. oleracea* and *B. rapa*. The process of repeated rounds of polyploidization, followed by rapid gene loss, slower diversification, and a subsequent return to diploid status has now been illustrated for many diploid plant species, notably soybean (Shoemaker et al. 2006) and in grasses (Paterson et al. 2004), and polyploidy is now recognized as a major force in the evolution of plants.

While many modern diploids have a polyploid origin, there are many other species which have retained their polyploid status. Differences in events occurring during meiosis between these and diploid species have ramifications for linkage analysis and trait mapping. This chapter intends to provide the reader with a basic introduction to some of the main features and problems that are likely to be encountered in linkage and QTL (quantitative trait loci) analysis in various types of polyploidy. It is written mainly for a readership with a strong background in molecular biology, and a basic understanding of Mendelian genetics, and deals with the principles and some of the practicalities involved in linkage and QTL analysis. Readers interested in gaining a deeper understanding of the statistical theory underlying the subject are directed to the many references cited in this chapter. The implications for plant breeding are also described.

1.2 Types of Polyploidy

Many examples of polyploidy can be found among crop plants (Smartt and Simmonds 1995), but most of them are allopolyploids which display disomic inheritance, e.g., wheat, cotton, and rapeseed. Examples of autopolyploids which display polysomic inheritance are rarer, with Gallais (2003) listing just 12 species. The two best known autotetraploids are the principal cultivated potato (*Solanum tuberosum*; $2n = 4x = 48$) and lucerne, which is also called alfalfa (*Medicago sativa*; $2n = 4x = 32$). Both of these species have relatively small chromosomes. As autopolyploids often have reduced fertility as a result of the production of genetically unbalanced gametes, it is not surprising that they can usually also reproduce vegetatively.

For the purposes of this chapter, we will differentiate between the two types of polyploidy mainly on the basis of pairing behavior of the chromosomes during meiosis and the resulting genetic segregation ratios. However, it is also useful to think of the manner in which the two modes of polyploidy have come about as this gives an insight into the pairing behavior. Thus, allopolyploidy can be considered to have effectively arisen from the merger of two or more chromosome sets that are sufficiently structurally different to pair independently during meiosis. The result is a functionally diploid plant. Although sufficiently structurally different to prevent inter-genome pairing, the progenitor genomes in allopolyploids are frequently similar in terms of basic chromosome number, and gene order and content, resulting in homoeolog sets that need to be distinguished from each other, preferably on the basis of their ancestral origin. For example, in allohexaploid bread wheat *Triticum aestivum* ($2n = 6x = 42$, AABBDD) there are three sets of seven pairs of chromosomes, referred to as the A, B, and D genomes, reflecting their proposed ancestral origins. It is worth noting that there may also be independent genetic mechanisms, such as the *Ph* loci in wheat, which are responsible for maintaining the allopolyploid behavior of these species (reviewed by Jenczewski and Alix 2004). In contrast, autopolyploidy can be viewed as effectively having arisen from the doubling of a single genome (for example, diploid AA to autotetraploid AAAA using the above nomenclature) giving rise to more than two copies of each homologous chromosome. Since all homologs are structurally similar, they are all equally capable of pairing with each other, and this pairing will occur at random during meiosis.

Allopolyploidy and autopolyploidy, as defined above, actually represent two extremes of behavior in terms of chromosome pairing as in some cases the situation is less clear cut. For example, sweet potato (*Ipomoea batatas*; $2n = 6x = 90$) is a hexaploid but there are doubts over whether it is an allohexaploid or an autohexaploid (Smartt and Simmonds 1995). The AFLP data of Kriegner et al. (2003) support predominance of polysomic inheritance with some degree of preferential pairing. Sugarcane (*Saccharum officinarum*; $2n = 8x = 80$) and strawberry (*Fragaria ananassa*; $2n = 8x = 56$) are octoploids, but again the extent of polysomic inheritance is not clear. Modern sugarcane cultivars are derivatives from the high-sucrose domesticated species *S. officinarum* and the hardier but lower-sugar content wild species *S. spontaneum* ($2n = 8x = 64$, but $2n$ can be 40-128) and have between 100 and 130 chromosomes (Aitken et al. 2005). In strawberry, there is evidence for mixed disomic and polysomic inheritance (Lerceteau-Kohler et al. 2003). This scenario, in which there is a mixture of autopolyploid and allopolyploid pairing behavior, may be referred to as segmental allopolyploidy.

For the purposes of subsequent discussions in this chapter, it is necessary that a basic understanding of the fundamental differences between the meiotic events that occur in autopolyploids and diploids are developed (for the moment it will be assumed that allopolyploids are functionally equivalent to diploids), and the effect that these events have on the inheritance patterns at marker loci. As autotetraploidy represents the simplest form of autopolyploidy, this scenario will be used to illustrate the points that need to be made.

2 MEIOSIS IN AUTOTETRAPLOIDS

2.1 Pairing Behavior

As mentioned previously, meiosis in allopolyploids proceeds in a functionally equivalent manner to that in diploid species. However, in autopolyploids, the presence of at least four (in autotetraploids) homologous chromosomes leads to diverse pairing scenarios amongst these homologs during prophase I of meiosis, as described in classical cytogenetics books (Lewis and John 1963).

In autotetraploids, as in diploids, DNA synthesis takes place before the start of the first division of meiosis so that each chromosome has doubled so far as DNA content is concerned. Hence each homologous chromosome comprises two sister chromatids which are joined at the centromere, although these individual chromatids are not initially visible.

The first meiotic division begins with competition between homologous chromosomes as they try to form pairs (synapsis), followed by chiasma formation and crossing over (recombination). As four chromosomes are involved, there are a number of possible outcomes depending upon whether or not there are changes of pairing partner along the lengths of the chromosomes. The simplest situation in autotetraploids is where four homologous chromosomes pair at random to give two bivalents, and the two cellular divisions of meiosis both proceed normally and produce equally viable diploid gametes, and male and female gametes unite at random to form tetraploid offspring. However, more complex pairing patterns may also occur, resulting in quadrivalents, trivalents, bivalents and univalents, or various mixtures thereof. Hence quadrivalent formation is evidence for autotetraploidy, but its absence does not prove disomic inheritance. Quadrivalents can be a chain (CIV) or a ring (RIV), with three chiasmata sufficient for a chain and four for a ring (Fig. 1). Five or more chiasmata would give a complex pattern but may not occur in practice, particularly with short chromosomes. Quadrivalent orientations on the spindle which result in 2 + 2 disjunction (i.e. the migration of two chromosomes to each pole) produce balanced diploid chromosome sets. As previously mentioned, formation of two bivalents (2II's) provides the simplest situation for tetrasomic inheritance. Where a trivalent and univalent (CIII + I) occur, the trivalent will usually disjoin two and one, and the univalent to one or another pole at random, or be left behind and get lost. The simplest trivalent has two chiasmata and forms a chain, but more complex arrangements are possible, particularly if there are three or more chiasmata. One bivalent and two univalents (1II + 2I's) and four univalents (4I's) are the remaining possibilities.

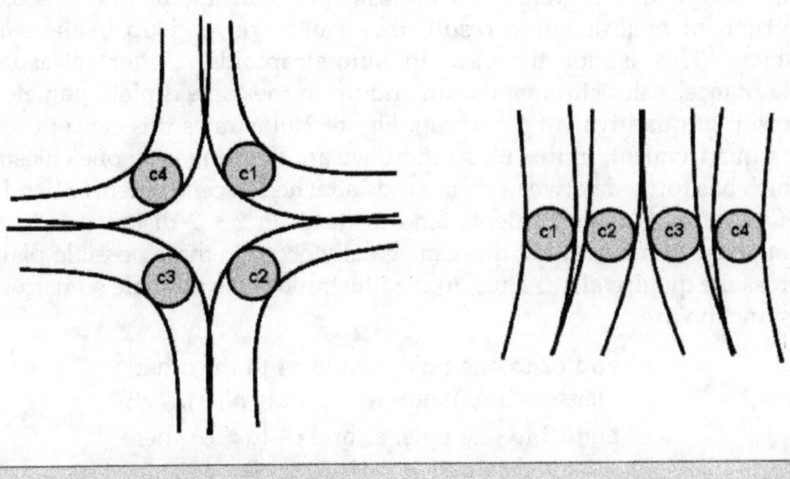

Fig. 1 Examples of possible configurations for a ring quadrivalent (left) and a chain quadrivalent (right) in an autopolyploid.

The second division of meiosis is straightforward in that it resembles two synchronized mitotic divisions, in which the centromeres lie on the spindle equator and then split into daughter centromeres which disjoin to opposite spindle poles. In this case, one just has to consider the two orientations for the disjunction of daughter centromeres. Overall, some of the possible outcomes of meiosis are more likely than others to lead to balanced diploid chromosome sets and hence viable and fertile diploid gametes.

In practice a mixture of these outcomes could occur. As genetic linkage mapping requires that at least some fixed assumptions can be made regarding chromosome pairing behavior during meiosis, it could be argued that it is not possible to develop a general theory of linkage analysis in autotetraploids. In practice, this depends largely on the species in question. In potato for instance, it has been observed that while low frequencies of quadrivalents, trivalents, and univalents occur, bivalents predominate (Swaminathan and Howard 1953). Furthermore, in alfalfa, Bingham and McCoy (1988) found that most cells have the full complement of 16 bivalents at metaphase I. Hence for these species good progress has been made by applying the theory that assumes that four homologous chromosomes pair at random to give two bivalents.

2.2 Double Reduction

One of the fundamental aspects of meiosis in diploids is the concept that sister chromatids migrate to the opposite poles during the first or second division of meiosis. As a result, they can never end up in the same gamete. This is not the case in autotetraploids. Under tetrasomic inheritance, sister chromatids can end up in the same diploid gamete as a result of quadrivalent formation. Figure 2 illustrates this concept for a ring quadrivalent. In this illustration, we are showing only one chiasma, which has formed between chromatids attached to centromeres c1 and c2 before locus A. Quadrivalents tend to result in 2 + 2 disjunction in the first phase of meiosis, but this can actually occur in three possible planes across the quadrivalent. Thus, in the illustration, the possible scenarios at disjunction are:

c1 and c2 to one pole, c3 and c4 to the other
(sister chromatids with A, together),
c1 and c3 to one pole, c2 and c4 to the other
(sister chromatids with A, separated),
c1 and c4 to one pole, c2 and c3 to the other
(sister chromatids with A, separated).

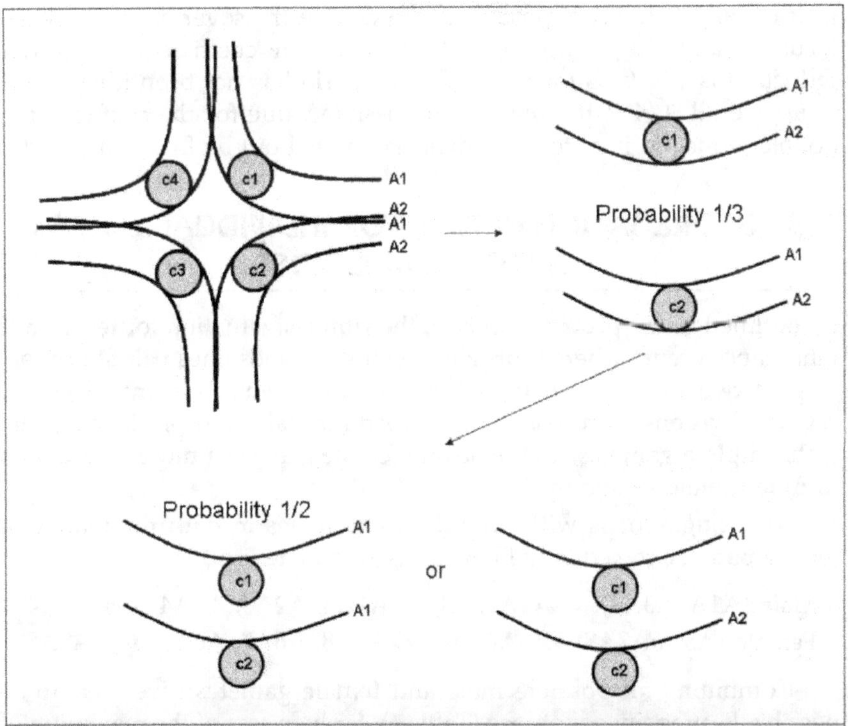

Fig. 2 Double reduction occurring at a ring quadrivalent. Homologous chromosomes c1 and c2 have homologous loci A1 and A2, respectively. A crossover between the centromeres of c1/c2 and locus A is shown. The first arrow indicates migration of c1 and c2 to the same pole at disjunction in the first division of meiosis. The second arrow indicates the two possible scenarios for daughter cells after the second division of meiosis. The scenario on the left indicates the occurrence of double reduction, where sister chromatids A1 end up in the same gamete.

As the orientation of the quadrivalent is random, all three scenarios are equally likely, but the first one results in sister chromatids with A alleles going to the same spindle pole. The probability of this occurring is 1/3. During the second division of meiosis, the probability that the sister chromatids that have disjoined together will end up in the same gamete is 1/2. The overall probability of the sister chromatids ending up in the same gamete, given quadrivalent formation and a chiasma, is the product of the probabilities during both phases of meiosis, namely 1/3 x 1/2 = 1/6. The probability of sister chromatids in the same gamete is referred to as the coefficient of double reduction (α).

In the example given here, the maximum value for the coefficient of double reduction is 1/6. However, in cases where the centromere occurs near the end of the chromosome and a double crossover occurs between locus A and the centromere, the maximum coefficient of double reduction is 1/4 (this more complex scenario has not been illustrated). Thus overall, 1/4 is the maximum possible value for the coefficient of double reduction in autotetraploids, as pointed out by Luo et al. (2006).

3 SEGREGATION RATIOS FOR INDIVIDUAL LOCI IN AUTOTETRAPLOIDS

As outlined in the previous section, the simplest situation for tetrasomic inheritance occurs where four homologous chromosomes pair at random to give two bivalents in both the male and female parents, the two cellular divisions of meiosis both proceed normally and produce equally viable diploid gametes, and male and female gametes unite at random to form tetraploid offspring.

At a single locus with four different alleles in both the male and female parents, the equally likely diploid gametes will be:

Male (A1A2A3A4): A1A2, A1A3, A1A4, A2A3, A2A4, and A3A4.
Female (A5A6A7A8): A5A6, A5A7, A5A8, A6A7, A6A8, and A7A8.

Combining all possible male and female gametes gives rise to 36 possible F_1 progeny genotypes (Table 1), far in excess of the maximum of four possible progeny genotype classes arising from the analogous situation in an outbreeding diploid (A1A2 x A3A4 = 4 possible F_1 progeny genotypes, A1A3, A1A4, A2A3, and A2A4).

Table 1 Progeny genotypes for a tetraploid cross A1A2A3A4 x A5A6A7A8.

	A1A2	*A1A3*	*A1A4*	*A2A3*	*A2A4*	*A3A4*
A5A6	A1A2 A5A6	A1A3 A5A6	A1A4 A5A6	A2A3 A5A6	A2A4 A5A6	A3A4 A5A6
A5A7	A1A2 A5A7	A1A3 A5A7	A1A4 A5A7	A2A3 A5A7	A2A4 A5A7	A3A4 A5A7
A5A8	A1A2 A5A8	A1A3 A5A8	A1A4 A5A8	A2A3 A5A8	A2A4 A5A8	A3A4 A5A8
A6A7	A1A2 A6A7	A1A3 A6A7	A1A4 A6A7	A2A3 A6A7	A2A4 A6A7	A3A4 A6A7
A6A8	A1A2 A6A8	A1A3 A6A8	A1A4 A6A8	A2A3 A6A8	A2A4 A6A8	A3A4 A6A8
A7A8	A1A2 A7A8	A1A3 A7A8	A1A4 A7A8	A2A3 A7A8	A2A4 A7A8	A3A4 A7A8

In reality, this level of diversity is not usually experienced for markers or genes at a single locus in most autotetraploid plant species. In addition, one is frequently dealing with dominant markers or traits. The expected segregation ratios for dominant alleles in the F_1 of an autotetraploid cross between genotypes with varying 'doses' of a single dominant allele are shown in Table 2 (these will become useful in some

Table 2 Segregation ratios for dominant markers in an autotetraploid.

Parent 1	Parent 2	Offspring configurations	
		Marker genotype	*Marker Phenotype A dominant*
Aaaa	aaaa	Aaaa:aaaa 1 : 1	A—:aaaa 1 : 1
AAaa	aaaa	AAaa:Aaaa:aaaa 1 : 4 : 1	A—:aaaa 5 : 1
AAAa	aaaa	AAaa:Aaaa 1 : 1	A— 1
Aaaa	Aaaa	AAaa:Aaaa:aaaa 1 : 2 : 1	A—:aaaa 3 : 1
AAaa	Aaaa	AAAa:AAaa:Aaaa:aaaa 1 : 5 : 5 : 1	A—:aaaa 11 : 1
AAaa	AAaa	AAAA:AAAa:AAaa:Aaaa:aaaa 1 : 8 : 18 : 8 : 1	A—:aaaa 35 : 1

of the subsequent discussions). It is important to remember that if one is referring to a dominant marker, such as an AFLP, it is not always possible to know what the marker genotype is by looking at the marker phenotype. For example, in a system where 'A' represents the presence of a marker band, and 'a' the absence, genotype aaaa will always result in band absence, but Aaaa, AAaa, and AAAa are all simply indicated by band presence, and so the marker genotype/phenotype is best referred to as A–. In practical terms it may be possible to determine the dosage state of the dominant allele by band intensity (or peak height in fluorescent systems) but this is not an entirely reliable guide to dosage. It is the 5:1, 11:1, and 35:1 ratios that are diagnostic of tetrasomic inheritance.

[*Aside - During the course of our subsequent discussions, we will have to refer to the marker phenotype quite regularly – for convenience, we may also use the phrase 'gel banding pattern'. Of course, this is analogous to the chromatogram output of many fluorescent based systems*]

There is a generally accepted nomenclature for describing the dosage states of the dominant allele of the marker genotypes illustrated above. Aaaa is referred to as simplex (or single dose), AAaa as duplex (or double dose), AAAa as triplex (or triple dose), aaaa as nulliplex, and AAAA as quadruplex (or quadruple dose). Frequently, this nomenclature is extended to describe the segregation arising from crosses involving markers in various dosage states, most frequently when one of the parents possesses the nulliplex configuration.

For example:

Aaaa x aaaa is referred to as simplex, implying 1:1 segregation

AAaa x aaaa is referred to as duplex, implying 5:1 segregation and

Aaaa x Aaaa is referred to as double simplex, implying a 3:1 ratio.

These examples are the most common, and the most useful in linkage mapping.

3.1 Distorted Segregation Ratios

Deviation from the expected segregation ratio for a marker is a well known phenomenon that occurs in diploid species. Autotetraploids are also subject to this phenomenon, although in this case the potential causes are more numerous:

1. In diploids, meiosis produces haploid gametes. In outbreeding species, the outcome can be the 'unmasking' of recessive alleles which have a negative impact on the fertility and the viability of the resulting gametes and zygotes. These are likely to be due to specific loci but will cause distortion at linked loci, although this will not have a large effect on the estimate of recombination frequency if recombinant chromosomes are 'distorted' in the same ratio as non-recombinant ones. Autotetraploids are subject to the same phenomenon, but, in theory, the fact that autotetraploid gametes are actually diploid should reduce the incidence of unmasking of recessive alleles, and as a consequence, reduce segregation distortion from this source. Nevertheless, autopolyploids are usually outbreeding species that display inbreeding depression and some segregation distortion is to be expected.
2. Double reduction is another cause of distorted segregation ratios because it disturbs the expected gametic output of the plant.
3. Chromosomes may not pair at random due to lack of complete homology. In the extreme case, the pairing amongst the four homologs results in the same pairs always being formed and disomic inheritance.

For example:

A1A2A3A4 gives gametes A1A2, A1A3, A1A4, A2A3, A2A4, and A3A4 (tetrasomic), whereas

A*1A*2A3A4 gives only gametes A*1A3, A*1A4, A*2A3, and A*2A4 (disomic).

If disomic inheritance is restricted to individual chromosomes (or indeed chromosome arms) then the result will be a mixture of inheritance patterns, indicating segmental allopolyploidy.

Care is required in the interpretation of segregation ratios. For example, in alfalfa, Julier et al. (2003) found segregation distortion at SSR loci that could be explained by double reduction, but all of the observed genotypes could have been formed without double reduction. Hence other causes of segregation distortion could not be ruled out.

3.2 Distinguishing Segregation Ratios and Predicting Parental Genotypes in Autotetraploids

The population size required for a mapping experiment is one of the first important practical decisions to be made on initiating a mapping study. The choice is frequently a trade-off between the practicalities associated with generating, maintaining, and phenotyping/genotyping a large population and the desire to have large numbers of individuals to increase the power to detect genuine linkage and QTLs and the precision of estimates. The lower limits of mapping population size for experiments involving polyploids, however, are governed by several requirements, the first of which involves the ability to successfully distinguish segregation ratios.

We can illustrate the point using a supposed autotetraploid. As disomic inheritance also gives a 1:1 ratio of Aa to aa in progeny derived from a cross Aa × aa, and a 3:1 ratio of A- to aa from Aa × Aa, it is important to be able to distinguish a 5:1 ratio from a 1:1 ratio, both to confirm tetrasomic inheritance and to confirm the parental genotype as AAaa rather than Aaaa. Likewise, one may want to distinguish a 3:1 ratio from an 11:1 ratio so that one knows both parents are Aaaa. But what is the population size required to be sure of doing this? Let us consider distinguishing a 1:1 ratio from a 5:1 ratio and a 3:1 ratio from an 11:1 ratio. The general formula for doing this can be found in Bailey (1961).

With a population size of 24, a 1:1 ratio would be 12:12 and a 5:1 ratio would be 20:4. An observed ratio of 16:8 would not be a statistically significant departure from a 1:1 ratio ($P > 0.05$ on chi-square test), but it would be a statistically significant departure from a 5:1 ratio. In contrast, the reverse would be true for an observed ratio of 17:7. Hence an observed ratio of 16:8 would be declared 1:1 and an observed ratio of 17:7 would be declared 5:1.

In contrast, such a small population size would be inadequate for distinguishing a 3:1 ratio from an 11:1 ratio (22:2 is not a statistically significant departure from 18:6). A population size of 72 would be adequate in the absence of double reduction as a ratio of 62:10 would be declared 11:1 and 61:11 would be declared 3:1, at the 5% level of significance. However, larger population sizes are required in the presence of double reduction and we shall see later that considerably larger population sizes are required for linkage analysis in autotetraploids.

4 THE IMPORTANCE OF LINKAGE PHASE AND MARKER TYPE IN AUTOTETRAPLOIDS

Until now we have been examining the events occurring at a single locus during tetrasomic inheritance. However, linkage analysis is based on observing the combined segregation patterns at two or more loci on the same chromosome. The first step of the linkage analysis is to test whether pairs of loci are segregating independently. This can be done for each pair of markers by representing their joint segregation in a two-way contingency table and testing for independent segregation. The observed and expected numbers may be compared by Pearson's chi-square statistic as described by Luo et al. (2001). Cluster analysis is then a suitable technique to partition the marker loci into linkage groups, so that a marker segregates independently of markers in different linkage groups and shows a significant association with at least some of the other markers within its linkage group. The basic idea of cluster analysis is simple. A similarity matrix is first calculated between all pairs of markers, based on the significance of the test of independence. The analysis then proceeds by a series of successive fusions of markers into groups, culminating at the stage where all markers are in one group. Luo et al. (2001) discuss different clustering methods and recommend comparing the dendrogram from single-linkage (also called nearest-neighbor) cluster analysis with that from average-linkage cluster analysis. The former can combine large groups on the strength of one marker from each subgroup, whereas the latter may not detect a genuine but loose linkage. The criterion for partitioning the dendrogram into different linkage groups can be determined as the distance measure by which significant linkage is inferred. However, the expected number of linkage groups is the haploid chromosome number of the species and this can be regarded as the ideal number of groups, though it is important not to force the markers into this number of groups if the data do not support this.

In a diploid, where there are two homologous copies of each chromosome, linked alleles (of different loci) may be on the same homologous chromosomes or opposite ones. This relationship is referred to as the linkage phase of the alleles. Alleles on the same homologous chromosome are in coupling phase, while those on different homologous chromosomes are in repulsion phase. For the purpose of the subsequent discussion, it is important to appreciate the difference in complexity between phase relationships in diploids and autotetraploids, and the importance of the marker type (dominant vs. co-dominant) in the context of this relationship. Figure 3 below illustrates five dominant markers (A/a-E/e) and a co-dominant marker (F) on four homologous chromosomes in a tetraploid. Allele A is linked in coupling phase with allele B, and in repulsion phase with alleles C and D. However, alleles C and D are also in repulsion phase with each other, something that would not be possible in a diploid, where there are only two possible homologs

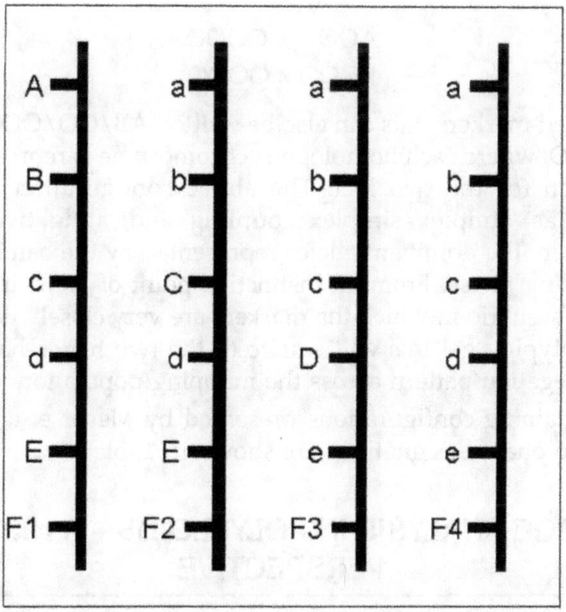

Fig. 3 Four homologous chromosomes in an autotetraploid with marker loci A-F.

for markers to occupy. Further complexity is added by the possibility of multiple dosage states, for example, marker E is duplex (EEee). Alleles A and B are in coupling phase with one E allele, allele C is in coupling phase with the other allele E, while allele D is in repulsion phase to both copies of allele E. Marker F is a co-dominant marker with four observable alleles, and individual alleles of F are in either coupling or repulsion phase with each of the dominant markers.

The importance of linkage phase relationships between markers in a tetraploid (which is discussed further below) requires a clear terminology to describe both dosage and phase of pairs of linked markers. Meyer et al. (1998) presented a system for dominant markers that has been widely adopted, and we will use this as the basis of further discussion. Consider marker loci A and B in Fig. 3. For a moment we will assume that both loci are involved in a cross in which the other parent is nulliplex at both loci:

Aaaa x aaaa
Bbbb x bbbb

Meyer et al. (1998) use a slightly different nomenclature, in which band presence is indicated by A and B, and band absence by O (rather than a or b):

AOOO x OOOO
BOOO x OOOO

For linked markers, this can also be written AB/OO/OO/OO x OO/OO/OO/OO, where each homologous chromosome is represented by its configuration for the two loci. The above configuration can also be referred to as 'simplex/simplex coupling' indicating two dominant markers where the dominant alleles represented by the bands on the gel are in coupling phase. From an instinctive point of view, it is useful to think of the scenario in which the markers are very closely linked so that on the genotyping gel this will equate to the two bands having a very similar segregation pattern across the mapping population.

The remaining configurations presented by Meyer et al. (1998) and their linkage phase designations are shown in Table 3.

5 LINKAGE ANALYSIS IN POLYPLOIDS – A HISTORICAL PERSPECTIVE

Several approaches to performing linkage analysis in polyploids were devised during the 1990s as various kinds of molecular markers became

Table 3 Marker pair configurations in an autotetraploid.

Allele dosage	*OOOO*	*AOOO*	*AAOO*	*AAAO*	*AAAA*
Designation	*Nulliplex*	*Simplex*	*Duplex*	*Triplex*	*Quadruplex*
Marker pair configuration			*Linkage phase designation*		
AB/OO/OO/OO × OO/OO/OO/OO			Simplex/simplex coupling SSC		
AO/OB/OO/OO × OO/OO/OO/OO			Simplex/simplex repulsion SSR		
AB/AO/OO/OO × OO/OO/OO/OO			Simplex/duplex coupling DSC		
AO/AO/OB/OO × OO/OO/OO/OO			Simplex/duplex repulsion DSR		
AB/AB/OO/OO × OO/OO/OO/OO			Duplex/duplex coupling DDC		
AB/AO/OB/OO × OO/OO/OO/OO			Duplex/duplex mixed DDM		
AO/AO/OB/OB × OO/OO/OO/OO			Duplex/duplex repulsion DDR		
AB/OO/OO/OO × AO/OO/OO/OO			Simplex/double-simplex coupling		
AO/OB/OO/OO × AO/OO/OO/OO			Simplex/double-simplex repulsion		
AB/OO/OO/OO × AB/OO/OO/OO			Double-simplex/double-simplex coupling		
AB/OO/OO/OO × AO/OB/OO/OO			Double-simplex/double-simplex mixed		
AO/OB/OO/OO × AO/OB/OO/OO			Double-simplex/double-simplex repulsion		

available. The simplest of these was proposed by Wu et al. (1992) and involved using markers in the simplex/simplex coupling configuration only. As outlined previously, simplex markers in an autopolyploid are equivalent to heterozygous markers (Aa x aa) exhibiting a 1:1 segregation ratio in diploids. The methodology for calculating recombination frequencies between heterozygous (or simplex) markers linked in coupling phase in diploids (Ritter et al. 1990) can also be applied to polyploids, regardless of their genetic constitution. The major advantage of using this approach at the time was that it allowed the use of widely available mapping software designed for diploids for mapping in polyploids. The technique does not extend to repulsion phase linkages amongst simplex markers in polyploids. As outlined above, these are more complex, and not analogous to the diploid scenario. This strategy was employed by Al-Janabi et al. (1993, 1994) to construct linkage maps using AP-PCR based markers in wild *Saccharum* (sugarcane), a polyploid with a complex genetic constitution which is thought to exhibit a combination of allopolyploid and autopolyploid pairing behavior. The disadvantage to this approach is that using only linkages in coupling, each resulting co-segregation group represents only one homolog of N homologous chromosomes (N = ploidy level). For instance, applying this technique to potato should, in theory, result in 48 (4 x 12) linkage groups per parent. In a diploid, coupling and repulsion linkages between simplex markers are estimated with equal precision and maps of the homologous chromosomes can generally be merged using markers linked in repulsion phase to produce maps with the same number of linkage groups as the basic chromosome number for the species (e.g., 12 linkage groups in potato). However, repulsion linkages between simplex markers in polyploids are harder to detect than simplex coupling linkages, and have larger standard errors, which hamper their use in the merging of homologous chromosomes into homology groups.

Two complementary approaches were originally devised to overcome this problem. One method involves the use of markers present in 'double dose' (duplex markers) in one parent and absent in the other (AAaa x aaaa), which should ideally be linked to two homologous groups in coupling, and two in repulsion. This strategy was originally adopted by Yu and Pauls (1993) to construct a very limited linkage map in tetraploid alfalfa using RAPD markers. It was also used in sweet potato by Ukoskit and Thompson (1997) and Kriegner et al. (2003). While it allows a certain amount of merging of homologous linkage groups, this strategy has one major shortfall, dominant markers such as RAPDs (and to a lesser extent AFLPs) are generally anonymous in nature, and may not allow comparison of maps produced in this manner to reference maps upon which the nomenclature of the linkage groups for that species is based. Thus, one can end up with a map comprising many unidentified, unorientated linkage groups. Nevertheless, in potato, if the parents SH and RH of a diploid ultra high density (UHD) reference population (*http://www.dpw.wageningen-ur.nl/uhd*) are included in the marker assays, one can identify AFLP alleles that co-migrate with ones in the UHD population, and hence confirm chromosomal identification of linkage groups as proposed by Rouppe van der Voort et al. (1997). They recommended as a rule of thumb that at least three putatively homologous markers per linkage group should be used for chromosome identification because they found 11% of co-migrating markers were not homologous.

The second method involves the use of co-dominant, multi-allelic markers revealing several bands segregating in the expected ratio for simplex markers. This would be the case for marker F in Fig. 3 above; each of the alleles F_1 - F_4 is represented by a band of different molecular weight, each acting in the same manner as a single dominant marker. Bands from the same assay are assumed to be allelic, and this acts as an anchor-point to allow the alignment of linkage groups representing homologous chromosomes. This strategy has the added advantage that co-dominant markers are much more portable between maps, and if one has used markers that have previously been placed on a reference map, it is possible to assign chromosomal identity and orientation to some or all of the linkage groups. This approach was originally deployed in mapping both wild (Da Silva et al. 1993) and cultivated sugarcane (Grivet et al. 1995) using RFLPs and was reviewed by Da Silva and Sobral (1996). Thus, if a single RFLP probe revealed two bands with different sizes, each segregating 1:1 for presence:absence in a progeny population, it was assumed that two homologous chromosomes had been detected. However, this approach will include duplicate loci on non-pairing chromosomes as well as true-pairing partners. More recently Aitken et al. (2005) used a combination of AFLP and SSR markers to provide extensive map coverage and identification of homo(eo)logous linkage groups in a sugarcane cultivar. They succeeded in forming 127 single-dose linkage groups into eight homo(eo)logy groups.

Brouwer and Osborn (1999) constructed a molecular marker linkage map of tetraploid alfalfa from 82 segregating single-dose RFLP restriction fragments in two backcross populations of 101 individuals. Four homologous coupling-phase co-segregation groups were detected for seven of the eight linkage groups of diploid alfalfa and aligned using probes in common. No co-segregation groups were found for linkage group 7 due to the lack of appropriate polymorphisms. More recent linkage maps of alfalfa (Julier et al. 2003; Robins et al. 2007a) did achieve eight groups of four homologous linkage groups for each parent in F_1 mapping populations of sizes 168 and 200, respectively. Julier et al. (2003) used a combination of SSR and AFLP markers and Robins et al. (2007a) a combination of RFLP and SSR markers. These papers demonstrated the benefits of using co-dominant markers for mapping, possibly helped by the apparent lack of double reduction in alfalfa. These mapping studies were also aided by advances in theory and computer software (Hackett and Luo 2003) which are explained in later sections.

6 LINKAGE MAPPING IN AUTOTETRAPLOIDS

In the construction of the first genetic linkage map of autotetraploid potato, Meyer et al. (1998) used AFLP markers in an F_1 population of 78 individuals derived from a cross between the cultivar Stirling and the advanced SCRI breeding clone 12601ab1. Markers exhibiting segregation distortion (approximately 17% of the 500 or so available) could not be used, since it was not possible to predict their probable configuration. Parallel theoretical and simulation studies by Hackett et al. (1998) showed that simplex-simplex coupling linkages are the most informative among dominant markers. Simplex-duplex pairs of markers are equally informative in coupling or repulsion phase in autotetraploids, and estimates of recombination frequency, though less informative than simplex-simplex coupling, are more precise than for simplex-simplex repulsion. Duplex-duplex pairs were only informative in coupling phase. Populations of 100 or less had particularly high standard errors associated with the estimates of recombination frequencies between markers in the least informative configurations (Fig. 4). These simulation studies demonstrated that a population size of at least 150, and preferably 250, should be used to identify homologous chromosomes using a combination of single and double dose markers. Recent mapping studies in alfalfa (Julier et al. 2003; Robins et al. 2007a), potato (Bradshaw et al. 2008), and sugarcane (Aitken et al. 2005) have used population sizes in the range 168 to 227.

Meyer et al. (1998) extended the calculation of recombination frequencies between marker pairs to those involving double simplex

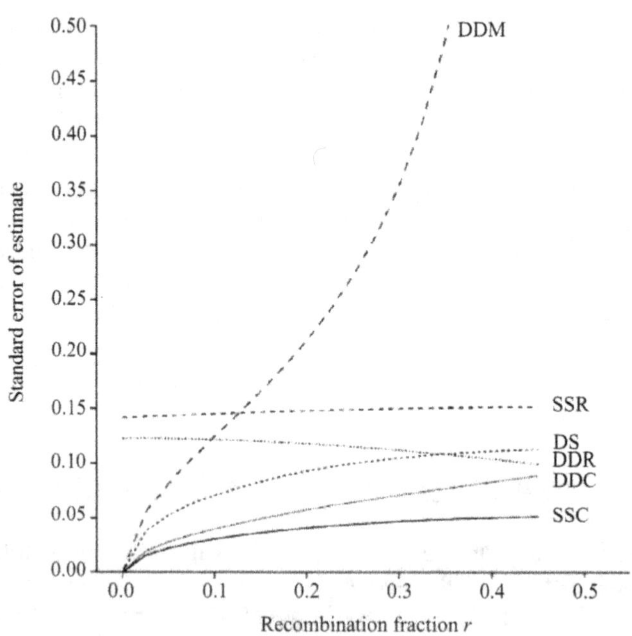

Fig. 4 Distribution of standard errors for the estimate of the recombination fraction in an autotetraploid, for various linkage phases, in a population size of 100. The abbreviations for marker configurations can be found in Table 3 reproduced from Hackett et al. (1998) with kind permission of Cambridge University Press.

markers. Dominant markers present in both parents were found to be extremely uninformative about recombination unless the segregation ratio is 3:1. Markers for which the most likely segregation ratio is 11:1 or 35:1 should not be used in linkage analysis. As with the previous examples, published formulae (Wu et al. 1992; Hackett et al. 1998; Meyer et al. 1998) were used to calculate pairwise recombination frequencies and LOD scores between all pairs of markers for the most likely phase. These were subsequently analyzed using a standard genetic mapping program to produce genetic maps. Figure 5 shows the largest single linkage group assembled during this study, and demonstrates the use of duplex markers to assemble homologous co-segregation groups (identified by simplex markers in coupling) into a single linkage group by exploiting the equally informative coupling and repulsion linkages between duplex and simplex markers. The limitations of the population size meant that, despite the availability of over 500 segregating AFLP markers for the population, the resulting maps covered only 12% and 25% of the paternal and maternal genomes, respectively.

Nevertheless, in a subsequent study of the 78 individuals from the Stirling by 12601ab1 cross (Bradshaw et al. 1998), use of SSR markers, which had been mapped in a diploid population, allowed the identification of a linkage group which was found to harbor a QTL for resistance to the white potato cyst nematode (*Globodera pallida*). However, SSRs were not as useful as expected in identifying homologous groups within each parent. Theoretically, if an SSR had a different simplex allele on each homologous chromosome of both parents as described above for marker F (Fig. 3), all homologous chromosomes could be identified. However, in practice, the number of alleles detected per SSR locus in the study was quite low, usually two to four alleles in a pair of parents with one to three present in each parent. This complicates the determination of the parental marker genotype because gel banding pattern may be compatible with more than one marker genotype. The impact on linkage mapping is significant since map construction relies on distinguishing parental and recombinant genotypes and thus knowledge of the parental genotype is essential.

In the mapping studies in tetraploid potato described above, SSR alleles had to be assessed as separate dominant markers, reducing the inherent advantage of this marker system. To address this problem, Luo et al. (2000) developed a theoretical model and a statistical analysis for predicting the genotypes of a pair of tetraploid parents at a co-dominant or dominant marker locus based on the marker phenotype of both the parents and the progeny. The theory allows for null alleles (the absence of one or more alleles due to a circumstance such as the failure of PCR

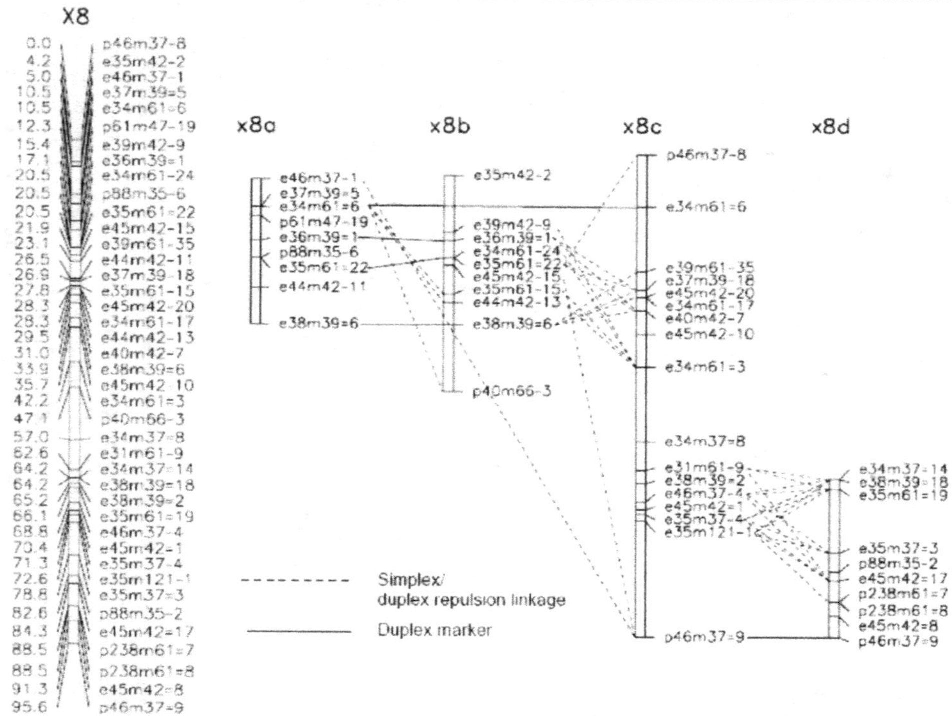

Fig. 5 Merging four homologous co-segregation groups of potato breeding line 12601ab1 using a combination of simplex/duplex repulsion linkages (dashed lines) and duplex markers (solid lines). The merged group is illustrated on the left. Reproduced and adapted from Meyer et al. (1998) with kind permission of Springer Science and Business Media.

primers to anneal to the relevant DNA templates properly), multiple dosage of alleles and double reduction, and compares the frequencies of the observed offspring phenotypes with those expected from the inferred parental genotypes. A significant coefficient of double reduction can, however, occur for several reasons, especially for dominant data and should be taken as no more than an indicator that the data does not follow any of the expected ratios in the absence of double reduction, given the parental genotypes. With co-dominant markers, it is easier to detect double reduction and individuals can have combinations of alleles that can only occur as a result of double reduction. These individuals need to be omitted from further analysis, or the alleles need to be analyzed separately, as current software (described later) cannot handle double reduction in its calculation of recombination frequencies and LOD scores.

The example given in Luo et al. (2000) is for SSR marker STM003 in potato cross 12601ab1 x Stirling (Fig. 6). The offspring had the following phenotypes with expected numbers in brackets for the most likely parental genotypes: 26 (28.8) were 0111, 22 (16.4) were 1111, 9 (10.3) were 1101, 6 (4.1) were 1011, 5 (6.2) were 0011, 3 (4.1) were 1110, 3 (2.1) were 0110, and none (2.1) were 1001. The probability of 12601ab1 being A1A3A3A4 (genotype 1334) and Stirling being A2A2A4A4 (2244) was estimated to be 0.999997, and the observed numbers of offspring phenotypes were not significantly different from those expected given

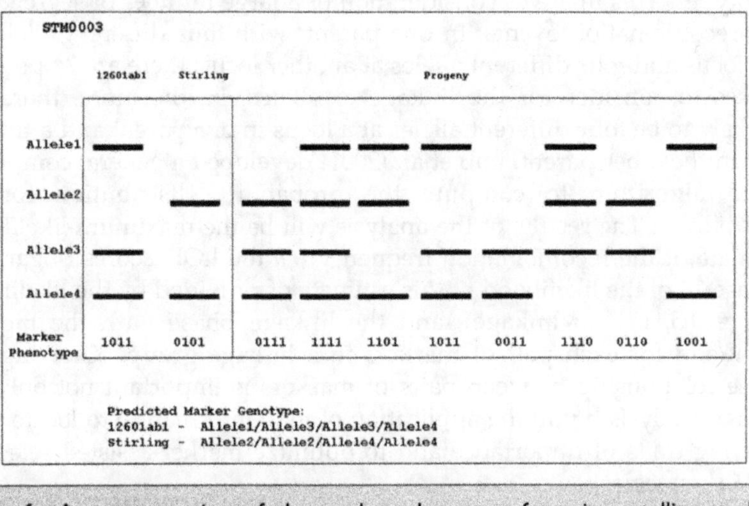

Fig. 6 A representation of the marker phenotype for microsatellite marker STM0003 on the parents and F_1 progeny of the cross 12601ab1 x Stirling.

these parental genotypes, based on a goodness of fit test ($\chi^2_{df=7}$ was 6.2, $P > 0.5$). There was no evidence of double reduction at this locus. Luo et al. (2000) also showed by simulation that in most cases both parental genotypes can be correctly recognized with a probability of nearly one when the progeny sample size is 100. However, as mentioned previously, a population size of 250 is desirable for constructing genetic linkage maps.

In a later simulation of linkage analysis and an analysis of experimental data, Luo et al. (2001) found examples of loci for which more than one genotype for the parents is possible. If the two parents have the same marker phenotypes but different genotypes are inferred, then information from linked markers is needed to resolve which parent has which genotype, for example to distinguish between configurations 1123 x 1233 and 1233 x 1123. Marker alleles which were present in all offspring and hence uninformative were not a problem as the choice between two possible genotypes had no consequence for the estimation of the linkage map. However, occasionally it may be necessary to exclude some marker loci from the linkage analysis.

Once the most likely marker genotypes for co-dominant and dominant markers have been established, linkage analysis can then be carried out using the most probable parental marker genotype(s). One of the major difficulties in linkage analysis with autotetraploid species is the calculation of the conditional distribution of the offspring genotypes, and hence phenotypes, at two linked loci for any given pair of parental genotypes. This involves consideration of a large number of segregation and recombination events. In one parent, with four different alleles at one locus and four different alleles at another locus, there are 24 possible phases to consider for the linkage of alleles. Furthermore, there are unlikely to be four different alleles at a locus in one parent and a further four in the other parent. Luo et al. (2001) developed a general computer-based algorithm to compute the probability distribution for all possibilities. The results of the analysis will be the maximum-likelihood estimate of the recombination frequency (r), the LOD scores (logarithm to base 10 of the likelihood of this estimate of r divided by the likelihood of r = 0.5, i.e. no linkage), and the linkage phase with the highest likelihood for each pair of markers in a linkage group. Knowing the phase relationship between pairs of markers is important not only for linkage analysis but in the application of map information to locate QTL affecting traits of importance and to optimize marker-assisted selection (MAS) strategies.

The recombination frequencies and LOD scores from the above analyses can be used to identify the order of the markers in each linkage group that has the minimum value of the weighted least squares criterion (Stam 1993) and to calculate map distances between markers. A linkage map with n markers has n!/2 possible orders, so an exhaustive search is feasible only for small n. Normally a very large number of possible orders presents a challenge to find the correct order. A seriation algorithm (Buetow and Chakravarti 1987) can be used for a rapid exploration to see whether any markers have been wrongly allocated to a linkage group. This can be followed with a slower but more thorough search by simulated annealing for a final ordering. Hackett et al. (2003) demonstrated the value of simulated annealing on real data as well as through computer simulation.

The methodology developed by Luo et al. (2000, 2001), and Hackett et al. (2001, 2003) assumes random, bivalent pairing during meiosis. While this is a simplification of the processes, it gives a tractable model that represents the available data on alfalfa (Julier et al. 2003; Robins et al. 2007a, b) and potato (Bradshaw et al. 2008) well. It may be less appropriate for autotetraploid species where there is evidence for a higher frequency of multivalents. More complicated models, including double reduction and/or preferential pairing have been proposed by Wu et al. (2001), Wu and Ma (2005) and by Luo et al. (2004, 2006). However, the practical advantages of such analyses are still to be thoroughly tested with real data.

7 QTL ANALYSIS IN AUTOTETRAPLOIDS

From a breeding perspective, the most important use of linkage maps is to locate major genes and QTLs for important traits. Ultimately, markers that are linked to (and preferably diagnostic for) the genes underlying these traits can be used in marker-assisted selection strategies in breeding and improvement of the target crop species. However, it is important to understand the genetic models that can be tested in a full-sib family.

7.1 Genetic Model for a QTL

In theory there could be four different QTL alleles in each of the two parents whose effects could be detected through their tight linkage to flanking markers. If there are no double crossovers between locus A and locus B, marker genotype A1B1 will be QTL genotype Q1:

Parent 1 ×	Parent 2
A1Q1B1	A5Q5B5
A2Q2B2	A6Q6B6

A3Q3B3 A7Q7B7
A4Q4B4 A8Q8B8

In the absence of double reduction, the 36 QTL genotypes are:

Q5Q6 Q5Q7 Q5Q8 Q6Q7 Q6Q8 Q7Q8

Q1Q2 Q1Q2Q5Q6 etc.
Q1Q3
Q1Q4
Q2Q3
Q2Q4
Q3Q4

In theory, one could make comparisons between the 36 QTL genotypes by comparing the trait means for the 36 corresponding marker genotypes (in the above Table Q1 is replaced by A1B1, etc.). If it is not possible to accurately align the marker maps of the two parents, then one has to do separate analyses on each parent, as is the current practice in potato. Comparisons are then made between the six sets of QTL genotypes from each parent separately (i.e., between Q1Q2 to Q3Q4 and between Q5Q6 to Q7Q8). In potato, it is also possible to analyze a tetraploid parent through its dihaploid (diploid) offspring which can be produced by parthenogenesis or anther culture. It must be remembered that Q1Q2 vs. Q3Q4 means Q1Q2(average of Q5Q6 to Q7Q8) vs. Q3Q4(average of Q5Q6 to Q7Q8) and there will be a loss of information. This can be seen in the simple example of Qqqq × Qqqq. The three offspring genotypes are QQqq, Qqqq, and qqqq in the ratio 1 : 2 : 1. However, the two comparisons that will be made are both Qq(QqQq + Qqqq) vs. qq(qqQq + qqqq) and information about dominance will be lost. It must also be remembered that whilst one may detect the segregation at a QTL locus in one parent, the other parent may be homozygous for a different allele, e.g. Qqqq × q′q′q′q′ giving offspring Qqq′q′ and qqq′q′, and this allele may be superior, inferior or intermediate to Q and q in its effect. The full model of six possible QTL genotypes inherited from one parent can be compared to a range of simpler models using a likelihood ratio test, to see if any of them is adequate to model the trait means. In a simplex model, the parental genotypes are Qqqq and qqqq, and one tests in turn for Q1, Q2, Q3 or Q4 being Q (e.g., Q1Q2 + Q1Q3 + Q1Q4 vs. Q2Q3 + Q2Q4 + Q3Q4 tests for Q1 = Q). In a duplex model, with complete dominance the parental genotypes are QQqq and qqqq, and one tests in turn for Q1Q2, Q1Q3, Q1Q4, Q2Q3, Q2Q4 or Q3Q4 being QQ (e.g., Q1Q2 to Q2Q4 vs. Q3Q4 tests for Q3Q4 = qq and hence Q1Q2 = QQ). Finally, in an additive model, one tests for Qq at the mid point of the trait values for QQ and qq, for example Q1Q2 = QQ, Q3Q4 = qq, Q1Q3 = Q1Q4 = Q2Q3 = Q2Q4 = Qq.

7.2 Marker-trait Associations

The earliest QTL mapping studies were based on testing for associations at single markers using regression-based approaches which are equivalent to single-factor analysis of variance. This type of analysis asks a simple question: is there a statistically significant difference between the mean trait values of the progeny that have the marker, and the progeny that do not have the marker? During the earliest of the mapping studies for autotetraploid potato described above, Meyer et al. (1998) and Bradshaw et al. (1998) adopted this approach to identify QTL for resistance to late blight and potato cyst nematode, respectively, in the F_1 progeny of the cross between 12601ab1 and Stirling. The advantage of this simple approach for tetraploids is that no modifications have to be made to take into account the dosage state of the marker. For instance, Meyer et al. (1998) found associations between late blight resistance (assessed using a 1-9 scale) and simplex, double-simplex and duplex markers, using a non-parametric Mann-Whitney U-test (Siegel 1956) to test the significance of the difference between the median values of the blight scores for progeny with the band versus the progeny without the band. The most significant association was attributed to a duplex marker. The authors proposed the presence of a duplex QTL for blight resistance with the two copies of the resistant allele of the QTL linked in coupling phase to the two copies of the duplex marker. This was supported by the fact that simplex markers on the remaining two homologs exhibited small negative associations with blight resistance (i.e. their presence was associated with susceptibility). Thus, this simple type of approach in combination with a linkage map is relatively useful for QTL detection for at least some scenarios in autotetraploids, and it can be extended equally well to more complex polyploids. It has been used in recently published papers on QTL detection in sugarcane (Aitken et al. 2006) and alfalfa (Robins et al. 2007a, b). The sugarcane results are presented in a later section. In alfalfa, multiple regression models with marker alleles as the independent variables were used to explain as much of the phenotypic variance as possible for yield, height, and regrowth at three harvest dates and for biomass production in five location-year combinations. Individual markers were found which explained from 3% to 20% of the variation and combinations of from two to five markers which cumulatively explained from 11% to 44% of the variation. Broad-sense heritabilities were in the range 45% to 87%.

There are several major drawbacks to the approach outlined above. The first of these is the fact that the performance of a separate regression test for each marker increases the overall Type I error rate (the probability of rejecting a true null hypothesis) associated with the experiment. More simply, this means that, when performing a separate ANOVA for each of several hundred marker/trait combinations some significant associations will arise by chance. In recognition of this fact a more stringent threshold value for declaring a QTL significant can be achieved by applying a permutation test (Churchill and Doerge 1994) to the data set. In the blight mapping study, subsets of markers corresponding to different dosage states were assembled from the total complement of segregating markers. For each subset, the resistance scores for the 78 clones were permuted and re-analyzed for QTLs at all markers, the largest U-value was stored, and the procedure was repeated up to 1,000 times. Stored values were ordered, and the 95% quantile was taken as the estimated critical value for which the overall Type 1 error rate for the experiment was 0.05 (5%) or less. However, the disadvantageous aspect of adopting this approach is that it tends to eliminate all but the most large-effect QTLs. In the study of Meyer et al. (1998), of 21 initial marker-trait associations, all but the largest (involving the duplex marker mentioned above, which explained ~30% of the observed phenotypic variation) were eliminated using this approach because of the high threshold U-values it generated.

The second drawback of this approach is that it is most likely to detect a QTL allele linked in coupling to the marker allele (e.g., M-Q in MQ/mq/mq/mq or MQ/MQ/mq/mq), but the magnitude of the effect will be reduced by any recombination between the marker and the QTL. Thus location and magnitude are confounded, and this could be a particular problem when the genetic map is not very densely populated with markers. Hence, one can not estimate the means of the QTL genotypes given in the previous section.

Thirdly, this approach is not very sensitive in relation to the relative effects of individual QTL alleles, and any interactions between them. The two parents in an autotetraploid cross could have up to eight different QTL alleles at an individual locus, some of which may be more or less effective than others and this information might be considered quite useful in a molecular breeding context. Ming et al. (2002), for example, used flanking single-dose RFLP markers for interval mapping in sugarcane, but this can only detect single-dose QTL alleles. The allele effect of each single-dose QTL was the average difference in phenotype of individuals differing by one copy of the indicated allele (single dose vs. zero dose).

7.3 Interval Mapping

In order to overcome some of these limitations, Hackett et al. (2001) developed a method for QTL interval mapping in autotetraploid species for a full-sib family derived by crossing two parents, and based on the simple model of tetrasomic inheritance in which the four homologous chromosomes pair at random to give two bivalents. The method assumes that the map of molecular markers has been produced as described earlier. As the offspring QTL genotypes are unknown, the conditional probabilities of QTL genotypes must be estimated from the marker information. The outcome is an estimate of the QTL location and the means of the QTL genotypes considered in the genetic model for a QTL.

For each offspring, the marker information on each homologous chromosome in a linkage group is used to identify possible configurations of chromosomes inherited from the two parents and the locations of crossovers on these chromosomes. A branch and bound algorithm is used to search for chromosome configurations for each offspring that are compatible with the marker phenotype information, arise from a possible bivalent pairing in each parent, and have the minimum number of crossovers. From these configurations (graphical genotypes), the conditional probability of each possible QTL genotype for a series of positions along the chromosome can be estimated. For some offspring a recombination event can be located precisely between two markers; for others a recombination event can have occurred anywhere in a longer interval. From the graphical genotypes, the set of possible genotypes at a putative QTL anywhere on the chromosome can be identified and their probabilities calculated. The homologous chromosomes from parent 1 will be referred to as h1-h4, and those from parent 2 as h5-h8. Assume that an offspring has a phenotype compatible with inheriting h1, h2, h5 and h6 at locus L1, say, and h1, h3, h5 and h6 at locus L2. This offspring will have two possible QTL genotypes between L1 and L2, Q1Q2Q5Q6 and Q1Q3Q5Q6. The probabilities of these two genotypes will be p and 1-p, respectively, where p = 1 at locus L1, p = 0.5 halfway between L1 and L2 and p = 0 at L2. An iterative weighted regression is then used to relate the trait values to the QTL genotype probabilities. The likelihood of a QTL at any location is expressed as the LOD score, the $\log_{10}$ of the ratio of the likelihood of the data given a QTL at this point to the likelihood of no QTL. The LOD score can be plotted against position on the linkage map, and the means and standard errors estimated for each QTL genotype at the position corresponding to the maximum LOD. A permutation test (Churchill and Doerge 1994) can be done to assess the significance of the QTL.

Hackett et al. (2001) performed simulation studies to assess the effects that parameters such as the types of marker used, population size and the heritability of the trait had on the ability to reliably detect QTLs using this method. The type of marker used for linkage mapping was found to have a very large effect on the ability to detect QTLs, mainly because it affected the ability to efficiently reconstruct the chromosomal configurations of progeny individuals. The most desirable scenario was the use of only co-dominant markers, the multi-allelic nature of which allow much better graphical genotype-based reconstruction of progeny individuals. However, the method also performed reasonably well when the more realistic scenario involving the use of a mixture of dominant and co-dominant markers was modelled. The effect of population size was also examined, and results suggested that populations of at least 200 individuals would be most desirable. Quantitative traits can have low heritabilities, implying a large environmental component relative to the stable, heritable genetic component for the trait. It is desirable, in these cases, to be able to identify the loci underlying this heritable genetic component, and a truly useful QTL mapping method should be able to detect QTL even in these scenarios. Although Hackett et al. (2001) found that the heritability of the trait did have an effect on the power of QTL detection, the method performed well with heritabilities as low as 10% when other parameters, such as the use of sufficiently large population sizes (ca. 200), and maps comprising mainly co-dominant markers, were optimal. As many autopolyploids can be clonally propagated, heritabilities can be increased by having a larger number of replicates of each offspring.

7.4 Experimental Results from Interval Mapping

Much of the methodology for linkage and QTL analysis described above was validated by extending the previous mapping experiments described for the cross between the advanced potato breeding line 12601ab1 and the potato cultivar Stirling. Bradshaw et al. (2004), Bryan et al. (2004), and Bradshaw et al. (2007) collectively carried out interval mapping of QTLs for blight and nematode resistance and 16 yield, agronomic and quality traits in potato on an expanded population comprising 227 clones from the cross between these genotypes. All of the traits scored had moderately high heritabilities with 54% to 94% of the variation in clone means being due to genetic differences. The disease tests had two or four replicates and the agronomic trials were conducted over three years in trials with two replicates. A total of 42 QTLs were identified. A well documented QTL for maturity was identified on chromosome V which explained 56% of the phenotypic variance. One for blight resistance was identified on chromosome IV which explained 37.2% of the variation and 2 for nematode resistance were identified on chromosomes IV and XI which explained 34.4% and 17.1%, respectively of the variation. The other QTLs individually explained between 5.4% and 16.5% of the variation. However, 6 QTLs were detected for after-cooking blackening and 4 for each of the traits: regularity of tuber shape, fry color (both after storage at 4°C and 10°C) and sprouting. Just two QTLs were found for each of the traits: yield, crop emergence, tuber size and common scab, and just one QTL was detected for each of the traits: dry matter content, keeping quality, growth cracks and internal condition. With a stringent permutation test, the smallest QTLs that could be detected were ones that explained just over 5% of the phenotypic variation. Although some QTLs of large effect may have been missed through inadequate marker coverage on some chromosomes, it is more likely that many QTLs of small effect went undetected. The 42 segregating QTLs can be classified as 14 simplex (Qqqq), 14 duplex with complete or partial dominance (QQqq where Qq acts more like more like QQ than qq), 4 duplex additive (QQqq where Qq is intermediate between QQ and qq) and 10 more complicated. Hence for 32 out of the 42 QTLs, a very simple gene model explains the variation. These results clearly demonstrate the feasibility of QTL analysis for a wide variety of traits in an autotetraploid, and pave the way for potential molecular breeding approaches based on marker-assisted selection in such species.

7.5 Computer Software

One factor limiting the routine use of all of the methodology described above was, for a period, the lack of a single analysis package that combined it all into a relatively user friendly format. Hackett et al. (2003) described the development of a software application called TetraploidMap, which could be used to construct genetic linkage maps in autotetraploid species. More recent versions of this software have been developed (Hackett et al. 2007) to run in a standard Microsoft Windows, facilitating the mapping process through a menu-driven graphical user interface that allows easy data entry (in standard spreadsheet format) and the alteration of several mapping parameters. The most recent version, at the time of writing, included a routine for

QTL analysis using the interval mapping-based approach described above (Hackett et al. 2007). The software package is available for download at: *http://www.bioss.ac.uk/knowledge/tetraploidmap/*.

8 LINKAGE ANALYSIS IN TRUE ALLOPOLYPLOIDS

Until now, we have focused our discussions largely on autotetraploids, using these as a model to demonstrate the feasibility of mapping in an autopolyploid species. At the beginning of the chapter, however, we defined two other types of polyploidy, namely allopolyploidy and segmental allopolyploidy. In terms of inheritance patterns true allopolyploids are essentially equivalent to diploids, and many are inbreeding species for which inbred lines are available and this simplifies genetic analysis. We will not discuss mapping and QTL analysis in these species in any great detail, since all of the tools developed for linkage and QTL analysis in diploids can equally well be applied to disomic allopolyploids. The interested reader should refer to the latest linkage maps of rapeseed (*B. napus*, an allotetraploid) and wheat (an allohexaploid) to see how they relate to the *Arabidopsis* and rice genomes which have been sequenced. A recent paper by Li et al. (2006) found that it is not easy to exploit *A. thaliana* information for *B. napus* based on synteny because rapeseed chromosomes are constituted as a patchwork of the homologous regions of different chromosomes of *Arabidopsis*. Landjeva et al. (2007) have provided a useful review of the actual and potential contributions of molecular markers to wheat genome characterization and breeding.

There are features of mapping in allopolyploids which can be problematic. For instance, while in most allopolyploid species, it is relatively easy to identify (using homoeologue specific markers) the ancestral genome to which different homoeologous linkage groups 'belong' (e.g., the A, B, and D genomes in wheat), in other species this is less trivial. White clover (*Trifolium repens*) is a good example of this. This species is an allotetraploid thought to have arisen from an ancient hybridization of two species that are represented by the extant species *T. occidentale* and *T. pallescens* (Ellison et al. 2006). Genetic mapping in white clover consistently reveals eight homoeolog pairs. Barrett et al. (2004) labelled the eight homoeologous pairs of linkage groups A to H, with subscript numerals (e.g., A_1 and A_2) to distinguish the homoeolog pairs. However, no homoeolog specific markers have been discovered to date in white clover, and, unlike other, more characterized allopolyploids, such as wheat, the subscripts do not designate the ancestral genomes of clover, but are arbitrarily assigned. This makes comparison of genetic linkage maps in different clover crosses somewhat problematic.

Dealing with co-dominant markers in allotetraploids can present some difficulties due to the presence of two homoeologous genomes. A single co-dominant marker may actually reveal two essentially independent loci (both homoeologs), or alternatively only one of the two homoeologs present. Several factors can make it difficult to derive the actual marker genotype present at a locus from the banding pattern or chromatogram (marker phenotype). For instance, independently segregating homoeologs may have one or more alleles of the same size. A particularly likely error would occur where two different homoeologs in the different parents possessed a marker allele of the same size. This would be scored as a biparental marker with an expected segregation ratio of 3 : 1, whereas in fact it is two independent markers. In addition, as with autopolyploids, null alleles may be present at one or both of the homoelogous loci. An extension of this idea is that it is not inconceivable that the marker is revealing both homoeologs in one parent, but only one homoeolog in the other. Arguably the easiest approach for dealing with these issues is to score each allele of problematic co-dominant markers independently, in a dominant fashion. Subsequent examination of the positions of these alleles on the linkage map may give the information required to re-score the markers as co-dominant.

9 MAPPING IN SEGMENTAL ALLOPOLYPLOIDS AND VERY COMPLEX POLYPLOIDS

Segmental allopolyploids present a greater challenge for linkage and QTL analysis because they exhibit a mixture of autopolyploid and allopolyploid inheritance patterns, meaning that some chromosomes pair randomly (as homologs, while others assort non-randomly (as homoeologs). This is thought to be the case in octoploid strawberry. In other plants, patterns of random and non-random pairing can be more complex than standard models of segmental allopolyploidy would imply, and these species can only really be described as complex polyploids. Modern sugarcane cultivars, which are derived from interspecific hybridizations between the domesticated species *Saccharum officianarum* (2*n* = 80) and the wild species *S. spontaneum* (2*n* = 40-128), are a good example of this latter class. Most sugarcane cultivars have between 100 and 130 chromosomes, which have been shown to exhibit a mixture of preferential and (mostly) non-preferential pairing (Aitken et al. 2005).

Despite the associated challenges, it is possible to describe a generalized strategy for mapping in these species using many of the techniques described already in this chapter, although one must understand that many of the specific details will vary from species to species. All of the below assumes that mapping is occurring in the F_1 of a cross between heterozygous parents (pseudotestcross).

While unpredictable pairing behavior of chromosomes in complex polyploids renders linkage analysis difficult, it is worth remembering that the detection and calculation of linkage between pairs of dominant simplex markers linked in coupling phase (simplex/simplex coupling) is unaffected by ploidy level and type. Thus, even in the presence of mixed autopolyploid and allopolyploid inheritance, it is possible to develop a map consisting entirely of linkage groups containing simplex markers linked in coupling phase (simplex-coupling LGs), using the same methodology and tools as used in diploids. This is generally the starting point for linkage mapping experiments in any scenario where the pairing behavior of the target species is unclear.

Most of the remaining efforts in the mapping process are directed towards 'merging' the simplex-coupling LGs comprising the framework map. As mentioned earlier, complex polyploids may possess chromosomes exhibiting both homologous and homoeologous pairing behavior. If co-dominant markers, such as SSRs or RFLPs have been used during the mapping process, the occurrence of individual simplex alleles of these markers on different simplex-coupling LGs may be an indicator of either homology or homoeology between the groups, depending on the ancestral genome relationships within the species. In the presence of both possible relationships, it may be useful to think of simplex-coupling LGs sharing these alleles as homo(eo)log groups, alluding to both possibilities. A recent illustration of this can be seen in the AFLP- and SSR-based map of sugarcane produced by Aitken et al. (2005).

In order to further refine the map, one can then scan the segregation data set for markers conforming to the expected segregation ratio for duplex markers associated with the most likely type/s of autopolyploidy expected to occur in the cross. These markers may then be incorporated into the simplex-coupling framework map using the appropriate formulae for estimation of recombination frequencies and LOD scores. Aitken et al. (2005) also successfully incorporated duplex markers into their map of sugarcane using this strategy.

Wu et al. (1992) outlined a method to distinguish between allo- and autopolyploid inheritance patterns based on the frequency of disomic repulsion phase linkages. Their approach was based on using diploid mapping software for phase-known markers, and involves taking the original marker segregation dataset and inverting it so that markers originally scored as '1' for band presence and '0' for band absence were rescored in the opposite way. The inverted data set is added to the original data set and linkage analysis performed again. Linkages detected between simplex markers in the original and recoded data sets represent repulsion phase linkages. More recent software that can analyze phase-unknown markers will calculate disomic repulsion phase linkages on the original data. If complete preferential pairing of homologous chromosomes (as in diploids and disomic polyploids) is occurring then there should be approximately equal numbers of coupling and repulsion phase linkages (1:1 repulsion:coupling). Alternatively, if completely random pairing is occurring (as in autopolyploids) then repulsion phase linkages should be much more difficult to detect. In autotetraploids, the ratio of detectable repulsion:coupling phase linkages is expected to be about 0.25:1 using this method, with the ratio dropping off towards 0:1 for higher order polyploids. Deviation from these ratios is an indication of mixed pairing behavior. In addition to being used as a tool to explore segregation type, this approach can also be used to further refine the simplex-coupling framework map for complex polyploids. For example, in their AFLP-based map of octoploid strawberry which exhibits mixed disomic and polysomic inheritance patterns, Lerceteau-Köhler et al. (2003) successfully identified pairs of simplex-coupling LGs representative of homoeologous linkage groups by virtue of multiple repulsion-phase linkages between these groups, while other simplex-coupling LGs exhibited no repulsion phase linkages to any other groups, indicating that these were probably acting under polysomic inheritance patterns. Aitken et al. (2005) also successfully used this approach in their sugarcane map, and in fact, built up 8 complex homo(eo)log groups composed of 127 linkage groups associated by all three of the methods (co-dominant markers, polysomic duplex markers, and repulsion-phase linkages between simplex markers) described above, accurately reflecting the complex structure of the genome of modern sugarcane varieties.

QTL analysis is also challenging in genomes as complex as those described above. In theory, the QTL mapping approach of Hackett et al. (2001) can be extended from autotetraploids to higher autopolyploids. However the main prerequisite for efficient detection of QTLs is good marker coverage, something that is frequently difficult to achieve in complex polyploids where there are limitations on the use of the marker segregation data, as we saw in earlier examples in the chapter. For instance, although the sugarcane map generated by Aitken et al. (2005) comprised over 1,300 markers, they estimated that this equated to only ~50% coverage of the genome in their cross. A similarly large mapping study in a different sugarcane cross was conducted by Hoarau et al. (2001) with over 1,000 AFLP markers, and the resulting map coverage was estimated at only 30%. Unlike the autopolyploid scenario, where the number of homologous chromosomes is known, and pairing behavior can be modelled, no well developed strategies exist for QTL mapping in species such as sugarcane, where the numbers of homologous chromosomes are variable, and pairing is irregular. It might be feasible to conduct interval mapping, using a diploid model, on the simplex-coupling framework maps in such species, but this would not allow the use of any multi-dose markers in the QTL analysis. In follow-up QTL mapping studies on their mapping populations, both Hoarau et al. (2002) and Aitken et al. (2006) adopted the single-marker ANOVA-based approach to locate QTLs for sugar-related traits and yield components respectively. Both experiments successfully identified QTLs at simplex and multiplex (mainly duplex) markers, although the percentage of phenotypic variation accounted for by individual QTLs was generally quite low. Multiple regression models with marker alleles as the independent variables were then used to explain as much of the phenotypic variance as possible. For sugar-related traits, each QTL explained from 3% to 9% of the phenotypic variation with models including all of the QTLs (8 to 15, including interactions) explaining from 37% to 66% of the total phenotypic variation. Broad-sense heritabilities were in the range of 59% to 93%. For yield components, 40 putative QTL alleles were found. Individually they explained between 3% and 7% of the variation and altogether, depending on the trait, they explained from 30% to 55% of the total phenotypic variance. Broad-sense heritabilities were high (77% to 91%) but consistency of the QTL alleles across crop-cycles appeared rather low. The authors in both studies attributed the small effects of individual QTLs to the high ploidy and heterozygosity in sugarcane, which means that many QTL alleles could be segregating at one locus, something that cannot be taken into account by following a marker that is linked in coupling phase to only one of the alleles of the QTL. Indeed, some apparent allelism was observable in both experiments. Aitken et al. (2006) found multiple QTLs on different linkage groups within the same homo(eo)logy group on several occasions, and postulated that these probably represented alleles of the same QTL. Despite the numerous limitations, at the conclusion of both studies, the authors were optimistic about the possibility of using the results of the studies for marker-assisted selection strategies in sugarcane.

10 IMPLICATIONS FOR BREEDING

The results of molecular mapping studies are confirming the models of classical quantitative genetics that explain the observed phenotypic variation in terms of Mendelian genetics and underpin the theory and practice of conventional plant breeding. In these models the effects of genes are measured in terms of their quantitative effects on the traits of interest, and not in terms of their effects at the levels of the transcriptome, proteome and metabolome. Examples can be found in the literature cited of different loci whose effects on a trait are additive; of dispersion between the parents of alleles with positive and negative effects on a trait; of loci displaying ambidirectional dominance for a trait; of loci displaying non-allelic interactions and of QTL allele by environment interaction. Examples can also be found of QTLs associated with effects on more than one trait. These associations may be due to the pleiotropic effects of a single gene or linkage of different genes on the same chromosome. Linkages can be broken through recombination, but in the short-term both pleiotropy and linkage will result in correlated responses to selection and breeders need to be aware of them. In future, breeders will be able to design their breeding programs based on knowledge of the distribution of desirable alleles among potential parents, the number of genes segregating, their chromosomal locations and the magnitudes of their effects. They will be able to choose the right parents, make the right crosses, and raise sufficient offspring to achieve their objectives, most likely in more than one cycle of crossing and selection.

However, perhaps the most important conclusion from the studies mentioned above is that the genetic variation in yield, agronomic and quality traits usually results from the segregation of alleles at many quantitative trait loci. These allelic differences usually result in small to intermediate differences in the traits of interest, but with the occasional larger difference such as the one found for maturity in potato (Bradshaw et al. 2004). More genes of large effect may be found as complex traits are dissected and candidate genes are sought. This has already happened for quantitative resistance to potato cyst nematodes (Bryan et al. 2004) and potato late blight (Bradshaw et al. 2004). For these genes the breeder will want to use molecular breeding methods (gene cloning and marker-assisted selection) to ensure that they are introduced into new cultivars as quickly as possible. However, efficient exploitation of most of the genetic variation in plant breeding programs will continue to rely on knowledge from classical quantitative genetics, complemented by the results of QTL analyses. Furthermore, efficient multi-trait genotypic recurrent selection will still be essential for continued steady progress in the genetic improvement of our major crops.

11 CONCLUSIONS

With continuing advances in both genotyping and computing, the rate at which linkage and QTL analysis for all plant species will advance over the next few years is likely to be high. Parallel avenues of endeavor, such as association mapping and high throughput genomics are likely to have both predictable and unpredictable impacts on our ability to dissect the genetics of quantitative traits for all plant species. Perhaps the real challenge for plant geneticists and breeders in the future will be to come up with imaginative routes to maximize the results of these advances by exploiting them to produce better varieties of crop plants, faster, and using fewer resources.

References

Aitken KS, Jackson PA, McIntyre CL (2005) A combination of AFLP and SSR markers provides extensive map coverage and identification of homoeologous linkage groups in a sugarcane cultivar. Theor Appl Genet 110: 789-801

Aitken KS, Jackson PA, McIntyre CL (2006) Quantitative trait loci identified for sugar related traits in a sugarcane (*Saccharum* spp.) cultivar × *Saccharum officinarum* population. Theor Appl Genet 112: 1306-1317

Al-Janabi S, Honeycutt R, McClelland M, Sobral B (1993) A genetic linkage map of *Saccharum spontaneum* L 'SES 208'. Genetics 134: 1249-1260

Al-Janabi S, Honeycutt R, Sobral B (1994) Chromosome assortment in Saccharum. Theor Appl Genet 89: 959-963

Bailey NTJ (1961) Introduction to The Mathematical Theory of Genetic Linkage. Claredon Press, Oxford, UK

Barrett B, Griffiths A, Schreiber M, Ellison N, Mercer C, Bouton J, Ong B, Forster J, Sawbridge T, Spangenberg G, Bryan G, Woodfield D (2004) A microsatellite map of white clover. Theor Appl Genet 109: 596-608

Bingham ET, McCoy TJ (1988) Cytology and cytogenetics of alfalfa. In: Hanson AA (ed) Agronomy Monographs 29, Alfalfa and Alfalfa Improvement. ASA, CSSA, SSSA, Madison, WI, USA, pp 737-776

Bradshaw JE , Hackett CA , Meyer RC , Milbourne D , McNicol JW, Phillips MS, Waugh R (1998) Identification of AFLP and SSR markers associated with quantitative resistance to *Globodera pallida* (Stone) in tetraploid potato

(*Solanum tuberosum* subsp. *tuberosum*) with a view to marker-assisted selection. Theor Appl Genet 97: 202-210

Bradshaw JE, Bryan GJ, Hackett CA, McLean K, Pande B, Stewart HE, Waugh R (2004) Dissection and analysis of quantitative disease resistance in tetraploid potato. Euphytica 137: 13-18

Bradshaw JE, Pande B, Bryan GJ, Hackett CA, McLean K, Stewart HE, Waugh R (2004) Interval mapping of quantitative trait loci for resistance to late blight [*Phytophthora infestans* (Mont) de Bary], height and maturity in a tetraploid population of potato (*Solanum tuberosum* subsp. *tuberosum*). Genetics 168: 983-995

Bradshaw JE, Hackett CA, Pande B, Waugh R, Bryan GJ (2008) QTL mapping of yield, agronomic and quality traits in tetraploid potato (*Solanum tuberosum* subsp. *tuberosum*). Theor Appl Genet 116: 193-211

Brouwer DJ, Osborn TC (1999) A molecular marker linkage map of tetraploid alfalfa (*Medicago sativa L*). Theor Appl Genet 99: 1194-1200

Bryan GJ, McLean K, Pande B, Purvis A, Hackett CA, Bradshaw JE, Waugh R (2004) Genetical dissection of H3-mediated polygenic PCN resistance in a heterozygous autotetraploid potato population. Mol Breed 14: 105-116

Buetow KH, Chakravarti A (1987) Multipoint gene mapping using seriation I: General methods. Am J Hum Genet 41: 180-188

Churchill GA, Doerge RW (1994) Empirical threshold values for quantitative trait mapping. Genetics 138: 963-971

Da Silva J, Sorrells ME, Burnquist WL, Tanksley SD (1993) RFLP linkage map and genome analysis of *Saccharum spontaneum* L. Genome 36: 782-791

Da Silva JAG, Sobral BWS (1996) Genetics of polyploids. In: Sobral BWS (ed) The Impact of Plant Molecular Genetics. Birkhauser, Boston, USA, pp 3-37

Ellison NW, Liston A, Steiner JJ, Williams WM, Taylor NL (2006) Molecular phylogenetics of the clover genus (*Trifolium*-Leguminosae). Mol Phylogenet Evol 39: 688-705

Gallais A (2003) Quantitative Genetics and Breeding Methods in Autopolyploid plants. INRA, Paris, France

Goldblatt P (1980) Polyploidy in angiosperms:Monocotyledons. In: Lewis WH (ed) Polyploidy: Biological Relevance. Plenum Press, New York, USA, pp 219-239

Grant V (1963) The Origin of Adaptations. Columbia Univ Press, New York, USA

Grant V (1981) Plant Speciation. Columbia Univ Press, New York, USA

Grivet L, D'Hont A, Roques D, Feldmann P, Lanaud C, Glaszmann J (1995) RFLP mapping in cultivated sugarcane (*Saccharum* spp.): Genome organisation in a highly polyploid and aneuploid interspecific hybrid. Genetics 142: 987-1000

Hackett CA, Luo ZW (2003) Tetraploid Map: Construction of a linkage map in autotetraploid species. J Hered 94: 358-359

Hackett CA, Bradshaw JE, Meyer RC, McNicol JW, Milbourne D, Waugh R (1998) Linkage analysis in tetraploid species: a simulation study. Genet Res Camb 71: 143-154

Hackett CA, Bradshaw JE, McNicol JW (2001) Interval mapping of quantitative trait loci in autoteraploid species. Genetics 159: 1819-1832

Hackett CA, Pande B, Bryan GJ (2003) Linkage mapping in autotetraploid species using simulated annealing. Theor Appl Genet 106: 1107-1115

Hackett CA, Milne I, Bradshaw JE, Luo ZW (2007) TetraploidMap for Windows: linkage map construction and QTL mapping in autotetraploid species. J Hered 98: 727-729

Hoarau JY, Offmann B, D'Hont A, Risterucci AM, Roques D, Glaszmann JC, Grivet L (2001) Genetic dissection of a modern sugarcane cultivar (*Saccharum* spp.). 1 Genome mapping with AFLP markers. Theor Appl Genet 103: 84-97

Hoarau JY, Grivet L, Offmann B, Raboin LM, Diorflat JP, Payet J, Hellmann M, D'Hont A, Glaszmann JC (2002) Genetic dissection of a modern sugarcane cultivar (*Saccharum* spp.). II Detection of QTLs for yield components. Theor Appl Genet 105: 1027-1037

Kriegner A, Cervantes JC, Burg K, Mwanga ROM, Zhang D (2003) A genetic linkage map of sweet potato [*Ipomoea butatos* (L.) Lam) based on AFLP markers. Mol Breed 11: 169-185

Landjeva S, Korzun V, Borner A (2007) Molecular markers: actual and potential contributions to wheat genome characterization and breeding. Euphytica 156: 271-296

Li Y, Ma C, Fu T, Yang G, Tu J, Chen Q, Wang T, Zhang X, Li C (2006) Construction of a molecular functional map of rapeseed (*Brassica napus* L) using differentially expressed genes between hybrid and its parents. Euphytica 152: 25-39

Jenczewski E, Alix K (2004) From diploids to allopolyploids: The emergence of efficient pairing control genes in plants. Crit Rev Plant Sci 23(1): 21-45

Julier B, Flajoulot S, Barre P, Cardinet G, Santoni S, Huguet T, Huyghe C (2003) Construction of two genetic linkage maps in cultivated tetraploid alfalfa (*Medicago sativa*) using microsatellite and AFLP markers. BMC Plant Biol 3: 9

Lagercrantz U (1998) Comparative mapping between *Arabidopsis thaliana* and *Brassica nigra* indicates that *Brassica* genomes have evolved through extensive genome replication accompanied by chromosome fusions and frequent rearrangements. Genetics 150: 1217-1228

Lagercrantz U, Lydiate DJ (1996) Comparative genomic mapping in *Brassica*. Genetics 144: 1903-1910

Lerceteau-Kohler E, Guerin G, Laigret F, Denoyes-Rothan B (2003) Characterization of mixed disomic and polysomic inheritance in the octoploid strawberry (*Fragaria* × *ananassa*) using AFLP mapping. Theor Appl Genet 107: 619-628

Lewis KR, John B (1963) Chromosome Marker. J and A Churchill Ltd, London, UK

Luo ZW, Hackett CA, Bradshaw JE, McNicol, JW, Milbourne D (2000) Predicting parental genotypes and gene segregation for tetrasomic inheritance. Theor Appl Genet 100: 1067-1073

Luo ZW, Hackett CA, Bradshaw JE, McNicol JW, Milbourne D (2001) Construction of a genetic linkage map in tetraploid species using molecular markers. Genetics 157: 1369-1385

Luo ZW, Zhang RM, Kearsey MJ (2004) Theoretical basis for genetic linkage analysis in autotetraploid species. Proc Natl Acad Sci USA 101: 7040-7045

Luo ZW, Zhang Z, Leach L, Zhang RM, Bradshaw JE, Kearsey MJ (2006) Constructing genetic linkage maps under a tetrasomic model. Genetics 172: 2635-2645

Masterson J (1994) Stomatal size in fossil plants: Evidence for polyploidy in the majority of angiosperms. Science 264: 421-423

Meyer RC, Milbourne D, Hackett CA, Bradshaw JE, McNicol JW, Waugh R (1998) Linkage analysis in tetraploid potato and association of markers with quantitative resistance to late blight (*Phytophthora infestans*). Mol Gen Genet 259: 150-160

Ming R, Wang YW, Draye X, Moore PH, Irvine JE, Paterson AH (2002) Molecular dissection of complex traits in autopolyploids: mapping QTLs affecting sugar yield and related traits in sugarcane. Theor Appl Genet 105: 332-345

Paterson AH, Bowers JE, Chapman BA (2004) Ancient polyploidisation predating divergence of the cereals and its consequences for comparative genomics. Proc Natl Acad Sci USA 101: 9903-9908

Ritter E, Gebhardt C, Salamini F (1990) Estimation of recombination frequencies and construction of RFLP linkage maps in plants from crosses between heterozygous parents. Genetics 125: 645-654

Robins JG, Bauchan GR, Brummer EC (2007b) Genetic mapping forage yield, plant height, and regrowth at multiple harvests in tetraploid alfalfa (*Medicago sativa* L). Crop Sci 47: 11-18

Robins JG, Luth D, Campbell TA, Bauchan GR, He C, Viands DR, Hansen JL, Brummer EC (2007a) Genetic mapping of biomass production in tetraploid alfalfa. Crop Sci 47: 1-10

Rouppe van der Voort JNAM, van Zandvoort P, van Eck HJ, Folkertsma RT, Hutten RCB, Draaistra J, Gommers FJ, Jacobsen E, Helder J, Bakker J (1997) Use of allele specificity of comigrating AFLP markers to align genetic maps from different potato genotypes. Mol Gen Genet 255: 438-447

Shoemaker RC, Schlueter J, Doyle JJ (2006) Paleopolyploidy and gene duplication in soybean and other legumes. Curr Opin Plant Biol 9: 104-109

Siegel S (1956) Nonparametric Statistics for the Behavioural Sciences. McGraw-Hill, New York, USA

Smartt J, Simmonds NW (1995) Evolution of Crop Plants. 2nd Edn. Longman Group, Harlow, UK

Stam P (1993) Construction of integrated genetic linkage maps by means of a new computer package: JOINMAP. Plant J 3: 739-744

Stebbins GL (1950) Variation and Evolution in Plants. Columbia Univ Press, Columbia, New York, USA

Swaminathan MS, Howard HW (1953) The cytology and genetics of the potato (*Solanum tuberosum*) and related species. Bibliogr Genet 16: 1-192

Ukoskit K, Thompson PG (1997) Autopolyploidy versus allopolyploidy and low-density randomy amplified polymorphic DNA linkage maps of sweet potato. J Am Soc Hort Sci 122: 822-828

Wu K, Burnquist W, Sorrells M, Tew T, Moore P, Tanksley S (1992) The detection and estimation of linkage in polyploids using single-dose restriction fragments. Theor Appl Genet 83: 294-300

Wu R, Ma CX (2005) A general framework for statistical linkage analysis in multivalent tetraploids. Genetics 170: 899 - 907

Wu R, Gallo-Meagher M, Littell RC, Zeng Z-B (2001) A general polyploid model for analyzing gene segregation in outcrossing tetraploid species. Genetics 159: 869-882

Wu SS, Wu R, Ma C-X, Zeng Z-B, Yang MCK, Casella G (2001) A multivalent pairing model of linkage analysis in autotetraploids. Genetics 159: 1339-1350

Yu K, Pauls K (1993) Segregation of random amplified polymorphic DNA markers and strategies for molecular mapping in tetraploid alfalfa. Genome 36: 844-851

11 Molecular Breeding Technologies for Forage and Turf Plants

John W. Forster[1,3*], Noel O.I. Cogan[1,3], Mark P. Dobrowolski[2,3], Eline van Zijll de Jong[1,3], Germán C. Spangenberg[1,3] and Kevin F. Smith[2,3]

[1]Department of Primary Industries, Biosciences Research Division, Victorian AgriBiosciences Centre, La Trobe Research and Development Park, Bundoora, Victoria 3083, Australia

[2]Department of Primary Industries, Biosciences Research Division, Hamilton Centre, Hamilton, Victoria 3300, Australia

[3]Molecular Plant Breeding Cooperative Research Centre, Australia

*Corresponding author: john.forster@dpi.vic.gov.au

1 INTRODUCTION

1.1 Major Temperate Forage and Turf Species

Forage species provide herbage for grazing, hay, and silage production servicing livestock production industries in both tropical and temperate regions of the world. The grazing industries are responsible for dairy, meat and fiber production. Temperate forage grasses include perennial ryegrass (*Lolium perenne* L.), Italian ryegrass (*Lolium multiflorum* Lam.), meadow fescue (*Festuca pratensis* Huds.), tall fescue (*Festuca arundinacea* Schreb.), cocksfoot or orchard grass (*Dactylis glomerata* L.), timothy (*Phleum pratense* L.), Kentucky bluegrass (*Poa pratensis* L.), smooth bromegrass (*Bromus inermis* L.), and harding grass (*Phalaris aquatica* L.). Temperate forage legumes include white clover (*Trifolium repens* L.), red clover (*Trifolium pratense* L.), subterranean clover (*T. subterraneum* L.), bird's foot trefoil (*Lotus corniculatus* L.), and lucerne/alfalfa (*Medicago sativa* L.). Tropical forage species include grasses such as buffelgrass (*Pennisetum ciliare* L.), and members of the genera *Brachiaria* and *Paspalum*, as well as legumes such as round-leafed cassia (*Chamaecrista rotundifolia*), siratro (*Macroptilium atropurpuerum*), and members of the genera *Stylosanthes*, *Centrosema* and *Desmodium*. A range of temperate and warm-season grasses are also important for non-forage applications such as turf and amenity cultivation. The temperate turf grasses include *Lolium*, *Festuca*, *Poa* and *Agrostis* (bentgrass) species, while warm-season and tropical turf grasses include switchgrass (*Panicum virgatum* L.), seashore paspalum and bahiagrass (*Paspalum vaginatum* Swartz and *Paspalum notatum* Flugge) and members of the *Zoysia* (zoysiagrass) and *Cynodon* (bermuda grass) genera (Forster et al. 2001a).

Perennial ryegrass and white clover are cultivated, generally in combination, in grassland-producing regions of Northern Europe, the Pacific North-West of the United States, Japan, South-Eastern Australia and New Zealand, and provide high-quality forage with superior palatability and nutrient content (Forster et al. 2001a). The major focus of this chapter is on the biology, genetics, genomic resources, and current progress in molecular breeding, using non-transgenic methods, of perennial ryegrass and white clover. The major challenges and opportunities defined by the biology of the target species are described, along with potential and actual solutions. Perennial ryegrass undergoes a symbiosis with the fungal endophyte species *Neotyphodium lolii* that provides important beneficial agronomic effects to the host (Prestidge and Gallagher 1988; Breen 1993). The status of endophyte genomics and genetics is consequently also assessed. The equivalent information for other temperate forage and turf grasses and forage legumes is briefly reviewed.

1.2 Taxonomy of Perennial Ryegrass and White Clover

Perennial ryegrass is a member of the Poaeae tribe of the Poodae super-tribe in the Pooideae sub-family of the grass and cereal family Poaceae (Soreng and Davis 1998). The *Lolium* and *Festuca* genera are closely allied, and the most nearly related major cereal species is cultivated oats (*Avena sativa* L.) within the Aveneae tribe of the Poodae. The Triticeae cereal tribe (wheat, barley, and rye) is located within the Triticodae super-tribe of the Pooidae. Rice (*Oryza sativa* L.), by contrast, is located in the Poaceae sub-family Bambusoideae. Translation genomics from rice to perennial ryegrass based on whole genome DNA sequence data consequently traverses a significant phylogenetic distance, but the use of partial genomic sequence and expressed sequence tag (EST) data from wheat (*Triticum aestivum* L.: Powell and Langridge 2004) and barley (*Hordeum vulgare* L.) exploits closer taxonomic affinities. Future whole genome sequencing of the model plant species *Brachypodium distachyon* (Draper et al. 2001; Vogel et al. 2006), which is more closely related to the Triticeae than the Oryzeae, is also likely to be valuable for perennial ryegrass.

White clover is a member of the Trifolieae tribe of the cool-season Galegoid clade in the Papilinoideae sub-family of the legume family Fabaceae (Doyle and Luckow 2003). The most closely related genus is *Melilotus* (sweet clovers) and the genus *Medicago*, including alfalfa, is also part of the Trifolieae. As a consequence, the model legume species barrel medic (*Medicago truncatula* Gaertn.) shares a common ancestor relatively recently in evolutionary time with white clover. Translational genomics based on whole genome sequencing of *M. truncatula* (Young et al. 2005; Zhu et al. 2005) is consequently anticipated to be highly efficient for members of the *Trifolium* genus. The other model legume species, *Lotus japonicus* Gifu, is also a Galegoid legume located in a separate tribe, Loteae.

1.3 Genome Structure of Perennial Ryegrass and White Clover

Members of the *Lolium* genus are diploids with a fundamental chromosome number of 7 ($2n = 2x = 14$). The genome size of perennial ryegrass has been estimated through measurements of nuclear DNA content by microdensitometry (Hutchinson et al. 1979; Seal and Rees 1982). A 2C value of 4.16 pg corresponds to a haploid genome size of c. 1.6×10^9 bp. The individual genome sizes of other *Lolium* species vary, so that inbreeding taxa such as *Lolium temulentum* exhibit nuclear DNA contents c. 50% larger than those of the outbreeding species. In common with other Poaceae family members, the genomes of *Lolium* species contain large numbers of dispersed repetitive sequences, frequently belonging to major retroelement families (Jenkins et al. 2000).

White clover is an allotetraploid species with a fundamental chromosome number of 8 ($2n = 4x = 32$). The evolutionary origin of white clover has been a matter of dispute over many years, although two diploid species, *T. occidentale* D.E. Coombe and *T. nigrescens* Viv., and another allotetraploid species, *T. uniflorum* L., were previously considered to be potential progenitors (Chen and Gibson 1970; Chen and Gibson 1971; Badr et al. 2002). More recent molecular phylogenetics studies based on chloroplast DNA and nuclear ribosomal DNA variation

(Ellison et al. 2006) have implicated *T. occidentale* and the diploid taxon *T. pallescens* Schreber as the paternal and maternal progenitors, respectively. Measurements of nuclear DNA content have been performed using Feulgen microdensitometry (Grime and Mowforth 1982; Campbell et al. 1999) and flow cytometry (Aramuganathan and Earle 1991). A range of values between 2.07 and 3.0 pg were reported. The lower value corresponds to a haploid genome size of ca. 8×10^8 bp, implying an average sub-genome size close to 400 Mb and comparable to that of *M. truncatula*. Molecular cytogenetic studies of 5S and 18S-26S rDNA organization have been performed (Ansari et al. 1999), and a highly repetitive tandemly repeated DNA sequence family (TrR350) has been isolated and localized to the centromeric chromosomal regions (Ansari et al. 2004).

1.4 Reproductive Behavior of Perennial Ryegrass and White Clover

Perennial ryegrass is an obligate outbreeding species, with a gametophytic self-incompatibility (SI) system controlled by two loci (*S* and *Z*). Incompatible matings occur when the alleles at both loci in the male gametophyte (pollen grain) match one of the two alleles at each locus in the female sporophyte (Cornish et al. 1979). This mechanism ensures a very low level of self-fertilization, and also limits the level of fertility in closely related individuals such as full-sibs. The proportions of compatible matings between related individuals depend on the degree of allelic complexity at the SI loci. Genetic studies of both natural and synthetic populations have revealed a large number of different alleles (ca. 20) each of the *S* and *Z* loci (Devey et al. 1994; Fearon et al. 1994). The *S*/*Z*-based SI system is present in a wide range of outbreeding Poaceae species. Genetic studies of SI have been performed in the reed canary grass species *Phalaris coerulescens* and cereal rye (*Secale cereale* L.), among others (Baumann et al. 2000).

White clover is also an obligate outbreeding species, with a gametophytic SI system controlled by a series of alleles at a single locus (*S*) (Attwood 1940, 1941, 1942a). Rare instances of self-compatibility have been reported (Attwood 1942b; Yamada et al. 1989), presumably due to the presence of self-fertile (S_f) alleles at the SI locus.

The three well-recognized outbreeding species within the *Lolium* genus [perennial ryegrass, Italian ryegrass and annual ryegrass (*Lolium rigidum* Gaud.)] chiefly differ in reproductive development and growth habit, with a continuous range of variation from short-lived annual ecotypes of annual ryegrass at one extreme to long-lived perennial ecotypes of perennial ryegrass at the other extreme. The different species show a high degree of interfertility, and are reproductively isolated from one another in nature through separate geographical locations and differing mean flowering time. The annual-type ryegrasses are more characteristic of Mediterranean environments and are early flowering, while the perennial-type ryegrasses are characteristic of cooler temperate environments and are generally later flowering (Forster et al. 2005a).

White clover also exhibits perennial growth behaviour, with known genetic variation for floral development traits (Connolly 1990; Williams et al. 1998). The stoloniferous growth habit of white clover strongly contributes to vegetative persistence. Morphogenetic traits are generally negatively correlated with flowering date (Cogan et al. 2006a), and such relationships may be related to the observed decline of vegetative growth following onset of flowering, due to inhibition of stolon elongation (Kawanabe et al. 1963).

1.5 Varietal Development for Perennial Ryegrass and White Clover

Due to the obligate outbreeding natures of both perennial ryegrass and white clover, both natural and synthetic populations are highly genetically heterogeneous. Varietal development is typically based on the following process: evaluation of base populations containing 2000-500 individuals; selection of c. 200 potential parental clones; and polycross between the parents to produce a synthetic 1 (Syn1) population (Vogel and Pedersen 1993). The number of foundation individuals may vary from as low as four for perennial ryegrass to 50-100 for polyploid species such as tall wheat grass and alfalfa (Bray and Irwin 1999).

1.6 Challenges and Opportunities for Molecular Breeding of Perennial Ryegrass and White Clover

In common with the majority of other forage and turf species, perennial ryegrass and white clover exhibit obligate outbreeding reproductive habits, and varietal development is based on the production of synthetic varieties through polycrosses of selected foundation genotypes. This breeding system presents important challenges for the implementation of molecular breeding strategies, using both transgenic and non-transgenic methods. In addition, the majority of target traits for improvement for outcrossing pasture plants are quantitative in nature (such as herbage quality, drought tolerance) and may show significant environmental variation and hence genotype x environment (G x E) interaction (such as herbage digestibility: Oram et al. 1974; Tyler and Hayward 1982; Cogan et al. 2005b). These features of the biology of perennial ryegrass and white clover present both challenges and opportunities for the development and implementation of molecular markers. Current strategies to address these issues are described in section 4.5.

2 GENOMIC RESOURCES FOR PERENNIAL RYEGRASS, FUNGAL ENDOPHYTES, AND WHITE CLOVER

2.1 EST Resources

High-throughput gene discovery by expressed sequence tag (EST) sequencing has generated substantial genomic resources for the temperate pasture species. A collection of 44,534 perennial ryegrass ESTs was generated from single pass sequencing of randomly selected clones from 29 cDNA libraries that represent a range of plant organs (leaf, root, seed, etc.), developmental stages (vegetative, reproductive, etc.) and environmental conditions (Sawbridge et al. 2003a). EST redundancy was resolved through assembly with the CAP3 application, leading to identification of 12,170 unigenes. Similarly, a collection of 42,017 white clover ESTs was generated from 16 cDNA libraries obtained from a broad range of plant organs, developmental stages, and environmental conditions, leading to identification of 15,989 unigenes (Sawbridge et al. 2003b). Each of the sequences was annotated by comparison to GenBank and SwissProt public sequence databases and automated intermediate Gene Ontology (GO) annotation was obtained (Spangenberg et al. 2005b). All sequences and annotation are maintained within ASTRA format MySQL databases, with web-based access for text searching, BLAST sequence comparison and GO hierarchical tree browsing. Each ryegrass sequence was mapped onto an EnsEMBL genome viewer for comparison with the complete genome sequence of rice and expressed sequences from related species. Similarly, each white clover sequence was mapped onto an EnsEMBL genome viewer for comparison with the complete *A. thaliana* genome sequence and expressed sequences from related legumes. Equivalent developments in the symbiogenomics of forage and turf grasses (large-scale gene discovery from grass endophyte species) (Spangenberg et al. 2001, 2005a) have led to the production of 13,964 ESTs collectively from the taxa *Neotyphodium lolii, N. coenophialum* (the endophyte of tall fescue) and *Epichloë festucae* (the putative sexual progenitor *of N. lolii*) hence identifying a total of 7,585 unigenes (Felitti et al. 2004, 2006; Spangenberg et al. 2005a).

2.2 BAC Libraries

Complementing the EST resources, large insert DNA libraries have been generated using bacterial artificial chromosome (BAC) vectors. A perennial ryegrass BAC library was constructed using *Hind*III-generated partial DNA fragments in the pBeloBACII vector, and consists of 50,304 clones with an average insert size of 113 kb, corresponding to 3.4 genome equivalents. Two other perennial ryegrass BAC libraries have been constructed and PCR-screened to identify clones containing candidate genes for disease resistance and forage quality (Farrar et al. 2007). The white clover BAC library was constructed in a similar fashion, and consists of 50,302 clones with an average insert size of 101 kb, corresponding to 6.3 genome equivalents (Spangenberg et al. 2005b). BAC library screening has been performed for both species using both macroarray hybridization and PCR analysis of microtitre plate pools. A BAC library for the *N. lolii* endophyte has also been established, consisting of ca. 6,000 clones with an average insert size of 120 kb, corresponding to 12 genome equivalents.

3 TRANSCRIPTOMICS ANALYSIS FOR PERENNIAL RYEGRASS, FUNGAL ENDOPHYTES, AND WHITE CLOVER

High-density cDNA microarrays representing approximately 15,000 unique genes for each of perennial ryegrass and white clover (Sawbridge et al. 2003a, b) and unigene microarrays allowing the interrogation of over 5,000 *Neotyphodium* and *Epichloe* genes (Nchip™ microarray and EndoChip™ microarray) (Felitti et al. 2004, 2006; Spangenberg et al. 2005a) have been developed. The pasture species-derived microarrays have been applied in hybridizations with labeled total RNA isolated from a variety of genotypes, plant organs, developmental stages, and growth conditions. The endophyte microarrays have been used to compare differential gene expression between saprophytic growth of *N. lolii* in solid and liquid culture, and between *N. lolii* and *N. coenophialum* grown in liquid culture (Felitti et al. 2006). Results from these studies have enabled validation of functions predicted through comparative sequence annotation, and also suggested roles for novel genes which lack comparative sequence annotation. As an example of the latter, unannotated genes which are co-regulated with the perennial ryegrass *Lp*CAD1 and *Lp*CAD3 (cinnamyl alcohol dehydrogenase) lignin biosynthesis genes have been identified and mapped as candidate gene-based markers to regions of the genome associated with herbage quality quantitative trait loci (QTL) (Cogan et al. 2006b). Microarrays may also be used for gene and promoter discovery when used in concert with the BAC libraries for each species. An *International Transcriptome Initiative for Forage and Turf* (ITIFT) has been proposed to facilitate international efforts in microarray-based transcriptome analyses for key forage and turf plants and their endosymbionts. Support of ITIFT will provide access to a range of platforms for microarray spotting, hybridization and scanning operationally integrated through a Scierra laboratory workflow system. An ITIFT database with a web-based front-end portal for secure access by the research community to appropriate data and information will be developed and maintained. The next generation of microarrays for ITIFT will be based on oligonucleotide arrays customized synthesized *in situ* using the CombiMatrix (*www.combimatrix.com*) CustomArray™ synthesizer system. Prototype arrays of 60-mer oligonucleotides corresponding to the perennial ryegrass, white clover and endophyte unigene sets have already been generated.

4 MOLECULAR MARKER TECHNOLOGY FOR IMPROVEMENT OF PERENNIAL RYEGRASS AND WHITE CLOVER

4.1 Molecular Marker Development, Genetic Map Construction, and Trait-Dissection in Perennial Ryegrass

The development of molecular genetic markers and associated genetic maps for perennial ryegrass has been comprehensively reviewed by Forster et al. (2001a, 2004), while the application of genetic marker analysis to trait dissection has been reviewed by Yamada and Forster (2005). A comprehensive set (ca. 400) of unique perennial ryegrass genomic DNA-derived simple sequence repeat (LPSSR) markers have been developed using enrichment library technology (Jones et al. 2001). This resource has been augmented by the results of similar studies on a smaller scale by Kubik et al. (1999), Lauvergeat et al. (2005) and Jensen et al. (2005a). The perennial ryegrass EST collection has also been used for the development of a set of 310 EST-SSR primer pairs (Faville et al. 2004). More recently, gene-associated SNP markers have been developed through the use of both *in vitro* and *in silico* discovery strategies (Shinozuka et al. 2005; Spangenberg et al. 2005b; Cogan et al. 2006b).

Development of the first generation reference genetic map of perennial ryegrass was performed through the use of public domain genetic markers, including restriction fragment length polymorphism (RFLP) and amplified fragment length polymorphism (AFLP). This was achieved through coordination of the International *Lolium* Genome Initiative (ILGI), using the p150/112 one-way pseudo-testcross population of 183 F_1 genotypes. The ILGI reference map, constructed through collaboration between Victorian DPI, Australia; Institute of Grassland and Environmental Research (IGER), UK; Yamanashi Prefectural Dairy Experiment Station (YPDES) and the National Agricultural Research Centre for Hokkaido Region (NARCH), Japan and the Institut National de la Recherche Agronomique, (INRA) France contained ca. 200 AFLP loci and 109 heterologous RFLP loci (detected by wheat, barley, oat and rice cDNA probes), spanning 811 cM across seven linkage groups (LGs), and allowing the inference of comparative relationships between perennial ryegrass and other Poaceae species (Jones et al. 2002a). A predominant relationship between each perennial ryegrass LG and one of the homoeologous groups of hexaploid wheat was observed, providing a standardized nomenclature system, although several major translocation results were subsequently observed (Sim et al. 2005). This analysis has supported translational genetics between perennial ryegrass and the Triticeae. The ILGI map was enhanced through the addition of 93 LPSSR loci, providing the basis of framework genetic mapping in other populations (Jones et al. 2002b).

A second generation reference genetic mapping family was developed based on the F_1 (NA_6 x AU_6) two-way pseudo-testcross family of 157 F_1 genotypes, generating two parental genetic maps. The consolidated genetic maps included 43 LPSSR loci, 88 EST-RFLP loci and 71 EST-SSR loci on the NA_6 parental map, with a total length of 963 cM; and 49 LPSSR loci, 67 EST-RFLP loci and 58 EST-SSR loci on the AU_6 parental map, with a total length of 779 cM (Faville et al. 2004). The F_1 (NA_6 x AU_6) population maps have consequently been enriched with functionally-associated genetic markers.

Trait dissection for perennial ryegrass has been performed in multiple populations to allow QTL analysis. The p150/112 population has been analyzed for traits such as vegetative and reproductive morphogenesis, reproductive development and winter hardiness, herbage quality, mineral content and gametophytic self-incompatibility (Thorogood et al. 2002; Yamada et al. 2004; Cogan et al. 2005a, b; Shinozuka et al. 2006), while the F_1 (NA_6 x AU_6) population has been studied for a range of root and shoot morphogenesis, photosynthetic efficiency, pseudostem water soluble carbohydrate content, mineral content and crown rust resistance characters (Forster et al. 2004; Cogan et al. 2005a). Other perennial ryegrass populations have been analyzed to detect genetic control of crown rust resistance (Dumsday et al. 2003; Muylle et al. 2005a, b), vernalization response (Jensen et al. 2005b) and flowering time variation (Armstead et al. 2002, 2004). Interspecific hybrid populations, based on crosses between *L. multiflorum* and *L. perenne* have been used to identify QTLs for flowering time variation (Warnke et al. 2004), fiber components and crude protein content (Xiong et al. 2007a) and winter hardiness (Xiong et al. 2007b).

4.2 Molecular Marker Development, Genetic Map Construction, and Trait-Dissection in White Clover

The development of molecular genetic markers for white clover has been reviewed by Forster et al. (2001a). A comprehensive set (ca. 400) of unique white clover genomic DNA-derived SSR (TRSSR) markers was developed using enrichment library construction technology (Kölliker et al. 2001a). The white clover EST library has also been used to develop 792 EST-SSR primer pairs (Barrett et al. 2004).

Genetic map development in white clover was performed using a combination of TRSSR and AFLP markers. The reference mapping population was the F_2 (I.4R x I.5J) family that was developed at IGER, Aberystwyth, UK, with parental genotypes from fourth and fifth generation inbred lines descended from plants containing the rare self-fertile (S_f) allele. A single F_1 plant was self-pollinated to generate an F_2 population of 150 individuals (Michaelson-Yeates et al. 1997). The level of genetic polymorphism between the inbred parents, as assessed with TRSSR markers, was 48% of those markers showing efficient amplification. The F_2 (I.4R x I.5J) map contained 135 loci (78 TRSSR and 57 AFLP) on 18 LGs (two more than the karyotypic number), with a total map length of 825 cM. The extent of map construction was limited by high levels of segregation distortion, affecting 39% of the TRSSR loci, with the majority distorted towards the heterozygous genotypic class (Jones et al. 2003). A higher-resolution genetic map largely based on EST-SSR markers was constructed using the F_1 (Sustain 6525-2 x NRS 364-7) mapping family (Barrett et al. 2004). A total of 335 EST-SSR and 30 genomic DNA-derived SSR primer pairs detected polymorphism and permitted assignment of 493 loci to a genetic map containing 16 LGs, with a total map length of 1,144 cM. The EST-SSR markers detected homoeologous locations between the ancestral genomes at high frequency, and provided the basis for standard chromosome nomenclature development.

The F_2 (I.4R x I.5J) genetic map has been exploited for QTL analysis of a number of vegetative morphogenesis, reproductive morphogenesis and reproductive development traits (Cogan et al. 2006a). Target traits were measured using clonally replicated sets of plants at different geographical sites in Wales and Scotland, UK and across a number of years of evaluation. Individual environment analyses detected a large number of QTLs for each trait, with QTLs clustering for correlated traits, especially on LGs 7 and 12. Multi-environment combined analysis revealed genomic locations that are relatively insensitive to genotype x environment effects. The F_1 (Sustain 6525-2 x NRS 364-7) population has also been used for QTL analysis, specifically targeting seed production traits such as inflorescence density, yield per inflorescence and thousand-seed weight (Barrett et al. 2005). Stability of QTL effect was observed across temporal replication, along with co-location of QTLs for correlated traits.

4.3 Molecular Marker Development for Grass Fungal Endophytes and Pathogens

Initial molecular genetic development for grass fungal endophytes was based on non-specific systems such as random amplified polymorphic DNA (RAPD) (Liu et al. 1995) and AFLP markers (Arroyo Garcia et al. 2002) that do not require any prior DNA sequence information. Genomic DNA-derived SSR markers were also developed (Groppe et al. 1995; Moon et al. 1999), but the accuracy of genetic diversity measurements by such markers was limited by the small available numbers. Analysis of the *Neotyphodium* EST unigene collection identified SSRs in 9.7% of the *N. coenophialum* sequences and 6.3% of the *N. lolii* sequences. Primer sequences were designed to a total of 107 SSR-containing ESTs, and were used for preliminary evaluation of interspecific and intraspecific genetic variation (van Zijll de Jong et al. 2003). Comparison with AFLP data confirmed that the EST-SSR markers were informative for intraspecific genetic diversity analysis. Phenetic analysis of SSR polymorphism and subsequent sequence analysis of selected SSR-flanking regions has confirmed and extended knowledge of genome affinities between different taxa (van Zijll de Jong et al. 2005).

EST-SSRs have also been generated to investigate population structure in *Puccinia coronata* f.sp. *lolii*, the causative organism of crown rust which is the most important foliar pathogen of ryegrasses. Analysis of 1,100 ESTs from a urediniospore-derived cDNA library detected 55 SSRs, of which 12 proved useful for genotyping of single pustule-derived DNA templates (Dracatos et al. 2006).

4.4 Population Structure of Perennial Ryegrass, Grass Fungal Endophytes and Pathogens, and White Clover

Genetic diversity analysis for perennial ryegrass has been performed using the RAPD (Bolaric et al. 2005a, b, c) and AFLP (Roldán-Ruiz et al. 2000, 2001; Forster et al. 2001b; Guthridge et al. 2001; van Treuren et al. 2005) marker systems, which lack sequence specificity. A particularly interesting application of AFLP-based profiling was the prediction of optimum genetic diversity in polycross combinations for herbage yield (Kölliker et al. 2005). SSR-based diversity analysis has also been performed (Kubik et al. 2001; Forster et al. 2005a). These studies have demonstrated larger levels of genetic variation within compared to between populations. Varieties based on small numbers of parental genotypes (restricted-base varieties) were found to show lower levels of intrapopulation diversity and to be more readily discriminated than those based on larger numbers of parental individuals (non-restricted base varieties) (Guthridge et al. 2001; Forster et al. 2001b).

RAPD (Gustine and Elwinger 2003) and AFLP profiling (Kölliker et al. 2001b; van Treuren et al. 2005) has also been used to determine levels of genetic variability within and between white clover populations. As for perennial ryegrass, the majority of genetic variation was detected within rather than between populations, and divergent varieties were largely discriminated on the basis of AFLP profile (Kölliker et al. 2001b). Bulking at the genotypic level followed by AFLP analysis was used to determine the level of congruence between morpho-physiological and genotypic variation in white clover. More recently, a study of 16 elite varieties using genomic DNA-derived SSR markers revealed very high levels of intrapopulation diversity and overlap between populations, despite small but significant interpopulation differentiation (George et al. 2006b).

Intraspecific genetic diversity in the perennial ryegrass fungal endophyte *N. lolii* has been determined by SSR polymorphism analysis. A sub-set of 18 EST-SSR markers were optimized for *in planta* detection from leaf sheath samples of infected plants (van Zijll de Jong et al. 2005). Tests performed using *in vitro*-composed admixtures of genomic DNA isolated from a cultured endophyte and an endophyte-free grass genotype revealed high levels of sensitivity. A survey of global germplasm diversity has been performed across 390 genotypes representing 41 perennial ryegrass accessions from 20 different countries. Genetic diversity shows non-random association with geographic origin, and major clades of diversity are correlated with variation for alkaloid content (van Zijll de Jong et al. 2008). *In planta* genotyping provides the basis for efficient quality control of endophyte content and identity in commercial varieties, especially those containing novel strains with reduced neurotoxic alkaloid production.

Preliminary analysis of genetic diversity in *P. coronata* f.sp. *lolii* was performed using 16 isolates from the UK, Japan, Australia, and New Zealand. High levels of variation within and between geographical areas were observed (Dracatos et al. 2006).

4.5 Functionally-associated Genetic Markers for Perennial Ryegrass and White Clover

Methods for marker-trait gene linkage analysis and selection based on donor-recipient introgression schemes which have been developed for inbreeding plant species are not readily adaptable to outcrossing species with complex breeding systems, except under specific artificial circumstances. To address these issues, diagnostic genetic markers based on functionally-associated candidate genes have been developed (Forster et al. 2004; Spangenberg et al. 2005b; Cogan et al. 2006b). The reproductive habit and presumptive population structure of outbreeding forage species would be expected to dispose towards limited linkage disequilibrium (LD), extending over relatively short molecular distances (Mackay 2001; Flint-Garcia et al. 2003; Forster et al. 2004; Dobrowolski and Forster 2007). This is especially relevant for long-established populations derived from a large number of founding parents, as expected for ecotypes and long-established varieties, in which many rounds of recombination have occurred. These factors would tend to favour the use of candidate gene-based functionally-associated marker analyses (Andersen and Lübberstedt 2003) rather than whole-genome scans, although newly synthesized populations with small numbers of parents may prove suitable for limited genome-wide marker-based analysis. On this basis, nucleotide variation in qualified candidate genes will be closely associated with the casual mutations that generate variability for key agronomic traits, and may be used as diagnostics for such variation. Successful correlation of gene haplotype structure and phenotypic variation will provide the basis for a new paradigm in pasture plant molecular breeding based on direct selection of superior allele content at target genetic loci, allowing highly effective exploitation of germplasm collections for identification of potential parental genotypes. The subsequent progeny selection process will permit accumulation, towards fixation, of favorable alleles. In order to minimize the effects of inbreeding depression, parental individuals will be selected based on genome-wide genetic diversity at otherwise unselected genetic locations, and variability at otherwise unselected genetic locations will be maximized during synthetic variety development (Forster et al. 2004; Spangenberg et al. 2005b) (Fig. 1).

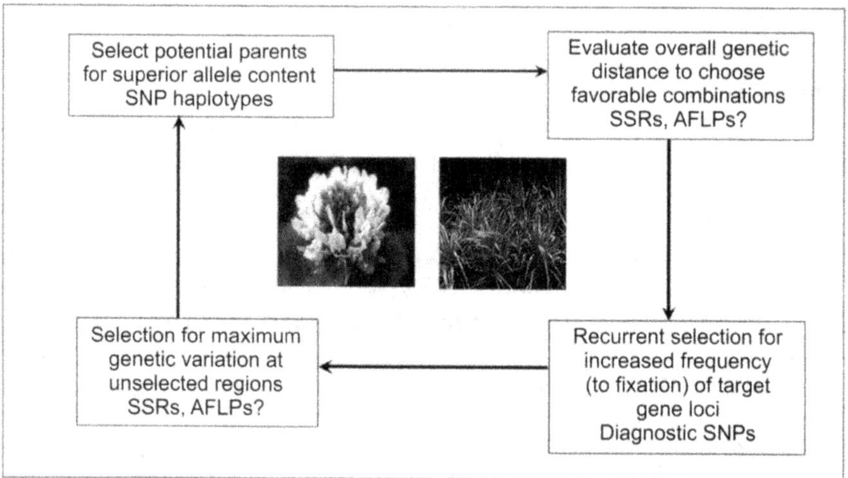

Fig. 1 Schematic representation of the concept for a molecular marker-based breeding cycle for outbreeding pasture species.

In order to determine the validity of the candidate gene-based molecular marker concept for outbreeding pasture species, major efforts have been devoted to the development and implementation of large-scale methods for single nucleotide polymorphism (SNP) discovery. This process has been based on a two-part strategy. An *in silico* discovery component has exploited computational detection of SNPs in sequence contigs from EST collections derived from multiple heterogeneous individuals from the white clover variety cultivar Grasslands Huia (Sawbridge et al. 2003a) and the perennial ryegrass cultivar Grasslands Nui (Sawbridge et al. 2002b), respectively (Spangenberg et al. 2005b). The *in vitro* discovery component has exploited a method based on cloning and sequencing of gene-specific amplicons from the heterozygous parents of two-way pseudo-testcross mapping families. Although this method is costly and time-consuming compared to direct sequencing of PCR products, it has several important advantages. Ambiguities in sequence traces due to heterozygous indels may be readily distinguished, and haplotype structures within amplicons are directly determined (Zhang and Hewitt 2003). In addition, paralogous sequences amplified by conserved primers are discriminated, and may be used to distinguish between members of multigene families (Edwards et al. 2007). The putative SNPs are then validated in the progeny set from the parental cross, and cross-validated in other sib-ships and diverse germplasm. The *in vitro* discovery method has been progressed in the form of three distinct intensity streams: the low intensity stream involved short ESTs, providing single SNP loci for structured map enhancement; the medium intensity stream involved involves full-length cDNAs, providing several SNP loci and partial haplotypic data; and the high intensity stream involved full-length genes with intron-exon structure, providing multiple SNP loci and determination of complete haplotype structures (Forster et al. 2005b; Cogan et al. 2006b).

For perennial ryegrass, the majority of SNPs were obtained by *in vitro* analysis from genes which were previously analyzed as RFLP markers (Faville et al. 2004). 'Proof-of-concept' for the *in vitro* discovery process was obtained with *Lp*ASRa2, an abscisic acid (ABA) inducible gene which is believed to be associated with osmotic stress tolerance in rice and maize (Vaidyanathan et al. 1999; Jeanneau et al. 2002). SNP haplotypes from the parents of the F_1 (NA_6 x AU_6) genetic mapping family were defined and correlated with amino acid changes which may affect the functionality of the derived protein. Alternatively, the characterized SNPs may be in LD with functionally-significant changes in the transcriptional control regions (Paran and Zamir 2003), given haplotype stability over gene-length distances. Validated SNPs were used to determine diversity and putative haplotype structure within diverse germplasm collections, along with preliminary determinations of LD (Cogan et al. 2006b). Data obtained from *Lp*ASRa2 and several full-length herbage quality (lignin biosynthesis, oligosaccharide metabolism) genes (Ponting et al. 2007) suggested that LD blocks in perennial ryegrass genes are unlikely to extend further than 1-2 kb.

Over 150 genes were introduced into the *in vitro* SNP discovery process for perennial ryegrass, of which 100 have been sequenced and aligned, with a total of 1,592 putative SNPs across 82 genes. SNPs were validated using the single nucleotide polymorphism (SNuPe) assay for 66 genes. Over a total of 87 kb of re-sequenced DNA, a relatively high SNP frequency (for four haplotypes) of 1 per 55 bp was observed, with higher incidence in intron compared to exon sequences, as anticipated (Cogan et al. 2006b). The *in vitro* discovery process has been augmented with limited *in silico* discovery, and continues with an emphasis on genes associated with resistance to plant diseases (Dracatos et al. 2008) such as crown rust, including defence response (DR) genes (chitinases, catalases, etc.) and resistance (R) genes (NBS-LRR class, receptor kinases, etc.), and genes associated with abiotic stress tolerance, such as CBF/DREB, dehydrin and late embryogenesis abundant (LEA) protein genes. Validated gene-associated SNP loci have been assigned to the perennial ryegrass p150/112 (Jones et al. 2002a, b) and F_1 (NA_6 x AU_6) (Faville et al. 2004) genetic maps and in many cases show co-location with pre-existing EST-RFLP loci (Cogan et al. 2006b). Co-location with QTLs for disease resistance, mineral content, flowering time and vegetative morphogenesis traits has been demonstrated (Cogan et al. 2005a; Shinozuka et al. 2005; Yamada et al. 2005).

The complementary SNP discovery process for white clover has been dominated by the *in silico* strategy, due to the presence of a relatively smaller number of well-characterized candidate genes. A sample of 236 EST contigs was selected for validation using the parents and progeny of the two-way pseudo-testcross mapping families F_1 ($Haifa_2$ x LCL_2) and F_1 ($S184_6$ x LCL_6). A total of 106 of the clusters (45%) were initially identified, and more detailed analysis confirmed 58 clusters as containing segregating SNPs. The SNP-containing genes belong to a range of predicted functional categories, including ribosomal proteins, heat-shock proteins, calmodulins, and lipid and organic acid biosynthesis enzymes (Cogan et al. 2007). The *in silico*-derived SNPs are being deployed along with genomic DNA-derived SSR and EST-SSR markers to generate second-generation trait-specific genetic maps for white clover. Polymorphic SNP and SSR-containing ESTs have also been used for comparative genomics analysis by comparison with whole genome data from the model legume species *M. truncatula* and *L. japonicus*, as well as *A. thaliana*. The homoeologous groups of white clover were aligned with the model species genomes, revealing a predominant correspondence between each group and one of the chromosomes of *M. truncatula* (George et al. 2006a).

The complementary *in vitro* SNP discovery process for white clover is expected to be influenced by the allopolypoid genetic constitution of this species ($2n = 4x = 32$). Close to 50 white clover cDNAs were selected from public databases, including the cyanogenesis-associated linimarase gene, *Tr*LIN, and from the unigene resource of Sawbridge et al. (2003a), including genes for flavonoid biosynthesis (relevant to bloat safety) and organic acid biosynthesis (relevant to aluminium tolerance and phosphorus acquisition). The F_1 ($Haifa_2$ x LCL_2) and F_1 ($S184_6$ x LCL_6) families provided the parental DNA templates for *in vitro* SNP discovery.

High levels of haplotypic complexity were observed, along with elevated failure rates for predicted SNP validation (Edwards et al. 2007; Lawless et al. 2008). Although intragenomic paralogy could account for these effects, homoeologous gene amplification is a more likely explanation. Nonetheless, individual SNP loci suitable for genetic mapping were validated for selected genes. Methods for improved gene and genome-specific SNP discovery for white clover have been developed, including protein domain identification, subtraction of haplotypes from putative progenitors and use of BAC libraries for gene- and genome-specific primer design.

Large-scale SNP development for perennial ryegrass and white clover has permitted the design of crucial 'proof-of-concept' experiments in association genetics (Thornsberry et al. 2001; Flint-Garcia et al. 2003). The emphasis is on perennial ryegrass herbage quality genes (Lidgett et al. 2002; Lynch et al. 2002; McInnes et al. 2002; Chalmers et al. 2003), including those subjected to high intensity SNP discovery. Several lignin biosynthesis genes have previously been identified as co-locating with QTLs for herbage digestibility on LG 7 (Cogan et al. 2005a). An association mapping panel (AMP) of 384 genotypes has been established, including both globally diverse and ecotypic materials. Population structure has been determined through genotyping with c. 150 SSR markers, followed by c. 150 SNPs from 20 genes. In parallel, phenotypic analysis is being performed through the use of calibrated near infra-red reflectance spectroscopy (NIRS) analysis. Successful identification of correlations between gene haplotype structure and phenotypic variation (Fig. 2) will validate the future use of this approach for other target agronomic traits, and generate diagnostic markers for implementation in practical breeding programs (Dobrowolski and Forster 2007).

5 MOLECULAR MARKER TECHNOLOGY FOR IMPROVEMENT OF OTHER FORAGE GRASSES

5.1 Molecular Marker Technology for Other *Lolium* Species

Genetic map development in Italian ryegrass has employed two-way pseudo-testcross populations and multiple marker systems. The first *Lolium multiflorum* genetic map was based on RFLPs detected by *Pst*I-generated genomic clones and heterologous anchor clones from other Poaceae species, in addition to AFLPs and telomeric repeat-associated sequence (TAS) markers. Seven LGs were identified, with a total map length of 1,244 cM (Inoue et al. 2004a), and comparative relationships

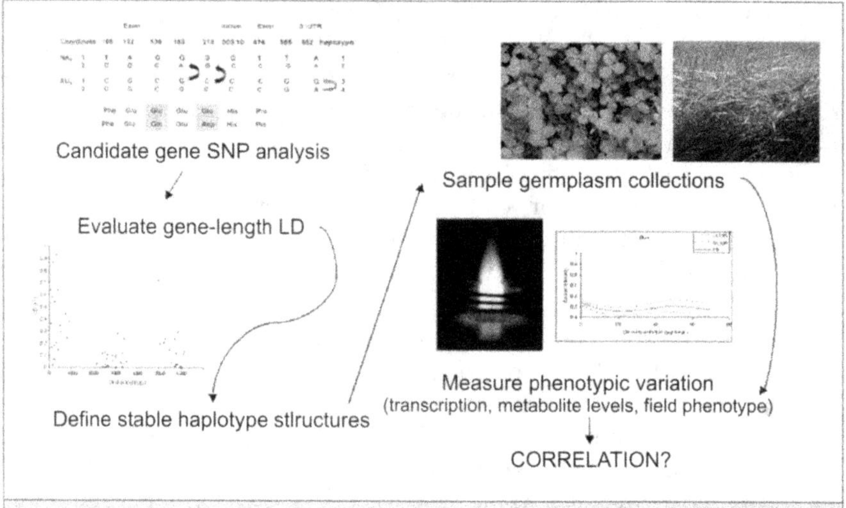

Fig. 2 Schematic representation of the process for confirmation of correlation between candidate gene-based SNP haplotype and phenotypic trait variation for outbreeding pasture species.

were inferred, which were in accordance with those predicted for perennial ryegrass. A sub-set of genomic DNA-derived RFLP probes were subsequently end-sequenced for conversion into PCR-based markers (SSRs and sequence tagged sites [STSs]) (Inoue et al. 2004b). Genomic DNA-derived SSR markers obtained from 3,965 unique clones were subsequently genotyped over the same population to generate comprehensive parental genetic maps (Hirata et al. 2006) providing the basis for framework mapping of other populations.

Trait-dissection in Italian ryegrass has permitted the detection of QTLs for lodging resistance and related traits such as heading date, plant height, culm weight, culm diameter, culm strength and tiller number (Inoue et al. 2004c). EST-derived cleaved amplified polymorphic sequence (CAPS) and AFLP markers were also used to map and tag a major gene locus (*LmPil*) for ryegrass blast (grey leaf spot) resistance (Miura et al. 2005). A large two-way pseudo-testcross F_1 progeny set based on two highly heterozygous individuals with contrasting resistance to bacterial wilt (*Xanthomonas translucens* pv. *graminis*) has been used to generate an SSR- and AFLP-based map and to locate a single large (accounting for 67% of total phenotypic variance) resistance QTL (Studer et al. 2006).

Annual ryegrass (*Lolium rigidum* Gaud.) has historically been cultivated as an annual pasture, especially for sheep production in the low rainfall regions of southern Australia. However, a shift to zero-tillage grain production in these regions has minimized the grazing role of annual ryegrass and created a new problem of weedy growth, coupled with resistance to multiple herbicides including glyphosate and paraquat (Powles et al. 1998; Powles and Preston 1999). Genomic DNA-derived SSR markers from perennial ryegrass were used to study the mode of inheritance of glyphosate resistance in the NLR70 ecotype. Phenotypic analysis of crosses involved parents from this ecotype had suggested the presence of a major semi-dominant nuclear resistance fixed in the homozygous state (Lorraine-Colwill et al. 2001), and this conclusion was supported by marker-association analysis (Forster et al. 2005a).

Darnel (*Lolium temulentum* L.) is an inbreeding relative of the ryegrasses and a weed species of cereal agriculture since ancient times: the parable of the tares and wheat told by Christ in the New Testament (Matthew 13: 24-30) probably refers to darnel. The autogamous nature of *L. temulentum* has generated interest in the use of the species as a model for aspects of pasture grass biology, such as floral development (Evans 1999). A small number of SSR markers have been generated from genomic libraries (Senda et al. 2004) and have been used to study geographical differentiation between populations of darnel (Senda et al. 2005).

5.2 Molecular Marker Technology for *Festuca* species

The *Festuca* genus is closely related to *Lolium* within the Poeae tribe of the Poaceae, and fertile hybrids may be generated between members of the two genera through the use of hyperploid bridges (Thomas et al. 1988; Humphreys 1989), permitting controlled introgression (King et al. 1998). Molecular markers for genetic mapping in *Festuca* species were developed as RFLP probes from the allohexaploid ($2n = 6x = 42$) taxon *F. arundinacea* (Xu et al. 1991). The clones were evaluated for genetic polymorphism in tall fescue, as well as for hybridization to other diploid and tetraploid species of *Festuca*. A first-generation tall fescue mapping population was developed by crossing two parents selected for maximal genetic divergence. One of the parents was also selected because of its use in a recurrent selection program to improve *in vitro* dry matter digestibility. F_1 plants were selfed to produce a mapping population consisting of 105 F_2 plants. The map covered 1,274 cM on 19 different LGs, with an average of 5 loci per LG and 17.9 cM between loci (Xu et al. 1995), and revealed homoeologous relationships between different LGs.

A total of 157 EST-SSR primer pairs were designed from tall fescue cDNA library sequences and tested for cross-species amplification in seven different grass and cereal species (Saha et al. 2004). A sub-set of these markers was used for more intensive phylogenetic analysis of cool-season grasses (Mian et al. 2005). EST-SSR and AFLP primer pairs were used to construct a tall fescue map based on PCR-generated marker genotyping of the F_1 (HD28-56 x R43-64) two-way pseudo-testcross population (Saha et al. 2005). Integration of the parental genetic maps produced a consensus structure of 17 LGs and a cumulative map length of 1,841 cM. Homoeologous relationships between six of the seven expected groups were observed. Further map enhancement would be expected to identify the remaining 5 homologous LGs, to the anticipated total of 22.

Tall fescue RFLP markers have also been used to efficiently detect RFLP in the diploid ($2n = 2x = 14$) taxon *F. pratensis*. Comparative genetic mapping between the two species of fescue (Chen et al. 1998) generated a map containing 66 markers on 7 LGs with a total map length of 280 cM. A high level of conservation for the orders of most of the genetic markers was observed, with 23 of 33 common markers located in corresponding LGs.

A comprehensive genetic map has been constructed for meadow fescue using a two-way pseudo-testcross F_1 population from the pair-cross of a single genotype from a Norwegian population that has been selected for frost tolerance (HF2/7) and a genotype from a Yugoslavian variety (B14/16). The combined data for homologous and heterologous RFLP, AFLP, SSR and isoenzyme markers from the two parental maps defined 466 loci with a total map length of 658 cM (Alm et al. 2003). Conserved synteny was analyzed through the use of heterologous RFLP anchor probes derived from perennial and Italian ryegrass, wheat, barley, oat, rice, maize, and sorghum, demonstrating a high degree of conserved synteny and colinearity with both perennial ryegrass and the Triticeae consensus map. As meadow fescue is a close taxonomic relative of perennial ryegrass, and high levels of genetic recombination are observed in interspecific F_1 hybrids (King et al. 1998), a high level of alignment with the *L. perenne* genetic map was anticipated. The LGs of meadow fescue have been designated 1F-7F to correspond with the Triticeae homoeologous chromosomes. The F_1 (B14/16 x HF2/7) mapping population has been used for QTL analysis of vernalization requirement, heading time and number of panicles (Ergon et al. 2006), following phenotypic evaluation (Fang et al. 2004). Molecular genetic markers have also been used to monitor the introgression of a crown rust resistance determinant from *F. pratensis* into *L. multiflorum* (Armstead et al. 2006).

5.3 Molecular Marker Technology for Other Temperate Pasture Grass Species

SSR enrichment genomic libraries were developed for timothy (*Phleum pratense* L.), permitting development of 355 unique markers, of which 90% detected polymorphism between the parents of a two-way pseudo-testcross population (Cai et al. 2003). These markers provide the basis for framework genetic mapping, QTL detection and marker-assisted selection (MAS) in this species.

Genetic maps based on AFLPs, SSRs and RFLPs were generated for full-sib populations obtained from interspecific crosses between the allotetraploid wild rye species *Leymus triticoides* and *Leymus cinereus* ($2n = 4x = 28$). Conserved synteny was demonstrated with each of the seven homoeologous groups of wheat and barley (Wu et al. 2003).

6 MOLECULAR MARKER TECHNOLOGY FOR IMPROVEMENT OF OTHER FORAGE LEGUMES

6.1 Molecular Marker Technology for Other *Trifolium* Species

A small set (27) of genomic DNA-derived SSR markers were obtained from enrichment libraries of red clover (*Trifolium pratense* L.), permitting assessment of high levels of genetic variability in red clover, as well as low level cross-species amplification in white clover (Kölliker et al. 2006). A much larger set of red clover SSR markers were derived from a large-scale EST discovery program (Sato et al. 2006), from which a total of 26,356 ESTs were generated and clustered into 9,339 unigenes. A total of 7,159 SSR locus primer pairs were designed from candidate sequences.

A comprehensive red clover genetic map has been constructed using the F_1 (HR x R130) two-way pseudo-testcross population, initially through the use of 157 cDNA-derived RFLP markers (Isobe et al. 2003), and more recently through incorporation of genomic DNA-derived and EST-SSR markers to generate a map containing 1,434 SSR loci across the 7 LGs with a cumulative map length of 869 cM. Macrosynteny was determined between red clover and both *Medicago truncatula* and *Lotus japonicus* (Sato et al. 2006). A second two-way pseudo-testcross population has been used to construct a map containing 216 AFLP and 42 SSR loci, with a cumulative length of 444 cM. This genetic map was utilized for QTL analysis of seed yield component traits, identifying 38 QTLs for eight traits, including two genomic regions with clustered QTLs suitable for marker-based selection (Herrmann et al. 2006).

6.2 Molecular Marker Technology for *Medicago* Species

Cultivated alfalfa (*Medicago sativa* L.) is an autotetraploid species ($2n = 4x = 32$) with a partial self-incompatibility system (Barnes et al. 1972). Due to the complexities inherent in genetic analysis of tetrasomic inheritance systems, initial genetic mapping of alfalfa was based on maps constructed using diploid sub-species (Brummer et al. 1993; Kiss et al. 1993; Echt et al. 1994; Tavoletti et al. 1996; Kalo et al. 2000) and populated by RFLP, RAPD, and isoenzyme markers.

Genetic mapping of tetraploid alfalfa was initially performed using RAPD markers (Yu and Pauls 1993) and a more detailed RFLP-based genetic map was developed by Brouwer and Osborn (1999). The most recent generation of genetic maps are predominantly based on SSR markers, initially derived from genomic enrichment libraries (Diwan et al. 1997). The availability of large numbers of ESTs from *M. truncatula* has permitted cross-species amplification and polymorphism assessment in other *Medicago* species (Eujayl et al. 2004). Genomic DNA-derived and EST-SSR markers have been used to construct a functionally-associated marker-based tetraploid genetic map (Sledge et al. 2005) in a pair-cross population varying for drought tolerance-associated traits. Other trait dissection studies in autotetraploid alfalfa have targeted resistance to root rot (*Phytophthora medicaginis*) (Musial et al. 2005).

6.3 Molecular Marker Technology for Other Temperate Pasture Legume Species

Bird's foot trefoil (*Lotus corniculatus* L.), like cultivated alfalfa, is a tetraploid species ($2n = 4x = 24$). Genetic linkage mapping using RFLP, RAPD, inter-simple sequence repeat (ISSR) and STS markers has demonstrated tetrasomic inheritance, confirming *L. corniculatus* as an autotetraploid. Six LGs were identified, with a cumulative map distance of 572 cM (Fjellstrom et al. 2003).

7 MOLECULAR MARKER TECHNOLOGY FOR IMPROVEMENT OF COOL-SEASON TURF GRASS SPECIES

The molecular genetic marker technologies developed for forage-type *Lolium* and *Festuca* species are clearly also important for turf-type varieties. The most important pathogens of turf-type ryegrasses are grey leaf spot (*Magnaporthe grisea*), dollar spot (*Sclerotinia homeocarpa*), and brown patch (*Rhizoctonia solani*). Trait dissection analysis of grey leaf spot resistance in an Italian ryegrass x perennial ryegrass interspecific hybrid population has identified QTLs on 4 LGs, including two (LGs 3 and 4) containing regions syntenic to resistance loci for the rice blast pathogen, which is also caused by *M. grisea* (Curley et al. 2005).

Genetic linkage mapping has also been performed for creeping bentgrass (*Agrostis stolonifera* L.), which is highly adapted to use on the fairways and putting greens of golf courses. *A. stolonifera* is an obligate outbreeding allotetraploid species ($2n = 4x = 28$). A genetic linkage map was constructed using AFLP, RAPD, and both homologous and heterologous RFLP markers, containing 14 LGs and spanning 1,130 cM. Comparative mapping analysis with wheat and perennial ryegrass demonstrated conserved synteny for each of the homoeologous groups (Chakraborty et al. 2005). Trait-dissection analysis for dollar spot resistance has been performed using the mapping population (Chakraborty et al. 2006) and similar studies for brown patch resistance are planned.

8 CONCLUSIONS

Historically, major efforts in the development of molecular genetic and genomic resources, genetic mapping and marker-assisted selection in crop plants have concentrated on obligate or facultative inbreeding species such as tomato, soybean, wheat, barley, rice, and maize. The relatively lower perceived economic importance of pasture species has delayed the establishment of equivalent activities, but large-scale resources have now been established for perennial ryegrass, tall fescue, white clover, and red clover. The obligate outbreeding reproductive habit exhibited by the majority of pasture species has also provided challenges for genetic map construction and trait-dissection, which have been effectively addressed. Although multiple parental genetic maps are generated for two-way pseudo-testcross derived populations, the individual maps may be effectively bridged with common markers, and

QTLs attributed to each map. The major remaining challenge is the successful implementation of molecular genetic marker-based technologies in varietal development. Diagnostic markers based on functionally-associated variation in candidate genes provide the means to achieve such success, and efficient methods for generation and validation of such markers based on *in silico* and *in vitro* SNP discovery have been developed. Selection based on superior allele content represented by SNP haplotype is a method of wide-applicability for outbreeding forage and turf species, and is capable of addressing the confounding effects of G x E during conventional selection processes. However, improved methods of candidate gene identification and selection are required to ensure the widespread adoption and impact of gene-associated marker-based breeding. Enhanced efforts in functional genomics, including transcriptomics, exploitation of comparative relationships and physical mapping leading to positional candidate identification are needed to achieve these objectives. Intensive developments are also required for methods for tactical implementation of molecular genetic marker technologies in pasture plant enhancement programs, especially to identify the optimal stages in varietal construction programs for intervention. Computer simulated models will provide important supporting information for this process.

Acknowledgments

This original research described here was supported by funding from the Victorian Department of Primary Industries (DPI), Dairy Australia Ltd., the Geoffrey Gardiner Dairy Foundation, Meat and Livestock Australia Ltd., and the Molecular Plant Breeding Cooperative Research Centre. The EST generation and characterization program was jointly funded by DPI and AgResearch New Zealand. The authors wish to thank all colleagues past and present within DPI for contributions, and Prof. Michael Hayward for scientific support and advice.

References

Alm V, Fang C, Busso CS, Devos KM, Vollan K, Grieg Z, Rognli OA (2003) A linkage map of meadow fescue (*Festuca pratensis* Huds.) and comparative mapping with other Poaceae species. Theor Appl Genet 108: 25-40

Andersen JR, Lübberstedt T (2003) Functional markers in plants. Trends Plant Sci 8: 554-560

Ansari HA, Ellison NW, Reader SM, Badaeva ED, Friebe B, Miller TE, Williams WM (1999) Molecular cytogenetic organisation of 5S and 18S-26S rDNA loci in white clover (*Trifolium repens* L.) and related species. Ann Bot 83: 199-206

Ansari HA, Ellison NW, Griffiths AG, Williams WM (2004) A lineage-specific centromeric satellite sequence in the genus *Trifolium*. Chrom Res 12: 357-367

Armstead IP, Turner LB, King IP, Cairns AJ, Humphreys MO (2002) Comparison and integration of genetic maps generated from F_2 and BC_1-type mapping populations in perennial ryegrass. Plant Breed 121: 501-507

Armstead IP, Turner LB, Farrell M, Skøt L, Gomez P, Montoya T, Donnison IS, King IP, Humphreys MO (2004) Synteny between a major heading-date QTL in perennial ryegrass (*Lolium perenne* L.) and the *Hd3* heading-date locus in rice. Theor Appl Genet 108: 822-828

Armstead IP, Harper JA, Turner LB, Skøt L, King IP, Humphreys MO, Morgan WG, Thomas HM, Roderick HW (2006) Introgression of crown rust (*Puccinia coronata*) resistance from meadow fescue (*Festuca pratensis*) into Italian ryegrass (*Lolium multiflorum*): genetic mapping and identification of associated molecular markers. Plant Pathol 55: 62-67

Arroyo Garcia R, Martinez Zapater JM, Garcia Criado B, Zabalgogeazcoa I (2002) Genetic structure of the grass endophyte *Epichloë festucae* in semiarid grasslands. Mol Ecol 11: 355-364

Arumuganathan K, Earle ED (1991) Nuclear DNA content of some important plant species. Plant Mol Biol Rep 9: 208-218

Attwood SS (1940) Genetics of cross-incompatibility among self-incompatible plants *of Trifolium repens*. J Am Soc Agron 32: 955-968

Attwood SS (1941) Controlled self- and cross-pollination of *Trifolium repens*. J Am Soc Agron 33: 538-545

Attwood SS (1942a) Oppositional alleles causing self-incompatibility in *Trifolium repens*. Genetics 27: 333-338

Attwood SS (1942b) Genetics of pseudo-self-incompatibility and its relation to cross-incompatibility in *Trifolium repens* L. J Agri Res 64: 699-709

Badr A, Sayed-Ahmed H, El-Shanshouri A, Watson LE (2002). Ancestors of white clover (*Trifolium repens* L.), as revealed by isozyme polymorphisms. Theor Appl Genet 106: 143-148

Barnes DK, Bingham ET, Axtell JD, Davis WH (1972) The flower, sterility mechanisms and pollination control. In: CH Hanson (ed) Alfalfa Science and Technology. Agronomy 15: 123-141 American Society of Agronomy, Madison, WI, USA

Barrett B, Griffiths A, Schreiber M, Ellison N, Mercer C, Bouton J, Ong B, Forster J, Sawbridge T, Spangenberg G, Bryan G, Woodfield D (2004) A microsatellite map of white clover (*Trifolium repens* L.). Theor Appl Genet 109: 596-608

Barrett BA, Baird IJ, Woodfield DR (2005) A QTL analysis of white clover seed production. Crop Sci 45: 1844-1850

Baumann U, Juttner J, Bian X-Y, Langridge P (2000) Self-incompatibility in the grasses. Ann Bot 85 (Suppl A): 203-209

Bolaric S, Barth S, Melchinger AE, Posselt UK (2005a) Genetic diversity in European perennial ryegrass cultivars investigated with RAPD markers. Plant Breed 124: 161-166

Bolaric S, Barth S, Melchinger AE, Posselt UK (2005b) Molecular genetic diversity within and among German ecotypes in comparison to European perennial ryegrass cultivars. Plant Breed 124: 257-262

Bolaric S, Barth S, Melchinger AE, Posselt UK (2005c) Molecular characterisation of genetic diversity in European germplasm of perennial ryegrass. Euphytica 146: 39-44

Bray RA, Irwin JAG (1999) *Medicago sativa* L. (lucerne) cv. Hallmark. Aus J Exp Agri 39: 643-644

Breen JP (1993) Enhanced resistance to three species of aphids (*Homoptera, Aphididae*) in *Acremonium* endophyte-infected turfgrasses. J Econ Entomol 86: 1279-1286

Brouwer DJ, Osborn TC (1999) A molecular linkage map of tetraploid alfalfa (*Medicago sativa* L.). Theor Appl Genet 99: 1194-1200

Brummer EC, Bouton JH, Kochert H (1993) Development of an RFLP map in diploid alfalfa. Theor Appl Genet 86: 329-332

Cai HW, Yuyama N, Tamaki H, Yoshizawa A (2003) Isolation and characterisation of simple sequence repeat markers in the hexaploid forage grass timothy (*Phleum pratense* L.). Theor Appl Genet 107: 1337-1349

Campbell BD, Caradus JR, Hunt CL (1999) Temperature responses and nuclear DNA amounts of seven white clover populations which differ in early spring growth rates. NZ J Agri Res 42: 9-17

Chakraborty N, Bae J, Warnke S, Chang T, Jung G (2005) Linkage map construction in allotetraploid creeping bentgrass (*Agrostis stolonifera* L.). Theor Appl Genet 111: 795-803

Chakraborty N, Curley J, Warnke S, Casler MD, Jung G (2006) Mapping QTL for dollar spot resistance in creeping bentgrass (*Agrostis stolonifera* L.). Theor Appl Genet 113: 1421-1435

Chalmers J, Johnson X, Lidgett A, Spangenberg G (2003) Isolation and characterisation of a sucrose:sucrose 1-fructosyltransferase gene from perennial ryegrass (*Lolium perenne* L.). J Plant Physiol 160: 1385-1391

Chen CC, Gibson PB (1970) Chromosome pairing in two interspecific hybrids of *Trifolium*. Can J Genet Cytol 12: 790-794

Chen CC, Gibson PB (1971) Karyotypes of fifteen *Trifolium* species in section Amoria. Crop Sci 11: 441-445

Chen C, Sleper DA, Johal GS (1998) Comparative RFLP mapping of meadow and tall fescue. Theor Appl Genet 97: 255-260

Cogan NOI, Vecchies AC, Yamada T, Dobrowolski MD, Smith KF, Forster JW (2005a) QTL analysis of mineral content in perennial ryegrass (*Lolium perenne* L.). In: Humphreys MO (ed) Molecular Breeding for the Genetic Improvement of Forage Crops and Turf. Academic Publ, Wageningen, The Netherlands, p 153

Cogan NOI, Smith KF, Yamada T, Francki MG, Vecchies AC, Jones ES, Spangenberg GC, Forster JW (2005b) QTL analysis and comparative genomics of herbage quality traits in perennial ryegrass (*Lolium perenne* L.). Theor Appl Genet 110: 364-380

Cogan NOI, Abberton MT, Smith KF, Kearney G, Marshall AH, Williams A, Michaelson-Yeates TPT, Bowen C, Jones ES, Vecchies AC, Forster JW (2006a) Individual and multi-environment combined analyses identify QTLs for morphogenetic and reproductive development traits in white clover (*Trifolium repens* L.). Theor Appl Genet 112: 1401-1415

Cogan NOI, Ponting RC, Vecchies AC, Drayton MC, George J, Dobrowolski MP, Sawbridge TI, Spangenberg GC, Smith KF, Forster JW (2006b) Gene-associated single nucleotide polymorphism (SNP) discovery in perennial ryegrass (*Lolium perenne* L.). Mol Genet Genom 276: 101-112

Cogan NOI, Drayton MC, Ponting RC, Vecchies AC, Bannan NR, Sawbridge TI, Smith KF, Spangenberg GC, Forster JW. (2007) Validation *of in silico*-predicted genic SNPs in white clover (*Trifolium repens* L.), an outbreeding allopolyploid species. Mol Genet Genom 277: 413-425

Connolly V (1990) Seed yield and yield components in ten white clover cultivars. Irish J Agri Res 29: 41-48

Cornish MA, Hayward MD, Lawrence MJ (1979) Self-incompatibility in ryegrass. I. Genetic control in diploid *Lolium perenne* L. Heredity 43: 95-106

Curley J, Sim SC, Warnke S, Leong S, Barker R, Jung G (2005) QTL mapping of resistance to grey leaf spot in ryegrass. Theor Appl Genet 111: 1107-1117

Devey F, Fearon CH, Hayward MD, Lawrence MJ (1994) Self-incompatibility in ryegrass. 11. Number and frequency of alleles in a cultivar of *Lolium perenne* L. Heredity 73: 262-264

Diwan N, Bhagwat AA, Bauchan GB, Cregan PB (1997) Simple sequence repeat DNA markers in alfalfa and perennial and annual *Medicago* species. Genome 40: 887-895

Dobrowolski MP, Forster JW (2007) Linkage disequilibrium-based association mapping in forage species. In: Oraguzie NC, Rikkerink E, Gardiner SE, De Silva NH (eds) Association Mapping in Plants. Springer, New York, USA, pp 197-209

Doyle JJ, Luckow MA (2003) The rest of the iceberg. Legume diversity and evolution in a phylogenetic context. Plant Physiol 131: 900-910

Dracatos PM, Dumsday JL, Olle RS, Cogan NOI, Dobrowolski MP, Smith KF, Forster JW (2006) Development and characterisation of EST-SSR markers from the crown rust pathogen of ryegrass (*Puccinia coronata* Corda f.sp. *lolii* Brown). Genome 49: 572-583

Dracatos PM, Cogan NOI, Dobrowolski MP, Sawbridge TI, Spangenberg GC, Smith KF, Forster JW (2008) Discovery and genetic mapping of single nucleotide polymorphisms in candidate genes for pathogen defense response in perennial ryegrass (*Lolium perenne* L.). Theor Appl Genet 117: 203-219

Draper J, Mur LAJ, Jenkins G, Ghosh-Biswas GC, Bablak P, Hasterok R, Routledge APM (2001) *Brachypodium distachyon*: a new model system for functional genomics in grasses. Plant Physiol 127: 1539-1555

Dumsday JL, Smith KF, Forster JW, Jones ES (2003) SSR-based genetic linkage analysis of resistance to crown rust (*Puccinia coronata* Corda f. sp. *lolii*) in perennial ryegrass (*Lolium perenne* L.). Plant Pathol 52: 628-637

Echt CS, Kidwell KK, Knapp SJ, Osborn TC, McCoy TK (1994) Linkage mapping in diploid alfalfa (*Medicago sativa*). Genome 37: 61-71

Edwards D, Forster JW, Cogan NOI, Batley J, Chagné D (2007) Single nucleotide polymorphism discovery in plants. In: Oraguzie NC, Rikkerink E, Gardiner SE, De Silva NH (eds) Association Mapping in Plants. Springer, New York, USA, pp 53-76

Ellison NW, Liston A, Steiner JJ, Williams WM, Taylor NL (2006) Molecular phylogenetics of the clover genus (*Trifolium* - Leguminosae). Mol Phylogenet Evol 39: 688-705

Ergon A, Fang C, Jorgensen O, Aamlid TS, Rognli OA (2006) Quantitative trait loci controlling vernalisation requirement, heading time and number of panicles in meadow fescue (*Festuca pratensis* Huds.). Theor App Genet 112: 232-242

Eujayl I, Sledge MK, Wang L, May GD, Chekhovskiy K, Zwonitzer JC, Mian MAR (2004) *Medicago truncatula* EST-SSRs reveal cross-species genetic markers for *Medicago* spp. Theor Appl Genet 108: 414-422

Evans LT (1999) Gibberellins and flowering in long day plants, with special reference to *Lolium temulentum*. Aust J Plant Physiol 26: 1-8

Fang C, Aamlid TS, Jorgensen O, Rognli OA (2004) Phenotypic and genotypic variation in seed production traits within a full-sib family of meadow fescue. Plant Breed 123: 241-246

Farrar K, Asp T, Lübberstedt T, Xu M, Thomas AM, Christiansen C, Humphreys MO, Donnison IS (2007) Construction of two *Lolium perenne* BAC libraries and identification of BACs containing candidate genes for disease resistance and forage quality. Mol Breed 19: 15-23

Faville M, Vecchies AC, Schreiber M, Drayton MC, Hughes LJ, Jones ES, Guthridge KM, Smith KF, Sawbridge T, Spangenberg GC, Bryan GT, Forster JW (2004) Functionally-associated molecular genetic marker map construction in perennial ryegrass (*Lolium perenne* L.). Theor Appl Genet 110: 12-32

Fearon CH, Cornish MA, Hayward MD, Lawrence MJ (1994) Self-incompatibility in ryegrass. 10. Number and frequency of alleles in a natural-population of *Lolium perenne* L. Heredity 73: 254-261

Felitti S, Shields K, Ramsperger M, Tian P, Webster T, Ong B, Sawbridge T, Spangenberg G (2004) Gene discovery and microarray-based transcriptome analysis in grass endophytes. In: Hopkins A, Wang Z-Y, Sledge M, Barker RE (eds) Molecular Breeding of Forage and Turf. Kluwer Academic Press, Dordrecht, The Netherlands, pp 145-153

Felitti S, Shields K, Ramsperger M, Tian P, Sawbridge T, Webster T, Logan E, Erwin T, Forster J, Edwards D, Spangenberg G (2006) Transcriptome analysis of *Neotyphodium* and *Epichloë* grass endophytes. Fungal Genet Biol 43: 465-475

Fjellstrom RG, Steiner JJ, Beuselinck PR (2003) Tetrasomic-linkage mapping of RFLP, PCR, and isozyme loci in *Lotus corniculatus* L. Crop Sci 43: 1006-1020

Flint-Garcia SA, Thornsberry JM, Buckler ESI (2003) Structure of linkage disequilibrium in plants. Annu Rev Plant Biol 54: 357-374

Forster JW, Jones ES, Kölliker R, Drayton MC, Dumsday J, Dupal MP, Guthridge KM, Mahoney NL, van Zijll de Jong E, Smith KF (2001a) Development and implementation of molecular markers for forage crop improvement. In: Spangenberg G (ed) Molecular Breeding of Forage Crops. Kluwer Academic Press, Dordrecht, The Netherlands, pp 101-133

Forster JW, Jones ES, Kölliker R, Drayton MC, Dupal MP, Guthridge KM, Smith KF (2001b) DNA profiling in outbreeding forage species. In: Henry R (ed) Plant Genotyping - the DNA Fingerprinting of Plants. CABI Press, Wallingford, Oxon, UK, pp 299-320

Forster JW, Jones ES, Batley J, Smith KF (2004) Molecular marker-based genetic analysis of pasture and turf grasses. In: Hopkins A, Wang Z-Y, Sledge M, Barker RE (eds) Molecular Breeding of Forage and Turf. Kluwer Academic Press, Dordrecht, The Netherlands, pp 197-239

Forster JW, Jones ES, Smith KF, Guthridge KM, Dupal MP, Howlett S, Hughes LJ, Garvie S, Preston C (2005a) Molecular marker technology for the study of molecular variation and comparative genetics in pasture grasses. In: Sharma AK, Sharma A (eds) Plant Genome: Biodiversity and Evolution. Vol 1, Pt B: Phanerogams. Science Publ, Enfield, New Hampshire, USA, pp 119-155

Forster JW, Cogan NOI, Vecchies AC, Ponting RC, Drayton MD, George J, Dumsday JL, Sawbridge TI, Spangenberg GC (2005b) Gene-associated single nucleotide polymorphism (SNP) discovery in perennial ryegrass (*Lolium perenne* L.). In: Humphreys MO (ed) Molecular Breeding for the Genetic Improvement of Forage Crops and Turf. Academic Publ, Wageningen, The Netherlands, p 199

George J, Cogan NOI, Smith KF, Spangenberg GC, Forster JW (2006a) Genetic map integration and comparative genome organisation of white clover (*Trifolium repens* L.) with model legume species. In: Plant, Animal and Microbe Genome XIV Conf, San Diego, CA, USA, p 542

George J, Dobrowolski MP, van Zijll de Jong E, Cogan NOI, Smith KF, Forster JW (2006b) Assessment of genetic diversity in cultivars of white clover (*Trifolium repens* L.) detected by simple sequence repeat polymorphism. Genome 49: 919-930

Grime JP, Mowforth MA (1982) Variation in genome size - an ecological interpretation. Nature 299: 151-153

Groppe K, Sanders I, Wiekman A, Boller T (1995) A microsatellite marker for studying the ecology and diversity of fungal endophytes (*Epichloë* species) in grass tissue. Appl Environ Microbiol 61: 3943-3949

Gustine DL, Elwinger GF (2003) Spatiotemporal genetic structure within white clover populations in grazed swards. Crop Sci 43: 337-344

Guthridge KM, Dupal MP, Kölliker R, Jones ES, Smith KF, Forster JW (2001) AFLP analysis of genetic diversity within and between populations of perennial ryegrass (*Lolium perenne* L.). Euphytica 122: 191-201

Herrmann D, Boller B, Studer B, Widmer F, Kölliker R (2006) QTL analysis of seed yield components in red clover (*Trifolium pratense* L.). Theor Appl Genet 112: 536-545

Hirata M, Cai H, Inoue M, Yuyama N, Miura Y, Komatsu T, Takamizo T, Fujimori M (2006) Development of simple sequence repeat (SSR) markers and construction of an SSR-based linkage map in Italian ryegrass (*Lolium multiflorum* Lam.) Theor Appl Genet 113: 270-279

Humphreys MW (1989) The controlled introgression of *Festuca arundinacea* genes into *Lolium multiflorum*. Euphytica 42: 105-116

Hutchinson J, Rees H, Seal AG (1979) An assay of the activity of supplementary DNA in *Lolium*. Heredity 43: 411-421

Inoue M, Gao ZS, Hirata M, Fujimori M, Cai HW (2004a) Construction of a high-density map of Italian ryegrass (*Lolium multiflorum* Lam.) using restriction fragment length polymorphism, amplified fragment length polymorphism and telomeric repeat-associated sequence markers. Genome 47: 57-65

Inoue M, Cai HW (2004b) Sequence analysis and conversion of genomic RFLP markers to STS and SSR markers in Italian ryegrass (*Lolium multiflorum* Lam.). Breed Sci 54: 245-251

Inoue M, Gao ZS, Cai HW (2004c) QTL analysis of lodging resistance and related traits in Italian ryegrass (*Lolium multiflorum* Lam.). Theor Appl Genet 109: 1576-1585

Isobe S, Klimenko I, Ivashuta S, Gau M, Kozlov NN (2003) First RFLP linkage map of red *clover (Trifolium pratense* L.) based on cDNA probes and its transferability to other red clover germplasm. Theor Appl Genet 108: 105-112

Jenkins G, Head J, Forster JW (2000) Probing meiosis in hybrids of *Lolium* (Poaceae) with a discriminatory repetitive genomic sequence. Chromosoma 109: 280-286

Jeanneau M, Gerentes D, Foueillassar X, Zivy M, Vidal J, Toppan A, Perez P (2002) Improvement of drought tolerance in maize: towards the functional validation of the *Zm-Asr1* gene and increase of water use efficiency by over-expressing C4-PEPC. Biochimie 84: 1127-1135

Jensen LB, Muylle H, Arens P, Andersen CH, Holm PB, Ghesquiere M, Julier B, Lübberstedt T, Nielsen KK, De Riek J, Roldán-Ruiz I, Roulund N, Taylor C, Vosman B, Barre P (2005a) Development and mapping of a public reference set of SSR markers in *Lolium perenne* L. Mol Ecol Notes 5: 951-957

Jensen LB, Andersen JR, Frei U, Xing Y, Taylor C, Holm PB, Lübberstedt T (2005b) QTL mapping of vernalisation response in perennial ryegrass (*Lolium perenne* L.) reveals co-location with an orthologue of wheat *VRN1*. Theor Appl Genet 110: 527-536

Jones ES, Dupal MP, Kölliker R, Drayton MC, Forster JW (2001) Development and characterisation of simple sequence repeat (SSR) markers for perennial ryegrass (*Lolium perenne* L.). Theor Appl Genet 102: 405-415

Jones ES, Mahoney NL, Hayward MD, Armstead IP, Jones JG, Humphreys MO, King IP, Kishida T, Yamada T, Balfourier F, Charmet C, Forster JW (2002a)

An enhanced molecular marker-based map of perennial ryegrass (*Lolium perenne* L.) reveals comparative relationships with other Poaceae species. Genome 45: 282-295

Jones ES, Dupal MD, Dumsday JL, Hughes LJ, Forster JW (2002b) An SSR-based genetic linkage map for perennial ryegrass (*Lolium perenne* L.). Theor Appl Genet 105: 577-584

Jones ES, Hughes LJ, Drayton MC, Abberton MT, Michaelson-Yeates TPT, Forster JW (2003) An SSR and AFLP molecular marker-based genetic map of white clover (*Trifolium repens* L.). Plant Sci 165: 531-539

Kálo P, Endre G, Zimnayi L, Csanádi G, Kiss GB (2000) Construction of an improved linkage map of diploid alfalfa (*Medicago sativa*). Theor Appl Genet 100: 641-657

Kawanabe S, Yoshihara K., Okada T, Ueno M, Hidaka M (1963) Studies on summer depression of pasture crops. 3. Influence of flower bud removal upon vegetative growth of Ladino clover. J Jap Soc Grassland Sci 9: 31-41

King IP, Morgan WG, Armstead IP, Harper JA, Hayward MD, Bollard A, Nash JV, Forster JW, Thomas HM (1998) Introgression mapping in the grasses. I. Introgression of *Festuca pratensis* chromosomes and chromosome segments into *Lolium perenne*. Heredity 81: 462-467

Kiss GB, Csanádi G, Kálmán K, Kálo O, Ökrész L (1993) Construction of a basic genetic map for alfalfa using RFLP, RAPD, isozyme and morphological markers. Mol Gen Genet 238: 119-127

Kölliker R, Jones ES, Drayton MC, Dupal MP, Forster JW (2001a) Development and characterisation of simple sequence repeat (SSR) markers for white clover (*Trifolium repens* L.). Theor Appl Genet 102: 6-424

Kölliker R, Jones ES, Jahufer MZZ, Forster JW (2001b) Bulked AFLP analysis for the assessment of genetic diversity in white clover (*Trifolium repens* L.). Euphytica 121: 305-315

Kölliker R, Boller B, Widmer F (2005) Marker-assisted polycross breeding to increase diversity and yield in perennial ryegrass (*Lolium perenne* L.). Euphytica 146: 55-65

Kölliker R, Enkerli J, Widmer F (2006) Characterisation of novel microsatellite loci for red clover (*Trifolium pratense* L.) from enriched genomic libraries. Mol Ecol Notes 6: 50-53

Kubik C, Meyer WA, Gaut BS (1999) Assessing the abundance and polymorphism of simple sequence repeats in perennial ryegrass. Crop Sci 39: 1136-1141

Kubik C, Sawkins M, Meyer WA, Gaut BS (2001) Genetic diversity in seven perennial ryegrass (*Lolium perenne* L.) cultivars based on SSR markers. Crop Sci 41: 1565-1572

Lauvergeat V, Barre P, Bonnet M, Ghesquiére M (2005) Sixty simple sequence repeat markers for use in the *Festuca-Lolium* complex of grasses. Mol Ecol Notes 5: 401-405

Lawless KA, Drayton MC, Hand MC, Ponting RC, Cogan NOI, Sawbridge TI, Smith KF, Spangenberg GC, Forster JW (2008) Interpretation of SNP haplotype complexity in white clover (*Trifolium repens* L.) an outbreeding alloteraploid species. In: Yamada T. Spangenberg G (eds) Molecular Breeding of Forage and Turf: Proc 5th Int Symp Molecular Breeding of Forage and Turf. Springer, New York, USA (in Press)

Lidgett A, Jennings K, Johnson X, Guthridge K, Jones E, Spangenberg G (2002) Isolation and characterisation of a fructosyltransferase gene from perennial ryegrass (*Lolium perenne*). J Plant Physiol 159: 1037-1043

Liu D, van Heeswijck G, Latch G, Leonforte T, Panaccio M., Langford C, Cunningham P, Reed R (1995) Rapid identification of *Acremonium lolii* and *Acremonium coenophialum* endophytes through arbitrarily primed PCR. FEMS Microbiol Lett 133: 95-98

Lorraine-Colwill DF, Powles SB, Hawkes TR, Preston C (2001) Inheritance of evolved glyphosate resistance in *Lolium rigidum* (Gaud.). Theor App Genet 102: 545-550

Lynch D, Lidgett A, McInnes R, Huxley H, Jones E, Mahoney N, Spangenberg G (2002) Isolation and characterisation of three cinnamyl alcohol dehydrogenase homologue cDNAs from perennial ryegrass (*Lolium perenne* L.). J Plant Physiol 159: 653-660

Mackay TFC (2001) The genetic architecture of quantitative traits. Annu Rev Genet 35: 303-309

McInnes R, Lidgett A, Lynch D, Huxley H, Jones E, Mahoney N, Spangenberg G (2002) Isolation and characterisation of a cinnamoyl-CoA reductase gene from perennial ryegrass (*Lolium perenne*). J Plant Physiol 159: 415-422

Mian MAR, Saha MC, Hopkins AA, Wang ZY (2005) Use of tall fescue EST-SSRs in phylogenetic analysis of cool-season forage grasses. Genome 48: 637-647

Michaelson-Yeates TPT, Marshall A, Abberton MT, Rhodes I (1997) Self-incompatibility and heterosis in white clover (*Trifolium repens* L.). Euphytica 94: 341-348

Miura Y, Ding CL, Ozaki R, Hirata M, Fujimori M, Takahashi W, Cai HW, Mizuno K (2005) Development of EST-derived CAPS and AFLP markers linked to a gene for resistance to ryegrass blast (*Pyricularia* sp.) in Italian ryegrass (*Lolium multiflorum* Lam.). Theor Appl Genet 111: 811-818

Moon CD, Tapper BA, Scott B (1999) Identification of *Epichloë* endophytes in planta by a microsatellite-based PCR fingerprinting assay with automated analysis. Appl Environ Microbiol 65: 1268-1279

Musial JM, Aitken KS, Mackie JM, Irwin JAG (2005) A genetic linkage map in autotetraploid lucerne adapted to northern Australia, and the use of the map to identify DNA markers linked to resistance to *Phytophthora medicaginis*. Aus J Agri Res 56: 333-344

Muylle H, Baert J, Van Bockstaele E, Moerkerke B, Goetghebeur E, Roldán-Ruiz I (2005a) Identification of molecular markers linked with crown rust (*Puccinia coronata* f.sp. *lolii*) resistance in perennial ryegrass (*Lolium perenne*) using AFLP markers and a bulked segregant approach. Euphytica 143: 135-144

Muylle H, Baert J, Van Bockstaele E, Petijs J, Roldán-Ruiz I (2005b) Four QTLs determine crown rust (*Puccinia coronata* f.sp. *lolii*) resistance in a perennial ryegrass (*Lolium perenne*) population. Heredity 95: 348-357

Oram RN, Clements RJ, McWilliam JR (1974) Inheritance of nutritive quality of summer herbage in *Phalaris tuberosa* L. Aus J Agri Res 25: 265-272

Paran I, Zamir D (2003) Quantitative traits in plants: beyond the QTL. Trends Genet 19: 303-306

Ponting RC, Drayton MD, Cogan NOI, Dobrowolski MD, Smith KF, Spangenberg GC, Forster JW (2007) SNP discovery, validation, haplotypic structure and linkage disequilibrium in full-length herbage nutritive quality genes of perennial ryegrass (*Lolium perenne* L.). Mol Genet Genom 278: 589-597

Powell W, Langridge P (2004) Unfashionable crop species flourish in the 21st century. Genome Biol 5: 233

Powles SB, Lorraine-Colwill DF, Dellow JJ, Preston C (1998) Evolved resistance to glyphosate in rigid ryegrass (*Lolium rigidum*) in Australia. Weed Sci 46: 604-607

Powles SB, Preston C (1999) Herbicide resistance: a case of rapid evolution. In: Attwell B, Kriedemann P, Turnbull C (eds) Plants in Action: Adaptation in Nature, Performance in Cultivation. Macmillan Ed, Australia, pp 607-613

Prestidge RA, Gallagher RT (1988) Endophyte fungus confers resistance to ryegrass: Argentine stem weevil larval studies. Ecol Entomol 13: 429-435

Roldán-Ruiz I, Dendauw J, van Bockstaele J, Depicker E, De Loose M (2000) AFLP markers reveal high polymorphic rates in ryegrasses (*Lolium* spp.) Mol Breed 6: 125-134

Roldán-Ruiz I, van Eeuwijk FA, Gilliland TJ, Dubreuil P, Dillmann C, Lallemand J, de Loose M, Baril CP (2001) A comparative study of molecular and morphological methods of describing relationships between perennial ryegrass (*Lolium perenne* L.) varieties. Theor Appl Genet 103: 1138-1150

Saha MC, Mian MAR, Eujayl I, Zwonitzer JC, Wang LJ, May GD (2004) Tall fescue EST-SSR markers with transferability across several grass species. Theor Appl Genet 109: 783-791

Saha MC, Mian R, Zwonitzer JC, Chekhovskiy K, Hopkins AA (2005) An SSR- and AFLP-based genetic linkage map of tall fescue (*Festuca arundinacea* Schreb.). Theor Appl Genet 110: 323-336

Sato S, Isobe S, Asamizu E, Ohmido N, Kataoka R, Nakamura Y, Kaneko T, Sakurai N, Okumura K, Klimenko I, Sasamoto S, Wada T, Watanabe A, Kohara M, Fujishiro T, Tabata S (2006) Comprehensive structural analysis of the genome of red clover (*Trifolium pratense* L.). DNA Res 12: 301-364

Sawbridge T, Ong E-K, Binnion C, Emmerling M, McInnes R, Meath K, Nguyen N, Nunan K, O'Neill M, O'Toole F, Rhodes C, Simmonds J, Tian P, Wearne K, Webster T, Winkworth A, Spangenberg G (2003a) Generation and analysis of expressed sequence tags in perennial ryegrass (*Lolium perenne* L.). Plant Sci 165: 1089-1100

Sawbridge T, Ong E-K, Binnion C, Emmerling M, Meath K, Nunan K, O'Neill O, O'Toole F, Simmonds J, Wearne K, Winkworth A, Spangenberg G (2003b) Generation and analysis of expressed sequence tags in white clover (*Trifolium repens* L.). Plant Sci 165: 1077-1089

Seal AG, Rees H (1982) The distribution of quantitative DNA changes associated with the evolution of the diploid Festuceae. Heredity 94: 179-190

Senda T, Kubo N, Hirai M, Tominaga T (2004) Development of microsatellite markers and their effectiveness in *Lolium temulentum*. Weed Res 44: 135-141

Senda T, Saito M, Ohsako T, Tominaga T (2005) Analysis of *Lolium temulentum* geographical differentiation by microsatellite and AFLP markers. Weed Res 45: 18-26

Shinozuka H, Hisano H, Ponting RC, Jones ES, Cogan NOI, Forster JW, Yamada T (2005) Molecular cloning and genetic mapping of perennial ryegrass protein kinase CK2a-subunit genes. Theor Appl Genet 112: 167-177

Shinozuka H, Hisano H, Yoneyama S, Shimamoto Y, Jones ES, Forster JW, Yamada T, Kanazawa A (2006) Cold-responsive gene expression, linkage mapping and protein localization in the nucleus of glycine-rich RNA-binding protein gene from perennial ryegrass suggest a role for cold adaptation. Mol Genet Genom 275: 399-408

Sledge MK, Ray IM, Jiang G (2005) An expressed sequence tag SSR map of tetraploid alfalfa (*Medicago sativa* L.). Theor Appl Genet 111: 980-992

Sim S, Chang T, Curley J, Warnke SE, Barker R, Jung G (2005) Chromosomal rearrangements differentiating the ryegrass genome from the Triticeae, oat and rice genomes using common heterologous RFLP probes. Theor Appl Genet 110: 1011-1019

Soreng RJ, Davis JI (1998) Phylogenetics and character evolution in the grass family (Poaceae): simultaneous analysis of morphological and chloroplast DNA restriction site character sets. Bot Rev 64: 1-85

Spangenberg G, Kalla R, Lidgett A, Sawbridge T, Ong EK, John U (2001) Breeding forage plants in the genome era. In: Spangenberg G (ed.) Molecular Breeding of Forage Crops. Kluwer Academic Publ, Dordrecht, The Netherlands, pp 1-39

Spangenberg G, Felitti S, Shields K, Ramsperger M, Tain P, Ong EK, Singh D, Logan E, Edwards D (2005a) Gene discovery and microarray-based transcriptome analysis of the grass-endophyte association. In: Roberts CA, West CP, Spiers DE (eds) *Neotyphodium* in Cool Season Grasses: Current Research and Applications. Blackwell Publ, Ames, Iowa, USA, pp 103-121

Spangenberg GS, Forster JW, Edwards D, John U, Mouradov A, Emmerling M, Batley J, Felitti S, Cogan NOI, Smith KF, Dobrowolski MP (2005b) Future directions in the molecular breeding of forage and turf. In: Humphreys MO (ed) Molecular Breeding for the Genetic Improvement of Forage Crops and Turf. Academic Publ, Wageningen, The Netherlands, pp 83-97

Studer B, Boller B, Herrmann D, Bauer E, Posselt UK, Widmer F, Kölliker R (2006) Genetic mapping reveals a single major QTL for bacterial wilt resistance in Italian ryegrass (*Lolium multiflorum* Lam.). Theor Appl Genet 113: 66-671

Tavoletti S, Veronesi F, Osborn TC (1996) RFLP linkage map of an alfalfa meiotic mutant based on a F_1 population. J Hered 87: 167-170

Thomas H, Morgan WG, Humphreys MW (1988) The use of a triploid hybrid for introgression in *Lolium* species. Theor Appl Genet 76: 299-304

Thornsberry JM, Goodman MM, Doebley J, Kresovich S, Nielsen D, Buckler IV ES (2001) *Dwarf8* polymorphisms associate with variation in flowering time. Nat Genet 28: 286-289

Thorogood D, Kaiser WJ, Jones JG, Armstead I (2002) Self-incompatibility in ryegrass 12: Genotyping and mapping the *S* and *Z* loci of *Lolium perenne* L. Heredity 88: 385-390

Tyler BF, Hayward MD (1982) A sampling technique for assessing digestibility in Italian ryegrass populations. Euphytica 31: 249-355

Vaidyanathan R, Kuruvilla S, Thomas G (1999) Characterisation and expression pattern of an abscisic acid and osmotic stress responsive gene from rice. Plant Sci 140: 25-36

van Zijll de Jong E, Guthridge KM, Spangenberg GC, Forster JW (2003) Development and characterisation of EST-derived simple sequence repeat (SSR) markers for pasture grass endophytes. Genome 46: 277-290

van Zijll de Jong E, Smith KF, Spangenberg GC, Forster JW (2005) Molecular genetic marker-based analysis of the forage grass host - *Neotyphodium* endophyte interaction. In: Roberts CA, West CP, Spiers DE (eds) *Neotyphodium* in Cool Season Grasses: Current Research and Applications. Blackwell Publ, Ames, Iowa, USA, pp 123-133

van Zijll de Jong, E, Bannan NR, Dobrowolski MP, Stewart AV, Smith KF, Spangenberg GC, Forster JW (2008) Global genetic diversity of the perennial ryegrass fungal endophyte *Neotyphodium lolii*. Crop Sci 48: 1487-1501

van Treuren R, Bas N, Goossens PJ, Jansen J, van Soest LJM (2005) Genetic diversity in perennial ryegrass and white clover among old Dutch grasslands as compared to cultivars and nature reserves. Mol Ecol 14: 39-52

Vogel KP, Pedersen JF (1993) Breeding systems for cross-pollinated forage grasses. Plant Breed Revs 11: 251-274

Vogel JP, Qiang Gu Y, Twigg P, Lazo GR, Laudencia-Chingcuanco D, Hayden DM, Donse TJ, Vivian LA, Stamova B, Coleman-Derr D (2006) EST sequencing and phylogenetic analysis of the model grass *Brachypodium distachyon*. Theor Appl Genet 113: 186-195.

Warnke SE, Barker RE, Jung G, Rouf Mian MA, Saha MC, Brilman LA, Dupal MD, Forster JW (2004) Genetic linkage mapping of an annual x perennial ryegrass population. Theor Appl Genet 109: 294-304

Williams TA, Abberton MT, Thornley WJ, Evans DR, Rhodes I (1998) Evaluation of seed production potential in white clover (*Trifolium repens* L.) varietal improvement programs. Grass Forage Sci 53: 197-207

Wu XL, Larson SR, Hu ZM, Palazzo AJ, Jones TA, Wang RRC, Jensen KB, Chatterton NJ (2003) Molecular genetic linkage maps for allotetraploid *Leymus* wildryes. Genome 46: 627-646

Xiong Y, Fei S-Z, Brummer EC, Moore KJ, Barker RE, Jung G, Curley J, Warnke SE (2007a) QTL analyses of fiber components and crude protein in an annual x perennial ryegrass interspecific hybrid population. Mol Breed 18: 327-340

Xiong Y, Fei S-Z, Arora R, Brummer EC, Barker RE, Jung G, Warnke SE (2007b) Identification of quantitative trait loci controlling winter hardiness in an annual x perennial ryegrass interspecific hybrid population. Mol Breed 19: 125-136

Xu WW, Sleper DA, Hoisington DA (1991) A survey of restriction fragment polymorphism in tall fescue and its relatives. Genome 34: 686-692

Xu WW, Sleper DA, Chao S (1995) Genome mapping of polyploid tall fescue (*Festuca arundinacea* Schreb.) with RFLP markers. Theor Appl Genet 91: 947-955

Yamada T, Higuchi A, Fukuoka A (1989) Recurrent selection of white clover (*Trifolium repens* L.) using self-compatible plants. I. Selection of self-compatible plants and inheritance of a self-compatibility factor. Euphytica 44: 167-172

Yamada T, Jones ES, Cogan NOI, Vecchies AC, Nomura T, Hisano H, Shimamoto Y, Smith KF, Forster JW (2004) QTL analysis of morphological, developmental and winter hardiness-associated traits in perennial ryegrass (*Lolium perenne* L.). Crop Sci 44: 925-935

Yamada T, Forster JW (2005) QTL analysis and trait dissection in ryegrasses (*Lolium* spp.). In: MO Humphreys (ed) Molecular Breeding for the Genetic Improvement of Forage Crops and Turf.). Academic Publ, Wageningen, The Netherlands, pp 43-53

Young ND, Cannon SB, Sato S, Kim D, Cook DR, Town CD, Roe BA, Tabata S (2005) Sequencing the genespaces of *Medicago truncatula* and *Lotus japonicus*. Plant Physiol 137: 1174-1181

Yu KF, Pauls KP (1993) Segregation of random amplified polymorphic DNA markers and strategies for molecular mapping in tetraploid alfalfa. Genome 36: 84-851

Zhang D-X, Hewitt GM (2003) Nuclear DNA analyses in genetic studies of populations: practice, problems and prospects. Mol Ecol 12: 563-584

Zhu H, Choi H-K, Cook DR, Shoemaker RC (2005) Bridging model and crop legumes through comparative genomics. Plant Physiol 137: 1189-1196

12 Transgenics for Genetic Improvement of Plants

M.R. Davey[1*], P. Anthony[1], J.B. Power[1] and K.C. Lowe[2]

[1]Plant Sciences Division, School of Biosciences, University of Nottingham, Sutton Bonington Campus, Loughborough LE12 5RD, UK

[2]School of Biology, University of Nottingham, University Park, Nottingham NG7 2RD, UK

*Corresponding author: Mike.Davey@nottingham.ac.uk

1 INTRODUCTION

Human existence and welfare rely upon plants for food, fiber, and fuel. During the last 10,000 years, these demands have been satisfied by the domestication of plants combined with breeding and selection. The next 25 years will necessitate a triple increase in food production because of the estimated 40% rise in the global population. However, only a 10% increase is possible in the area of land available for cultivation (Khachatourians 2002). Environmental issues, such as global warming, drought, climate change and increased salinity of agricultural land, combined with political instability, are additional factors that determine the success or failure of agricultural practices. Traditional plant breeding will continue to provide the baseline for crop improvement, directed towards increased productivity and quality. In order to meet new demands, it is imperative that breeding programs are underpinned by biotechnological approaches. The latter involve gene insertion into plants by transformation and somatic hybridization through protoplast fusion to circumvent pre- and post-zygotic incompatibility barriers since the latter frequently hinder sexual hybridization and gene transfer.

Genetic manipulation technology permits novel approaches to studies of plant genetics and physiology, with the ability to exploit gene manipulation procedures to generate 'designer' crops.

The last two decades have witnessed major advances in biotechnology, particularly the ability to isolate genetic material from plants, animals and microorganisms, to sequence DNA, and to reconfigure it prior to introduction into target species. Recombinant DNA technology and the ability to culture plant cells in the laboratory and to induce such cells to express their totipotency by developing into fertile individuals, have provided the basis for the genetic modification of plants. Since weeds, insects, fungi, and viruses are the main causes of crop losses, it is understandable that increased tolerance to these agents has been the prime objective of most of the initial genetic manipulation technology. Undoubtedly, this will continue, accompanied by a wider range of targets. For example, there is already considerable effort directed to the biosynthesis of pharmaceutically-important products in plants, so-called 'molecular pharming', increasing nutritional value of plant products ('biofortification'), and plant-based production of antibodies ('plantibodies'). Other targets include genetic modification for tolerance to biotic stress, increased shelf-life, and modifying the aesthetic appeal of amenity species, including manipulation of floral characteristics and plant stature.

Many aspects of crop plants ranging from their physiology, interactions with microorganisms and biodiversity, are discussed in a compendium of comprehensive articles edited by Goodman (2004). In addition, there is an extensive literature relating to the procedures that have been developed to generate transgenic plants and the characteristics that have been introduced into many species. In general, this ability to reduce the time to obtain homozygous plants to within two to three years provides a strong incentive to generate transgenic plants. Much of the original literature has been summarized in several excellent reviews and books to which the reader is directed. Consequently, this chapter provides an overview of fundamental aspects of plant genetic manipulation with selected examples throughout the text.

2 TRANSFORMATION PROCEDURES FOR PLANTS

Genes for introduction into target plants can originate from plants, animals or microorganisms, with recombinant DNA technology enabling chimaeric genes to be constructed using promoters, coding sequences and terminators from different sources. Additionally, it is possible to control tissue-specificity, the timing of expression and the extent of gene expression. Several plant transformation procedures have been developed to introduce genes into plants, some of which are more reproducible and user-friendly than others. All demand reliable DNA delivery combined with efficient regeneration of fertile transgenic individuals from transformed cells. Most rely upon explant-, cell-, or protoplast-to-plant regeneration systems, involving plant recovery by organogenesis (shoot formation) and/or somatic embryogenesis, irrespective of whether the end products are exploited in basic studies or have commercial application. The establishment of robust protocols for shoot regeneration generally demands considerable effort and resources, since the conditions required for reproducible plant regeneration are often species and genotype-specific. Typical culture protocols for specific crops are provided by Loyola-Vargas and Vázquez-Flota (2005), while Veltcheva et al. (2005) cite the common bean (*Phaseolus vulgaris*) as an example of a plant that is difficult to regenerate and transform. Other critical parameters that influence the generation of transgenic plants include systems for the selection of tissues and shoots carrying the gene(s) of interest and the micropropagation, where appropriate, of selected transgenic plants, followed by their phenotypic and molecular characterization. Khachatourians et al. (2002) have compiled a volume with contributions of topic areas ranging from basic aspects of cell culture to transgene expression in fruit, vegetables, grain, and other seed crops.

The most successful procedure and one that is exploited most extensively for plant transformation in many laboratories world-wide, utilizes the natural gene transfer system of *Agrobacterium*. Gene delivery by particle bombardment is exploited for plants that remain recalcitrant to *Agrobacterium* infection. Other procedures that are less reproducible and that are not universally applicable are also considered in this discussion. Newell (2000) and Sharma et al. (2005) have provided excellent reviews of the general procedures of plant transformation.

2.1 *Agrobacterium*-Mediated Gene Delivery to Plants

Agrobacterium, a Gram-negative soil bacterium, is a natural genetic engineer that induces galls in plants by introducing and expressing genes controlling the biosynthesis of growth regulators (auxins and cytokinins) in host tissues. The increased synthesis of growth regulators results in uncontrolled plant cell proliferation. Unusual amino acids (opines) are also secreted by transformed plant cells and are used by *Agrobacterium* as a source of carbon and nitrogen. *Agrobacterium tumefaciens* induces crown gall disease, *A. rhizogenes* causes proliferation of roots (hairy roots), while *A. rubi* incites cane gall disease. Galls on grape and some other plants are induced by *A. vitis*. Of these bacteria, *A. tumefaciens* and *A. rhizogenes* have been studied most extensively, particularly the former. Considerable research has been focused on attempting to elucidate the molecular biology of *A. tumefaciens*-mediated plant transformation, although some of the precise details of the process still remain unclear.

Oncogenic strains of *Agrobacterium* each carry a megaplasmid, 200-800 kb in size, designated the Ti or tumour-inducing plasmid in the case of *A. tumefaciens*, and the Ri or root-inducing plasmid in *A. rhizogenes*. As Gelvin (2003) indicated, the species designation of *Agrobacterium* is almost meaningless, since *A. tumefaciens* can be converted to *A. rhizogenes* by replacing, in the laboratory, the Ti plasmid by an Ri plasmid. The host range of *Agrobacterium* is strain dependent, with genetic control of the transformation process involving the chromosomal background of both the bacterium and plant hosts. During transformation of plant cells, a defined piece of bacterial DNA of 10-30 kb in size delimited by 25 bp direct repeat left and right border regions, the T-DNA, is transferred from the Ti or Ri plasmid to the plant genome. Wounding of plant cells is essential to initiate tumour induction. Phenolic compounds, such as acetosyringone, and sugars released from wounded cells act initially as bacterial chemoattractants. Subsequently, these compounds induce the expression of virulence (*vir*) genes on the Ti plasmid, the *vir* genes being located outside the T-DNA region. Kuta and Tripathi (2005) have indicated the importance of exogenously supplied acetosyringone in *Agrobacterium*-mediated plant transformation. Stimulation of a two component *virA* plus *virG* activator initiates a cascade of events that result through a strand-replication mechanism in the generation of a single-stranded copy of the T-DNA, the T-strand, corresponding to the non-transcribed strand of T-DNA. Excellent detailed summaries of the complex activity of the *vir* genes and their products, and the way in which the T-strand becomes protected and is transferred to plant cells, are provided by Gelvin (2003), Valentine (2003), Tzfira et al. (2004), and Lacroix et al. (2006a). Subsequently, movement of a T-complex involving bacterial and host plant proteins through the plant cell cytoplasm to the nucleus, possibly by microtubular and dynein-mediated transport, is discussed by Tzfira and Citovsky (2006). Plant cyclophilins and karyopherin-α cooperate with bacterial factors to translocate the T-complex through nuclear pores into the karyoplasm. Uncoating of the T-complex within the nucleus may be coupled with the synthesis of a second strand of the T-DNA. Tzfira et al. (2004) provided an explanation for the way in which full length, truncate, and multiple T-DNAs may be found in the genomes of plant cells. These authors also indicated possible ways in which the T-DNA may integrate into plant chromosomes by double-stranded-break repair, single-strand-gap repair or microhomology. Lacroix et al. (2006a) favored the integration of double-stranded T-DNA at double-stranded breaks with the DNA of the plant genome, a view supported by Tzifira and Citovsky (2006). These three reviews also contain detailed sketches of the mechanisms proposed by the authors. What is clear is that *Agrobacterium* exploits the host cellular mechanisms to effect the transformation of plant cells. However, aspects have not been elucidated of the targeting of T-DNA to its integration site(s) and the precise mechanism of integration.

2.1.1 Vectors for Plant Transformation: Exploiting the Molecular Biology of Bacterial-Plant Interactions

Since bacterial genes on the T-DNA become integrated and expressed in plant cells, the logical step in exploiting this natural gene transfer was to introduce foreign genes into the T-DNA. Insertion of genes into the T-DNA was not a trivial task because the lack of DNA restriction sites on the T-region and the size of the Ti or Ri plasmid made the introduction of foreign genes into the T-region extremely difficult. Whilst the T-DNA is defined by its left and right borders, its transfer and integration into plant cells is not dependent upon the DNA sequence between these borders. Consequently, removal of the oncogenicity genes between the T-DNA borders, to 'dis-arm' the T-DNA, followed by cointegration of foreign DNA between the left and right borders of the T-DNA, enabled foreign genes to be delivered by *Agrobacterium* carrying engineered Ti or Ri plasmids, with Ti plasmids being used most extensively. Importantly, it was demonstrated that the T-region and *vir* genes could be separated within *Agrobacterium,* with the *vir* gene products still acting on the T-region to effect T-strand formation and its transfer to plant cells. This binary vector system offered the advantage that the dis-armed T-DNA, with unique restriction sites between the T-DNA borders for the insertion of foreign genes, could be cloned on small plasmids capable of being manipulated in *Escherichia coli* and then introduced into *Agrobacterium*. Binary vectors have been developed in a number of laboratories. Some of the plasmids and *Agrobacterium* strains have been discussed by Hellens et al. (2000), while Goderis et al. (2002) constructed binary vectors that permit insertion of up to six expression units.

An important advance in the development of *Agrobacterium*-based vector systems was recognition of the fact that extra copies of the *vir* genes on a small replicon within *Agrobacterium* increased bacterial virulence and, consequently, the efficiency of gene transfer to plant cells. These extra *vir* genes work in conjunction with the *vir* genes on the resident Ti plasmid from which the T-DNA has been removed (this Ti plasmid acts as a 'helper' for gene transfer), to promote reproducible and efficient T-strand formation and T-DNA transfer from *Agrobacterium* to plant cells. Van der Fits et al. (2000) described a ternary system in which the introduction of a plasmid carrying a constitutive *virG* mutant gene (*vir*GN54D) in addition to a binary vector with a T-DNA and the resident Ti plasmid lacking its T-DNA, dramatically increased plant transformation in the case of *A. tumefaciens* strain LBA4404. A further advance in vector technology was the incorporation of *vir* genes onto the binary plasmid carrying the T-DNA, to create 'super-binary' vectors. The relevance of these super-virulent systems was that they enabled *Agrobacterium* to be used as a gene transfer system for monocotyledons, especially cereals. Indeed, the transformation of monocotyledons had been questionable prior to the development of these vectors. Thus, efficient transformation systems based on *Agrobacterium* are available for the most important cereals such as rice, wheat, maize, barley, and sorghum, together with some ornamental species (Cheng et al. 2004). These authors have provided an excellent summary of some of the parameters that influence the transformation of monocotyledons by *Agrobacterium*, with a comprehensive list of target plants, *Agrobacterium* strains, transgenes, explants, transformation frequency, and relevant notes relating to the transformation system. Likewise, Jones et al. (2005) described the methodologies reported for *Agrobacterium*-mediated transformation of wheat and the protocol used by these authors to transform spring and winter wheat varieties.

GATEWAY™ technology, based on site-specific recombination between two DNA molecules carrying complementary recombination sites that permit the cloning of gene(s) of interest into specifically designed vectors without the need for restriction, was also introduced into *Agrobacterium* vectors (Karimi et al. 2002). Later, Chung et al. (2005) described the new and versatile vectors for transformation that are available, with several promoters, terminators, and autofluorescent tags. Such vectors permit the expression of multiple genes from a single construct. Chung et al. (2005) highlighted the generation of easy-to-use vector systems incorporating GATEWAY™ technology and reviewed the information relating to the development of pAUX and pSAT systems for multiple gene expression in plants.

An interesting concept was proposed by Pappas and Winans (2003) in which they transformed tobacco by coinoculating leaf explants with a disarmed strain of *A. tumefaciens* and *E. coli* carrying the IncN-type conjugation system. This system mobilized a shuttle plasmid from *E. coli* to *A. tumefaciens* and the latter transferred the T-DNA to recipient plant cells. The size of the transformation vector was minimized as the mobilization plasmid in *E. coli* did not require a broad host range origin of replication. Gene transfer was enhanced by incubating bacterial strains on medium optimized for bacterial conjugation prior to inoculation of leaf explants. Transgenic plants were regenerated from selected, transformed tissues. As Pappas and Winans (2003) emphasized, this approach negates the requirement to construct *Agrobacterium* strains carrying binary vectors and could simplify the generation of transgenic plants. Undoubtedly, further improvements will be made in the future in vector construction, especially when more information is available relating to the precise mechanism of T-DNA integration into plant genomes.

Immersion in a suspension of *Agrobacterium* of stem, root and leaf explants, especially the latter, that are capable of plant regeneration remains a simple procedure for interacting engineered strains of *Agrobacterium* with totipotent plant cells. Several parameters influence the efficiency of transformation, including the temperature, the density and age of the bacterial suspension, and the period of co-cultivation of the bacteria with plant cells, the latter being about 48 hours in several plant systems (Gelvin 2003). Other approaches have been adopted to promote the transformation process. Wu et al. (2003) reported some of the parameters that influence *Agrobacterium*-mediated transformation of bread wheat, including the size of the embryo, duration of pre-culture, inculation and co-cultivation with bacteria, and the presence of acetosyringone and the surfactant Silwet-L77 in the culture medium. Exposing target tissues to brief periods of ultrasound in the presence of *Agrobacterium* stimulates transformation through sonication-assisted *Agrobacterium* transformation (SAAT), or so-called sonoporation (Liu et al. 2006). The latter increased the transformation efficiency by the formation of micro-wounds, facilitating access of *Agrobacterium* to regions of DNA-receptive cells. At the single cell level, ultrasound is believed to alter the transient permeability of the plasma membrane, facilitating DNA transfer. The optimum time of SAAT depends on the tissue, but should be kept to a minimum (e.g., 0.5-2.0 sec). The equipment required is simple to use, inexpensive and multifunctional. Interestingly, the procedure was developed initially for animal cells, both *in vivo* and *in vitro*, and extended to plant cells.

2.1.2 *In Planta Transformation by* Agrobacterium

An important aspect of plant transformation technology has been elimination of the requirement to culture totipotent plant cells and tissues in order to regenerate transgenic plants. In this respect, an *in planta* floral dip procedure has been developed, first for the model molecular plant, *Arabidopsis thaliana,* in attempts to simplify the *Agrobacterium* transformation system. The technique involves the immersion of developing flowers in a suspension of *Agrobacterium,* generally in a vacuum with a surfactant, such as Silwet L-19. Subsequently, *Agrobacterium*-treated plants are grown to maturity, seed is harvested and seedlings are selected carrying the foreign gene(s) of interest. Most progeny are non-chimaeric, hemizygous for the transgene(s) and any somaclonal variation resulting from tissue culture is negated. Studies of the transformation process with a β-glucuronidase (*gus*) reporter gene showed the female gametophyte to be the target for gene delivery. However, it is unclear whether *Agrobacterium*-mediated transformation is facilitated by pollen tube growth following pollination and whether agrobacteria reach the ovules to effect gene delivery (Bechtold et al. 2003). This procedure is very effective in *A. thaliana* that produces large numbers of seeds. Chung et al. (2000) modified the procedure by spraying developing flowers with *Agrobacterium,* rather than immersing the developing flowers in a bacterial suspension. Although the floral dip technique is effective in other plants, such as radish, a member of the Brassicaceae (Curtis 2004), extension of this procedure to a range of crop species has yet to be demonstrated. Certainly, it would simplify some of the current tissue culture-based transformation systems. Detailed protocols have been reported for transferring genes into crop species. Such procedures are described and in many cases illustrated, for example, in volumes edited by Khachatourians et al. (2002), Curtis (2004), and Loyola-Vargas and Vázquez-Flota (2005).

2.2 Direct Gene Transfer to Plant Cells

2.2.1 *Particle Bombardment*

Transformation procedures that do not involve *Agrobacterium* to mediate gene transfer into recipient plant genomes have been discussed by Newell (2000) and Sharma et al. (2005). Of these approaches, particle bombardment (biolistics, microprojectile bombardment, particle acceleration, gene gunning) has been exploited most extensively in laboratories world-wide. It involves the coating of metal particles, usually gold, with DNA and their insertion, at high velocity, into totipotent target cells. Acceleration of the particles normally involves helium gas or compressed air, with the helium driven HE-1000 instrument being used in several laboratories for a number of years. The use of this instrument has facilitated exchange of protocols between laboratories. Several parameters influence the success of DNA delivery, including the speed of flight of the particles, their size and shape, the binding of the DNA to the particles, the ploidy of the target cells and the semi solid medium supporting the target tissues during gene delivery. Other critical factors that influence transformation include the characteristics of the cells in the plant tissues, their totipotency and the pretreatment of tissues prior to bombardment. Importantly, the technique requires less technical expertise and input of labour than several other gene delivery approaches (Taylor and Fauquet 2002).

A critique of particle bombardment has been presented by Altpeter et al. (2005). Considerable criticism has been leveled at particle bombardment for DNA delivery because of rearranged and truncated copies of transgenes in recipient cells, with associated transgene silencing and instability. *Agrobacterium*-based vectors with T-DNA borders, are claimed to give 'cleaner' DNA integration at low copy number into plant genomes. However, in reviewing the evidence, Altpeter et al. (2005) concluded that there was little to substantiate these claims. In fact, they indicated that the regeneration of transgenic plants containing intact single copy integration events with high-level gene expression was relatively easy by particle bombardment. As explained by these authors, DNA transfer into plant cells using this technology is a physical process, negating any dependence on *Agrobacterium* and any naturally occurring bacterial-host interactions and restraints. This versatile transformation procedure is independent of plant cell type, species or genotype. Indeed, it has facilitated the transformation of some of the most recalcitrant plant species, including cereals and legumes. Transgenes of any size can be introduced into target cells, while co-transformation with two or more genes is facile. The technique is particularly useful for transient gene expression studies for the rapid evaluation of the efficiency of DNA transfer, and to verify the expression of genes delivered by specific constructs. Stable transgene integration into plant genomes probably occurs by DNA repair mechanisms.

A range of cells and tissue explants have been used for particle bombardment, including embryos, shoot apices, leaf discs, microspores, and immature pollen grains and callus. Only the gene expression cassette, comprising a promoter, open reading frame and polyadenylation site, are required for transformation. T-DNA borders are superfluous, although Hansen and Chilton (1996) described a novel system that combined aspects of the molecular biology of *Agrobacterium* with particle bombardment. In their approach, plant expression cassettes for *VirD1* and *VirD2* genes were co-delivered with a vector carrying the gene(s) of interest flanked by T-DNA border sequences. As a result of T-DNA border cleavage by the products of *VirD1* and *VirD2* gene expression, transformed plants were generated lacking any extraneous vector DNA. Multiple gene insertion is possible with particle bombardment, which has application in several areas of research, particularly metabolic engineering.

Transgene integration is random and often into transcriptionally active regions of plant genomes, particularly the sub-terminal regions of chromosomes. Shearing of genomic DNA by the microprojectiles used for DNA delivery may facilitate foreign DNA integration into the nuclei of recipient plant cells. Whilst particle bombardment may lead to large, multicopy complex transgenic loci, plants can also be regenerated that have simple transgenic loci. Multicopy transgenic plants generally have all transgenes at a single locus, in contrast to *Agrobacterium*-transformed plants where transgenes tend to be dispersed to more than one locus, making transgenic plants more difficult to exploit in breeding programs. Additionally, the premise that low transgene copy number in biolistics-transformed plants is preferable to ensure maximal gene expression does not hold true. Clearly, particle bombardment remains of major importance for delivering DNA to cells of those plants that do not respond readily to gene transfer by *Agrobacterium*. It is ironical that as particle bombardment was developed for species that were considered recalcitrant to *Agrobacterium*-mediated gene delivery, success was achieved with *Agrobacterium* transformation of previously difficult-to-transform species, particularly cereals. This parallel development of two techniques initiated considerable rivalry between workers, the views of those who favoured an *Agrobacterium* approach and those who supported particle bombardment undoubtedly promoting considerable scientific debate and experimentation.

2.2.2 Uptake of DNA into Isolated Protoplasts

The fluid mosaic characteristics of the plasma membrane of living plant cells enables DNA to be taken up by chemical and physical procedures once the cell walls have been removed as a physical barrier. As T-DNA borders are not essential for DNA integration into plant genomes, simple *E. coli*-based cloning vectors can be used for gene delivery. Supercoiled and linearized plasmids can effect transformation, together with single and double stranded DNA. The application of enzymatically isolated protoplasts (naked cells) to plant genetic manipulation through transformation has been described by Davey et al. (2005a, b). Isolated protoplasts are exploitable in both transient and stable transformation investigations, particularly where protoplast-to-plant systems are available for target crops.

The development of protoplast-to-plant systems represents the most labour intensive part of the transformation procedure. Protoplast isolation generally involves enzymatic digestion of the walls of cells of primary tissues, followed by culture in liquid or semi-solid media. Following protoplast isolation, protoplast-plasmid mixtures are treated with polyethylene glycol (PEG), usually in the presence of cations such as Ca^{2+} or Mg^{2+} and/or electroporation involving high voltage (~ 2000 volts) electrical pulses of short duration to induce DNA uptake. Since the frequency of stable transformation is low (ca. 1 in 10^4 protoplasts), the requirement for reproducible protoplast-to-plant systems, accompanied by rigorous selection, is of paramount importance. Protoplasts can be transformed with more than one gene carried on the same or on separate plasmids. The stage of the cell cycle at the time of transformation of the recipient protoplasts influences the success of DNA integration, while heat shock and irradiation treatments of recipient protoplasts also stimulate transformation. The presence of carrier DNA mixed with the plasmid carrying the foreign genes of interest, and the nature of the plant genome itself, influence DNA integration and expression. Temperature and pH affect transformation, with high pH (9.0) and reduced temperature (0°C) stimulating the process. An important parameter that must be considered for electroporation, includes the type and intensity of the electric field, and the conductivity of the solution in which the protoplasts are suspended, since such factors must be adapted to the protoplast system under investigation. Liposomes, microscopic spherical vesicles formed by the hydration of lipids, can be loaded with DNA and used to deliver the DNA by fusion and endocytosis with the plasma membranes of isolated protoplasts. This procedure, which was developed originally for the transformation of animal cells, has not become routine because it adds more labour to protoplast-to-plant systems.

Direct DNA uptake into protoplasts was the system of choice particularly in monocotyledons, such as cereals, and was exploited immediately protoplast-to-plant systems became available. However, the use of protoplasts for transformation was superseded by *Agrobacterium*-based gene delivery once this procedure had been established for cereals

(Cheng et al. 2004). Nevertheless, protoplasts still provide useful experimental systems for short-term transient gene expression studies.

2.2.3 Other Procedures for DNA Delivery to Plants

Several other approaches for plant transformation, together with the supporting literature, have been described by Rakoczy-Trojanowska (2002) and Sharma et al. (2005). Microinjection using glass needles, initially developed for transforming animal cells, demands sophisticated micromanipulation equipment, considerable manual dexterity and embryogenic single cells or small cell clusters as targets. This approach is somewhat limited for plant cells because of their cell walls, although this difficulty can be circumvented using as targets isolated protoplasts, or protoplasts that have recently commenced wall synthesis. Stringent culture conditions are required to regenerate injected cells to intact, fertile plants. This exacting procedure has seen limited application, although it is useful in studies of short-term cell-specific gene expression, in assessing gene constructs and for inserting chromosomes into cells. Macroinjecting large quantities of DNA into developing inflorescences, particularly the leaf sheath enclosing immature panicles of cereals, has also received limited attention. Other procedures include vortexing of plant cells with DNA in the presence of silicon carbide fibers, the electrophoretic transfer of DNA into immature embryos, and the introduction of DNA into intact cells and tissues by electroporation. A pollen-tube pathway has also been reported. Whilst some of these approaches, particularly vortexing with silicon carbide fibers, are simple and require the minimum of equipment, reagents, labour, and expertise, the questionable reproducibility of these procedures has limited their exploitation.

Patent issues associated with *Agrobacterium*-mediated gene transfer have inspired workers to assess alternative procedures to insert genes into plants (Chung et al. 2006). For example, virus-based vectors are useful for transient expression because of their ability to replicate in host cells, and to move from cell to cell, giving the potential for high yields of recombinant proteins. These authors have illustrated the way in which such viral vectors may deliver genes to plant cells. The speed of expression compared to the time required to generate transgenic plants, is also relevant. Major advances have been made by Broothaerts et al. (2005) by demonstrating that *Rhizobium* sp. NGR234, *Sinorhizobium meliloti* and *Mesorhizobium loti* can transform plants if the bacteria are provided with Ti plasmid virulence genes and the T-DNA on a binary plasmid. Genes for hygromycin resistance and the *gus* reporter gene were on the T-DNA used in the experiments initiated by these authors.

Inoculation of tobacco leaf discs resulted in transgenic tobacco plants. Importantly, *Rhizobium* sp. and *S. meliloti* could effect floral dip transformation of *A. thaliana*.

3 NUCLEAR VERSUS PLASTID TRANSFORMATION

Whilst foreign genes are normally targeted to the nucleus of recipient cells, plastid transformation has received increasing attention in recent years as plastids have their own genomes and transcription-translation machinery. Importantly, plastids are transferred to the progeny through the maternal parent and not through pollen, which has advantages in terms of biosafety involving the containment of transgenes in genetically manipulated plants (Ruf et al. 2001). There are exceptions to this rule. Additionally, Stegemann et al. (2003) demonstrated that escape of genetic information from plastids to the nucleus is possible. The genomes of plastids (plastomes) are 120-180 kb in size of double stranded DNA with duplication of a region about 25 kb in size in an inverted orientation. The plastome is specific to each type of plastid although plastid number and, consequently, plastome copies per cell depend upon the cell type. Transformation involves a few plastome copies, followed by replacement of non-transformed plastomes. The generation of transplastomic plants involves the transformation of plastids in leaf cells, followed by shoot regeneration from cells on selection medium. DNA delivery has been affected by particle bombardment of leaf explants, or by PEG-induced gene delivery into isolated protoplasts, followed by plant regeneration from selected cells. The recovery of homoplastic plants is facilitated if target cells form undifferentiated callus and shoot regeneration is delayed until plastid segregation is complete.

Maliga (2004) provided a detailed discussion of vector design, genetic markers for selection and transplastomic plants reported to the date of submission of his review. More recently, Lelivelt et al. (2005) contributed to the literature by describing the generation of fertile, homoplastic plastid-transformed lettuce following PEG-induced DNA uptake into isolated leaf protoplasts, prior to plant regeneration. Previously, transplastomic plants had been reported only in some members of the Solanaceae, namely tobacco, tomato and petunia, and in soybean, cotton and *Lesquerella fendeleri*, the latter being a member of the Brassicaceae. The approach in lettuce involved a vector targeting genes to the *trnA*/ *trnI* integenic region of the plastome, with the *aadA* gene as a selectable marker conferring resistance to spectinomycin. Transmission of the plastid-encoded transgene was demonstrated in the first seed generation plants. Importantly, crossing transplastomic plants with male sterile wild type lettuce confirmed that spectinomycin resistance was not transmitted via pollen. An important application of plastid transformation is in the genetic engineering of cytoplasmic male sterility (CMS) (Chase 2006) through introduction of the β-ketothiolase gene from *Acinetobacter* into plastids of *Nicotiana tabacum*. This approach may permit engineering of CMS in other species.

The adaptation of plastid transformation protocols for the major food crops is, undoubtedly, the next logical step in this technology (Bock and Khan 2004), the high gene expression in plastids possibly being directed to more efficient photosynthesis through manipulation of Rubisco in the plastids themselves and to metabolic engineering directed to more nutritious food. Walmsley and Arntzen (2003) also extolled the virtues of plastid transformation to maximize the biosynthesis of recombinant pharmaceuticals in plants, the main challenge being to achieve cost-effective large-scale production. Undoubtedly, major advances in the production of metabolites, or the synthesis of entirely novel compounds, will draw upon the most recent knowledge of plant genomics and metabolite profiling coupled, in some cases, to the use of cell cultures as production systems (Oksman-Caldentey and Inzé 2004). Léon-Bañares et al. (2004) also compared the characteristics of nuclear and plastid transformation and described the potential of exploiting microalgae, such as *Chlamydomonas reinhardtii*, as green cell factories to generate economically important food products, pharmaceuticals and biomass for animal feed. Another interesting concept proposed by Qin et al. (2005) is the possible exploitation of marine algae as bioreactors. Adaptation of the biolistic transformation technology developed for land plants enabled these authors to establish a transformation system for the most commonly cultivated brown seaweed, *Laminaria japonica* (kelp). Clearly, there is the potential to achieve plastid transformation of kelp, again exploiting the procedures developed for model species such as tobacco.

4 MARKER GENES IN TRANSFORMATION TECHNOLOGY

In any transformation procedure adopted for a specific plant, the efficient selection of transformed cells and tissues is a key step in the generation of nuclear and plastid transformants. Reporter genes that can be assayed rapidly following the transformation procedure are essential in monitoring gene transfer, in providing confidence of the reproducibility of specific protocols and in monitoring the expression of new gene constructs. Whilst some reporter genes can be exploited to facilitate the manual selection of transgenic from non-transformed material, in general, they do not provide cells and tissues with an advantage at the culture level over their non-transformed counterparts. In contrast, selectable marker genes enable transformed cells to be isolated from populations of non-transformed cells, generally following the inclusion of specific substrates in the culture medium.

Miki and McHugh (2004) reviewed the extensive literature relating to marker gene technology. They tabulated the marker genes and their sources reported in the literature and concluded that about 50 genes have been exploited in nuclear and plastid transformation strategies, although only a limited number have been used routinely for transforming crop plants. The *uidA* (*gusA*) gene for β-glucuronidase has featured extensively as a reporter gene in both fluorometric and histochemical assays, the latter effectively localizing transgene expression within tissues following exposure to the substrate 5-bromo-4-chloro-3-indolyl-β-D-glucuronide (X-Gluc) and its cleavage to an indole compound. The substrate 4-methylumbelliferyl-β-D-glucuronide (MUG) is a substrate in the fluorometric assay. A limitation of these assays is their destructive nature to confirm expression of the *gus* gene. Consequently, this has encouraged the development of *luc* and *lux A, B,* and *F* genes for luciferase. However, the need for specialized equipment to assess luciferase activity prompted the development and exploitation of non-destructive assays based on expression of the *gfp* gene for green fluorescent protein, cloned from the jellyfish *Aequorea victoria*. Importantly, GFP is extremely stable in plant cells. Genes for phytoene synthase and maize R, C1, P1, and B transcription factors, have also been developed as reporter genes. They alter carotenoid and anthocyanin biosynthetic pathways in transformed cells, resulting in changes in pigmentation, permitting transformed cells to be visualized against a background of non-transformed cells, prior to manual selection.

Selection systems are in common use involving the addition to the culture medium of a specific agent, the latter being an antibiotic, herbicide, a drug or a metabolite analogue, a carbon source, or a phytohormone precursor. In all cases, the enzymes encoded by the marker genes in transformed plant cells encourage the growth of the transformed cells in preference to the growth of their non-transformed counterparts. However, some non-transformed cells may escape selection, depending upon the 'tightness' of the system. Antibiotics for selection include a range of aminoglycosides, in particular kanamycin, tolerance to this agent being conferred by the neomycin phosphotransferase (*npt*II) gene. Resistance to hygromycin, conferred by the hygromycin phosphotransferase (*hph, hpt, aph*IV) gene, has been extremely effective in selection systems. In both cases, the antibiotics are phosphorylated by transgene activity, rendering them harmless within transformed cells. Herbicides used for selection are dominated by phosphinothricin and glyphosate, with resistance to phosphinothricin being conferred by the phosphinothricin acetyltransferase (*pat, bar*) gene. Normally, phosphinothricin inhibits glutamine synthase, resulting in accumulation of ammonia leading to cell death, but following acetylation, the herbicide no longer inhibits cell metabolism. The genes for 5-enolpyruvylshikimate-3-phosphate synthase (EPSP synthase, *aroA, cp4, epsps*) and glyphosate oxidoreductase (*gox*) confer tolerance to glyphosate. Pline-Srnic (2006) provided a detailed description of the the physiology of glyphosate resistance.

Kanamycin, followed by hygromycin and phosphinothricin, were the agents most used in selection of transformed cells in the investigations described in the 450 publications that Miki and McHugh (2004) assessed in 2002. These authors also summarized the traits introduced into a range of transgenic plants from *Beta vulgaris* to *Zea mays*, together with the selectable marker genes. They emphasized that the selection systems employed often depend upon the plant species, since no single system is adequate for all plants. As these authors indicated, the transformation frequency may be sub-optimal in many systems because dying, non-transformed cells may inhibit the development of transformed cells in their immediate vicinity.

Toxic drugs and metabolite analogues exploited in selection systems include 2-deoxyglucose, betaine aldehyde, S-aminoethyl L-cysteine, 4-methyltryptophan and methotrexate, while D-xylose, D-mannose, benzyladenine-*N*-3-glucuronide provide examples of non-toxic agents for the selection of transformed cells. Mannose, for example, is normally phosphorylated to non-metabolizable mannose-6-phosphate that accumulates and inhibits growth. However, in cells transformed with the phosphomannose isomerase (*manA; pmi*) genes, mannose-6-phosphate is converted to fructose-6-phosphate that is readily utilized. The majority of genes used as selectable markers are of bacterial origin, with a limited number from other sources, including plants such as *A. thaliana, Catharanthus roseus, Petunia hybrida, Spinacea oleracea* and *Z. mays*, and fungi (e.g., *Candida albicans, Myrothecium verrucaria*).

Some selection systems do not require exogenously supplied substrates in the culture medium, as selection is based on the growth and differentiation of transformed cells. An excellent example is in the case of isopentenyl transferase (*ipt, pga22*) genes from *A. tumefaciens* and *Arabidopsis thaliana*, respectively, that modify endogenous growth regulator concentrations, promoting shoot regeneration from transformed cells. There are also selection systems in which the transgenes convert non-toxic agents to toxic compounds within the transformed cells, resulting in the death of the latter. Miki and McHugh (2004) provided an in-depth description of the common selection systems for nuclear and plastid transformation, with extensive reference to examples cited in the literature.

5 FACTORS INFLUENCING GENE EXPRESSION IN TRANSGENIC PLANTS

Several parameters influence the expression of transgenes in plants, particularly their sites of integration into the recipient genome and the promoter(s) driving the gene(s) of interest. Currently, none of the procedures for gene integration provide a reliable mechanism for inserting DNA at a defined locus. Consequently, foreign DNA is inserted at random locations into host chromosomes. Thus, even different plants generated using the same cassettes of foreign gene(s) have different gene integration patterns and behave differently. Transgenes may be down- or up-regulated depending on whether they integrate into heterochromatic regions, or in the vicinity of enhancer elements. The number of copies of the gene will also influence expression. Vector 'backbone' sequences may also be transferred with transgenes; truncation, inversion and deletion of gene sequences are not uncommon. Thus, the generation of large populations of transgenic plants is essential to confirm the effect of the expression of a transgene in any tissue (Bhat and Srinivasan 2002). As explained by these authors, the evaluation of transgenic plants and their progeny demands an integrated approach involving cell culture, transformation, and conventional plant breeding.

Lyznik et al. (2003) discussed the combination of homologous and site-specific recombination for large-scale chromosome engineering in mammals and insects and the application of this technology to plant transformation, particularly involving the *Cre/LoxP* and *FLP/FRT* systems. Although *Cre*- and *FLP*-mediated processes are not common at present in plants, such recombinase-based procedures will increase, undoubtedly, in the future. More recently, Iida and Terada (2004) proposed models for gene targeting in rice by homologous recombination, including procedures involving double-strand break repair, synthesis-dependent strand annealing and break-induced replication. As these authors indicated, homologous recombination will provide not only a powerful approach for characterizing genes in rice, but has application to other cereals. Cotsaftis and Guiderdoni (2005) also described the challenge of gene targeting as a routine procedure in plants. Indeed, the importance of this technology cannot be over-emphasized since plant breeders, in particular, require the means to modify specific loci in genomes. At present, gene targeting by homologous recombination remains very low, in the order of 10^{-3} to 10^{-6}.

Promoters play a key role in governing gene expression in transgenic plants. Those that control constitutive expression of transgenes featured in most of the initial transformation technology, particularly the 35S promoter from cauliflower mosaic virus (CaMV), because of its ready availability. Subsequently, other viral promoters were exploited for plant transformation, including those from cassava vein mosaic virus, Australian streak virus, mirabilis mosaic virus and figwort mosaic virus. Promoters from plants that have been used extensively include those from actin and ubiquitin genes. A disadvantage of constitutive gene expression is that it may interfere with aspects of plant development. In order to circumvent this difficulty, promoter technology has focused on the isolation and use of regulatory elements that control the temporal and spatial expression of transgenes. Indeed, promoters are available that are specific to most plant tissues and plastids. Synthetic promoters have also been constructed, increasing the application of transformation technology. Potenza et al. (2004) discussed the promoters that are available from fruit, seeds/grain, tubers and storage organs, flowers, roots and root nodules, leaves, and other green tissues. These authors provided a very useful annotated sketch of the source tissues used for the isolation of such promoters, together with supporting references. Trichome, guard cell/stomata, root hair and companion cell-specific promoters have been reported, while plastid-specific promoters assume increasing importance as more effort is directed to the generation of transplastomic plants. Several inducible/ repressible promoters have been described that can be activated by externally supplied agents (Padidam 2003).

Despite the range of promoters available, variation in transgene expression is common. Eike et al. (2005) showed that transcripts from cryptic promoters may affect transgene silencing, while Butaye et al. (2005) reviewed the factors that contribute to unpredictable transgene expression, including the integration of multiple gene copies, often with complex rearrangements and/or deletions of the foreign DNA, into the chromosomes of target plants. Gene silencing at the transcriptional or post-transcriptional levels, the position of integration of transgenes into the host genome, the activity of promoters and terminators and the presence of matrix attachment sites flanking transgene cassettes, all influence foreign DNA expression in transgenic plants. As Butaye et al. (2005) emphasized, the correct tailoring of constructs is crucial in maximizing gene expression and, in the longer term, the way of minimizing variation in transgene expression may be through the creation of plant artificial chromosomes. Frequently, genetic engineering technology is not directed to maximizing gene expression. In fact, the ability to silence genes is essential in some cases, with virus-induced gene silencing (VIGS) being an important tool for down-regulating expression (Robertson 2004). This author has provided comprehensive tables of examples of virus-induced gene silencing in plants, together with details of the relevant delivery procedure, the resulting plant phenotype and the criteria for assessing gene silencing.

6 EXPRESSION OF USEFUL GENES IN HIGHER PLANTS: A BRIEF RESUME

The increasing availability of genes and their integration and expression in plants have progressed to the stage at which transformation technology is now directed not only to understanding gene function, but as importantly, engineering plants for specific traits. A survey of the global status revealed that genetically manipulated crops were grown on 81 million hectares by more than 8 million farmers in 17 countries in 2004 (James 2005). These areas of cultivation and the range of crops are likely to increase significantly during the next decade. However, classical plant breeding will never be replaced by genetic manipulation technology; the latter will merely underpin conventional approaches, reducing the time required to generate new varieties. The range of genes introduced into plants has escalated to such an extent during the last five years that it is impossible to summarize all the relevant literature. Consequently, only a limited number of specific examples are cited in the present discussion. In most cases, the literature cited tabulates those agronomically useful genes and the target plants into which such genes have been introduced and expressed.

Wenzel (2006) listed 50 important crop plants, ranging from alfalfa to wheat, that have been transformed, together with countries where the main transgenic crops are cultivated and the area of land under cultivation. Mohan Babu et al. (2003) provided an overview of the range of chimaeric genes introduced into plants, from those conferring tolerance to biotic and abiotic stresses to adapt to changing environmental conditions, through to changes in floral pigmentation. It is not unexpected that because of crop losses caused by weeds, insects, viruses and fungi, that increased tolerance to these agents, constitutes a prime target for genetic manipulation. Of the 50 transgenic plants listed by Wenzel (2006), attention has been focused on cotton, soybean, maize and rapeseed, not only as 'models' *per se*, but because of their importance in global economy. Herbicide tolerance in soybean and insect resistance in maize have dominated this area of research. Popelka et al. (2004) discussed the relevance of genetically manipulating legumes to challenge potential food shortages. The demand for these genetically manipulated crops is likely to continue in the immediate future. In addition to maize, rice and wheat have also been targeted as major cereal crops, with manipulation of the β-carotene to generate 'Golden Rice' and 'Golden Rice 2' undoubtedly representing the most significant advance in transgenic cereal research in recent years (Paine et al. 2005). Importantly, complete sequencing of the rice genome (Matsumoto et al. 2005) will aid in understanding the development of rice and other cereals and their genetic improvement through transformation. The simplicity of particle bombardment for transient gene expression studies has facilitated molecular analysis of the interaction of mildew with cereal leaves and the effect of over-expressing or silencing single host genes (Panstruga 2004). Shrawat and Lörz (2006) reviewed the literature relating to *Agrobacterium*-mediated gene delivery to rice and cereals, indicating the achievements, particularly in rice, but also emphasizing some of the inherent difficulties associated with the application of this technology to barley, wheat, maize, and sorghum.

Tolerance to insect pests has been achieved by expressing the *Bt* and *cry* genes from *Bacillus thuriengensis* in transgenic plants, this approach following the widespread use of the bacterium itself as a biological insecticide. The protein crystal toxins produced by the bacterium disrupt cells of the insect midgut, the toxins being active, depending on their structure, against Coleoptera, Diptera, and Lepidoptera. An interesting scenario was reported by Zaidi et al. (2006) in which the Bt toxin Cry1Ab produced in the leaves of transgenic rice was extracted and applied to cabbage (*Brassca juncea*), resulting in 78 to 100% mortality of the larvae of *Pieris rapae*. Certainly, it may be useful to extend this strategy to plants that remain difficult to transform directly to insect resistance using *Agrobacterium* and biolistics-mediated gene delivery. A second category of genes, including proteinase inhibitors, α-amylase inhibitors and lectin genes, normally contribute to natural plant defence systems by interfering with insect digestion. Insect-derived chitinase genes have also been assessed for their ability to confer insect tolerance on transgenic plants. A comprehensive list of target crops and transgenes for insect resistance, with the supporting references, has been compiled by Mohan Babu et al. (2003). Clearly, field production has already benefited from the introduction of insect-resistant crops, accompanied by a reduction in the use and cost of insecticides. However, as Mohan Babu et al. (2003) indicated, many of the genes used are only mildly effective against the target insect pests, with the pyramiding of genes being an essential strategy for achieving long-term durable insect resistance. François et al. (2002) reported the procedures that are available for stacking transgenes in plants, through conventional sexual crossing and genetic manipulation approaches. The latter include sequential transformation, cotransformation, the exploitation of internal ribosome entry sites, a polyprotein approach and, importantly, the use of transplastomic technology, plastid transformation being used either alone or combined with nuclear transformation. Halpin and Boerjan (2003) listed the potential traits for stacking in forest trees, including those for herbicide, insect and disease resistances, biomass and wood content, and the control of flowering. Likewise, Petri and Burgos (2005) considered the progress achieved in genetically manipulating fruit trees, particularly in terms of disease resistance, while Gómez-Lim and Litz (2004) focused attention on the transformation of tropical fruits, particularly citrus, banana, mango, pineapple, avocado, longan, litchi, and passion fruit.

The genetic manipulation of storage products is a key area of experimentation, with substantial progress being reported in rapeseed in terms of fatty acid composition (Kinney et al. 2002). Cardoza and Stewart (2004) provided an overview of the advances in cellular and molecular biology of the genus *Brassica,* both in terms of transformation for introducing durable genes and also gene introgression by somatic hybridization. The relevance of somaclonal variation, together with marker-assisted selection and breeding, was also described.

Potato remains a prime target of the tuber crops for genetic manipulation, since it is cultivated in about 125 countries and consumed by more than a billion persons daily. Mullins et al. (2006) evaluated the role of molecular approaches for potato improvement. Transformation of potato is well established and despite genotype-dependent variation, genes introduced into potato include those with resistance to Colorado beetle, the fungi *Verticillium dahliae, Phytophthora infestans,* and *Alternaria, solani,* and the bacterial soft-rot pathogen, *Erwinia carotovora*. Similarly, resistance to potato viruses X and Y has also been accomplished.

Of the environmental stresses experienced by plants, drought poses one of the greatest constraints world-wide for agriculture. Approaches for manipulating plants to increase drought tolerance include the introduction of genes that encode enzymes associated with the synthesis of osmotically active compounds, transporters, chaperones, and scavengers of reactive oxygen species (Umezawa et al. 2006). Importantly, drought, cold, and salt tolerance are often linked, facilitating genetic manipulation of these traits. Dita et al. (2006) have indicated the biotechnological approaches that may be exploited to overcome biotic and abiotic stresses in legumes.

Since floriculture makes a significant contribution to the economy of several countries, it follows that effort has been directed to the genetic improvement of ornamentals especially in relation to flower color, fragrance, vase life, insect and disease resistances and plant stature. The *rol* genes from the T-DNA of *A. rhizogenes* have featured in many of these transformation studies (Casanova et al. 2005).

Undoubtedly, the longer-term focus of plant genetic manipulation is towards metabolite engineering for the biosynthesis of pharmaceuticals (Fischer et al. 2004) and other specialist compounds. For example, oral vaccines synthesized in plants have the potential to overcome some of the limitations of traditional vaccines, including the cost of production and the ease of storage prior to distribution (Robert and Kirk 2006). Lettuce, for example, a palatable species that exhibits rapid growth, has been engineered to express measles virus haemaglutinin protein (Webster et al 2006). Likewise, Mullins et al. (2006) stressed the potential of transgenic potato for the biosynthesis of edible plant-based vaccines for protection, for example, against HIV-1 infection, and also the possible use of this crop from industrial and veterinarian perspectives. Alongside genetic manipulation for biosynthesis of pharmaceuticals is the possibility of fortifying crops, particularly those consumed in the raw state such as leafy vegetables, with micronutrients. The latter include vitamins A and E and folate, and minerals such as zinc, iron, and selenium (Poletti and Sautter 2005).

7 BIOSAFETY: IMPACT OF GM TECHNOLOGY ON SOCIETY AND HUMAN HEALTH

Whilst more emphasis has focused on the interaction of *Agrobacterium* with plant cells, it is interesting to note that this bacterium does not transfer DNA only to plant cells. Indeed, it can affect gene transfer to fungi and human cells (Lacroix et al. 2006b), raising concerns that the release into the environment of *Agrobacterium* carrying recombinant DNA could be followed by gene transfer to soil and plant-borne fungi and other non-plant organisms. There is some evidence that *Agrobacterium* can be a rare but opportunistic human pathogen capable of infecting immune-compromised patients.

Goldstein et al. (2005) described genetically manipulated plants in relation to human safety. As they emphasized, the inefficiency of plant cell transformation necessitates the use of markers to select transformed cells from their non-transformed counterparts. Antibiotic resistance markers include genes for resistance to kanamycin (*npt*II), hygromycin (*aph, hpt*), and streptomycin and spectinomycin (*aad*A). When introducing herbicide resistance into plants, the gene of interest itself can be used for selection. Currently, kanamycin, streptomycin, and spectinomycin have limited utility in clinical medicine, whilst hygromycin has no clinical use presently or in the immediate future. Thus, in reviewing the literature, Goldstein et al. (2005) concluded that antibiotic resistance markers pose no health risks to humans or animals and that gene products have no unique hazards compared to those of dietary proteins. There is no documentation of intact functional hygromycin gene transfer from plants to bacteria in nature. Likewise, there is no evidence for regular incorporation of plant or bacterial genes into human cells, inspite of the fact that because all foods contain DNA, humans have been exposed to plant and bacterial DNA throughout evolution.

The transfer of antibiotic resistance genes from plants to bacteria is not an important mechanism of bacterial resistance. Furthermore, antibiotic release into the environment through manufacture, disposal and therapeutic use encourages the presence of antibiotic resistance genes in bacteria. Clearly, antibiotic resistance markers and their products in plants will be present in the early generation of transgenic plants for some time, although legislation will demand the removal of such markers from future generations of transgenic plants. Three procedures are available for the removal of marker genes, these involving cotransformation followed by segregation of the resistance marker, homologous recombination (Puchta 2003) and recombinase-mediated excision, the latter based on the *Cre/Lox, FLP/FRT,* and *R-RS* systems, all of which are described by Goldstein et al. (2005). Transposition-mediated repositioning of genes has been summarized by Miki and McHugh (2004), the maize *Ac/Ds* transposable element system being exploited to construct novel T-DNA vectors. Whilst marker-free transgenic plants can be generated using this approach, repositioning of genes of interest may also occur, sometimes accompanied by their enhanced expression. Goldstein et al. (2005) concluded that the use of antibiotic resistance markers is an inexpensive, readily available, widely accepted technology that has not provided any documented harmful effects. They posed the question. Why replace it with a more costly, less understood system?

Bennett et al. (2004) and Gay and Gillespie (2005) also considered the risks of antibiotic resistance genes. The latter authors concluded that DNA from genetically manipulated plants must be degraded in the field, taken up by potential recipient soil microorganism, integrated into their genomes and expressed. Although fragments of DNA large enough to contain antibiotic resistance genes may survive in the environment, the barriers to transfer, incorporation, and transmission are so considerable that any contribution to antibiotic resistance made by genetically manipulated plants must be overwhelmed by the contribution made by antibiotic prescription in clinical practice. In terms of tissue culture technology, Zuo et al. (2002) questioned the use of some of the selectable marker gene systems, since the antibiotics and herbicides used are known to inhibit growth and regeneration of transformed cells, decreasing the transformation frequency. Consequently, they favour the use of genes that promote shoot regeneration. An area that has received relatively little attention, is that of the risk of allergenicity from transgenic plants. In this respect, Moneret-Vautrin (2006) suggested that the lead being taken by France in screening transgenic plants for their possible ability to induce allergenic reactions should be extended to international networks in order to identify and possibly preempt any difficulties associated with transgenic crops before the latter are commercialized.

As a means to circumvent some of the difficulties associated with the use of genes from microorganisms, Rommens (2004) advocated more extensive use of plant-based genes for plant transformation, since genes for fungal and insect resistance, fruit ripening, drought and salinity tolerances, for example, are already available in plants. Some genes, such as a modified plant protoporphrinogen gene, can be used as selectable markers. Antisense expression of plant genes can also be exploited to control the biosynthesis of antinutritional compounds, such as alkaloids, and processes such as fruit ripening. In addition, sequences have been identified in plant DNA that resemble *Agrobacterium* T-DNA borders. Since these plant DNAs support gene transfer from *Agrobacterium* to plant cells, they may be used more extensively in the future in the construction of plant transformation vectors.

8 GENE TRANSFER TO OTHER ORGANISMS – RISKS

As gene flow from transgenic plants to wild species is a major impediment to the cultivation of genetically manipulated plants in some countries, several investigators have tracked the spread of genes from transgenic to wild type plants. For example, Alibert et al. (2005) studied the transmission of the *epsps* gene for resistance to the herbicide glyphosate and the *gus* gene from transgenic beet to ruderal beet under glasshouse and field conditions. Pollen was capable fertilizing target plants at 1 km from donor plants. Transgenic seed was recovered in these experiments, indicating that care is required in rotation cropping of genetically manipulated beet. As expected, the proportion of transgenic hybrids in the recipient plant population decreased with distance from the transgenic source being about 0.5% for ruderal beets and 1.5% for male sterile beet at 200 m, the values always being greater when the recipient plants were downwind of the transgenic plants. Alibert et al. (2005) quoted several relevant experiments to track gene transfer under field conditions. An interesting concept was one proposed by Stewart (2005) using bio- and nano-technologies to monitor the presence and expression of transgenes. The latter tagged with green fluorescent protein could be imaged using instrumentation designed to detect fluorescent signals, while nano-technologies employing aptamers, quantum dots and molecular beacons may also be exploited for *in vivo* monitoring.

Other authors have stressed their concerns of the impact of genetically manipulated crops on the environment, not only in terms of gene flow to other species but also their effects on biodiversity, all of which are interdisciplinary and highly complex (Saito and Miyata 2005). In order to investigate transgenic plants in the natural habitat of their related wild-type species, Celis et al. (2004) introduced potatoes, genetically modified for resistance to nematodes, into the central Andes, the centre of diversity of this crop. These authors demonstrated that there was no harm to non-target organisms, although gene flow to wild relatives in the vicinity of the transgenic crops could affect the fitness of the wild species if stable introgression occurred. Consequently, in order to avoid this possibility, transformation was targeted to the male sterile potato cultivar Revolucion. Indeed, this approach should not prevent the cultivation of genetically manipulated potato or, indeed, other crops, in their centers of crop biodiversity. As Conner et al. (2003) stressed, a difficulty in assessing genetically manipulated crops in terms of their impact under field conditions is in defining the appropriate baseline for comparison and decision. They emphasized that the most appropriate baseline for comparison and reference is the impact that may occur by plants developed through traditional breeding, since these are accepted as part of present-day agricultural systems.

9 CONCLUSIONS

World food security to satisfy an increasing population necessitates continued intensive crop production, matched against a background of a reduction in arable land because of deterioration of top-soil quality and erosion, lack of water aggravated by climatic change, disease and insect attack, and political instability in some regions. The green revolution of the 1960s and 1970s that exploited conventional breeding to generate cereals with improved yields, has been followed by the biotechnological era in which conventional breeding is underpinned by the ability to manipulate genetic traits in a meaningful way, based on improved understanding of plant genetics and genome organization. Nevertheless, it must not be overlooked that the genetic manipulation of many plants is not feasible, or remains highly inefficient, and the technology is still not available as a routine plant breeding tool (Arias et al. 2006). Indeed, these authors have indicated that since transformation is associated positively with cell division and differentiation, with optimum transformation at the S and G2 phases of the cell cycle, manipulating genes that control the cell cycle may facilitate transformation in several plants.

Biotechnology, of which genetically manipulated crops are only one aspect, has already induced major restructuring in the agrifood sector, seen, for example, in the early generation of plants expressing herbicide tolerance and insect resistance. The ability to manipulate the genetic code of plants has caused the major agrochemical companies to change their marketing programs. Shifting crop production between countries has altered world patterns of trade which will benefit some parties but not others. The ability to breed and grow crops expressing specific traits has provided the opportunity for innovation that, undoubtedly, will alter the economic foundation of agriculture, especially in the next decade. Clearly, there are relevant issues of intellectual property and technology transfer (Khachatourians et al. 2002). Whilst some countries, producers, and the public have accepted the need, especially longer-term, for biotechnological approaches to underpin conventional breeding, others have focused on the perceived risks of genetically manipulated crops, accompanied by national legislation, rather than following global trends (Singh et al. 2006). Consequently, in some countries, the perceived risks of transgenic crops outweigh their potential advantages, probably because of poor communication and inadequate discussion between scientists, politicians and consumers. Clearly, there is a requirement to rebuild trust in agrifood systems, to address relevant issues associated with genetically manipulated plants, particularly their effect upon ecology and biodiversity, and to educate farmers, politicians and the general public in this important area of research and development.

References

Alibert B, Sellier H, Souvré A (2005) A combined method to study gene flow from cultivated sugar beet to ruderal beets in the glasshouse and open field. Eur J Agron 23: 195-208

Altpeter F, Baisakh N, Beachy R, Bock R, Capell T, Christou P, Daniell H, Datta K, Datta S, Dix PJ, Fauquet C, Huang N, Kohli A, Mooibroek H, Nicholson L, Nguyen TT, Nugent G, Raemakers K, Romano A, Somers DA, Stoger E, Taylor N, Visser R (2005) Particle bombardment and the genetic enhancement of crops: Myths and realities. Mol Breed 15: 305-327

Arias RS, Filichkin SA, Strauss SH (2006) Divide and conquer: development and cell cycle genes in plant transformation. Trends Biotechnol 24: 267-273

Bennett PM, Livesay CT, Nathwani D, Reeves DS, Saunders JR, Wise R (2004) An assessment of the risks associated with the use of antibiotic resistance genes in genetically modified plants: Report of the working party of the British Society of Antimicrobial Chemotherapy. J Antimicrob Chem 53: 418-431

Bechtold N, Jolivet S, Voisin R, Pelletier G (2003) The endosperm and the embryo of *Arabidopsis thaliana* are independently transformed through infiltration by *Agrobacterium tumefaciens*. Transgenic Res 12: 509-517

Bhat SR, Srinivasan S (2002) Molecular and genetic analyses of transgenic plants: Considerations and approaches. Plant Sci 163: 673-681

Bock R, Khan MS (2004) Taming plastids for a green future. Trends Biotechnol 22: 311-318

Broothaerts W, Mitchell HJ, Weir B, Kaines S, Smith LMA, Yang W, Mayer JE, Roa-Rodriguez C, Jefferson RA (2005) Gene transfer to plants by diverse species of bacteria. Nature 433: 629-633

Butaye KMJ, Cammue BPA, Delauré SL, De Bolle MFC (2005) Approaches to minimize variation of transgene expression in plants. Mol Breed 16: 79-91

Cardoza V, Stewart CN (2004) *Brassica* technology: Progress in cellular and molecular biology. In Vitro Cell Dev Biol-Plant 40: 542-551

Casanova E, Trillas MI, Moysset L, Vainstein A (2005) Influence of *rol* genes in floriculture. Biotechnol Adv 23: 3-39

Celis C, Scurrah M, Cowgill S, Chumbiauca S, Green J, Franco J, Main G, Kiezebrink D, Visser RGF, Atkinson HJ (2004) Environmental safety and transgenic potato in a centre of diversity of this crop. Nature 432: 222-225

Chase CD (2006) Genetically engineered cytoplasmic male sterility. Trends Plant Sci 11: 7-9

Cheng M, Lowe BA, Spencer TM, Ye X, Armstrong CL (2004) Factors influencing *Agrobacterium*-mediated transformation of monocotyledonous species. In Vitro Cell Dev Biol-Plant 40: 31-45

Chung MH, Chen MK, Pan SM (2000) Floral spray transformation can efficiently generate *Arabidopsis* transgenic plants. Transgenic Res 9: 471-476

Chung S-M, Frankman EL, Tzifira T (2005) A versatile vector system for multiple gene expression in plants. Trends Plant Sci 10: 357-361

Chung S-M, Vaidya M, Tzfira T (2006) *Agrobacterium* is not alone: gene transfer to plants by viruses and other bacteria. Trends Plant Sci 11: 1-4

Conner AJ, Glare TR, Nap J-P (2003) The release of genetically modified crops into the environment. Part II. Overview of ecological risk assessment. Plant J 33: 19-46

Cotsaftis O, Guiderdoni E (2005) Enhancing gene targeting efficiency in plants: rice is on the move. Transgenic Res 14: 1-14

Curtis IS (2004) Transgenic Crops of the World. Essential Protocols. Kluwer Academic Publ, Dordrecht, The Netherlands.

Davey MR, Anthony P, Power JB, Lowe KC (2005a) Plant protoplasts: status and biotechnological perspectives. Biotechnol Adv 23: 131-171

Davey MR, Anthony P, Power JB, Lowe KC (2005b) 2004 SIVB Congress Symposium Proceedings "Thinking outside the cell": Plant protoplast technology: Status and applications. In Vitro Cell Dev Biol-Plant 41: 202-212

Dita MA, Rispail N, Prats E, Rubiales D, Singh KB (2006) Biotechnology approaches to overcome biotic and abiotic stress constraints in legumes. Euphytica 147: 1-24

Eike MC, Mercy IS, Aalen RB (2005) Transgene silencing may be mediated by aberrant sense promoter sequence transcripts generated from cryptic promoters. Cell Mol Life Sci 62: 3080-3091

Fischer R, Stoger E, Schillberg S, Christou P, Twyman RM (2004) Plant-based production of biopharmaceuticals. Curr Opin Plant Biol 7: 152-158

François IEJA, Broekaert WF, Cammue BPA (2002) Different approaches for multi-transgene-stacking in plants. Plant Sci 163: 281-295

Gay PB, Gillespie SH (2005) Antibiotic markers in genetically modified plants: A risk to human health? Lancet 5: 637-646

Gelvin S (2003) *Agrobacterium*-mediated plant transformation: the biology behind the "Gene-Jockeying" tool. Microbiol Mol Biol Rev 67: 16-37

Goderis IJWM, De Bolle MFC, François IEJA, Wouters PFJ, Broekaert WF, Cammue BPA (2002) A set of modular plant transformation vectors allowing flexible insertion of up to six expression units. Plant Mol Biol 50: 17-27

Goldstein DA, Tinland B, Gilbertson LA, Staub JM, Bannon GA, Goodman RE, McCoy RL, Silvanovich A (2005) Human safety and genetically modified plants: a review of antibiotic resistance markers and future transformation selection technologies. J Appl Microbiol 99: 7-23

Goodman RM (2004) Encyclopedia of Plant and Crop Science. Marcel Dekker, New York, USA

Gómez-Lim MA, Litz RE (2004) Genetic transformation of perennial tropical fruits. In Vitro Cell Dev Biol-Plant 40: 442-449

Halpin C, Boerjan W (2003) Stacking genes in forest trees. Trends Plant Sci 8: 363-365

Hansen G, Chilton MD (1996) "*Agrolistic*" transformation of plant cells: integration of T-strands generated *in planta*. Proc Natl Acad Sci USA 93: 14978-14983

Hellens R, Mullineaux P, Klee H (2000) A guide to *Agrobacterium* binary Ti vectors. Trends Plant Sci 5: 446-451

Iida S, Terada R (2004) A tale of two integrations, transgene and T-DNA: Gene targeting by homologous recombination in rice. Curr Opin Biotechnol 15: 132-138

James C (2005) Global Status of Commercialized Biotech/GM Crops: 2004: *http:/ /www.isaaa.org.Kc/CBTNewa/press_release/brieff32*

Jones HD, Doherty A, Wu H (2005) Review of methodologies and a protocol for the *Agrobacterium*-mediated transformation of wheat. Plant Methods: *http:// www.plantmethods.com/content/1/1/5*

Kinney AJ, Cahoon EB, Hitz WD (2002) Manipulating desaturase activities in transgenic crop plants. Biochem Soc Trans 30: 1099-1103

Khachatourians GG (2002) Agriculture and food crops: Development, science and society. In: Khachatourians GG, McHughen A, Scorza R, Nip W-K, Hui YH (eds) Transgenic Plants and Crops. Marcel Dekker, New York, USA, pp 1-27

Khachatourians GG, McHughen A, Scorza R, Nip W-K, Hui YH (2002) Transgenic Plants and Crops. Marcel Dekker, New York, USA

Karimi M, Inzé D, Depicker A (2002) GATEWAY™ vectors for *Agrobacterium*-mediated plant transformation. Trends Plant Sci 7: 193-195

Kuta DD, Tripathi L (2005) *Agrobacterium*-induced hypersensitive necrotic reaction in plant cells: A resistance response against *Agrobacterium*-mediated DNA delivery. Afr J Biotechnol 4: 752-757

Lacroix B, Li J, Tzfira T, Citovsky V (2006a) Will you let me use your nucleus? How *Agrobacterium* gets its T-DNA expressed in the host plant cell. Can J Physiol Pharmacol 84: 333-345

Lacroix B, Tzfira T, Vainstein A, Citovsky V (2006b) A case of promiscuity: *Agrobacterium's* endless hunt for new partners. Trends Genet 22: 29-37

Lelivelt CLC, McCabe MS, Newell CA, Bastiaan de Snoo C, van Dun KMP, Birch-Machin I, Gray JC, Mills KHG, Nugent JM (2005) Stable plastid transformation in lettuce (*Lactuca sativa* L.). Plant Mol Biol 58: 763-774

León-Bañares R, González-Ballester D, Galván A, Fernández E (2004) Transgenic microalgae as green cell-factories. Trends Biotechnol 22: 45-52

Loyola-Vargas VM, Vázquez-Flota (2005) Methods in Molecular Biology™. Plant Cell Culture Protocols. 2nd edn. Humana Press, Totowa, New Jersey, USA

Liu Y, Yang H, Sakanishi A (2006) Ultrasound: Mechanical gene transfer into plant cells by sonoporation. Biotechnol Adv 24: 1-16

Lyznik LA, Gordon-Kamm WJ, Tao Y (2003) Site-specific recombination for genetic engineering in plants. Plant Cell Rep 21: 925-932

Maliga P (2004) Plastid transformation in higher plants. Annu Rev Plant Biol 55: 289-313

Matsumoto et al. (2005) The map-based sequence of the rice genome. Nature 436: 793-800

Miki B, McHugh S (2004) Selectable marker genes in transgenic plants: Applications, alternatives and biosafety. J Biotechnol 107: 193-232

Mohan Babu R, Sajeena A, Seetharaman K, Reddy MS (2003) Advances in genetically engineered (transgenic) plants in pest management - An over view. Crop Protect 22: 1071-1086

Moneret-Vautrin AD (2006) Transgenic plants (GM plants): What we do and don't know about the risks of allergenicity. Rev Franc Allergol Immunol Clin 46: 85-91

Mullins E, Milbourne D, Petti C, Doyle-Prestwich BM, Meade C (2006) Potato in the age of biotechnology. Trends Plant Sci 11: 254-260

Newell C (2000) Plant transformation technology. Developments and applications. Mol Biotechnol 16: 53-65

Oksman-Caldentey K-M, Inzé D (2004) Plant cell factories in the post-genomic era: New ways to produce designer secondary metabolites. Trends Plant Sci 9: 433-440

Padidam M (2003) Chemically regulated gene expression in plants. Curr Opin Plant Biol 6: 169-177

Paine JA, Shipton CA, Chaggar S, Howells RM, Kennedy MJ, Vernon G, Wright SY, Hinchliffe E, Adams JL, Silverstone AL, Drake R (2005) Improving the nutritional value of Golden Rice through increased provitamin A content. Nat Biotechnol 23: 482-487

Panstruga R (2004) A golden shot: How ballistic single cell transformation boosts the molecular analysis of cereal-mildew interactions. Mol Plant Pathol 5: 141-148

Pappas KM, Winans SC (2003) Plant transformation by coinoculation with a disarmed *Agrobacterium tumefaciens* strain and an *Escherichia coli* strain carrying mobilizable transgenes. Appl Environ Microbiol 69: 6731-6739

Petri C, Burgos L (2005) Transformation of fruit trees. Useful breeding tool or continued future prospect? Transgenic Res 14: 15-26

Pline-Srnic W (2006) Physiological mechanisms of glyphosate resistance. Weed Technol 20: 290-300

Poletti S, Sautter C (2005) Biofortification of the crops with micronutrients using plant breeding and/or transgenic strategies. Minerva Biotechnol 17: 1-11

Popelka JC, Terryn N, Higgins TJV (2004) Gene technology for grain legumes: Can it contribute to the food challenge in developing countries? Plant Sci 167: 195-206

Potenza C, Aleman L, Sengupta-Gopalan C (2004) Targetted gene expression in research, agricultural, and environmental applications: Promoters used in plant transformation. In Vitro Cell Dev Biol-Plant 40: 1-22

Puchta H (2003) Towards the ideal GMP: Homologous recombination and marker gene excision. J Plant Physiol 160: 743-754

Qin S, Jiang P, Tseng C (2005) Transforming kelp into a marine bioreactor. Trends Biotechnol 23: 264-268

Rakoczy-Trojanowska M (2002) Alternative methods of plant transformation - a short review. Cell Mol Biol Lett 7: 849-858

Robert JS, Kirk DD (2006) Ethics, biotechnology, and global health: The development of vaccines in transgenic plants. Am J Bioethics 6: W29-W41

Robertson D (2004) VIGS vectors for gene silencing: Many targets, many tools. Annu Rev Plant Biol 55: 495-519

Rommens CM (2004) All-native DNA transformation: a new approach to plant genetic engineering. Trends Plant Sci 9: 457-464

Ruf S, Hermann M, Berger IJ, Carrer H, Bock R (2001) Stable genetic transformation of tomato plastids and expression of a foreign protein in fruit. Nat Biotechnol 19: 870-875

Saito T, Miyata T (2005) Situation and problems on transgenic technology for insect pest control. Jap J Appl Entomol Zool 49: 171-185

Sharma KK, Bhatnagar-Mathur P, Thorpe TA (2005) Genetic transformation technology: Status and problems. In Vitro Cell Dev Biol-Plant 41: 102-112

Shrawat AK, Lörz H (2006) *Agrobacterium*-mediated transformation of cereals: A promising approach to crossing barriers. Plant Biotechnol J 4: 575-603

Singh OV, Ghai S, Paul D, Jain RK (2006) Genetically modified crops: Success, safety assessment, and public concern. Appl Microbiol Biotechnol 71: 598-607

Stegemann S, Hartmann S, Ruf S, Bock R (2003) High-frequency gene transfer from the chloroplast genome to the nucleus. Proc Natl Acad Sci USA 100: 8828-8833

Stewart Jr CN (2005) Monitoring the presence and expression of transgenes in living plants. Trends Plant Sci 8: 390-396

Taylor NJ, Fauquet CM (2002) Microparticle bombardment as a tool in plant science and agricultural biotechnology. DNA Cell Biol 21: 963-977

Tzfira T, Citovsky V (2006) *Agrobacterium*-mediated genetic transformation of plants: Biology and biotechnology. Curr Opin Biotechnol 17: 147-154

Tzfira T, Li J, Lacroix B, Citovsky V (2004) *Agrobacterium* T-DNA integration: molecules and models. Trends Genet 20: 375-383

Umezawa T, Fujita M, Fujita Y, Yamaguchi-Shinozaki K, Shinozaki K (2006) Engineering drought tolerance in plants: discovering and tailoring genes to unlock the future. Curr Opin Biotechnol 17: 113-122

Valentine L (2003) *Agrobacterium tumefaciens* and the plant: The David and Goliath of modern genetics. Plant Physiol 133: 948-955

van der Fits L, Deakin EA, Hoge JHC, Memelink J (2000) The ternary transformation system: Constitutive *virG* on a compatible plasmid dramatically increases *Agrobacterium*-mediated plant transformation. Plant Mol Biol 43: 495-502

Veltcheva M, Svetleva D, Petkova S, Perl A (2005) *In vitro* regeneration and genetic transformation of common bean (*Phaseolus vulgaris* L.) - Problems and progress. Sci Hort 107: 2-10

Walmsley AM, Arntzen CJ (2003) Plant cell factories and mucosal vaccines. Curr Opin Biotechnol 14: 145-150

Webster DE, Smith SD, Pickering RJ, Strugnell RA, Dry IB, Wesselingh SL (2006) Measles virus hemagglutinin protein expressed in transgenic lettuce induces neutralizing antibodies in mice following mucosal vaccination. Vaccine 24: 3538-3544

Wenzel G (2006) Molecular plant breeding: Achievements in green biotechnology and future perspectives. Appl Microbiol Biotechnol 70: 642-650

Wu H, Sparks C, Amoah B, Jones HD (2003) Factors influencing successful *Agrobacterium*-mediated genetic transformation of wheat. Plant Cell Rep 21: 659-668

Zaidi MA, Cheng XY, Xu HR, Altosaar I (2006) *Ex situ* application of foliar-produced lepidoptericides from transgenic rice to *Pieris rapae* and Cry1Ab stability *in vivo*. Crop Protect 25: 748-752

Zuo J, Niu Q-W, Ikeda Y, Chua N-H (2002) Marker-free transformation: Increasing transformation frequency by the use of regeneration-promoting genes. Curr Opin Biotechnol 13: 173-180

13 Intellectual Property Rights and Protection of Agbiotechnological Inventions

M. Blakeney[1*], R.S. Kulkarni[2], K. Chalapathy Reddy[3] and A. Mohan Rao[2]

[1]Queen Mary Intellectual Property Research Institute, University of London, 67-69 Lincoln's Inn Fields, London, WC2A 3JB, UK

[2]Department of Genetics and Plant Breeding, University of Agricultural Sciences, Bangalore 560006, India

[3]Mission Biofuels India Pvt. Ltd., 608, Powai Plaza, Hiranandani Business Park, Powai, Mumbai 400076, India

*Corresponding author: m.blakeney@qmul.ac.uk

1 INTRODUCTION

The key determinant of the possibility in agricultural research in both developed and developing countries is the funding which is made available for that research. The availability of that funding is in part a function of the legal environment for research investment. An important feature of that environment is the Intellectual Property Rights (IPR) system, which provides a mechanism both for encouraging investment and for protecting innovations. The principal determinant of the IPR systems which prevails in most countries is the World Trade Organization (WTO) Agreement on Trade-Related Aspects of Intellectual Property Rights 1994 (TRIPs). The implementation of this Agreement is an obligation of all WTO members. The TRIPs Agreement lays down basic rules (norms) which countries have to adopt in their IPR laws. A number of other international agreements also contain IPR prescriptions. The most significant of these are: (i) the Convention on Biological Diversity 1992 (CBD), which contains obligations of consent of source countries to access to their biological resources and the sharing of benefits when those resources are exploited; and (ii) the International Treaty on Plant Genetic Resources for Food and Agriculture 2001, which contains rules about IPRs arising from the exploitation of the biological resources which are conserved within the Consulting Group for International Agricultural Research (CGIAR).

Countries can be divided into (a) exploiters of biological resources; and (b) source countries. As a general rule, the principal exploiters are industrialized countries with an advanced research capacity. These countries are invariably signatories of the TRIPs Agreement, which provides legal support for the protection of their research investments. Source countries are invariably those biologically diverse countries in the tropics which are looking for a share of the benefits which exploiting countries make from their biological diversity. Unfortunately, the exploiting countries have not yet all subscribed to the CBD or the International Treaty.

The current international negotiating agenda, whether within the TRIPs Agreement, or the Conference of Parties of the CBD, or within the negotiations in the World Intellectual Property Organization for a Substantive Patent Law Treaty, is to introduce the access and benefit sharing principles as a mandatory principle of intellectual property protection. Research programs in developing and developed countries, therefore, take account of the legal infrastructure surrounding the rights of and access to research innovations and genetic resources.

This chapter demonstrates that IPR protection in agricultural research is a fact of life. Public research institutions have to operate within this reality and have to develop their own responses, particularly a 'defensive' strategy. It cautions that there are difficulties in documenting any significant gains from using IPR protection as a strategy for generating new, external funds for research. The chapter underlines the fact that patents simply protect innovation and secure the potential rights for future development. The chance for national programs to earn financial benefits from research comes mainly from working with the private sector and by providing for technology transfer.

The most important mechanisms for legally protecting agricultural innovations are Plant Variety Rights (PVR) and patents. Other forms of protection can be provided through trademarks, trade secrets, and copyright. Alternatives to these include Material Transfer Agreements (MTAs) of a private contractual nature. If no form of protection is taken, then research results are generally placed in the public domain, mostly in the form of publications, making results available to all without restrictions on use.

Also it is important to note that agricultural research institutes which require access to proprietary genetic material, are obliged to introduce systems for the management of third party IP which they might wish to use and for the conservation of any IPR generated by public/private collaborative research.

2 INTELLECTUAL PROPERTY RIGHTS

Human beings are distinguished from animals by the intellectual faculty. This is probably reflected by the guarantee in Article 27(2) of the Universal Declaration of Human Rights that "Everyone has the right to protection of the moral and material interest resulting from any scientific, literary or artistic production of which he is the author". This is also reflected in the similarly-worded Art.15.1.c of the International Convention on Economic, Social and Cultural Rights. A guarantee of this human right is the legal protections which are made available to protect intellectual creations from misappropriation. According to the World Intellectual Property Organization (WIPO), the specialized agency of the United Nations, Intellectual Property, "refers to creations of the mind: inventions, literary and artistic works, and symbols, names, images, and designs used in commerce". Intellectual property is split into two main categories. Firstly there is industrial property, which includes patents, trademarks, industrial designs and trade secrets, and secondly copyright, which includes literary, musical, artistic and audio-visual works (Fig. 1; source Ravishankar and Archak 2000).

Intellectual property is neither a new nor a static concept. The grant by a state of some form of exclusive rights in their discoveries to inventors originated in the early part of the fifteenth century in Venice and spread rapidly during the sixteenth century to Germany, France, the Netherlands and England. The Gutenberg printing press is often cited as the main catalyst in the evolution of intellectual property rights. Further technological developments over the centuries propelled the need for stricter measures in protecting people's creative works. The rules governing intellectual property in traditional media are well established. The Internet however, is forcing governments, companies and individuals to re-address the way in which intellectual property; particularly copyright law should be governed (Singh 2003).

Intellectual property surrounds us in nearly everything we do, at home, in school, at work, during rest and play, no matter what we—do the fruits of human creativity and invention surround us. Intellectual property refers to certain kinds of exclusive rights to intellectual capital,

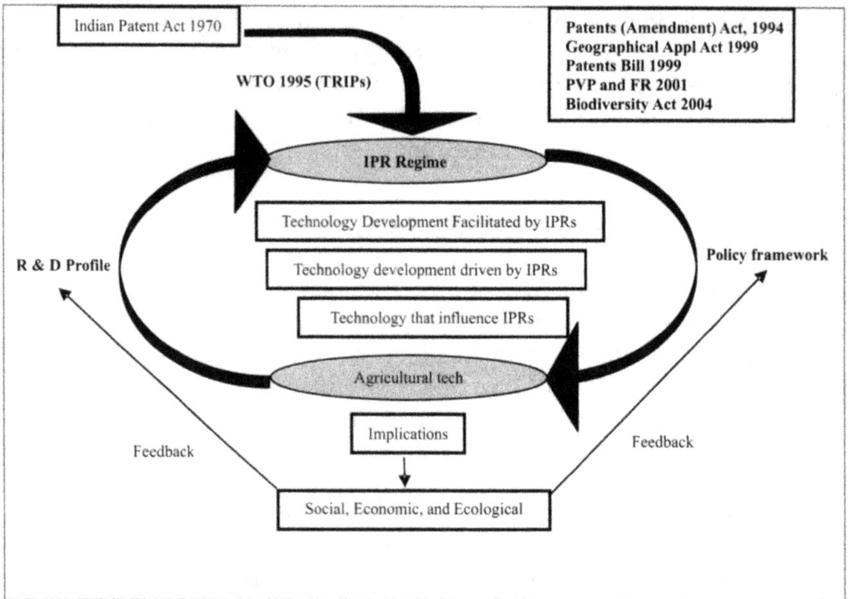

Fig. 1 Agriculture and IPRs in India: Interplay and interaction.

some forms of which can expire after a set period of time, and other forms of which can last indefinitely. The World Intellectual Property Organization defines Intellectual Property as legal rights that result from intellectual activity. The intellectual activity may include any activity in the industrial, scientific, literary and artistic fields. According to the Centre for Intellectual Property Rights in India, the major Indian intellectual properties typically fall into four major buckets; Copyright, Patent, Trademark, and Design Protection.

2.1 Protection of IPRs

The protection of IPRs may take several forms depending mainly on the type of intellectual property sought; each form of protection has its own advantages and pitfalls. The main forms of IPR protection are as follows:

1. Trademarks
2. Trade secrets
3. Patents
4. Copyright
5. Industrial designs

6. Geographical indications
7. Plant Breeder's Rights (PBRs)

2.2 Trade Secret Laws Can Be Used for Protection

Plants can be protected provided that reasonable efforts have been made to keep the plant characteristics confidential, secret and out of the public domain. In the United States, trade secrets have been used for decades to protect parental lines of *hybrid corn*. Recently, a US court recognized that '*genetic messages*' of inbred plant varietal lines might be protected by trade secret law subject to the provision that reasonable effort has been taken to preserve the secrecy of the gene sequence (Pioneer Hi-Bred case). In the Pioneer Hi-Bred case it was held that the accidental inclusion of inbred seed in bags of commercially available first generation seed by Pioneer Hi-Bred was not enough to destroy the secrecy surrounding its '*genetic message*'. Reasonable precaution had been taken to keep the '*genetic message*' contained in the inbred seed out of the public domain.

2.3 Patents

> **What kind of protection does a patent offer?**
>
> Patent protection means that the invention cannot be commercially made, used, distributed, or sold without the patent owner's consent. These patent rights are usually enforced in a court, which, in most systems, holds the authority to stop patent infringement. Conversely, a court can also declare a patent invalid upon a successful challenge by a third party.

An invention is a product or process that provides a new way of doing something, or offers a new technical solution to a problem. Inventions are protected by *patents*, which provide exclusive rights to the owner of the patent. This means that the patented invention cannot be made, used, distributed or sold without the patent owner's consent. Patent protection is granted for a limited time usually 20 years from the filing date of the patent application. Once a patent expires, the protection ends, and an invention enters the public domain, available for commercial exploitation by others. A patent provides protection for the invention to the owner of the patent. Patent area is granted for (1) an invention (including product), (2) innovation/improvement in an invention, (3) the process of an invention/product and (4) a concept (Subbaram 2000).

2.3.1 *Patent Requirements*

The main requirements for granting a patent are as follows:

1. **Novelty**: The invention must be new and should not be already known to the public
2. **Inventiveness**: The invention should not be obvious to a person skilled in the art and should represent an innovation.
3. **Industrial application and usefulness**: The subject matter of the patent must have an industrial application, either immediate or in the future and this application should be useful to the society/ nation.
4. **Patentability**: The subject matter of the patent must be patentable under the existing law and its current interpretation. This criterion at present varies from country to country and with time within the same country.
5. **Disclosure**: The inventor is required to describe his invention in sufficient details so that a person of normal skill is able to reproduce it. In case of biological organisms, already known organisms may be simply named.

A patent owner has the right to decide who may - or may not - use the patented invention for the period in which the invention is protected. The patent owner may give permission, or license to, other parties to use the invention on mutually agreed terms. The owner may also sell the right to the invention to someone else, who will then become the new owner of the patent. Once a patent expires, the protection ends, and an invention enters the public domain, that is, the owner no longer holds exclusive rights to the invention, which becomes available to commercial exploitation by others.

2.3.2 *Importance of Patents*

Patents provide incentives to individuals by offering them recognition for their creativity and material reward for their marketable inventions. These incentives encourage innovation, which assures that the quality of human life is continuously enhanced. Patented inventions have, in fact, pervaded every aspect of human life, from electric lighting (patents held by Edison and Swan) and plastic (patents held by Baekeland), to ballpoint pens (patents held by Biro) and microprocessors (patents held by Intel, for example). All patent owners are obliged, in return for patent protection, to publicly disclose information on their invention in order to enrich the total body of technical knowledge in the world. Such an ever-increasing body of public knowledge promotes further creativity and innovation in others. In this way, patents provide not only protection for the owner but valuable information and inspiration for future generations of researchers and inventors.

2.4 Patent Protection for Plants

In some countries (including United States, Australia and Europe) plants can be protected by patents provided that they are able to meet all of the necessary standards and requirements that exist in that country for patentability (Baenziger et al. 1993).

Under the TRIPs Agreement, member countries that choose not to provide protection for plant varieties under their national patent system must provide an alternative *sui generis* form of protection (Rai 1994).

> **Plant patents** may be granted to anyone who has invented or discovered and asexually reproduced any distinct and new variety of plants, including cultivated sports, mutants, hybrids, and newly found seedlings, other than a tuber-propagated plant or a plant found in an uncultivated state.

Utility and Standard Patents May be Used to Protect (not exhaustive)

- Transgenic plants,
- Plant components (e.g., specific genes or chromosomes),
- Vectors and processes involved in the production of transgenic plants.
- Plant culture cells,
- Plant material used in industrial processes (e.g., cell lines used in cultivation methods),
- Plant groups,
- New varieties of plants (US only),
- Individual plants and their descendants,
- Particular plant traits,
- Plant parts,
- Plant products (e.g., fruit, oils, pharmaceuticals),
- Reproductive material (e.g., seeds or cuttings),
- Plant breeding methodologies.

2.5 Plant Variety Rights Protection

The obligation of countries to protect plant varieties is imposed by Article 27.3(b) of the TRIPs Agreement. This requires countries to protect plant varieties by patent laws or by sui generic laws or by a combination of both. Most Plant Variety Protection (PVP) laws are modelled on the international Convention for the Protection of Plant Varieties (UPOV), which originated in 1961, but which has undergone revisions in 1978 and 1991. Generally, under this legislation the plant breeder is conferred an exclusive right to do or to license the following acts in relation to propagating material of the variety:

- produce or reproduce the material;
- condition the material for the purpose of propagation;
- offer the material for sale;
- sell the material;
- import the material;
- export the material;
- stock the material for the purposes described above.

Generally excepted from these rights are acts done privately and for non-commercial purposes, or for experimental purposes, or for the purpose of breeding other plant varieties. Under the 1978 and 1991 versions of UPOV, seed saved by a farmer from harvested material and treated for the purpose of sowing a crop on that farmer's own land may be permitted under legislation as a non-infringing use. In some countries plant variety rights might be exempted from infringement when propagating material is used as a food, food ingredient or fuel, or for any other purpose not leading to or involving the production or reproduction of propagating material.

The general duration of PBR under the UPOV Convention is 25 years in the case of trees and vines and 20 years for any other variety.

As with patents, PVP is established after a registration process. A plant variety is considered to be registrable, if it has a breeder, is distinct, uniform, stable and has not been or has only recently been exploited. A plant variety is considered distinct if it is clearly distinguishable from any other variety whose existence is a matter of common knowledge. It is uniform if, subject to the variation which may be expected from the particular features of its propagation, it is uniform in its relevant characteristics on propagation. A plant variety is stable if its relevant characteristics remain unchanged after repeated propagation. A plant variety is taken not to have been exploited if it or propagating material has not been sold to another person by or with the consent of the breeder. The test of no commercial exploitation is easier to satisfy than the test for novelty under patent law and the choice between the two forms of intellectual property right is a matter to be considered by the agricultural research institute.

Legislation based on the UPOV Convention generally provides for the grantee of PVP to take all reasonable steps to ensure reasonable public access to the plant variety. This requirement is taken to be satisfied if propagating material of reasonable quality is available to the public at reasonable prices, or as gifts to the public, in sufficient quantities to meet demand. An appropriate person may be licensed to sell or produce propagating material of plants of that variety on reasonable terms and conditions. Generally an exception to the grant of a compulsory license applies in the case of a plant variety which has no direct use as a consumer product.

3 PROTECTION OF AGBIOTECHNOLOGICAL INVENTIONS

> Biotechnology, the boom technology of this century, is a phenomenally fast-growing area of science, encompassing a massive range of disciplines and technologies: it promises rapid advancement in medicine, agriculture, nutrition, and the environment. But harnessing these rapid leaps in science to obtain commercial outcomes requires sound intellectual property advice and guidance.

If one were to forecast the areas involving the greatest technological breakthroughs for the new millennium, genetic engineering made possible by biotechnology would be at or near the top of the list. It is entirely possible that over the next 50 years thousands of genetically modified bacteria, viruses, plants and animals could be developed and released into the global environment for pharmaceutical, agricultural, medical, environmental remediation, alternative fuel, and other purposes. There is something special about IPR and biotechnology and which is why biotechnological products and processes maybe easier to copy than in other areas of technology. Seed can be replaced, genes can be cloned based on sequence information and methods can be copied following established protocols. This makes enforcement of IPR harder to keep up with. The most common IPRs in agbiotech are Patents, MTAs and Plant Breeders' Rights (PBRs).

The development of recombinant DNA technology - the joining of DNA from different organisms - has resulted in greatly increased understanding of the genetic and molecular basis of life. Following the first successful directed insertion of microorganism in 1973, scientific research began to recognize the potential for directing the cellular machinery to develop new and improved products and processes in a wide variety of industrial products. With the development of recombinant DNA technology, the potential of patenting the living organism from this type technology arose (Van Wijk et al. 1993).

In recent times due to dramatic developments in the ability to select and manipulate genetic materials, have generated interests in the commercial uses of living organisms. The rapid progress being made in the area of biotechnology points out that in the very near future, the existing industries, particularly the food and pharmaceutical industry is likely to be influenced by the unique developments in this area. Similarly in the fields of agriculture, biotechnology is likely to play an important role in the production of biocides, growth promoting hormones and the like? Other areas in which biotechnology will be applied are the fermentation industry, food industry, treatment of waste for the protection of environment, etc., accordingly IPRs have come to be viewed by many as a way to protect their commercial interest.

It is a well-recognized fact that there is some specific problems in patenting of inventions relating to biotechnology including agriculture as the inventions in these fields invariably involve living entities. The problem arises mainly due to the fact that the living materials are very hard/complicated to describe due to which difficulty arises for making a distinction as to whether it as an invention or a discovery (present in the nature), and if it is an invention the difficulty in repeating the invention. The ability to patent biological inventions is central to protecting scientists' work and the development of the biotechnology industry. What can be patented, for how long and the extent of global protection are critical issues. Economists' question whether patenting is the quickest and best way to diffuse new knowledge throughout the market place.

The decision to adopt an IPRs system for agricultural research hinges on several factors affecting the use of genetic material and germplasm. Research programs in developing and developed countries, therefore, seek clarification on the rights of and access to research innovations and genetic resources. The Agreement on TRIPs and the CBD are meant to provide such clarifications. Countries pursuing commitments under both the TRIPs agreement and the CBD must also address the possible conflict between provisions regarding IPR in the two agreements. In addition, decision-makers want to improve their understanding of innovations produced by scientists.

Simultaneously with the development of biotechnology resulting in the creations of new inventions resulting in the production of a variety of useful products, such as vaccines, antibiotics, enzymes, etc., protection of such inventions by patents has also gained importance. Biotechnology

> GE: A heritable alteration or a change that can be carried from one generation to the next. GE is performed by modifying an organism's own DNA or introducing new DNA to perform desired functions.

involves use of living forms and patenting of living forms which raises many issues in favour and against the issuance of patents for such forms. The debate on the question as to whether to allow patent protection for living forms goes beyond simple consequences of commercial use of the patented materials or the underlying merits of biotechnology. What makes many people feel uncomfortable about IPR in agbiotech is that agriculture was perceived until now as 'the last stronghold of the free'. Farmers have had the freedom to replant their own seed and to sell it to others farmers since the dawn of agriculture.

For many, agbiotech implies gene, genetically modified plants and the processes involved in their production. But other tools and technologies are equally important, including tissue culture, diagnostic tools, molecular marker technologies and bioinformatics. While holding great promise for the future of agriculture, transgenic crops are not yet playing a major role in most developing countries. Various factors have contributed to this delay; one being that the first generation of transgenics has been geared toward herbicide and insect resistance in crops more relevant to a few developed countries. Another reason is the low level of public acceptance of transgenic crops in many developed countries that are the traditional importers of crop produce from the developing world.

Genetic engineering (GE) provides materials and methods to introduce genes into many crop varieties of one species or into completely different species. Development of GE products requires several technologies and/or raw materials. For example, production of transgenic plant would require the following:

1. Outstanding variety of the concerned crop.
2. The gene specifying the concerned trait.
3. Specific restriction enzymes.
4. Suitable promoters or enhancer sequences.
5. Appropriate reporter genes.
6. Suitable transformation method/protocol.
7. A specific technology.

Patents provide the strongest protection for know-how and genetically modified plants. A patent may protect a process used to obtain transgenic plants as well as the plants themselves and uses thereof. While PBRs are generally the best way to protect individual plant varieties, they are not applicable to Essential Derived Varieties (EDV). For example, a variety, into which one gene construct has been introduced by genetic manipulation, retains the rest of the genome practically unscathed.

Products developed using a patented technology/raw material cannot be commercialized without a license. Patent/license holders generally do not raise questions of infringement during the research and development phase and prefer to wait till the product (in the case of transgenic plant) is ready to be marketed. At that time, the patent/ license holders initiate legal proceedings claiming infringement (Singh 2003). When developing 'golden rice', its authors (Prof. Potrykus and Dr. Beyer, succeeded in installing several genes in rice, including genes from bacteria and daffodils) had to use genes and procedures developed by other companies with great expenses. The Golden Rice™ case is the leading example of the novelty of IPR complexity involved in the engineering of a single trait; in this case the enhanced production of Vitamin A in a crop plant. Those companies wanted to get back the money and which is why they had the genes and the procedures patented. The common practice is to require royalties as soon as a product is obtained with a help of an invention that is commercialized.

Golden Rice™ has become a *cause celebre* in the struggle between the need to protect and motivate creativity and the basic needs of the developing world. Modification of one trait in plant requires the use of several key steps, from cloning the genes of interest into vectors to transformation using *Agrobacterium* and regeneration of fully fertile plants. Many patents can swamp such a process, the UK with 35 or France with 37.

In developing countries the number can fluctuate between zero and just a few, such as in Indonesia with six or in South Africa with five. The situation becomes even more advantageous when we consider that some of these patents will not be enforced by their owners or are not worth enforcing at all. And another important example involving the precise idea of the IPR positions of the various materials/processes being used by the inventor in developing '*Endless Summer*', [a GE tomato (delayed-ripening) developed by DNA Plant Technologies] tomato is typical of the IPR situation in agbiotechnological innovations.

In 1998 the Board of Appeals of the European Patent Office (EPO) had to decide whether patents containing claims to processes and products that could result in multiple genetically modified varieties were allowable. The Enlarged Board of Appeals finally decided that genetically modified plants were patentable subject matter as long as the claims were not restricted to a single variety (EPO case number G 0001/98).

IPRs do not prevent access to traditional technologies used in plant breeding. The real problem stems from underfunded or non-existent breeding programs. To deliver efficiently, a good breeding program should employ state-of-the-art technologies such as marker-assisted selection (MAS), anther culture, embryo rescue and genetic transformation. Only competent, interdisciplinary teams can identify bottlenecks in their programs and propose entry points for novel technologies to overcome those bottlenecks (Jorge 2003).

A major change occurred in 1980, particularly in the USA, for patenting inventions in the area of biotechnology and agriculture, after the decision of the US Supreme Court. The Supreme Court in the USA, in the case of Chakrabarty vs. Diamond ruled that a 'man made' microorganism through genetic engineering could be patented. The court ruled that the definition of invention under the US patent Status is broad enough to include 'anything under the sun' provided there is 'human intervention'.

The main biotechnological inventions are concerned with life forms and involve one or more of the following.

1. A process of genetically altering or otherwise inducing a single or multi-celled organism to:
   a. Express an exogenous nucleotide sequence.
   b. Inhibit, eliminate, augment, or alter expression of an endogenous nucleotide sequence.
   c. Express a specific physiological characteristic not normally associated with that organism.
2. A process of generating useful biotechnological products (e.g., vaccines).
3. Various biotechnological products, e.g., antibiotics, purified vitamins, etc.
4. Cell fusion procedures yielding a cell line that expresses a specific protein (e.g., monoclonal antibodies).
5. Application of various processes/products (biocontrol agents).
6. Various microorganisms, cell lines, plant/animal lines obtained through biotechnology approaches.
7. Various DNA sequences and the proteins encoded by them.
8. Approaches used to modify the different properties of organisms.

3.1 Protection of Genes and DNA Sequences

The patenting of genes is the subject of major international debate, around the question of whether genes can be considered novel inventions or merely discoveries, and whether patents' applications for gene sequences adequately prove practical usefulness. Patent law evolved to protect and encourage technological innovation. But the rise of biotechnology has blurred the boundary between invention and discovery. According to patent law, it is not possible to patent raw DNA sequences. Nor is it is not possible to patent a naturally occurring or conventionally bred new animal or plant variety, nor any part of the human body at any stage of development (Crespi 1995).

However, isolated genes, gene sequences and gene products may be patentable - no distinction is made between DNA and other chemicals. And, where the creation of a new variety of organism involves an 'inventive step', in the laboratory rather than the field, it too may be patentable. Patents are now allowed on such genes in the USA; the first patent was awarded for the gene *aroA* isolated from a mutant bacterial strain and intended for transfer into plants to confer glyphosate resistance. Calgene, Inc., USA, holds the patent for *aroA* in terms of a DNA sequence containing this gene.

According to the 'product of nature' theory, "any kind of structure made by human hand", is patentable, but things that exist in substantially the same form in nature (i.e., not made by human hand), namely 'products of nature' are not patentable. This is a theory that has often been employed in US patent case laws.

4 OTHER PROTECTIVE MECHANISMS

4.1 Trademarks and Registrable Marks

A trademark is a sign used to indicate the origin of goods or services. Legal protection is provided for trademarks through a system of registration. To be registered as a trademark, a sign must be represented in a visible form. In some countries, audible or olfactory 'signs' may be registered if they can be recorded in an appropriate manner. Visible signs typically include names, invented or existing words, letters, numbers, pictures, and symbols, or combinations of these signs. The TRIPs agreement envisages that the shape of goods or their containers, as well as colours and smells may be registered as marks. Exempted from registration in most countries are marks that are not distinctive or that are deceptively similar to existing marks and marks that violate public order or morality (Blakeney et al. 1999).

4.1.1 Registration

The application process usually includes an examination by the granting office to ensure that (1) the application complies with the formal registration requirements, (2) the mark meets the substantive requirement of distinctiveness, and (3) the mark is not in conflict with prior rights. After an application is published, most countries provide for an opposition process whereby interested third parties may protest the registration of a mark. Common objections include prior rights or deceptive similarity with another mark.

Upon acceptance by the registering office, registration is conferred for a term of between 10 and 20 years, with a possibility for renewal. A registered mark will expire if a renewal is not sought. Cancellation of a mark may also be sought where its use becomes deceptive or where the mark becomes generic of goods or services. Vaseline and gramophone are two examples of marks that became generic descriptions of the type of goods to which they were appended.

A requirement of some trademark laws is that registration is contingent on its use with the classes of goods or services for which it is registered. A similar requirement provides for the removal of the registration after a prescribed period of non-use. Protection without registration may be extended to 'well-known marks' i.e., those with a significant reputation in a country. Such marks invariably have a substantial international reputation through advertising and use.

4.1.2 Trademark Rights

Registration of a mark confers protection against emulation by traders using identical or substantially similar marks. Most systems of registration permit assignment or licensure. A system of registered user may be provided to record trademark licenses. In the event of infringement of a registered mark, a trade mark proprietor may seek relief in the form of injunction, compensation orders, and seizure of infringing goods.

The use of a trademark is a warranty of the quality of the goods or services supplied under that mark. The name, acronym or logo of a research institute is often a warranty of the quality of the services supplied by that institute. Its designation is worthy of protection, particularly because unauthorized traders may falsely represent an affiliation with the institute. Similarly, the products which are produced by an agricultural research institute may also be worthy of protection, e.g., the IRRI prefix for rice types developed at that institute. Similarly, the research institute may wish to protect its Internet domain name as a trade mark.

4.2 Geographic Marks and Appellations of Origin

A specialized form of trademark, which has been identified as the subject of a separate system of protection, includes those marks which identify that a product or service originates in a country, region, or particular place. The false or deceptive indication of a source is actionable. An appellation of origin is a mark that indicates that in addition to the geographic source of goods, the place of origin decisively influences the character or quality of the goods. For example, the soil and climatic influences in a wine-producing district, such as Burgundy or Champagne, can be demonstrated to produce a wine of a particular quality which it would be deceptive to permit other wine producers to use those appellations of origin.

4.3 Confidential Information and Trade Secrets

IPR cannot be effectively used without the deployment of substantial know-how. However, there is no system of registration for know-how. It can be protected only if it is regarded as confidential or if a restriction is placed on its unauthorized communication. In the case of contracts with employees or researchers, most common law systems imply a contractual term obliging employees not to divulge information that is considered to be an employer's property. This contractual term is enforceable only against an employee. In determining whether the information is protectable as a trade secret, the courts have regard to (1) the extent to which the information is known by employees and by persons outside the relevant business, (2) the extent of measures made to guard its secrecy, (3) the value of the information to the business and to competitors, (4) the amount of effort or money expended in developing the information, and (5) the ease or difficulty with which the information could be properly acquired or duplicated by others. The TRIPs agreement, article 39, permits restrictions on the disclosure of confidential information.

4.4 Copyright

Copyright law is concerned with the protection and exploitation of the expression of ideas in a tangible form. The law developed in response to the invention of the printing press, which made possible the mass production of print works. The technological context in which the law developed explains the way in which copyright concepts have been formulated. For example, the law focuses on the rights of authors and those who claim through the author; it protects 'original works' and only if those works are fixed in a material form. The central right that the law confers is to prevent unauthorized persons from copying a work. To be protected as copyright, ideas have to be expressed in an original way; i.e., they must have their origin in the labour of the creator. Works are protected irrespective of their quality (De Biswas 2004).

4.4.1 Subject Matter of Protection

Originally, the subject matter of copyright protection was printed literary artistic and literary works. Research notes and reports will be protectable as copyright works. More recently, copyright protection has been extended to computer programs and databases.

4.4.2 Rights Comprized in Copyright

The owner of a copyrighted work may exclude others from using it without authorization.

Authorization of the copyright owner is usually required for copying or reproducing the work, performing the work in public, making a sound recording of the work, making a motion picture of the work, broadcasting a work through the electromagnetic spectrum or through cable diffusion, and translating or adapting the work. In addition to these rights, the Berne Convention for the Protection of Literary and Artistic Works also recognizes certain immoral rights. These include the right to claim authorship of a work and the right to object to any distortion, mutilation, or other modification of, or other derogatory action in relation to a work that would be prejudicial to an authorized honor or reputation. These moral rights usually remain with an author, even after the transfer of the various economic rights mentioned above.

5 MATERIAL TRANSFER AGREEMENTS

Contractual in nature, MTAs offer a form of intellectual property protection that can cover material not generally protected by patents (Barton and Siebeck 1994). Most international agricultural research centers use such agreements for the genetic resources they hold in trust for the world community. They are also widely used among public-sector research organizations in industrialized countries. MTAs provide interim protection for material sent to collaborating organizations for advanced research. They can thus be used until more formal IPR is sought (Barton and Siebeck 1994). MTAs are becoming especially important in the exchange and use of plant genetic resources, particularly since open access to such material is essential for the development of food and agriculture. In effect, these agreements can help clear the way for research and breeding by setting out the conditions that govern each germplasm exchange.

Under the International Treaty on Plant Genetic Resources for Food and Agriculture, a Standard Materials Transfer Agreement (SMTA) is being developed to deal with the transfer of germplasm from CGIAR Centres. At the FAO Conference which adopted the Treaty, an Expert Group was established to prepare and analyze possible options for the terms of the SMTA. At its second meeting, the following questions were addressed:

- "What constitutes commercialisation in terms of Article 13.2d(ii) of the Treaty?
- What constitutes a product?
- What constitutes incorporation of material accessed from the Multilateral System?
- When would a product be considered to be available without restriction to others for further research and breeding?
- What should be the level, form and manner of payments in line with commercial practice?
- Whether different levels of payment should be established for various categories of recipients who commercialize such products or for different sectors and, if so, what those levels, various categories of recipients and sectors should be?
- Whether to exempt small farmers in developing countries and in countries with economies in transition from the payments, and if so, who qualifies as such as a small farmer?
- How will monetary and other benefits be defined, for the purposes of the standard MTA?
- What terms should be included in the MTA, so that recipients are bound by it on acceptance of the material from the Multilateral System?
- By what means will the MTA ensure the application of Article 12.3 of the Treaty?"

Table 1 Intellectual property rights available for crop improvement and protection technologies (Source Ravishankar and Archak 2000).

Sl. No.	*Technologies*	*IPR Applicable*
A	**Crop Improvement Technologies**	
1	Conventional Breeding	Patents, Plant Breeders' Rights, Trademarks, and Trade secrets
2	Agbiotech	Patents, Plant Breeders' Rights, Trademarks, and Trade secrets
3	Agro-Chemicals	Patents, Trademarks, and ID
4	Knowledge Base Technologies	Patents, and Copyright
B	**Crop Protection Technologies**	
1	Conventional Breeding	Patents, Plant Breeders' Rights, and Trademarks
2	Ag Biotech	Patents, Plant Breeders' Rights, and Trademarks
3	Agro-Chemicals	Patents and Trademarks
4	Knowledge Base Technologies	Patents and Copyright

A definitive *Second Draft Standard Material Transfer Agreement* was adopted on 28 April 2006.

6 CONCLUSION

Selecting among the types of protection to be applied to innovations arising from agricultural research is a complex management decision. In many public organizations, research liaison officers or offices of intellectual property help with these decisions. These individuals will help assess the accountability requirements and public expectations regarding innovations produced with public funds. Such offices can also help the national research program anticipate means to license, develop, and move its innovations into production. Since such technology transfer agreements are especially important for new products serving farmers who rely on purchased inputs and make capital expenditures, technology transfer will usually include the licensing of some proprietary right.

A patent system is an effective way of achieving co-ordination and sharing of the risks and benefits of R&D among the various players in the agbiotech value chain. Stated alternatively, it is critical to design IP regimes that adequately reward the inventor for her efforts and provide economic stability to promote R&D investment by understanding and capturing the downstream economic benefits of innovation. On the other hand, as the fruits of biotechnology are becoming an integral part of our daily living, the social welfare benefits of the intellectual property system for the consumer population and society at large must also be carefully defined. Satisfying both these objectives is the central policy challenge in the relationship between intellectual property protection and agbiotechnology.

Opportunities arising from an increased availability of information and from little obstruction by patents in the developing world most often do not compensate and are needed in the scientific as much as in the legal aspects of agbiotech in order to take advantage of those opportunities. IPRs were never designed to prevent development but to promote it and thus the process of co-evolution between technology and society will continue. Fine-tuned IPR laws will be instrumental in the acquisition of agricultural biotechnologies for developing countries ready to embrace them.

References

Baenziger SP, Kleese RA, Barnes RF (1993) Intellectual Property Rights: Protection of Plant Materials. Executive Summary and Work Group Reports. CSSA Spl Publ No 21. Crop Sci Soc Am, Madison, WI, USA

Barton JH, Siebeck WE (1994) Material Transfer Agreements in Genetic Resources Exchange: The Case of the International Agricultural Research Centers. Issues in Genetic Resources No 1. International Plant Genetic Resources Institute, Rome, Italy

Blakeney M, Cohen JI, Crespi S (1999) Intellectual property rights and agricultural biotechnology. In: Cohen JI (ed) Managing Agricultural Biotechnology-Addressing Research Program Needs and Policy Implications. CABI Publ, Wallingford, Oxon, UK, pp 209-227

Crespi RS (1995) Intellectual Property in Agricultural Biotechnology: Issues for Developing Countries. In: Turning Priorities into Feasible Programs: Proceedings of a Regional Seminar on Planning, Priorities and Policies for Agricultural Biotechnology in Southeast Asia. Singapore, 25-29 Sept 1994. Intermediary Biotechnology Service, The Hague and Nanyang Technological University, Singapore

De Biswas SK (2004) Copyrightability of characters. J Intellect Prop Rights 9: 148-156

Jorge Niosi (2003) Alliances are not enough explaining rapid growth in biotechnology firms. Res Pol 32 (5): 737-750

Mayer JE (2003) Intellectual property rights and access to agbiotech by developing countries. AGBiotechNet, Vol 5 March, ABN 108

Rai M (1994) The current situation of the plant breeding and seed industries in India: The policy on the protection of new varieties. In: Seminar on the

Nature of and Rationale for the Protection of Plant Varieties under the UPOV Convention, UPOV, Beijing, China

Ravishankar A, Archak S (2000) IPRs and agricultural technology: Interplay and implications for India. In: Policy Brief. National Centre for Agricultural Economics and Policy Research and NRC on DNA Fingerprinting, NBPGR, ICAR, New Delhi, India

Singh BD (2003) Plant Breeding (Chapter on Intellectual Property Rights). Kalyani Publ, New Delhi, India

Subbaram NR (2000) Intellectual Property. Chapter in Patent protection of the Results of R and D. National Bank for Agriculture and Rural Development, Mumbai, India

Van Wijk J, Cohen JI, Komen J (1993) Intellectual Property Rights for Agricultural Biotechnology: Options and Implications for Developing Countries. ISNAR Research Report No. 3. The Hague: International Service for National Agricultural Research

Index

www.ingramcontent.com/pod-product-compliance
Lightning Source LLC
LaVergne TN
LVHW010536300926
849162LV00002B/154
* 9 7 8 1 5 7 8 0 8 5 3 7 8 *